LAMB—Elements of Soliton Theory
LAY—Convex Sets and Their Applications
LINZ—Theoretical Numerical Analysis: An Introduction to Advanced Techniques
LOVELOCK and RUND—Tensors, Differential Forms, and Variational Principles
MARTIN—Nonlinear Operators and Differential Equations in Banach Spaces
MELZAK—Companion to Concrete Mathematics
MELZAK—Invitation to Geometry
NAYFEH—Perturbation Methods
NAYFEH and MOOK—Nonlinear Oscillations
ODEN and REDDY—An Introduction to the Mathematical Theory of Finite Elements
PASSMAN—The Algebraic Structure of Group Rings
PETRICH—Inverse Semigroups
PRENTER—Splines and Variational Methods
RIBENBOIM—Algebraic Numbers
RICHTMYER and MORTON—Difference Methods for Initial-Value Problems, 2nd Edition
RIVLIN—The Chebyshev Polynomials
RUDIN—Fourier Analysis on Groups
SAMELSON—An Introduction to Linear Algebra
SCHUMAKER—Spline Functions: Basic Theory
SHAPIRO—Introduction to the Theory of Numbers
SIEGEL—Topics in Complex Function Theory
  Volume 1—Elliptic Functions and Uniformization Theory
  Volume 2—Automorphic Functions and Abelian Integrals
  Volume 3—Abelian Functions and Modular Functions of Several Variables
STAKGOLD—Green's Functions and Boundary Value Problems
STOKER—Differential Geometry
STOKER—Nonlinear Vibrations in Mechanical and Electrical Systems
STOKER—Water Waves
TURÁN—On A New Method of Analysis and Its Applications
WHITHAM—Linear and Nonlinear Waves
WOUK—A Course of Applied Functional Analysis
ZAUDERER—Partial Differential Equations of Applied Mathematics

ON A NEW METHOD OF ANALYSIS AND ITS APPLICATIONS

ON A NEW METHOD OF ANALYSIS AND ITS APPLICATIONS

PAUL TURÁN

Published posthumously, with the assistance of Gábor Halász and János Pintz

A Wiley-Interscience Publication
JOHN WILEY & SONS
New York • Chichester • Brisbane • Toronto • Singapore

Library of Congress Cataloging in Publication Data:

Turán, P. (Paul), 1910–1976.
On a new method of analysis and its applications.

(Pure and applied mathematics)
Includes index.
I. Mathematical analysis. I. Title. II. Series: Pure and applied mathematics (John Wiley & Sons)
QA300.T87 1983 515 83-16742
ISBN 0-471-89255-6

Printed in the United States of America

10 9 8 7 6 5 4 3 2 1

To my wife,
Vera

FOREWORD

This book is one of the principal efforts of the last 20 years of Paul Turán's life. He saw in its contents a new unifying method in analysis, providing a powerful technique and wide applications in both pure and applied mathematics. The first paper expounding this approach appeared in 1949, and a systematic book was published in 1953 in German and Hungarian editions. Deepening his insight into the method and extending the scope of its applications occupied him constantly. In 1959 he contracted to publish an entirely new edition, deeper and wider in its scope. His unremitting labors led to constant enrichments—and constant rewriting. For example, by 1965 he had completed a manuscript that his new ideas then led him to abandon. At his death on September 26, 1976 he left a carefully organized, but not entirely complete, manuscript comprising 57 sections in two parts. Part I (Sections 1–16) deals with the fundamental theory of the method, Part II (Sections 17–57) with its manifold applications.

Sections 1–37 (except for Section 26) were in finished form. For the remaining 20 sections, he had indicated the intended contents, chiefly by precise references to the relevant mathematical literature and by section headings.

Generous colleagues, formerly his students, later his co-workers, undertook to develop these indications into a completed version. G. Halász wrote Sections 26, 37–39, 43–46, and 57. J. Pintz wrote Sections 40–42, 47–56, and in these, following Paul Turán's intentions specifically expressed to me, he incorporated some of his own results. All this required expert knowledge and sound judgment together with enormous devoted labor. Their exceptional competence guarantees that this book truly follows the intentions of Paul Turán. In addition, M. Szalay checked the entire manuscript in detail, making many useful remarks which Paul Turán incorporated in the 36 sections he had completed and for which he was very grateful, as am I.

The work of checking the manuscript and making grammatical revisions was done by T. Kővári and J. Pipher.

I am deeply indebted to all of them for their selfless aid, which makes it possible to offer now to the mathematical public a scientific work that represents such a great portion of my husband's life, a work so close to his heart that he carried it forward during his last days.

Special thanks are due to Ms. Erzsébet Szlatki for the devotion and skill with which she typed the long manuscript and to Mrs. Éva Várnai for typing the first version with an enthusiasm typical of the warm friendship she has shown over the years.

I consider it my duty to express my most heartfelt thanks to Beatrice Shube, my first editor at Wiley, for her understanding and concern, the steady interest she has shown for almost 20 years in the completion of the book and to D. B. Kaplan, my second editor, for his care and professionalism in the publication of the book.

Finally, in offering this work, I cannot help but think of what my husband wrote about another mathematician: "Taking in hand any work finished shortly before the death of the author one cannot suppress—I think generally—a sort of emotion, a feeling of the tragic. Being aware of a grave illness and still being able to concentrate on questions beyond life shows passion, devotion, energy, and one gets the impression of a heroic last ditch fight against death." This was his case.

VERA T. SÓS

Budapest, Hungary
March 1984

PREFACE

A number of important problems of complex analysis and number theory can be reduced to showing that a sum $(z_1 + \cdots + z_n)$ of complex numbers is not zero or to giving a lower bound for its absolute value. For p-adic analysis the difficulties are much diminished by the general equality $|a_1 + \cdots + a_n|_p = |a_1|_p$, provided $|a_1|_p > |a_\nu|_p$ $(\nu = 2,\ldots,n)$. This raises the naive question whether or not one can have lower bounds for a sum of complex numbers given some bounds on the absolute values of either the single terms or of some combinations of terms. Generally, of course, only trivialities can be stated. Nevertheless, the aim of this book is to develop systematic methods—in the form of the solution or partial solution of minimax problems concerning power sums of complex numbers—that lead to some general lower bounds (Part I) and then in Part II to show to how many different situations these methods can be successfully adapted to yield new results in important areas. One should also mention the unifying power of the method. (As to the new results: in some cases they were later superseded by others; these were obtained however in the psychologically favorable position of having prior knowledge of the newly explored situation. Today, record holders are more honored than pioneers.) I speak of a *method* since the results of Part I arose mainly from a problem in analysis or analytic number theory which then was reduced gradually to an appropriate "power sum problem."

The results of Part I could also be formulated as Banach space inequalities. Occasionally we use these formulations.

It will also emerge that the theory can be considered as a natural development of the theory of diophantine approximation. How readily the family of theorems of diophantine approximation will accept the newcomers is questionable; some time elapsed until the family embraced H. Weyl's estimations of exponential sums. How great a role was played in the appeasement of the family by the successes of Weyl's estimations, I do not know; rich relatives are generally welcomed. If so, one can entertain hopes that any possible resentment will disappear soon.

This is the fourth form of a book that appeared in 1953 in Hungarian and German and then in 1956—in much improved form—in Chinese. Apart from the fundamentals and the historical background (which cannot be changed although this last circumstance is not always recognized, even in mathematics), the present form has very little resemblance even to the Chinese edition. I wrote in 1953 in the preface: "...Ich muss daher mit dem Ausdruck meiner Hoffnung schliessen, dass zahlreiche Resultate dieses Buches bald überholt sein mögen." I can now say that this hope is—thanks also to the results of several other authors—largely fulfilled. However, even now the book is far from perfection, which is indicated by the rather incomplete set of open problems at the end of Part I.

I am much indebted to a number of friends for their remarks; these are acknowledged at their respective places. Particular thanks are due to Dr. S. Dancs who read all previous versions and made a number of improvements, some of which were the starting points of our joint papers. Thanks are due to Dr. G. Halász and Dr. R. Tijdeman for several discussions and to my colleague Dr. M. Szalay, who read the last version of this book and made a number of valuable remarks.

PAUL TURÁN

GENERAL REMARKS, NOTATION, AND CONVENTIONS

As mentioned in the preface, the book consists of minimax problems (Part I) and their applications (Part II). The *exact* solution of the minimax problems is generally very difficult. Most frequently we have only inexact lower bounds which, however, are sufficient for the intended application or for a weaker form of it. I had to choose whether or not to investigate only a few minimax problems, solve them (perhaps) completely, and have only a few applications, or else to get the not-best-possible lower bounds for more minimax problems and thus have many more applications. Sinful greediness and excessive curiosity led me to follow the second course. As a consequence, the book generally makes no attempt to squeeze out the strongest possible results (even if I *had* the means) when these required a longer discussion. Another general feature is the great number of appendixes. The aim of this book is to show how these minimax bounds can be applied in various forms. For these applications, however, other auxiliary considerations are necessary too. To make the treatment more transparent these auxiliary considerations are *often* (not always) relegated to appendixes.

Positive, explicitly calculable constants whose numerical values do not matter are denoted by $c_1, c_2, \ldots$; their numeration starts anew in each section. The dependence of constants on some important parameters is stated explicitly. Lemma 5.4 means that the lemma is fourth in Section 5, Lemma C.2 that it is the second in Appendix C. Theorems are treated in the same way. Formula (5.1.3) means that the formula is to be found in Section 5, Formula (D.3.1) that it can be found in Section D.3 of Appendix D. The letter p is reserved for primes. The generalized power sum $\sum_{j=1}^{n} b_j z_j^{\nu}$ of complex numbers z_j with complex coefficients b_j and integer ν is denoted by $g(\nu)$. $\operatorname{Re} z$ (resp. $\operatorname{Im} z$) denotes the real (resp. imaginary) part of z, $\operatorname{arc} z$ resp. $\bar{z}$ the argument resp. the complex conjugate of z,

$$\begin{aligned} &\exp x = \exp_1 x = e^x, \qquad &&\exp_k x = \exp(\exp_{k-1} x), \\ &\log_1 x = \log x, &&\log_{k+1} x = \log(\log_k x). \end{aligned} \tag{0.1}$$

The symbol $\doteq$ means "equal by definition." The expression $\|\sum_{\nu=0}^{k} \mu_\nu z^\nu\|$ stands for

$$\sum_{\nu=0}^{k} |\mu_\nu|; \tag{0.2}$$

the symbol $\{x\}$, as usual, stands for the distance of x from the nearest integer.

The numbers in brackets after names of authors refer to the bibliography at the end of the book. Referring to classical results, we quote them in the text. Results without quotation are first published in this book.

We follow throughout the convention that empty sums mean 0 *and empty products* 1.

CONTENTS

I

MINIMAX PROBLEMS

1

INTRODUCTION

1.1

The problems we are dealing with cannot be properly understood without knowing their background. The story started with a paper of H. Bohr in 1911 (see Bohr [1]), which gave the first solution of a problem raised by E. Lindelöf that asked whether or not the Riemann zeta function $\zeta(s) = \zeta(\sigma + it)$ defined for $\sigma > 1$ by

$$\zeta(s) = \sum_{j=1}^{\infty} j^{-s} \tag{1.1.1}$$

is bounded for $\sigma > 1$, $|t| \geq 1$. Bohr's negative answer was based on Dirichlet's classical theorem from the theory of diophantine approximation, which states that for arbitrary real numbers $\vartheta_1, \ldots, \vartheta_n$ and integer $q \geq 3$ there exists an integer ν_0 with

$$1 \leq \nu_0 \leq q^n \tag{1.1.2}$$

so that ($\{x\}$ being the distance of x from the nearest integer)

$$\{\nu_0 \vartheta_j\} \leq \frac{1}{q} \tag{1.1.3}$$

holds simultaneously for $j = 1, 2, \ldots, n$.

Briefly, Bohr's ingenious proof runs as follows. For an arbitrary integer n

$$\operatorname{Re} \zeta(\sigma + it) \geq \sum_{j=1}^{n} \frac{\cos(t \log j)}{j^{\sigma}} - \sum_{j=n+1}^{\infty} \frac{1}{j^{\sigma}}.$$

Applying (1.1.2)–(1.1.3) with

$$q = 5, \qquad \vartheta_j = \frac{\log j}{2\pi} \qquad (j = 1,\ldots,n)$$

we get an (integer) ν_0 with

$$1 \le \nu_0 \le 5^n \tag{1.1.4}$$

so that

$$\nu_0 \frac{\log j}{2\pi} = \text{integer} + \frac{\Theta_j}{5}$$

where $|\Theta_j| \le 1$.

Hence for $j = 1,\ldots,n$

$$\cos(\nu_0 \log j) = \cos\frac{2\pi\Theta_j}{5} \ge \cos\frac{2\pi}{5}$$

and hence

$$\operatorname{Re}\zeta(\sigma + \nu_0 i) \ge \cos\frac{2\pi}{5} \sum_{j=1}^{n} \frac{1}{j^\sigma} - \sum_{j=n+1}^{\infty} \frac{1}{j^\sigma}$$

$$> \cos\frac{2\pi}{5} \cdot \zeta(\sigma) - 2 \sum_{j=n+1}^{\infty} \frac{1}{j^\sigma}.$$

Given arbitrarily large ω we may choose $\sigma = \sigma_0 > 1$ so close to 1 that

$$\zeta(\sigma_0) > 5\omega$$

and then choose n so large that

$$\sum_{j=n+1}^{\infty} \frac{1}{j^{\sigma_0}} < \frac{\omega}{4}$$

to obtain

$$|\zeta(\sigma_0 + \nu_0 i)| \ge \omega, \qquad \nu_0 \ge 1.$$

We remark that in the proof the fact that ν_0 is an integer was not essential, and neither was the localization (1.1.4) of ν_0. This last fact is essential, however, to the slightly modified proof of the much stronger inequality

$$\varlimsup_{\substack{t \to +\infty \\ \sigma > 1}} \frac{|\zeta(\sigma + it)|}{\log\log t} > \frac{1}{10}.$$

Let us further observe that the essence of *this* proof is embodied in the inequality

$$\max_{\substack{1\le t\le 5^n \\ t \text{ integer}}} \frac{\left|\sum_{j=1}^{n} b_j e^{i\lambda_j t}\right|}{\sum_{j=1}^{n} b_j} \ge \cos\frac{2\pi}{5} \tag{1.1.5}$$

where $b_n > 0$ and λ_j are real numbers.

1.2

In the same year Bohr proved (Bohr [2]) a very surprising theorem using analogous reasoning, namely, that

$$\inf_{\sigma>1} |\zeta(\sigma + it)| = 0.$$

The interest is obvious in light of Riemann's unproven conjecture according to which $\zeta(s) \neq 0$ for $\sigma > \frac{1}{2}$. For this one must show that for arbitrarily large ω and for suitable $s_1 = \sigma_1 + it_1$, $\sigma_1 > 1$ the inequality

$$\left|\frac{1}{\zeta(s_1)}\right| \ge \omega \tag{1.2.1}$$

holds. Bohr's proof of (1.2.1) was based on Kronecker's classical theorem from the theory of diophantine approximation. This—in a specialized form—says that if $\vartheta_1, \ldots, \vartheta_n$ are linearly independent over the rationals, $\beta_1, \ldots, \beta_n$ are real, and $\varepsilon > 0$ is arbitrarily small, then for a suitable real t_0 the inequalities

$$\{t_0\vartheta_j - \beta_j\} \le \varepsilon \qquad j = 1, 2, \ldots, n \tag{1.2.2}$$

hold simultaneously. Note that this time no localization of t_0 *is* given (or *can* be given in general). Then Bohr's reasoning was based on the representation

$$\frac{1}{\zeta(s)} = \prod_p \left(1 - \frac{1}{p^s}\right) \qquad \sigma > 1$$

or

$$\log\frac{1}{|\zeta(s)|} = -\sum_p \frac{\cos(t\log p)}{p^\sigma} + O(1) \tag{1.2.3}$$

$$> -\sum_{p\le p_N} \frac{\cos(t\log p)}{p^\sigma} - \sum_{p>p_N} \frac{1}{p^\sigma} - O(1)$$

where $N \geq 10$ is an otherwise arbitrary integer and $O(1)$ is uniform for $\sigma \geq 1$. Since the numbers $(1/2\pi)\log p_\nu$ $(\nu = 1,\ldots,N)$ are linearly independent over the rationals, Kronecker's theorem gives the existence of a t^* such that

$$\frac{t^*}{2\pi}\log p_\nu = \left[\frac{t^*}{2\pi}\log p_\nu\right] + \frac{1}{2} + \frac{\Theta_\nu}{10}$$

$$\nu = 1,2,\ldots,N, \qquad |\Theta_\nu| \leq 1.$$

Hence the expression in (1.2.3) is larger than

$$\cos\frac{\pi}{5}\sum_{p\leq p_N}\frac{1}{p^\sigma} - \sum_{p>p_N}\frac{1}{p^\sigma} - O(1),$$

and the proof of (1.2.1) can be completed as before.

We observe that the essence of this proof is embodied in the inequality

$$\sup_t \frac{\left|\sum_{j=1}^n b_j e^{i\lambda_j t}\right|}{\sum_{j=1}^n |b_j|} = 1 \tag{1.2.4}$$

where $\lambda_1,\ldots,\lambda_n$ are real, linearly independent over the rationals, and the b_j's are arbitrary complex numbers.

1.3

Bohr made several applications of the inequalities (1.1.5) and (1.2.4) (though he never stated them explicitly) to the general theory of Dirichlet series and almost periodic functions. Furthermore, Hardy and Littlewood based several applications on it in analytic number theory. I observed several years ago that the fields of applications of Bohr's idea could be essentially enlarged if the following obstacles were removed:

1. In (1.1.5) we have some weak localization of t, in (1.2.4) none. In both cases strong localization would be desirable.
2. Both inequalities only hold under restrictions that are in most cases either not fulfilled or difficult to verify. It is desirable to remove them.
3. In (1.2.4) the t-value is some real number. For some applications here one would need to know that t is an integer.

4. Inequality (1.1.5) can be written as

$$\max_{\substack{1 \le \nu \le 5^n \\ \nu \text{ integer}}} \frac{\left|\sum_{j=1}^{n} b_j z_j^{\nu}\right|}{\sum_{j=1}^{n} |b_j|\,|z_j|^{\nu}} \ge \cos\frac{2\pi}{5} \tag{1.3.1}$$

if $|z_j| = 1$ and $b_j > 0$ for $j = 1,\dots,n$. It is desirable to extend to generalized power sums $g(\nu) = \sum_{j=1}^{n} b_j z_j^{\nu}$ with complex z_j and b_j.

Partly heuristic arguments soon showed that most of these requirements cannot be fulfilled. But three observations helped. First, for most of the intended applications the lower bounds in (1.1.5) and (1.2.4) are unnecessarily strong, much weaker lower bounds are sufficient. Second, instead of comparing $|g(\nu)|$ with the "norm"

$$M_0(\nu) = \sum_{j=1}^{n} |b_j|\,|z_j|^{\nu} \qquad \text{(Bohr-norm)} \tag{1.3.2}$$

it is enough to compare it with other norms such as

$$M_1(\nu) = \min_j |z_j|^{\nu} \qquad \text{(minimum norm)} \tag{1.3.3}$$

$$M_2(\nu) = \max_j |z_j|^{\nu} \qquad \text{(maximum norm)}$$

$$M_3(\nu) = \left(\sum_{j=1}^{n} |b_j|^2 |z_j|^{2\nu}\right)^{1/2} \qquad \text{(Wiener norm)}$$

$$M_4(\nu) = \min_{1 \le j_1 \ne j_2 \le n} |z_{j_1} - z_{j_2}| \qquad \text{(separation norm)}$$

$$M_5 = \max_j |b_j| \qquad \text{(Cauchy norm)}$$

$$M_6 = \min_j |\operatorname{arc} z_j| \qquad \text{(argument norm)}$$

(with the convention that $0 \le |\operatorname{arc} z_j| \le \pi$).

Third, the necessary localization of ν in $g(\nu)$ should be treated elastically, making it depend on the intended application. With all these concessions, the problems could be reformulated in terms of certain minimax problems referring to $g(\nu)$ which could either be solved, or where useful lower bounds could be attained. Their classification is discussed in Section 2.

1.4

When tracing the necessity of such investigations concerning $g(\nu)$ one can go rather far back (newcomers like to flaunt not only their wealth, but also their ancestry). D. Bernoulli found (in 1728) the following method for determining $|z_1|$ where $z_1,\ldots,z_n$ with

$$(1.4.1) \qquad |z_1| > |z_2| \geq \cdots \geq |z_n|$$

are zeros of the polynomial $H(z)$. With an arbitrary polynomial $h(z)$ of degree $\leq n-1$ we have, for all sufficiently large $|z|$,

$$\frac{h(z)}{H(z)} = \sum_{\nu=1}^{\infty} \frac{q_{\nu-1}}{z^{\nu}}.$$

He then remarked that when $\nu \to \infty$

$$\frac{q_\nu}{q_{\nu-1}} \to z_1.$$

Many mathematicians of the eighteenth and nineteenth centuries, among them Euler, Fourier, and Jacobi, tried to settle the more general case when, instead of (1.4.1), we have only

$$|z_1| = |z_2| = \cdots = |z_k| > |z_{k+1}| > \cdots.$$

After realizing that in the case of exclusively simple zeros of $H(z)$ one has

$$q_\nu = c_1 z_1^\nu + \cdots + c_n z_n^\nu$$

(c_j independent of ν) Jacobi wrote (Jacobi [1]) as follows, "In expressione generali ipsius q_ν prae terminis ductis in k radices maximas ad ν *tam* dignitatem elatas, negligimus reliquos terminos omnes, quod eo maiore iure licet, quo maior nemerus ν. Hinc statuimus proxime

$$q_\nu = c_1 z_1^\nu + \cdots + c_k z_k^\nu."$$

It is difficult to understand in which sense the last relation can be true for *all* sufficiently large ν; even to see that it is true for a "dense" set of ν-values we need some mild statement on the lower bound of $\max_{m+1\leq\nu\leq m+k}|q_\nu|$ when $\max_{j=1,\ldots,k}|z_j| = 1$, that is, an (easier) problem of our type.

2

CURSORY CLASSIFICATION OF THE PROBLEMS

2.1

A good theory usually must classify its problems, but with the present one this is not the case. Any rigid classification would include a number of uninteresting problems and at the same time leave out important ones. So we enumerate only those groups of problems that have already led to interesting applications, or that were suggested by the possibility of further applications. We also deal later with a few problems suggested only by the inner dynamics of the subject. There are examples, however, of theorems whose necessity was indicated only by inner reasons of the subject and whose applicability was established later.

2.2. EXTREMAL PROBLEMS OF THE FIRST TYPE

Let S be a fixed finite set of integers and d an integer with $0 \leq d \leq 3$. Then for fixed b_j we wish to determine [see (1.3.3.)]

$$\inf_{z_j} \max_{\nu \in S} \frac{\left| \sum_{j=1}^{n} b_j z_j^{\nu} \right|}{M_d(\nu)} \tag{2.2.1}$$

and all minimizing $z_1, \ldots, z_n$ systems ("two-sided direct problems").

In some cases S can be infinite; in such cases inf max must be replaced by inf sup. Also in all further groups of problems we restrict S to be finite, but throughout it should be understood that S could be infinite, in which case inf max would be replaced by inf sup. We do not formulate separately the problems arising by fixing the z_j and treating the b_j as variables with $3 \leq d \leq 5$ in (1.3.3).

In the problems in (2.2.1) the z_j are free variables. In some applications the z_j are no longer free variables but are restricted by some geometric conditions [the most important being, in the notation of (1.3.3), the restrictions

$$M_4 \geq \delta \qquad (\delta > 0)$$

and

$$M_6 \geq \kappa \qquad (\kappa > 0)]$$

These constitute *extremal problems of second type* ("two-sided conditional problems").

2.3. EXTREMAL PROBLEMS OF THE THIRD TYPE

Again let S be a fixed finite set of integers, d be an integer with $0 \leq d \leq 3$, and the variables z_j be restricted to a domain U. Then we wish to determine

$$\text{(2.3.1)} \qquad \inf_{z_1,\ldots,z_n \in U} \max_{\nu \in S} \frac{\operatorname{Re}\left(\sum_{j=1}^{n} b_j z_j^{\nu} \right)}{M_d(\nu)}$$

and all minimizing $z_1,\ldots,z_n$ systems ("one-sided conditional problems").

It is apparent that the group of unconditional one-sided problems is missing from our list. The reason is our selection principle. Namely, the applications require a *positive* value for the quantity in (2.3.1), and in the case $d = 2$ where

$$b_1 = b_2 = \cdots = b_n = -1$$

the choice of the system $z_1 = \cdots = z_n = 1$ shows that our quantity is $(-n)$, however large the set S is. There will be a similar reason each time we omit a plausible type of problem.

2.4. EXTREMAL PROBLEMS OF THE FOURTH TYPE

Let S be a fixed set of integers, and to each $\nu \in S$ a weight $\psi(\nu, n) > 0$ is attached. Further, let d be an integer with $0 \leq d \leq 3$. Then we wish to determine

$$\text{(2.4.1)} \qquad \inf_{z_j} \max_{\nu \in S} \left(\frac{\left| \sum_{j=1}^{n} b_j z_j^{\nu} \right| \psi(\nu, n)}{M_d(\nu)} \right)^{1/\nu}$$

and all minimizing systems ("weighted two-sided direct problems").

Next we state *extremal problems of the fifth type*. Let S be a finite set of integers, d be an integer with $0 \le d \le 3$, and let the variables $z_1, \ldots, z_n$ be restricted to a domain U. Then we wish to determine

$$(2.4.2) \qquad \sup_{z_1,\ldots,z_n \in U} \min_{\nu \in S} \frac{\left|\sum_{j=1}^{n} b_j z_j^\nu\right|}{M_d(\nu)}$$

and all extremal systems ("dual conditional problems").

Obviously, unconditionally, nothing better than the trivial $\sqrt{n}$ can be attained in (2.4.2) if $d = 3$.

2.5

So far all of our problems refer to the generalized power sums $g(\nu) = \sum_{j=1}^{n} b_j z_j^\nu$. For some applications we need results of the preceding type for expressions of the more general form

$$(2.5.1) \qquad G_1(\nu) = \sum_{j=1}^{n} \varphi_j(\nu) z_j^\nu$$

and

$$(2.5.2) \qquad G_2(\nu) = \sum_{j=1}^{n} b_j z_j^\nu \varphi(z_j)$$

where the $\varphi_j(x)$ are polynomials with fixed coefficients (independent of the b_j and z_j) and $\varphi(x)$ is a polynomial with fixed coefficients (independent of the b_j and ν). For these we formulate *extremal problems of the sixth type*. Let S be a finite set of integers, d any integer with $0 \le d \le 3$. Then determine

$$(2.5.3) \qquad \inf_{z_j} \max_{\nu \in S} \frac{|G_1(\nu)|}{M_d(\nu)}$$

and

$$(2.5.4) \qquad \inf_{z_j} \max_{\nu \in S} \frac{|G_2(\nu)|}{M_d(\nu)}$$

and the corresponding extremal systems ("two-sided direct operator problems"). *Extremal problems of seventh type*: Let S be a finite set of integers,

d an integer with $0 \leq d \leq 3$, and let the z_j's be restricted to a domain U. Then determine

$$\inf_{z_1,\ldots,z_n \in U} \max_{\nu \in S} \frac{\operatorname{Re} G_1(\nu)}{M_d(\nu)} \tag{2.5.5}$$

and

$$\inf_{z_1,\ldots,z_n \in U} \max_{\nu \in S} \frac{\operatorname{Re} G_2(\nu)}{M_d(\nu)} \tag{2.5.6}$$

and the corresponding extremal systems ("one-sided conditional operator problems").

The name "operator" refers to the fact that for the rational function

$$f(z) = \sum_{h=1}^{n} \frac{b_h}{z - \alpha_h}$$

for fixed z_0 and variable a, the expression

$$\frac{(-1)^{a-1}}{(a-1)!} f^{(a-1)}(z_0)$$

is the generalized power sum

$$\sum_{h=1}^{n} b_h \left(\frac{1}{z_0 - \alpha_h} \right)^a$$

of the quantities $(z_0 - \alpha_h)^{-1}$ with the differential operator

$$\sum_{a=\nu}^{\nu+k} (-1)^{a-1} \frac{\beta_{a-\nu}}{(a-1)!} \frac{d^{a-1}}{dz^{a-1}} \qquad (\beta_j \text{ independent of } \nu)$$

acting on $f(z)$ (for fixed z_0 and variable ν) and is of type $G_2(\nu)$ in (2.5.2).

Problems of the sixth (resp. seventh) type are more general than those of the first (resp. third) type; so it is justifiable to ask why they are considered separately. There are several reasons. The proofs of results of the first (resp. third) type are much simpler than those of the sixth (resp. seventh) type, and most of the applications are derived from them. Hence the simplicity and elegance of these applications would suffer were they deduced from the much more difficult theorems of the sixth (resp. seventh) type. (The elegance of some applications suffers at any rate from the fact that, according to the etiquette of books, theory and applications are treated separately). Moreover, some of the results of the sixth (resp. seventh) type can be *deduced* (by suitable passage to the limit) from the corresponding results of the first (resp. third) type. Furthermore, *not all* theorems referring to generalized power sums, which are applied in the second part, have been extended to $G_j(\nu)$ expressions. Further reasons could be mentioned.

2.6

The types of problems we have listed are those for which we already have noteworthy results *and* applications, or have at least clear prospects of applications. Next we turn to other types of problems for which we have no significant results at this time but are aware of effective applications such results would have. Some of these problems occur explicitly among the open problems in Section 16, and some possible applications are discussed in Section 47. Here we indicate only their general types.

1. *The Simultaneous Problems.* Given a finite set S of integers, two generalized power sums

$$g_1(\nu) = \sum_{j=1}^{n} b_j' z_j'^{\nu}$$

(2.6.1)

$$g_2(\nu) = \sum_{j=1}^{n} b_j'' z_j''^{\nu}$$

(the b_j and z_j being arbitrary complex numbers), and d', d'' integers between 0 and 3, then for fixed b_j how large can the quantities

$$\frac{|g_1(\nu)|}{M_{d'}(\nu)} \quad \text{and} \quad \frac{|g_2(\nu)|}{M_{d''}(\nu)}$$

be made *simultaneously* depending only on n, S, and the b_j?

2. *The Several Variable Problems.* Given S_1 and S_2, two finite sets of integers,

$$h(\mu, \nu) = \sum_{j=1}^{n} b_j z_j^{\nu} w_j^{\mu},$$

$$|z_1| \geq |z_2| \geq \cdots \geq |z_n|$$

$$|w_1| \geq |w_2| \geq \cdots \geq |w_n|$$

and $0 \leq d', d'' \leq 3$, then for fixed b_j what is

$$\inf_{z_j, w_j} \max_{\substack{\nu \in S_1 \\ \mu \in S_2}} \frac{|h(\mu, \nu)|}{M_{d'}(\nu) M_{d''}(\mu)}$$

and what are the extremal systems?

Now we turn to the treatment of extremal problems of the first type of (2.2.1). The case

$$b_1 = b_2 = \cdots = b_n$$

that is, the "pure" power sums are important in some applications and so we commence with them.

3

PURE POWER SUMS

3.1

We start with the simplest of these problems, choosing

(3.1.1) $$S = (1, 2, \ldots, n)$$

and as norm the $M_1(\nu)$ norm defined in (1.3.3). We use throughout the notation

(3.1.2) $$s_\nu = z_1^\nu + \cdots + z_n^\nu.$$

Equivalently we want to determine

$$\min \max_{\nu=1,\ldots,n} |s_\nu|,$$

where the z_j variables are subject to the normalization

(3.1.3) $$\min_{j=1,\ldots,n} |z_j| = 1.$$

Then we assert (see Turán [32])

Theorem 3.1. *Under condition* (3.1.3) *the relation*

$$\min_{z_j} \max_{\nu=1,\ldots,n} |s_\nu| = 1$$

holds, and there is equality if and only if the z_j *are the* n *vertices of a regular* $(n+1)$-gon *on the unit circle.*

Before proving this theorem we formulate a corollary. Let $\xi_1,\ldots,\xi_n$ be arbitrary complex numbers with

$$|\xi_1| \geq \cdots \geq |\xi_n| > 0. \tag{3.1.4}$$

Then, choosing in Theorem 3.1

$$z_j = \frac{\xi_j}{\xi_n} \qquad j = 1,\ldots,n,$$

condition (3.1.3) is satisfied, and hence we get, for a suitable integer $1 \leq \nu_0 \leq n$,

$$|\xi_1^{\nu_0} + \cdots + \xi_n^{\nu_0}| \geq |\xi_n|^{\nu_0},$$

that is,

$$|\xi_n| \leq |\xi_1^{\nu_0} + \cdots + \xi_n^{\nu_0}|^{1/\nu_0} \leq \max_{\nu=1,\ldots,n} |\xi_1^{\nu} + \cdots + \xi_n^{\nu}|^{1/\nu};$$

we have established the following consequence.

Corollary. *For arbitrary complex numbers $\xi_1,\ldots,\xi_n$ the inequality*

$$\max_{\nu=1,\ldots,n} |\xi_1^{\nu} + \cdots + \xi_n^{\nu}|^{1/\nu} \geq \min_j |\xi_j| \tag{3.1.5}$$

holds. Equality holds if and only if $\xi_1,\ldots,\xi_n$ are n vertices of a regular $(n + 1)$-gon *with center in the origin.*

3.2. PROOF OF THEOREM 3.1

Fix a system $(z_1,\ldots,z_n)$ satisfying (3.1.3) and let

$$z^n + a_1 z^{n-1} + \cdots + a_n = 0 \tag{3.2.1}$$

be the equation with $z_1,\ldots,z_n$ as roots. Equation (3.1.3) gives

$$|a_n| \geq 1, \tag{3.2.2}$$

and hence

$$|a_l| = \max_j |a_j| \geq 1. \tag{3.2.3}$$

We apply the Newton–Girard formulas

$$s_\nu + a_1 s_{\nu-1} + \cdots + a_{\nu-1} s_1 + \nu a_\nu = 0 \qquad \nu = 1, 2,\ldots,n. \tag{3.2.4}$$

Taking the lth formula (3.2.4) and using (3.2.3) we get

$$l|a_l| = |s_l + a_1 s_{l-1} + \cdots + a_{l-1}s_1|$$
$$\leq (1 + |a_1| + \cdots + |a_{l-1}|) \max_{1\leq\nu\leq l} |s_\nu| \leq l|a_l| \max_{1\leq\nu\leq n} |s_\nu|.$$

Hence

$$\max_{\nu=1,\ldots,n} |s_\nu| \geq 1 \tag{3.2.5}$$

which proves the first half of the theorem.

3.3

Now we investigate when equality holds in (3.2.5). Marking with stars the quantities belonging to an extremal system we have

$$1 = |a_1^*| = \cdots = |a_l^*|.$$

But due to (3.2.2) we also have

$$|a_n^*| = 1.$$

That is, a_n^* can serve as a_l^*, so we get

$$1 = |a_1^*| = \cdots = |a_n^*| \tag{3.3.1}$$

and, of course,

$$|s_1^*| = \cdots = |s_n^*| = 1. \tag{3.3.2}$$

Obviously, together with $(z_1^*, \ldots, z_n^*)$, the n-tuple $(z_1^* e^{i\alpha}, z_2^* e^{i\alpha}, \ldots, z_n^* e^{i\alpha})$—with real α—is also an extremal system; hence we may choose α so that

$$s_1^* = -1 \quad \text{and} \quad a_1^* = 1. \tag{3.3.3}$$

We assert that

$$s_1^* = \cdots = s_n^* = -1$$
$$a_1^* = \cdots = a_n^* = 1. \tag{3.3.4}$$

Proof by induction: according to (3.3.3) we may suppose that

$$s_1^* = \cdots = s_\mu^* = -1$$

$$a_1^* = \cdots = a_\mu^* = 1,$$

with $1 \le \mu \le n - 1$ already proved. Using the $(\mu + 1)$-th equation from (3.2.4) we get

$$s^*_{\mu+1} = \mu - (\mu + 1)a^*_{\mu+1} \tag{3.3.5}$$

and hence from (3.3.2)

$$1 = |\mu - (\mu + 1)a^*_{\mu+1}|.$$

But this means, on the one hand, that $a^*_{\mu+1}$ is on a circle with radius $1/(\mu + 1)$ around $\mu/(\mu + 1)$. On the other hand, (3.3.1) gives $|a^*_{\mu+1}| = 1$ and hence $a^*_{\mu+1} = 1$; then from (3.3.5) we get $s^*_{\mu+1} = -1$. This completes the proof of Theorem 3.1.

For an analogous question, also interesting in its application, see Problem 3.

3.4

The situation with

$$S = (1, 2, \ldots, n) \tag{3.4.1}$$

and with the $M_2(\nu)$ norm, which amounts to

$$\max_{j=1,\ldots,n} |z_j| = 1, \tag{3.4.2}$$

is far more complex and far from being settled as nicely as with the $M_1(\nu)$ norm. Therefore we deal first with the case

$$S = (1, 2, \ldots, 2n - 1), \tag{3.4.3}$$

attributable to Cassels (Cassels [1] and also Ostrowski [1]).

Theorem 3.2 (Cassels). *Under condition* (3.4.2) *we have*

$$\min_{z_j} \max_{\nu=1,\ldots,2n-1} |s_\nu| = 1.$$

Equality can be attained, for example, for

$$z_1 = e^{i\alpha}, \qquad z_2 = z_3 = \cdots = z_n = 0 \quad (\alpha \text{ real}). \tag{3.4.4}$$

Once again we formulate a corollary of this theorem. With ξ_j as in (3.1.4) we can apply this theorem with $z_j = \xi_j/\xi_1$ $(j = 1, \ldots, n)$, thereby obtaining as in

3.1 the following:

Corollary. *For arbitrary complex* $\xi_1,\ldots,\xi_n$ *the inequality*

$$\max_{\nu=1,\ldots,2n-1} |\xi_1^\nu + \cdots + \xi_n^\nu|^{1/\nu} \geq \max_j |\xi_j| \tag{3.4.5}$$

holds. Equality can be attained.

The first result in this direction was due to N. Schweitzer (oral communication) with $2n$ instead of $2n - 1$ and $\frac{1}{2}$ instead of 1. We shall see that the quantity $(2n - 1)$ in (3.4.5) cannot be replaced by $(2n - 2)$.

Inequality (3.4.5) implies several other inequalities (which are not sharp). For example, for arbitrary complex ξ_j the inequality

$$\max_{\nu=1,\ldots,2n-1} |\xi_1^\nu + \cdots + \xi_n^\nu|^{1/\nu} \geq \frac{1}{n}(|\xi_1| + \cdots + |\xi_n|) \tag{3.4.6}$$

holds. Later we do not display such consequences unless we have applications.

3.5

Cassels' proof is based on an elegant lemma. We call a system $(w_1, w_2,\ldots,w_k)$ a *real system* if it is symmetric to the real axis. Then all power sums

$$\sigma_\nu = w_1^\nu + \cdots + w_k^\nu$$

are real. He needs the following.

Lemma 3.1. *For a real* $(w_1,\ldots,w_k)$*-system the inequality*

$$\max_{\nu=1,2,\ldots,k+1} \sigma_\nu \geq 0$$

holds.

Proof. Let

$$w^k + a_1 w^{k-1} + \cdots + a_k = 0 \tag{3.5.1}$$

be the equation with $w_1,\ldots,w_k$ as roots. The a_ν are obviously real. If

$$\max_{\nu=1,\ldots,k} \sigma_\nu \geq 0$$

we have nothing to prove. If

$$\max_{\nu=1,\ldots,k} \sigma_\nu < 0 \tag{3.5.2}$$

then the formulas (3.2.4) give successively

$$a_1 > 0, a_2 > 0, \ldots, a_k > 0.$$

But because

$$\sigma_{k+1} + a_1\sigma_k + \cdots + a_k\sigma_1 = 0$$

we have $\sigma_{k+1} > 0$. ■

Proof of Theorem 3.2. Suppose, without loss of generality, that

$$z_1 = \max_j |z_j| = 1. \tag{3.5.3}$$

Adding to $(z_2, z_3, \ldots, z_n)$ the points $(\overline{z_2}, \ldots, \overline{z_n})$, the new system forms a real system with $k = 2n - 2$. Thus Lemma 3.1 is applicable and hence

(3.5.4)

$$2 \max_{\nu=1,\ldots,2n-1} \mathrm{Re}(z_2^\nu + \cdots + z_n^\nu) = \max_{\nu=1,\ldots,2n-1} \left(\sum_{j=2}^{n} z_j^\nu + \sum_{j=2}^{n} \left(\overline{z_j}\right)^\nu \right) \geq 0.$$

If the maximum in (3.5.4) is realized for $\nu = \nu_0$ then—using the notation (3.1.2)—we have

$$\max_{\nu=1,\ldots,(2n-1)} |s_\nu| \geq |s_{\nu_0}| = \left|1 + \left(z_2^{\nu_0} + \cdots + z_n^{\nu_0}\right)\right|$$

$$\geq 1 + \mathrm{Re}\left(z_2^{\nu_0} + \cdots + z_n^{\nu_0}\right) \geq 1. \quad ■$$

3.6

In contrast to Theorem 3.1 we see that Theorem 3.2 made no claim to account for *all* extremal systems. The matter was cleared up by S. Dancs (Dancs [1]). He found the surprising fact that there are many essentially different extremal systems in Theorem 3.2 and even the cardinality of the different systems with

$$\max_{\nu=1,\ldots,3n-4} |s_\nu^*| = 1, \qquad n \geq 3$$

is that of the continuum. He showed further that Theorem 3.2 is the best possible result in the sense that the *only* systems with

$$\max_{\nu=1,\ldots,3n-3} |s_\nu^*| = 1$$

are those in (3.4.4). For details we refer to his paper.

3.7

Applications to numerical analysis raise the question whether or not the set $(1,2,\ldots,2n-1)$ in Theorem 3.2 can be replaced by $S=(1,2,\ldots,2n-2)$ while keeping the value 1 as minimax value. This was first refuted by M. Bowen (written communication). A simpler example was given in Cassels [1]. He exhibited a system $(z_1^*,\ldots,z_n^*)$ with $\max_j|z_j^*|=1$ and

$$\max_{\nu=1,2,\ldots,(2n-2)}|s_\nu^*|<1-\frac{1}{2^{2n}n^{4n-4}}. \tag{3.7.1}$$

His construction was improved by M. Szalay [1] who gives a system (for $n\geq 2$) $(z_1^{**},\ldots,z_n^{**})$ with $\max_j|z_j^{**}|=1$ and

$$\max_{\nu=1,\ldots,2n-2}|s_\nu^{**}|<1-\frac{1}{16n^4}. \tag{3.7.2}$$

Namely, let

$$z_1^{**}=1,\qquad z_j^{**}=\alpha\exp\frac{2\pi i(2j-3)}{2n-1}\qquad (j=2,\ldots,n) \tag{3.7.3}$$

where $0<\alpha<1$ is to be determined. One easily obtains

$$s_\nu^{**}=1-\frac{\alpha^\nu}{2}+i\frac{\alpha^\nu}{2}\operatorname{tg}\frac{\nu\pi}{2n-1},$$

that is,

$$|s_\nu^{**}|^2=1-\alpha^\nu+\frac{\alpha^{2\nu}}{4\cos^2\dfrac{\nu\pi}{2n-1}}\qquad (\nu=1,2,\ldots,2n-2). \tag{3.7.4}$$

Choosing

$$\alpha=\left(n\sqrt{2}\right)^{-2/(n-1)}, \tag{3.7.5}$$

(3.7.4) yields, for $1\leq\nu\leq n-1$,

$$|s_\nu^{**}|\leq 1-\frac{1}{8n^2}<1-\frac{1}{16n^4}$$

and for $n\leq\nu\leq 2n-2$

$$|s_\nu^{**}|^2\leq 1-\alpha^\nu+n^2\alpha^{2\nu}\leq 1-\frac{\alpha^\nu}{2}$$

$$|s_\nu^{**}|\leq 1-\frac{\alpha^\nu}{4}\leq 1-\frac{1}{16n^4}.$$

3.8

The determination of

$$R_n = \min \max_{\nu=1,\ldots,n} |s_\nu|, \tag{3.8.1}$$

under the normalization (3.4.2) (where Weierstrass' theorem yields the existence of the minimum) is much more difficult. One can easily see that

$$R_2 = \frac{\sqrt{5}-1}{\sqrt{2}} \sim 0.87403$$

with equality only for the systems

$$\left(1, \frac{\sqrt{5}-1}{2} e^{\pm(2\pi i/3)}\right).$$

In two papers, Ławrynowicz ([1, 2]) succeeded in proving by variational methods that

$$R_3 \sim 0.82478,$$

the only extremal systems being

$$\left(1, z_1^*, z_2^*\right) \quad \text{and} \quad \left(1, \overline{z_1^*}, \overline{z_2^*}\right)$$

where

$$z_1^* \sim -0.583346 + i0.091465$$

$$z_2^* \sim 0.089975 - i0.742306.$$

Hence R_n shows a decreasing tendency—at least at the beginning. The first lower bound, which is currently the simplest and is still useful, is given by the following.

Theorem 3.3. *The inequality*

$$R_n > \frac{\log 2}{\frac{1}{1} + \frac{1}{2} + \cdots + \frac{1}{n}}$$

holds.

Proof. Suppose again that

$$z_1 = 1. \tag{3.8.2}$$

Let our $(z_1,\ldots,z_n)$-system be the zeros of

$$z^n + a_1 z^{n-1} + \cdots + a_n = 0 \tag{3.8.3}$$

and let

$$\max_{\nu=1,\ldots,n} |s_\nu| = \mu. \tag{3.8.4}$$

The formulas (3.2.4) give successively

$$|a_1| \leq \mu = \binom{\mu}{1}$$

$$2|a_2| \leq \mu + \mu^2, \qquad |a_2| \leq \binom{\mu+1}{2}$$

and an easy induction shows

$$|a_\nu| \leq \binom{\mu+\nu-1}{\nu} \qquad (\nu = 1,2,\ldots,n). \tag{3.8.5}$$

Now (3.8.2) and (3.8.3) give

$$1 = -(a_1 + \cdots + a_n).$$

This and (3.8.5) give

$$1 \leq \binom{\mu}{1} + \binom{\mu+1}{2} + \cdots + \binom{\mu+n-1}{n} = \binom{\mu+n}{n} - 1$$

or

$$2 \leq \left(1+\frac{\mu}{n}\right)\left(1+\frac{\mu}{n-1}\right)\cdots\left(1+\frac{\mu}{1}\right) < \exp\left\{\mu\left(\frac{1}{1}+\cdots+\frac{1}{n}\right)\right\}. \quad \blacksquare$$

3.9

Theorem 3.3 made plausible the conjecture that $R_n \geq A$, for some constant A independent of n. The only known proof of this conjecture is due to Atkinson (Atkinson [1]) as follows:

Theorem 3.4 (Atkinson). *With the notation of* (3.8.1) *the inequality*

$$R_n > \tfrac{1}{6} \tag{3.9.1}$$

holds.

Remarks. As before, this gives the inequality

$$\max_{\nu=1,\ldots,n} \left|6(\xi_1^\nu + \cdots + \xi_n^\nu)\right|^{1/\nu} > \max_{j=1,\ldots,n} |\xi_j| \tag{3.9.2}$$

for arbitrary complex ξ_j.

Atkinson proved further (Atkinson [2]) that

$$R_n > \frac{\pi}{8} \quad \text{for} \quad n < 1.6 \cdot 10^3 \tag{3.9.3}$$

and, with an unspecified n_0,

$$R_n > \vartheta \quad \text{for} \quad n > n_0; \tag{3.9.4}$$

where

$$T(x) = \int_0^x \frac{|\sin r|}{r}\, dr, \tag{3.9.5}$$

ϑ is the (only) root in $(0, \frac{1}{2})$ of the transcendental equation

$$\frac{x^2}{2\pi} \int_0^\infty \left(\frac{-1 + \exp(2xT(r))}{r} \right)^2 dr = 1. \tag{3.9.6}$$

For details of the intricate proof we refer to the original paper and confine ourselves to proving (3.9.1).

Let us remark—and in the applications this is *not* inessential—that for $n < \exp 3.54$ Theorem 3.3 gives a better lower bound than Theorem 3.4.

3.10. PROOF OF THEOREM 3.4

Suppose again that

$$z_1 = 1 \qquad \left(\text{and also } R_n < \tfrac{1}{4}\right). \tag{3.10.1}$$

Since, due to (3.4.2), we have

$$\prod_{j=1}^n (1 - z_j z) = \exp\left\{ -\sum_{\nu=1}^\infty \frac{s_\nu}{\nu} z^\nu \right\}, \quad \text{for} \quad |z| < 1$$

and

$$\exp\left\{ \sum_{\nu=n+1}^\infty \frac{s_\nu}{\nu} z^\nu \right\} = 1 + \sum_{\nu=n+1}^\infty c_\nu z^\nu$$

where

$$c_\nu = c_\nu(z_1,\dots,z_n),$$

we get with

$$G(z) = -\sum_{\nu=1}^{n} \frac{s_\nu}{\nu} z^\nu \tag{3.10.2}$$

the relation

$$\begin{aligned} \exp G(z) &= \prod_{j=1}^{n} (1 - z_j z)\left\{1 + \sum_{\nu=n+1}^{\infty} c_\nu z^\nu\right\} \\ &= \prod_{j=1}^{n} (1 - z_j z) + \sum_{\nu=n+1}^{\infty} c'_\nu z^\nu \end{aligned} \tag{3.10.3}$$

with some $c'_\nu = c'_\nu(z_1,\dots,z_n)$. This last identity obviously holds for all complex z. Thus for $z = 1$ we get

$$\exp G(1) = \sum_{\nu=n+1}^{\infty} c'_\nu. \tag{3.10.4}$$

Since for $\nu \geq n + 1$ and $r > 1$

$$c'_\nu = \frac{1}{2\pi i}\int_{|z|=r} z^{-\nu-1}\exp G(z)\,dz$$

we get from (3.10.4)

$$\begin{aligned} \exp G(1) &= \frac{1}{2\pi i}\int_{|z|=r} \exp G(z)\left(\sum_{\nu=n+1}^{\infty} z^{-\nu-1}\right) dz \\ &= \frac{1}{2\pi i}\int_{|z|=r} G'(z)\exp G(z)\left\{\sum_{\nu=n+1}^{\infty} \frac{z^{-\nu}}{\nu}\right\} dz \end{aligned}$$

or

$$1 = \frac{1}{2\pi i}\int_{|z|=r} G'(z)\exp(G(z) - G(1))\left\{\sum_{\nu=n+1}^{\infty} \frac{z^{-\nu}}{\nu}\right\} dz.$$

Passage to the limit $r \to 1$ gives, with the notation

$$-\sum_{\nu=1}^{n} \frac{s_\nu}{\nu} e^{\nu i\vartheta} = G_1(\vartheta)$$

$$\sum_{\nu=n+1}^{\infty} \frac{\exp(-\nu i\vartheta)}{\nu} = h(\vartheta) \tag{3.10.5}$$

the important identity

$$(3.10.6)\quad 1 = -\frac{1}{2\pi}\int_{-\pi}^{\pi}\left(\sum_{1}^{n} s_\nu e^{\nu i\vartheta}\right)\exp(G_1(\vartheta) - G_1(0))h(\vartheta)\,d\vartheta.$$

3.11

Cauchy's inequality then gives

$$(3.11.1)\quad 1 \le \frac{1}{2\pi}\left(\sum_{\nu=1}^{n}|s_\nu|^2\right)\left(\int_{-\pi}^{\pi}|h(\vartheta)|^2\exp\{2\operatorname{Re}(G_1(\vartheta) - G_1(0))\}\,d\vartheta\right)$$

$$\le \frac{nR_n^2}{2\pi}\left\{\int_{-\pi}^{-\pi/n}\cdots + \int_{-\pi/n}^{\pi/n}\cdots + \int_{\pi/n}^{\pi}\cdots\right\}.$$

First consider the middle term. Since

$$(3.11.2)\quad |G_1(\vartheta) - G_1(0)| \le \sum_{\nu=1}^{n}\frac{|s_\nu|}{\nu}|e^{\nu i\vartheta} - 1| \le |\vartheta|\sum_{\nu=1}^{n}|s_\nu| \le nR_n|\vartheta|$$

holds, this integral cannot exceed the value

$$(3.11.3)\quad e^{2\pi R_n}\int_{-\pi/n}^{\pi/n}|h(\vartheta)|^2\,d\vartheta < \frac{2\pi}{n}e^{2\pi R_n}$$

by (3.10.5) and Parseval's formula.

Now consider the last integral in (3.11.1). First we remark that, due to the inequality

$$|(1 - e^{-i\vartheta})h(\vartheta)| = \left|\frac{e^{-(n+1)i\vartheta}}{n+1} - \sum_{\nu=n+2}^{\infty}\left(\frac{1}{\nu-1} - \frac{1}{\nu}\right)e^{-\nu i\vartheta}\right| \le \frac{2}{n+1},$$

the inequality

$$(3.11.4)\quad |h(\vartheta)| \le \frac{|(1 - e^{-i\vartheta})h(\vartheta)|}{(2/\pi)\vartheta} < \frac{\pi}{n\vartheta}$$

also holds in our range. Furthermore, we have

$$|G_1(\vartheta) - G_1(0)| = \left|\sum_{\nu\le\pi/\vartheta}\cdots + \sum_{\pi/\vartheta<\nu\le n}\frac{s_\nu}{\nu}(e^{\nu i\vartheta} - 1)\right|.$$

Using inequality (3.11.2) this is

$$\le R_n \vartheta \frac{\pi}{\vartheta} + 2R_n \sum_{\pi/\vartheta < \nu \le n} \frac{1}{\nu} < \pi R_n + 2R_n\left(\log\frac{n\vartheta}{\pi} + 1\right).$$

Hence

$$\int_{\pi/n}^{\pi} |h(\vartheta)|^2 \exp\{2\,\mathrm{Re}(G_1(\vartheta) - G_1(0))\}\, d\vartheta$$

$$\le \exp\{2R_n(\pi + 2)\} \int_{\pi/n}^{\pi} \left(\frac{n\vartheta}{\pi}\right)^{4R_n - 2} d\vartheta$$

$$< \frac{\pi}{n}\frac{1}{1 - 4R_n} \exp\{2R_n(\pi + 2)\}$$

by the assumption (3.10.1). Collecting (3.11.1), (3.11.3), and (3.11.5) we obtain

$$1 \le \frac{nR_n^2}{2\pi}\left\{\frac{2\pi}{n}\exp(2\pi R_n) + \frac{2\pi}{n}\frac{\exp(2R_n(\pi + 2))}{1 - 4R_n}\right\}$$

$$= R_n^2 \exp(2\pi R_n)\left\{1 + \frac{e^{4R_n}}{1 - 4R_n}\right\}.$$

Since the function

$$f(x) = x^2 e^{2\pi x}\left\{1 + \frac{e^{4x}}{1 - 4x}\right\}$$

increases monotonically for $0 \le x < \frac{1}{4}$ and because $f(\frac{1}{6}) < 1$ Theorem 3.4 is proved.

3.12

It follows from (3.7.1) that

$$R_n < 1 \quad \text{for} \quad n \ge 2. \tag{3.12.1}$$

As concerns the *exact* value of $\min \max_{\nu=1,\ldots,n} |s_\nu|$ with the normalization (3.4.2), the difficulty is emphasized by two facts. Superficial reasoning, based on the maximum principle for analytic functions, would imply that for the extremal system $(1, z_2^*, \ldots, z_n^*)$ we have

$$|z_j^*| = 1. \tag{3.12.2}$$

But then Theorem 3.1 would contradict (3.12.1). So (3.12.2) is not true. The further guess that the extremal system is a real system in the sense of 3.5 is refuted by Cassel's Lemma 3.1, applied to $(z_2^*,\ldots,z_n^*)$. This would give

$$\max_{\nu=1\ldots n} |1 + z_2^{*\nu} + \cdots + z_n^{*\nu}| \geq 1$$

in contradiction to (3.12.1).

An important property of the family of extremal systems is that *some* of the extremal systems will satisfy

$$|s_1^*| = |s_2^*| = \cdots = |s_n^*|, \tag{3.12.3}$$

where s_ν^* denotes their power sum. This fact due to Ławrynowicz (Ławrynowicz [1]), can be proved as follows. First we observe that

$$R_n = \min_{z_1=1} \max_{\nu=1,\ldots,n} |s_\nu|. \tag{3.12.4}$$

For this, one must only show that for every extremal system of the right side

$$\max_{j=1,\ldots,n} |z_j^*| \leq 1$$

is automatically satisfied. But if we had

$$\max_{j=2,\ldots,n} |z_j^*| = |z_2^*| > 1$$

then for the system

$$z_j^{**} = \frac{z_j^*}{z_2^*} \qquad (j = 1,\ldots,n)$$

we obviously have

$$|s_\nu^{**}| < |s_\nu^*|, \qquad R_n \leq \max_{\nu=1,\ldots,n} |s_\nu^{**}| < \max_{\nu=1,\ldots,n} |s_\nu^*|,$$

for $\nu = 1, 2,\ldots,n$ which contradicts the minimum property.* Hence we can deal with the form of the extremal problem given by the right side of (3.12.4). Second, it follows from the Newton–Girard formulas (3.2.4) that—apart from a constant factor—the only polynomial of nth degree with prescribed first-

*It is obvious that every extremal system for R_n yields an extremal system for the right side of (3.12.4) by a suitable rotation.

power sums is given by the determinant

$$f(z)=\begin{vmatrix} z^n & z^{n-1} & z^{n-2} & \cdots & z & 1 \\ s_1 & 1 & 0 & \cdots & 0 & 0 \\ s_2 & s_1 & 2 & \cdots & 0 & 0 \\ \vdots & \vdots & \vdots & & \vdots & \vdots \\ s_{n-1} & s_{n-2} & s_{n-3} & \cdots & (n-1) & 0 \\ s_n & s_{n-1} & s_{n-2} & \cdots & s_1 & n \end{vmatrix}.$$

Hence the fact that $z_1 = 1$ means that

$$\text{(3.12.5)} \qquad F(s_1,\dots,s_n)=\begin{vmatrix} 1 & 1 & 1 & \cdots & 1 \\ s_1 & 1 & 0 & \cdots & 0 \\ s_2 & s_1 & 2 & \cdots & 0 \\ \vdots & \vdots & \vdots & & \vdots \\ s_n & s_{n-1} & s_{n-2} & \cdots & n \end{vmatrix} = 0.$$

Thus the new form of our extremal problem requires us to find the points $(s_1^*,\dots,s_n^*)$ in the n-complex dimensional space which are on the algebraic surface (3.12.5) and for which

$$\text{(3.12.6)} \qquad \max_{\nu=1,\dots,n} |s_n^*|$$

is minimal. We now show that (3.12.3) holds for *some* of these points. Suppose, to the contrary, that

$$(s_1^*,\dots,s_n^*)$$

is such an extremal system where the number k of the coordinates s_ν^* with

$$\text{(3.12.7)} \qquad |s_\nu^*| < R_n$$

is minimal and ≥ 1. Let s^* be such a coordinate and consider the polynomial

$$\text{(3.12.8)} \qquad F(s_1^*,\dots,s_{l-1}^*, s_l, s_{l+1}^*,\dots,s_n^*)$$

in s_l. If this is identically zero, then s_l can be chosen as R_n, and this settles the case $k = 1$. In case $k > 1$ the system

$$(s_1^*,\dots,s_{l-1}^*, R_n, s_{l+1}^*,\dots,s_n^*)$$

would satisfy $F = 0$ in contradiction to the minimality requirement of (3.12.7).

If the polynomial in (3.12.8) is not identically zero then, because the zeros of an algebraic equation depend continuously on the coefficients, a sufficiently small change of the quantities

$$s_1^*, \dots, s_{l-1}^*, s_{l+1}^*, \dots, s_n^*$$

to

$$s_1^{**}, \dots, s_{l-1}^{**}, s_{l+1}^{**}, \dots, s_n^{**} \quad \text{with} \quad |s_j^{**}| < |s_j^*| \qquad (j \neq l)$$

yields a quantity s_l^{**} determined by

$$F\left(s_1^{**}, \dots, s_{l-1}^{**}, s_l^{**}, \dots, s_n^{**}\right) = 0$$

which still satisfies

$$|s_l^{**}| < R_n.$$

Thus

$$\max_{\nu=1,\dots,n} |s_\nu^{**}| < R_n = \max_{\nu=1,\dots,n} |s_\nu^*|,$$

a contradiction.

3.13

The values R_2, R_3 given together with Theorem 3.4 suggest that R_n is between $\frac{1}{6}$ and 0.88 and tends to a limit as $n \to \infty$. As to the upper *bound* of R_n, the best results are due to Komlós–Sárközy–Szemerédi [1] according to which

$$R_n < 1 - \frac{1}{250n} \quad \text{for} \quad n > n_0$$

and

$$R_n < 1 - \frac{1}{3}\frac{\log n}{n}, \quad \text{for infinitely many } n.$$

For the proofs we refer to their paper.

3.14

The system $(1, 0, \dots, 0)$ excludes the possibility of having *any* $N = N(n)$ such that

$$\min_{z_j} \max_{\nu=1,\dots,N} |s_\nu| > 1 \tag{3.14.1}$$

under the restriction $\max_j |z_j| = 1$. To get such lower bounds one must exclude this system, making appropriate restrictions on the z_j. The only known case of a *complete* solution is the following.

Theorem 3.5. *Let $n = 2k + 1$. In addition to the requirement*

$$z_1 = \max_j |z_j| = 1 \tag{3.14.2}$$

we subject the z_j to the restriction

$$s_1 = s_2 = \cdots = s_k = 0.$$

Then

$$\min_{z_j} \max_{\nu=1,\ldots,2k+1} |s_\nu| = \left(\sum_{j=k+1}^{2k+1} \frac{1}{j} \right)^{-1}$$

($\sim 1/\log 2$, *if k is large*) *with equality for, and only for, the roots of the equation**

$$z^{2k+1}\left(\sum_{j=k+1}^{2k+1} \frac{1}{j} \right) - \sum_{l=0}^{k} \frac{z^l}{2k+1-l} = 0. \tag{3.14.3}$$

Proof. Let

$$z^{2k+1} + a_1 z^{2k} + \cdots + a_{2k+1} = 0$$

be the equation whose roots are our z_j. The first k of the formulas (3.2.4) give successively

$$a_1 = \cdots = a_k = 0.$$

The $(k+1)$th formula gives

$$a_{k+1} = -\frac{s_{k+1}}{k+1}$$

and

$$a_j = -\frac{s_j}{j} \qquad \text{for } j = k+1,\ldots,2k+1.$$

These, together with $z_1 = 1$, at once give

$$\sum_{j=k+1}^{2k+1} \frac{s_j}{j} = 1,$$

*The proof shows that an analogous theorem can be proved mutatis mutandis for even n too.

that is,

$$\max_{\nu=1,\ldots,2k+1} |s_\nu| \geq \left(\sum_{j=k+1}^{2k+1} \frac{1}{j} \right)^{-1}.$$

Equality holds if and only if

$$s_l = \left(\sum_{j=k+1}^{2k+1} \frac{1}{j} \right)^{-1}; \qquad l = k+1,\ldots,2k+1$$

which shows that the only extremal system [after the normalization (3.14.2)] is given by the roots of the equation in* (3.14.3).

3.15

Obtaining a substantial improvement of some existing applications in Section 31 led Buchholtz ([1, 2]) to the consideration of an extremal problem of the fourth type. Choose, in (2.4.1),

$$d = 2, \qquad b_1 = \cdots = b_n = 1, \qquad \psi(\nu, n) = \frac{1}{n}.$$

Theorem 3.6 (Buchholtz). *For all systems* $(z_1,\ldots,z_n)$ *with* $\max_j |z_j| = 1$ *the inequality*

(3.15.1) $$M \doteq \max_{\nu=1,\ldots,n} \left(\frac{|s_\nu|}{n} \right)^{1/\nu} > \frac{1}{5}$$

holds.

Perhaps it is worth mentioning the following consequence.

Corollary. *For arbitrary complex numbers* $\xi_1,\ldots,\xi_n$ *we have the inequality*

(3.15.2) $$\max_{\nu=1,\ldots,n} \left(\frac{|\xi_1^\nu + \cdots + \xi_n^\nu|}{n} \right)^{1/\nu} \geq \frac{1}{5} \max_{j=1,\ldots,n} |\xi_j|.$$

In spite of appearances, for some purposes (3.15.2) is better than (3.9.2).

*One can also see directly that the roots of the equation (3.14.3) are in $|z| \leq 1$.

3.16. PROOF OF (3.15.1)

Suppose that

(3.16.1) $$z_1 = 1.$$

Using (3.10.4) we have for all $r > 1$

(3.16.2) $$\exp\left(-\sum_{\nu=1}^{n} \frac{|s_\nu|}{\nu}\right) \le \left|\exp\left(-\sum_{\nu=1}^{n} \frac{s_\nu}{\nu}\right)\right|$$

$$= \left|\sum_{\nu=n+1}^{\infty} c'_\nu\right| \le \frac{1}{r^n} \sum_{\nu=n+1}^{\infty} |c'_\nu| r^\nu.$$

Writing (3.10.3) in the form

(3.16.3) $$\exp G(z) = \sum_{\nu=0}^{\infty} c'_\nu z^\nu$$

we get

$$\sum_{\nu=0}^{\infty} |c'_\nu| r^\nu \le \exp\left(\sum_{\nu=1}^{n} \frac{|s_\nu|}{\nu} r^\nu\right)$$

and hence from (3.16.2)

$$\exp\left(-\sum_{\nu=1}^{n} \frac{|s_\nu|}{\nu}\right) \le \frac{1}{r^n} \exp\left(\sum_{\nu=1}^{n} \frac{|s_\nu|}{\nu} r^\nu\right).$$

In other words,

(3.16.4) $$\sum_{\nu=1}^{n} \frac{|s_\nu|}{\nu}(1 + r^\nu) \ge n \log r$$

or

(3.16.5) $$\sum_{\nu=1}^{n} \left\{\left(\frac{|s_\nu|}{n}\right)^{1/\nu}\right\}^{\nu} \frac{1 + r^\nu}{\nu} \ge \log r.$$

Under the assumption

$$\max_{\nu=1,\ldots,n} \left(\frac{|s_\nu|}{n}\right)^{1/\nu} \le \frac{1}{5},$$

(3.16.5) would imply

$$\sum_{\nu=1}^{n} \frac{1}{5^\nu} \frac{1+r^\nu}{\nu} \geq \log r, \quad \text{for } r > 1. \tag{3.16.6}$$

Restricting r to $1 < r < 5$, this implies

$$\log\left(\frac{1}{\left(1-\frac{r}{5}\right)r}\frac{5}{4}\right) \geq \sum_{\nu=n+1}^{\infty} \frac{1}{\nu}\frac{1}{5^\nu},$$

which is a contradiction for $r = \frac{5}{2}$.

3.17

Buchholtz also showed the inequality

$$\max_{\nu=1,\dots,n} \left(\frac{|s_\nu|}{n}\right)^{1/\nu} \geq \frac{1}{2(1+\sqrt{2})}, \tag{3.17.1}$$

and this is the best possible result. To prove this he looks for the system satisfying

$$s_1 = s_2 = \cdots = s_n = -n;$$

this is given—according to formulas (3.2.4)—by the roots of the equation

$$z^n + \binom{n}{1} z^{n-1} + \binom{n+1}{2} z^{n-2} + \cdots + \binom{2n-1}{n} = 0. \tag{3.17.2}$$

Asymptotic representation of this polynomial shows that its zeros cluster to each point of the curve

$$\left|\frac{z^2}{z-1}\right| = 4,$$

and the point of this curve farthest from the origin is the point $z = -2(1+\sqrt{2})$.

4

PURE POWER SUMS (CONTINUATION)

4.1

In the case when the set S of integers consists of

(4.1.1) $$(2, 3, \ldots, n, n+1) = S_2$$

an unexpected phenomenon was discovered by P. Erdős. Knowing Theorem 3.4, one might expect for

(4.1.2) $$M_2 \doteq \min_{z_j} \max_{\nu \in S_2} |s_\nu|$$

—with the restriction $\max_j |z_j| = 1$—a lower bound like $\log^{-2}(n)$ or even $1/n$. But he found the following.

Theorem 4.1 (Erdős). *For the quantity M_2 in (4.1.2) the inequality*

$$M_2 \leq 2(n+1)^2 e^{-\vartheta_0 n}$$

holds where ϑ_0 approximately equal to 0.2784 is the (positive) root of tne equation

(4.1.3) $$xe^{x+1} = 1.$$

Since

$$e^{-\vartheta_0} \sim \frac{1}{1.321}$$

this gives

$$M_2 < \frac{1}{1.32^n}, \quad \text{for } n > n_0. \tag{4.1.4}$$

4.2. PROOF

Consider the equation

$$z^n + a_1 z^{n-1} + \cdots + a_n = 0 \tag{4.2.1}$$

whose roots satisfy

$$z_1^* = 1 \tag{4.2.2}$$

$$s_1^* = \vartheta n \tag{4.2.3}$$

($0 < \vartheta < 1$ to be determined) and

$$s_2^* = \cdots = s_{n-1}^* = 0. \tag{4.2.4}$$

Formulas (3.2.4) give

$$a_\nu = \frac{\left(-s_1^*\right)^\nu}{\nu!} \qquad (\nu = 1, 2, \ldots, n-1).$$

Furthermore, (4.2.2) gives

$$a_n = -(1 + a_1 + \cdots + a_{n-1}) = -V_{n-1} \tag{4.2.5}$$

where

$$V_k = \sum_{j=0}^{k} \frac{\left(-s_1^*\right)^j}{j!}. \tag{4.2.6}$$

To give an upper bound for M_2 from (4.1.2) we need s_n^* and s_{n+1}^*. Equations (3.2.4) and (4.2.4) give

$$s_n^* + a_{n-1}s_1^* + na_n = 0.$$

With the notation (4.2.5) we have

$$s_n^* = nV_n. \tag{4.2.7}$$

Furthermore, one can see that

$$s_{n+1}^* + a_1 s_n^* + a_n s_1^* = 0,$$

that is,

$$s_{n+1}^* = s_1^*(nV_n + V_{n-1}). \tag{4.2.8}$$

Since

$$V_k = \exp(-s_1^*) - \sum_{j=k+1}^{\infty} \frac{(-s_1^*)^j}{j!}$$

and for $j > \vartheta n$ the terms $(\vartheta n)^j/j!$ are monotonically decreasing, we get

$$|s_n^*| \le n\left(e^{-\vartheta n} + \frac{(\vartheta n)^n}{n!}\right)$$

$$\begin{aligned} |s_{n+1}^*| &\le \vartheta n\left\{(n+1)e^{-\vartheta n} + n\frac{(\vartheta n)^{n+1}}{(n+1)!} + \frac{(\vartheta n)^n}{n!}\right\} \\ &< \vartheta n(n+1)\left(e^{-\vartheta n} + \frac{(\vartheta n)^n}{n!}\right), \end{aligned}$$

that is,

$$\begin{aligned} M_2 \le M^* = \max(|s_n^*|, |s_{n+1}^*|) &< (n+1)^2\left\{e^{-\vartheta n} + \frac{(\vartheta n)^n}{n!}\right\} \\ &< (n+1)^2\{e^{-\vartheta n} + (\vartheta e)^n\}. \end{aligned}$$

Choosing ϑ as the positive solution of Equation (4.1.3), the assertion follows if $\max|z_j^*| = 1$ is satisfied. Otherwise we have

$$\max_j |z_j^*| = |z_2^*| > 1,$$

and the system $(z_1^*/z_2^*, z_2^*/z_2^*, \ldots, z_n^*/z_2^*)$ satisfies the assertion a fortiori.

4.3

The importance of the problem of determining

$$\min_{\substack{z_j \\ (\max_j |z_j| = 1)}} \quad \max_{\substack{m+1 \le \nu \le m+n \\ \nu \text{ integer}}} |s_\nu| \tag{4.3.1}$$

(see the whole of Part II) emphasizes the importance of examples limiting the possibilities of improvements. For the construction of such examples it is useful to study nontrivial systems (those excluding $z_1 = \cdots = z_n = 0$) that for a given integer $m \geq 0$ satisfy

$$s_{m+1} = s_{m+2} = \cdots = s_{m+n-1} = 0. \tag{4.3.2}$$

Having a command of all such systems can be useful in another context too. Let ε_j denote the nth roots of unity and let $f(z) = \sum_{k=0}^{\infty} a_k z^k$ be an entire function. Then the formula

$$\frac{1}{n} \sum_{j=1}^{n} f(z\varepsilon_j) = \sum_{k=0}^{\infty} a_{nk} z^{nk}$$

is useful in a variety of ways. Having a system $(\xi_1, \ldots, \xi_n)$ with (4.3.2) we then have the formula

$$\sum_{j=1}^{n} f(z\xi_j) = \sum_{k=0}^{m} a_k s_k z^k + \sum_{k=m+n}^{\infty} a_k s_k z^k, \tag{4.3.3}$$

useful in the study of the effect of the coefficients $a_{m+1}, \ldots, a_{m+n-1}$. The case $m = 0$ is trivial by formulas (3.2.4). All such systems are derived from all the zeros of $z^n - 1 = 0$ by rotation and dilatation. For the case $m = 1$ one can see (as in 4.2) that all systems with $s_2 = \cdots = s_n = 0$ can be derived from the zeros of

$$z^n + \frac{z^{n-1}}{1!} + \frac{z^{n-2}}{2!} + \cdots + \frac{1}{n!} = 0 \tag{4.3.4}$$

by rotation and dilatation. The case $m = 2$ is more intricate. Defining as usual the νth Hermitean polynomial $H_\nu(x)$ by

$$e^{-x^2} H_\nu(x) = (-1)^\nu \left(e^{-x^2}\right)^{(\nu)} \tag{4.3.5}$$

we assert the following:

Theorem 4.2. *All nontrivial systems $(z_1, \ldots, z_n)$ with the property*

$$s_3 = s_4 = \cdots = s_{n+1} = 0 \tag{4.3.6}$$

can be obtained by rotation and dilatation from the ($[n/2] + 1$) *systems given by the roots of the equations*

$$\sum_{\nu=0}^{n} \frac{H_\nu(\lambda_k)}{\nu!} z^{n-\nu} = 0 \tag{4.3.7}$$

where λ_k stands for one of the nonnegative solutions of $H_{n+1}(\lambda) = 0$.

4.4. PROOF

Write

$$s_1 = -2a, \qquad s_2 = 2b^2.$$

We may suppose that

$$b \neq 0$$

and let

$$z^n + c_1 z^{n-1} + \cdots + c_n = 0$$

be the equation with our z_j as zeros. We have

$$c_1 = -s_1 = 2a = \frac{b}{1!} H_1\left(\frac{a}{b}\right)$$

$$2c_2 = -s_2 - c_1 s_1 = -2b^2 + 4a^2 = b^2 H_2\left(\frac{a}{b}\right),$$

that is,

$$c_2 = \frac{b^2}{2!} H_2\left(\frac{a}{b}\right).$$

Using the recursion formula

$$H_m(x) = 2xH_{m-1}(x) - 2(m-1)H_{m-2}(x) \tag{4.4.1}$$

the formulas (3.2.4) and (4.3.6) give

$$c_l = \frac{b^l}{l!} H_l\left(\frac{a}{b}\right) \qquad (l = 1, 2, \ldots, n).$$

Using the requirement $s_{n+1} = 0$ by way of the formula

$$s_{n+1} + c_1 s_n + \cdots + c_n s_1 = 0$$

and by (4.4.1), we get

$$\frac{b^{n+1}}{(n+1)!} H_{n+1}\left(\frac{a}{b}\right) = 0.$$

Since $b \neq 0$, we conclude that a/b is a zero of H_{n+1}, that is,

$$\frac{a}{b} = \lambda_k \qquad (k = 1, 2, \ldots, n).$$

Since the evenness or oddness of $n + 1$ implies the same parity of $H_{n+1}(x)$, the proof is complete. ■

4.5

To deduce from our result an upper bound for

$$M_3 \doteq \min_{z_j} \max_{\substack{3 \le \nu \le n+2 \\ \nu \text{ integer}}} |s_\nu| \tag{4.5.1}$$

where the $\min_{z_j}$ refers to all z_j-systems with

$$\max_j |z_j| = 1 \tag{4.5.2}$$

We need the value of s_{n+2} for the systems in 4.3. Since $s_{n+2} = -c_n s_2$ we get

$$M_3 \le 2 \min_{k,b} \frac{|b|^{n+2}}{n!} |H_n(\lambda_k)|$$

where b is to be determined so that (4.5.2) is fulfilled. Our z_j are zeros of an equation of the form

$$\sum_{\nu=0}^{n} \frac{H_\nu(\lambda_k)}{\nu!} b^\nu z^{n-\nu} = 0;$$

denoting for a fixed λ the minimal absolute value of the zeros of

$$\sum_{\nu=0}^{n} \frac{H_\nu(\lambda)}{\nu!} w^\nu = 0 \tag{4.5.3}$$

by $D_n(\lambda)$ we must choose

$$b = D_n(\lambda_k)$$

and

$$M_3 \le \frac{2}{n!} \min_k \left| H_n(\lambda_k) D_n(\lambda_k)^{n+2} \right|. \tag{4.5.4}$$

The difficult task of determining the minimum in (4.5.4) asymptotically for $n \to \infty$ was accomplished by E. Makai [1] using delicate techniques involving Hermitean polynomials. Referring to his paper for details, we state his result.

Theorem 4.3 (Makai). *For M_3 defined in (4.5.1) and (4.5.2) the inequality*

$$M_3 < \frac{1}{1.473^n}$$

holds for $n > n_0$.

4.6

A complete characterization of all nontrivial $(z_1,\ldots,z_n)$-systems with (4.3.2) has been given by S. Uchiyama [1]. Defining the $C_\nu(\lambda_1,\ldots,\lambda_m)$-polynomials by

$$\exp\left(-\sum_{j=1}^{m}\frac{\lambda_j}{j}z^j\right)=\sum_{\nu=0}^{\infty}C_\nu(\lambda_1,\ldots,\lambda_m)z^\nu \tag{4.6.1}$$

he showed that all systems in question are given by the zeros of the polynomials

$$\sum_{\nu=0}^{n}C_\nu(\lambda_1,\ldots,\lambda_m)z^{n-\nu}=0 \tag{4.6.2}$$

where the λ_j are restricted by the system

$$C_\nu(\lambda_1,\ldots,\lambda_m)=0 \qquad \nu=n+1,\ldots,n+m-1. \tag{4.6.3}$$

To use this even for the improvement of Makai's Theorem 4.3 seems to be difficult.

4.7

Thus the important problem of determining the dependence of the minimax in (4.3.1) on m cannot be solved at present along these lines. So we must confine ourselves at present to the assertion that an inequality

$$\max_{\nu=n+1,\ldots,m+n}|s_\nu|>\left(\frac{\pi n}{2m}\right)^{(\log n/\log 2)} \tag{4.7.1}$$

cannot be generally true. To show this let $n=2^k$, and for $l=0,1,\ldots,k$ let $\binom{k}{l}$ of the z_j be given by

$$\exp\left\{\frac{\pi i(k-2l)}{2(m+n/2)}\right\}.$$

Then we have

$$s_\nu^*=\left(2\cos\frac{\pi\nu}{2m+n}\right)^k \quad \text{for all integers } \nu>0, \tag{4.7.2}$$

and hence

$$\max_{m+1\le\nu\le m+n}|s_\nu^*|\le\left(-2\cos\frac{\pi(m+n)}{2m+n}\right)^k=\left(2\sin\frac{\pi n}{4m+2n}\right)^k<\left(\frac{\pi n}{2m}\right)^{(\log n/\log 2)}$$

4.8

So far the set S has consisted of integers of a single interval of length $\geq n$. For some possible applications one would need lower bounds of similar type for sets S consisting of several disjoint intervals each of length $\leq n-1$. Clearly, the system consisting of the nth roots of unity already shows some limitation of the possibilities. It indicates that any results must have an arithmetical character. Results have been obtained which, in spite of their simplicity, are much more than arithmetical curiosities. H. Jager [1] found that even for the generalized power sum

$$g(\nu) = \sum_{j=1}^{n} b_j z_j^{\nu} \tag{4.8.1}$$

with complex numbers b_j and distinct z_j, the existence of infinitely many positive integers m_ν with

$$g(m_\nu + 1) = g(m_\nu + 2) = \cdots = g\left(m_\nu + \left[\frac{n}{2}\right]\right) = 0, \tag{4.8.2}$$

and $\max_j |z_j| \neq 0$ implies the existence of an integer A such that

$$z_1^A = z_2^A = \cdots = z_n^A. \tag{4.8.3}$$

The length $[n/2]$ is a certain barrier. This is shown in the case $b_1 = \cdots = b_n = 1$ (and even n) by the system

$$z_l = \exp\frac{4\pi i l}{n}$$

$$z_{n/2+l} = 2\exp\frac{4\pi i l}{n} \quad \text{for } l = 1, 2, \ldots, \left(\frac{n}{2} - 1\right), \frac{n}{2} \tag{4.8.4}$$

which is, though not on *one* circle, still a "regular" system. In this direction R. Tijdeman found [1] that even the existence of *two* integers $m_1 < m_2$ with

$$g(m_1 + 1) = \cdots = g(m_1 + n - 1) = 0$$

$$g(m_2 + 1) = \cdots = g(m_2 + n - 1) = 0 \tag{4.8.5}$$

and $z_1 = 1$ implies that the z_j are $(m_2 - m_1)$-th roots of unity, while in the case $b_1 = \cdots = b_n$ (pure power sums) and *odd* n the z_j are in fact the nth roots of unity. This last statement does not hold for even n. Therefore what one may expect in this direction are results yielding lower bounds for $|g(\nu)|$ on such

S-sets if the $(z_1, \ldots, z_n)$-systems are "not too near a regular point-system." Many applications can be reduced to such situations.

Interesting results on the structure of integers ν which satisfy the more general equation $\sum_{j=1}^{n} P_j(\nu) z_j^{\nu} = 0$ (some of which were used by Jager in his proof) can be found in the papers of Mahler ([1, 2]), Lech [1], and H. N. Shapiro [1]. Their results are probably more closely connected to our problems than has been currently explored.

5

SOME LEMMAS

5.1

For further discussions of Part I we shall need some lemmas. We collect them in this section. We quote the classical theorem of Chebychev:

Lemma 5.1. *For $a < b$ and*

$$\pi_n(z) = a_0 z^n + a_1 z^{n-1} + \cdots + a_n \tag{5.1.1}$$

the inequality

$$\max_{a \le x \le b} |\pi_n(x)| \ge 2|a_0| \left(\frac{b-a}{4} \right)^n \tag{5.1.2}$$

holds.

For a proof see, for instance, G. Pólya and G. Szegő's book *Aufgaben und Lehrsätze aus der Analysis*, Bd. II., p. 85.

We need a generalization of this lemma which was conjectured by K. Mahler in slightly different form.

Lemma 5.2. *If $\xi_1, \ldots, \xi_n$ are the zeros of the polynomial $\pi_n(z)$ in* (5.1.1) *then the inequality*

$$\max_{-1 \le x \le 1} |\pi_n(x)| \ge \frac{|a_0|}{2^{n-1}} \Pi' |\xi_j| \tag{5.1.3}$$

holds, where the product is extended to all ξ_j with $|\xi_j| > 1$.

Proof. (Turán [36]). Consider the auxiliary function

$$\varphi_n(z) = \pi_n(z) \Pi' \left(\frac{z - \xi_j}{1 - \bar{\xi}_j z} \right)^{-1} \tag{5.1.4}$$

where Π' has the same meaning as in (5.1.3). $\varphi_n(z)$ is also a polynomial of degree n and has the form

$$\varphi_n(z) = a_0\left[\Pi'(-\bar{\xi}_j)\right] z^n + \cdots . \tag{5.1.5}$$

Let us observe that for our ξ_j the inequality

$$\left|\frac{z - \xi_j}{1 - \bar{\xi}_j z}\right| \geq 1, \quad \text{for} \quad |z| \leq 1 \tag{5.1.6}$$

holds and thus (putting $z = x + iy$) also

$$\min_{-1 \leq x \leq +1} \left|\frac{x - \xi_j}{1 - \bar{\xi}_j x}\right| = 1.$$

Hence from (5.1.4) we get the inequality

$$\max_{-1 \leq x \leq 1} |\pi_n(x)| \geq \max_{-1 \leq x \leq 1} |\varphi_n(x)|.$$

Applying Chebychev's Lemma 5.1 to $\varphi_n(x)$ and using (5.1.5), we complete the proof of Lemma 5.2. ∎

Actually we need a form of this lemma concerning an arbitrary real interval (a, b). Let

$$q_n(z) = a_0 \prod_{j=1}^{n} (z - \eta_j). \tag{5.1.7}$$

Then

$$\max_{a \leq x \leq b} |q_n(x)| = \max_{-1 \leq t \leq 1} \left| q_n\left(\frac{b+a}{2} + \frac{b-a}{2} t\right)\right| =$$

$$= |a_0| \left(\frac{b-a}{2}\right)^n \max_{-1 \leq t \leq 1} \left| \prod_{j=1}^{n} \left(t - \left(\frac{2\eta_j - (b+a)}{b-a}\right)\right)\right|.$$

Application of Lemma 5.2 gives the following.

Corollary. *For $a < b$ and $q_n(z)$ in (5.1.7) the inequality*

$$\max_{a \leq x \leq b} |q_n(x)| \geq 2|a_0| \left(\frac{b-a}{4}\right)^n \Pi'' \frac{2\eta_j - (b+a)}{b-a} \tag{5.1.8}$$

holds where the product Π'' is extended to all η_j satisfying the inequality

$$\left|\eta_j - \frac{b+a}{2}\right| > \frac{b-a}{2}. \tag{5.1.9}$$

5.2

Let $\xi_1, \xi_2, \ldots, \xi_N$ be distinct complex numbers, and let $y_1, y_2, \ldots, y_N$ be arbitrary complex numbers. Then there is a unique polynomial

$$K(w) = e_0 + e_1(w-\xi_1) + e_2(w-\xi_1)(w-\xi_2) \cdots + e_{N-1}(w-\xi_1)\cdots(w-\xi_{N-1}) \tag{5.2.1}$$

with the property that

$$K(\xi_j) = y_j \qquad (j = 1, 2, \ldots, N). \tag{5.2.2}$$

Let L be a Jordan curve consisting of a finite number of analytic arcs and let the points $\xi_1, \ldots, \xi_N$ lie outside L. Let $h(w)$ be analytic outside L and let

$$\lim_{w \to \infty} h(w) = 0.$$

If the values y_j are chosen to satisfy

$$y_j = h(\xi_j) \qquad j = 1, \ldots, N, \tag{5.2.3}$$

we assert the following.

Lemma 5.3 (Nörlund). *The coefficients e_j in* (5.2.1) *are given by*

(5.2.4)

$$e_j = -\frac{1}{2\pi i}\int_{(L)} \frac{h(z)\,dz}{(z-\xi_1)(z-\xi_2)\cdots(z-\xi_{j+1})}; \qquad j = 0, 1, \ldots, (N-1).$$

Proof. We first observe the identity

$$\frac{1}{z-\xi_1} + \frac{\xi_\nu - \xi_1}{(z-\xi_1)(z-\xi_2)} + \frac{(\xi_\nu-\xi_1)(\xi_\nu-\xi_2)}{(z-\xi_1)(z-\xi_2)(z-\xi_3)}$$

$$+ \cdots + \frac{(\xi_\nu-\xi_1)\cdots(\xi_\nu-\xi_{\nu-1})}{(z-\xi_1)\cdots(z-\xi_\nu)} = \frac{1}{z-\xi_\nu}; \qquad \nu = 2, 3, \ldots, N \tag{5.2.5}$$

which is easy to prove. Construct the polynomial $K_1(w)$ of the form (5.2.1) with the integrals in (5.2.4) as coefficients. This can be written as

$$-\frac{1}{2\pi i}\int_{(L)} h(z)\left\{\frac{1}{z-\xi_1} + \frac{w-\xi_1}{(z-\xi_1)(z-\xi_2)} + \cdots + \frac{(w-\xi_1)(w-\xi_2)\cdots(w-\xi_{N-1})}{(z-\xi_1)(z-\xi_2)\cdots(z-\xi_N)}\right\} dz.$$

Hence

$$K_1(\xi_\nu) = \frac{-1}{2\pi i}\int_{(L)} h(z)\left\{\frac{1}{z-\xi_1} + \frac{\xi_\nu - \xi_1}{(z-\xi_1)(z-\xi_2)} + \cdots + \frac{(\xi_\nu - \xi_1)\cdots(\xi_\nu - \xi_{\nu-1})}{(z-\xi_1)\cdots(z-\xi_\nu)}\right\} dz \qquad \nu = 1,2,\ldots,N;$$

and, from the identity (5.2.5),

$$K_1(\xi_\nu) = \frac{-1}{2\pi i}\int_{(L)} \frac{h(z)}{z-\xi_\nu}\, dz = h(\xi_\nu) \qquad \nu = 1,2,\ldots,N.$$

This completes the proof of the lemma. ■

The same proof yields another result.

Lemma 5.4. *Let L be a smooth Jordan curve, $\xi_1,\ldots,\xi_n$ lie inside L, $h(w)$ regular inside L, and $K_2(w)$ the unique polynomial of degree $\le N-1$ with*

$$K_2(\xi_j) = h(\xi_j) \qquad j = 1,2,\ldots,N.$$

Then, writing, $K_2(w)$ in the form (5.2.1), *the representation*

$$e_j = \frac{1}{2\pi i}\int_L \frac{h(z)\, dz}{(z-\xi_1)(z-\xi_2)\cdots(z-\xi_{j+1})} \tag{5.2.6}$$

holds for $j = 0,1,\ldots,(N-1)$.

5.3

We also need the following.

Lemma 5.5. *Let $0 < \kappa \le \dfrac{\pi}{2}$ and let*

$$F(z) = 1 + a_1 z + \cdots + a_N z^N \tag{5.3.1}$$

be a polynomial with real coefficients having no roots in the angle

$$|\text{arc } z| < \kappa. \tag{5.3.2}$$

Then one can multiply $F(z)$ by a suitable polynomial

$$\varphi(z) = 1 + \cdots$$

with real coefficients so that

1. $\operatorname{gr} F(z)\varphi(z) \le \frac{N}{2}\left(1 + \left[\frac{\pi}{\kappa}\right]\right)$,
2. *The coefficients of*

$$F(z)\varphi(z) = 1 + \cdots$$

are nonnegative.

Proof. Assume that

$$F(z) = 1 + \lambda z, \quad \text{with} \quad \lambda \text{ real}. \tag{5.3.3}$$

Then $\lambda > 0$, that is, $\varphi(z) \equiv 1$ is suitable. Next, let

$$F(z) = 1 - 2\cos\alpha \cdot z + z^2 \tag{5.3.4}$$

with α satisfying

$$\kappa \le \alpha \le \pi. \tag{5.3.5}$$

For $\pi/2 \le \alpha \le \pi$ we can take $\varphi(z) \equiv 1$ again, so assume

$$\kappa \le \alpha \le \frac{\pi}{2}. \tag{5.3.6}$$

For arbitrary nonnegative integers k one has the identity

$$(1 - 2\cos\alpha \cdot z + z^2) \sum_{l=0}^{k} \frac{\sin(l+1)\alpha}{\sin\alpha} \cdot z^l \tag{5.3.7}$$

$$= 1 - \frac{\sin(k+2)\alpha}{\sin\alpha} \cdot z^{k+1} + \frac{\sin(k+1)\alpha}{\sin\alpha} z^{k+2}.$$

If

$$k = \left[\frac{\pi}{\alpha}\right] - 1 \tag{5.3.8}$$

then the coefficients on the right side of (5.3.7) are nonnegative.

From (5.3.3) and (5.3.4) one can turn to the general case. We then write

$$F(z) = \prod_{j=1}^{n_1}\left(1 + \frac{z}{\lambda_j}\right)\prod_{j=1}^{n_2}\left\{1 - \frac{2\cos\alpha_j}{\rho_j} z + \frac{z^2}{\rho_j^2}\right\} \tag{5.3.9}$$

$$\cdot \prod_{j=n_2+1}^{n_2+n_3}\left\{1 - \frac{2\cos\alpha_j}{\rho_j} z + \frac{z^2}{\rho_j^2}\right\} = \Pi_1\Pi_2\Pi_3$$

where $n_1 + 2n_2 + 2n_3 = N$, the negative zeros are $-\lambda_j$, the complex zeros with $\pi/2 \le \alpha_j \le \pi$ are

$$\rho_j e^{\pm i\alpha_j} \qquad (j = 1,\ldots,n_2),$$

and the zeros with $\kappa \le \alpha_j < \pi/2$ are

$$\rho_j e^{\pm i\alpha_j} \qquad (j = n_2 + 1, \ldots, n_2 + n_3).$$

The factors of Π_1 and Π_2 already have nonnegative coefficients. Let

$$\sum_{l=0}^{[\pi/\alpha]-1} \frac{\sin(l+1)\alpha}{\sin\alpha} z^l \doteq \varphi(z, \alpha). \tag{5.3.10}$$

Then the polynomial

$$\varphi(z) = \prod_{\alpha_j < \pi/2} \varphi\left(\frac{z}{\rho_j}, \alpha_j\right) \tag{5.3.11}$$

clearly satisfies requirement 2. Requirement 1 follows from

$$\operatorname{gr} F(z)\varphi(z) = \sum_{j=1}^{n_3} \left(1 + \left[\frac{\pi}{\alpha_j}\right]\right) + n_1 + 2n_2 \le \frac{N}{2}\left(1 + \left[\frac{\pi}{\kappa}\right]\right). \quad \blacksquare$$

5.4

We also need the following.

Lemma 5.6. *If the polynomial $F(z)$ of the previous lemma satisfies the additional condition that all of its zeros are in $|z| \ge R$ (≥ 1) and if*

$$F(z)\varphi(z) = e_0 + e_1 z + \cdots, \qquad e_0 = 1,$$

then the inequality

$$\sum_\nu e_\nu \le \left(1 + \frac{1}{R}\right)^N$$

follows.

Proof. Write out $F(z)\varphi(z)$ explicitly. Equations (5.3.9) and (5.3.7) give

$$F(z)\varphi(z) = \prod_{j=1}^{n_1}\left(1 + \frac{z}{\lambda_j}\right) \prod_{j=1}^{n_2}\left\{1 + 2|\cos\alpha_j|\left(\frac{z}{\rho_j}\right) + \left(\frac{z}{\rho_j}\right)^2\right\} \tag{5.4.1}$$

$$\cdot \prod_{j=n_2+1}^{n_2+n_3}\left\{1 + \frac{|\sin(1 + [\pi/\alpha_j])\alpha_j|}{\sin\alpha_j}\left(\frac{z}{\rho_j}\right)^{[\pi/\alpha_j]} + \frac{\sin[\pi/\alpha_j]\alpha_j}{\sin\alpha_j}\left(\frac{z}{\rho_j}\right)^{1+[\pi/\alpha_j]}\right\}.$$

Since $R \geq 1$, we immediately obtain that

$$\sum_\nu e_\nu = F(1)\varphi(1) \leq \left(1 + \frac{1}{R}\right)^{n_1+2n_2}$$

$$\cdot \prod_{j=n_2+1}^{n_2+n_3} \left\{1 - \frac{\sin(1+[\pi/\alpha_j])\alpha_j}{\sin \alpha_j} \frac{1}{R} + \frac{\sin[\pi/\alpha_j]\alpha_j}{\sin \alpha_j} \frac{1}{R^2}\right\}.$$

Set $[\pi/\alpha_j] = \pi/\alpha_j - \vartheta_j$, $0 \leq \vartheta_j < 1$, then the expression in the curly brackets is equal to

$$1 + \frac{\sin(1-\vartheta_j)\alpha_j}{\sin \alpha_j} \frac{1}{R} + \frac{\sin \vartheta_j \alpha_j}{\sin \alpha_j} \frac{1}{R^2} < 1 + \frac{1}{R} + \frac{1}{R^2} < \left(1 + \frac{1}{R}\right)^2.$$

Then

$$\sum e_\nu \leq \left(1 + \frac{1}{R}\right)^N$$

as required. ■

5.5

Suppose again that $0 < \kappa \leq \pi/2$, but now

$$F_1(z) = z^N + a_1 z^{N-1} + \cdots + a_N \tag{5.5.1}$$

with real coefficients is such that all its zeros are in

$$(|z| \leq R) \cap (\kappa \leq |\arc z| \leq \pi), \qquad (0 < R < 1). \tag{5.5.2}$$

Then,

Lemma 5.7. *One can multiply $F_1(z)$ by a suitable polynomial $\varphi_1(z)$ with real coefficients so that setting*

$$F_1(z)\varphi_1(z) = e_0' z^{N_1} + e_1' z^{N_1-1} + \cdots + e_{N_1}' \tag{5.5.3}$$

we have

1. $e_0' = 1$
2. $e_\nu' \geq 0$
3. $N_1 \leq (N/2)(1 + [\pi/\kappa])$
4. $\Sigma_\nu e_\nu' \leq (1 + R)^N$.

For the proof it suffices to note that $z^N F_1(1/z)$ satisfies the requirements of the previous lemma.

5.6

We need another property of the Hermitean polynomials (4.3.5), namely, that on the real axis $M_\nu(x) \neq 0$ for $|x| \geq \sqrt{2\nu + 1}$. This follows at once from the fact that the function $e^{-x^2/2} H_\nu(x) = y$ satisfies the differential equation

$$y'' + (2\nu + 1 - x^2)y = 0.$$

We also need the ingenious observation of Sonin that for the foregoing y the function

$$y'^2 + (2\nu + 1 - x^2)y^2$$

decreases monotonically when $0 \leq x \leq \sqrt{2\nu + 1}$, and hence for the nonnegative zeros

$$x_1 < x_2 < \cdots < x_{[(\nu+1)/2]}$$

of $H_\nu(x)$ the sequence

$$|H_\nu'(x_j)| \exp\left(-\tfrac{1}{2}x_j^2\right), \qquad j = 1, 2, \ldots, \left[\frac{\nu + 1}{2}\right] \tag{5.6.1}$$

decreases monotonically.

Let $z_1, \ldots, z_n$ be pairwise distinct complex numbers and let

$$w(z) = \prod_{j=1}^{n} (z - z_j). \tag{5.6.2}$$

Then we need Euler's observation that

$$\sum_{j=1}^{n} \frac{z_j^\nu}{w'(z_j)} = \begin{cases} 0 & \text{for} \quad \nu = 0, 1, \ldots, (n-2) \\ 1 & \text{for} \quad \nu = n - 1. \end{cases} \tag{5.6.3}$$

If we denote by $l_j(z)$ the fundamental functions of the Lagrange interpolation, based on the points $z_1, \ldots, z_n$, we have the identity

$$\sum_{j=1}^{n} z_j^\nu l_j(z) \equiv z^\nu, \qquad \nu = 0, 1, \ldots, (n-1). \tag{5.6.4}$$

The proof of (5.6.3) follows by comparing the coefficients of z^{n-1}.

5.7

We also need the next result.

Lemma 5.8 (G. Halász). *For each $k \geq 2$ there is a polynomial $\pi_k^*(x)$ of degree k so that*

1. $\pi_k^*(z) \neq 0$ *for* $|z| < 1$
2. $\pi_k^*(0) = 1$
3. $\pi_k^*(1) = 0$ *and*
4. $\max_\vartheta |\pi_k^*(e^{i\vartheta})| < e^{2/k}$.

The proof we use is attributable to Q. I. Rahman and F. Stenger (written communication) and is based on the classical theorem of Fejér–F. Riesz*: If $t(\vartheta) \not\equiv 0$ is a nonnegative trigonometric polynomial of order k, then there is a uniquely determined rational polynomial $f_k(z)$ of degree k such that

1. $|f_k(e^{i\vartheta})|^2 = t(\vartheta)$
2. $f_k(z) \neq 0$ *for* $|z| < 1$
3. $f_k(0) > 0$.

Moreover, for $|z| < 1$, G. Szegő has given the representation ($z = re^{i\varphi}$)

$$(5.7.1) \qquad f_k(z) = \exp\left(\frac{1}{4\pi} \int_{-\pi}^{\pi} \log t(\vartheta) \frac{1 + ze^{-i\vartheta}}{1 - ze^{-i\vartheta}} d\vartheta \right).^{\dagger}$$

We apply this theorem with

$$(5.7.2) \qquad t_0(\vartheta) = 1 - \left(\frac{\sin[\vartheta(k+1)/2]}{(k+1)\sin(\vartheta/2)} \right)^2$$

and assert that

$$(5.7.3) \qquad \pi_k^*(z) = \frac{f_k(z)}{f_k(0)}$$

satisfies requirements 1, 2, 3, and 4 of Lemma 5.8. It is evident that 1, 2 and 3 are satisfied; it remains to show 4. From condition 1 of the Fejér–Riesz theorem and (5.7.2) we have

$$\max_\vartheta |f_k(e^{i\vartheta})| = 1$$

*L. Fejér, "Über trigonometrische Polynome," *J. reine Angew. Math.*, **146**, 53–82 (1915).
†G. Szegő, "Über die Randwerte analytischer Funktionen," *Math. Ann.*, **84**, 232–244 (1921).

and — using (5.7.1) —

$$\max_{\vartheta}|\pi_k^*(e^{i\vartheta})| = \frac{1}{|f_k(0)|} = \exp\left\{-\frac{1}{4\pi}\int_{-\pi}^{\pi}\log t_0(\vartheta)\,d\vartheta\right\} \tag{5.7.4}$$

$$= \exp\left\{-\frac{1}{2\pi}\int_0^{\pi}\log t_0(\vartheta)\,d\vartheta\right\}.$$

As a result of the inequality (A.1.4) (see Appendix A) this cannot exceed

$$\exp\left\{-\frac{1}{2\pi}\int_0^{\pi}\log\left(1-\frac{1}{k+1}\right)d\vartheta\right.$$

$$\left.-\frac{1}{2\pi}\int_0^{\pi}\log\left\{1-\left(\frac{\sin[\vartheta(k+1)/2]}{\vartheta(k+1)/2}\right)^2\right\}d\vartheta\right\}$$

$$= \exp\left\{-\frac{1}{2}\log\left(1-\frac{1}{k+1}\right)-\frac{1}{\pi(k+1)}\int_0^{(k+1)\pi/2}\log\left(1-\left(\frac{\sin x}{x}\right)^2\right)dx\right\}$$

$$< \exp\left\{\frac{1}{2k}+\frac{1}{k\pi}\int_0^{\infty}\log\frac{1}{1-(\sin x/x)^2}dx\right\} \leq \exp\left(\frac{2}{k}\right),$$

due to (A.1.5).

It would be important to significantly improve requirement 4 of Lemma 5.8. However, this is not possible since one can easily see that for $\pi_k(z) = 1 + c_1 z + \cdots + c_k z^k$ satisfying $\pi_k(1) = 0$ the inequality

$$\frac{1}{2\pi}\int_{-\pi}^{\pi}|\pi_k(e^{i\vartheta})|^2\,d\theta \geq 1+\frac{1}{k}$$

holds, with equality only for

$$1-\frac{z+z^2+\cdots+z^k}{k}.$$

Hence, for such $\pi_k(z)$

$$\max_{|z|=1}|\pi_k(z)| \geq \left\{\frac{1}{2\pi}\int_{-\pi}^{\pi}|\pi_k(e^{i\vartheta})|^2\,d\vartheta\right\}^{1/2} \geq \left(1+\frac{1}{k}\right)^{1/2} > e^{1/3k}.$$

5.8

We need an easy corollary of Lemma 5.8. The function

$$H(z) = \log \pi_k^*(z)$$

with $H(0) = 0$ is regular for $|z| < 1$, and from Lemma 5.8 we have

$$\mathrm{Re}H(z) \le \frac{2}{k}.$$

But then the theorem of Hadamard–Carathéodory* yields the inequality

$$\max_{\vartheta} |\log \pi_k^*(re^{i\vartheta})| \le \frac{2r}{1-r} \cdot \frac{2}{k}, \quad \text{for} \quad 0 < r < 1.$$

A fortiori, we have the following consequence.

Corollary. *The polynomial* $\pi_k^*(z)$ *of Lemma 5.8 satisfies the inequality*

$$\left| \frac{1}{\pi_k^*(z)} \right| \le \exp \frac{4r}{k(1-r)} \tag{5.8.1}$$

in the disk $|z| \le r$ $(0 < r < 1)$.

5.9

We require the following simple lemma.

Lemma 5.9. *For integer* $P \ge 2$ *and*

$$J_n(P) \doteq \sum_{i_1 + \cdots + i_n = P} \left(\frac{P!}{i_1! i_2! \cdots i_n!} \right)^2 \tag{5.9.1}$$

the inequality

$$J_n(P) \le (2nP)^P$$

holds.

Proof. We use the easily demonstrated recurrence formula

(5.9.2)

$$J_n(P) = \sum_{j=0}^{P} \binom{P}{j}^2 J_{n-1}(P-j) = \sum_{j=0}^{P-2} \binom{P}{j}^2 J_{n-1}(P-j) + (n-1)P^2 + 1.$$

For $n = 1$ the assertion is trivial. If it is true for $n \le N$ then

$$J_{N+1}(P) \le \sum_{j=0}^{P-1} \binom{P}{j}^2 (2N(P-j))^{P-j} + 1. \tag{5.9.3}$$

*See, for example, A. E. Ingham, "The distribution of prime numbers," *Cambridge Tracts* 1932, p. 50.

But for $0 \le j \le P$

$$\frac{1}{j!2^j} \le 1 \le \left(\frac{P}{P-1}\right)^{P-j}$$

$$\frac{1}{j!2^j}\left(\frac{P}{P-1}\right)^j \le \left(\frac{P}{P-1}\right)^P$$

$$\frac{\binom{P}{j}}{(P-1)^j 2^j} \le \frac{(P/(P-1))^j}{j!2^j} \le \left(\frac{P}{P-1}\right)^P$$

$$\binom{P}{j}(P-1)^{P-j} \le 2^j P^P = P^{P-j}(2P)^j$$

$$\binom{P}{j}(2N(P-1))^{P-j} \le (2NP)^{P-j}(2P)^j$$

and hence the right side of (5.9.3)

$$\le 1 + \sum_{j=0}^{P-1}\binom{P}{j}(2NP)^{P-j}(2P)^j \le (2NP+2P)^P = \{(2N+2)P\}^P. \qquad \blacksquare$$

6

GENERALIZED POWER SUMS
THE CASE OF THE MINIMUM NORMS

First Main Theorem

6.1

The results of this section have many applications demonstrated in Part II. The main result is stated in several forms, but this first is the most suitable for applications (Turán [1] from 1941).

Theorem 6.1. *For arbitrary nonnegative integer m and complex b_j and z_j with*

$$\min_j |z_j| = 1 \tag{6.1.1}$$

the inequality

$$\max_{\nu=m+1,\ldots,m+n} |g(\nu)| \geq \left(\frac{n}{2e(m+n)}\right)^n |b_1 + \cdots + b_n| \tag{6.1.2}$$

holds.

This implies, as before, that for arbitrary complex b_j and ξ_j with

$$|\xi_1| \geq |\xi_2| \geq \cdots \geq |\xi_n| \tag{6.1.3}$$

and any nonnegative integer m we have an integer ν_0 with

$$m + 1 \le \nu_0 \le m + n, \tag{6.1.4}$$

so that the inequality

$$|b_1 \xi_1^{\nu_0} + \cdots + b_n \xi_n^{\nu_0}| \ge \left(\frac{n}{2e(m+n)} \right)^n |b_1 + \cdots + b_n| \, |\xi_n|^{\nu_0} \tag{6.1.5}$$

holds.

As we shall see, the dependence upon the b_j in theorem (6.1) is very useful. For some other applications, however, it would be useful to have—under the restriction (6.1.1)—an inequality of the type

$$\max_{\substack{m+1 \le \nu \le m+n \\ \nu \text{ integer}}} |g(\nu)| \ge \left(\frac{n}{c(m+n)} \right)^n \min_j |b_j| \tag{6.1.6}$$

with a positive number c. Unfortunately no such *general* theorem can exist, as the next example shows. Take

$$z_1 = \cdots = z_n = 1, \qquad b_1 = \cdots = b_{n-1} = \frac{1}{n-1}, \qquad b_n = -1. \tag{6.1.7}$$

Inequalities of the type (6.1.6) can be obtained, if the z_j are subject to certain geometric restriction [as in (11.1.2)].*

6.2. PROOF OF THEOREM 6.1

Let

$$f_1(z) = \prod_{j=1}^{n} \left(1 - \frac{z}{z_j} \right) = \sum_{\nu=0}^{n} c_\nu^{(1)} z^\nu \tag{6.2.1}$$

Since $f_1(z)^{-1}$ is regular in $|z| < 1$ we have

$$\frac{1}{f_1(z)} = \prod_{j=1}^{n} \frac{1}{1 - z/z_j} = \sum_{\nu=0}^{\infty} c_\nu^{(2)} z^\nu. \tag{6.2.2}$$

Let

$$s_m\left(\frac{1}{f_1} \right) = \sum_{\nu=0}^{m} c_\nu^{(2)} z^\nu \tag{6.2.3}$$

*Of course in case the b_j are in an angle $\le \pi - \delta$, $\delta > 0$, inequality (6.1.6) follows trivially from Theorem 6.1.

and let

$$f_2(z) = 1 - f_1(z)s_m\left(\frac{1}{f_1}\right). \tag{6.2.4}$$

This is a polynomial of degree $\leq m + n$. Since for $|z| < 1$

$$f_2(z) = 1 - f_1(z)\left\{\frac{1}{f_1(z)} - \sum_{\nu=m+1}^{\infty} c_\nu^{(2)} z^\nu\right\} = f_1(z) \sum_{\nu=m+1}^{\infty} c_\nu^{(2)} z^\nu$$

we have

$$f_2(z) = \sum_{\nu=m+1}^{m+n} c_\nu^{(3)} z^\nu. \tag{6.2.5}$$

This gives, using (6.2.4),

$$\sum_{\nu=m+1}^{m+n} c_\nu^{(3)} z_j^\nu = 1 \qquad j = 1, 2, \ldots, n. \tag{6.2.6}$$

Multiplication with b_j and summation for $j = 1, \ldots, n$ yield the fundamental identity

$$\sum_{\nu=m+1}^{m+n} c_\nu^{(3)} g(\nu) = b_1 + \cdots + b_n. \tag{6.2.7}$$

This immediately gives [notation (0.2)]

$$\max_{\nu=m+1,\ldots,m+n} |g(\nu)| \geq \frac{\left|\sum_{j=1}^{n} b_j\right|}{\|f_2\|}. \tag{6.2.8}$$

To estimate the foregoing denominator we remark first that (6.2.5)–(6.2.4) show that

$$\|f_2\| \leq \|f_1\| \cdot \left\|s_m\left(\frac{1}{f_1}\right)\right\|. \tag{6.2.9}$$

From (6.2.1) we get

$$\|f_1\| \leq 2^n. \tag{6.2.10}$$

Furthermore, we obtain from (6.2.2)

$$c_\nu^{(2)} = \sum_{i_1 + \cdots + i_n = \nu} \frac{1}{z_1^{i_1} z_2^{i_2} \cdots z_n^{i_n}},$$

that is,

$$(6.2.11)\qquad |c_\nu^{(2)}| \le \sum_{i_1+\cdots+i_n=\nu} 1 = \binom{\nu+n-1}{n-1},$$

from (6.1.1) and hence

$$\left\|s_m\left(\frac{1}{f_1}\right)\right\| \le \binom{m+n}{n} \le \left(e\frac{m+n}{n}\right)^n.$$

And so,

$$\|f_2\| \le \left(2e\frac{m+n}{n}\right)^n,$$

from (6.2.9). This and (6.2.8) completes the proof of Theorem 6.1. ■

This form of the proof gives the impression that any proof requires much analysis. However, the proof is actually based on the rational identity (6.2.6) among the variables z_j, and this can be proved—as J. Surányi showed—by elementary algebraic arguments (although it takes a bit longer). So Theorem 6.1 is really a theorem from the algebra of the complex field.

6.3

To get another useful form of Theorem 6.1 let m_1 be an arbitrary nonnegative number and apply Theorem 6.1 with $[m_1]$ instead of m. This at once gives the inequality

$$(6.3.1)\qquad \max_{\substack{m_1\le\nu\le m_1+n\\ \nu\,\text{integer}}} |g(\nu)| \ge \left(\frac{n}{2e(m_1+n)}\right)^n |b_1+\cdots+b_n|.$$

The next form of Theorem 6.1 will give the first glimpse of the applicability of such results. Let $a>0$, $d>0$, $\alpha_1,\ldots,\alpha_n$ be arbitrary complex numbers with

$$(6.3.2)\qquad \min_j \operatorname{Re}\alpha_j = 0$$

and let b_j be any complex numbers. We apply the (6.3.1) form of Theorem 6.1 with

$$(6.3.3)\qquad m_1 = a\frac{n}{d},\qquad z_j = \exp\left(\frac{d}{n}\alpha_j\right)\qquad j=1,\ldots,n.$$

This implies the existence of an integer ν_0 satisfying

$$(6.3.4)\qquad a\frac{n}{d} \le \nu_0 \le a\frac{n}{d} + n$$

for which the inequality

$$\left|\sum_{j=1}^{n} b_j \exp\left(\frac{d\nu_0}{n}\alpha_j\right)\right| \geq \left(\frac{d}{2e(a+d)}\right)^n |b_1 + \cdots + b_n|$$

holds. In view of (6.3.4), we obtain a fortiori the inequality

$$\max_{a \leq t \leq a+d} \left|\sum_{j=1}^{n} b_j e^{\alpha_j t}\right| \geq \left(\frac{d}{2e(a+d)}\right)^n \left|\sum_{j=1}^{n} b_j\right| \tag{6.3.5}$$

provided that (6.3.2) holds (which is simply normalization). Or, if we set

$$\sum_{j=1}^{n} b_j e^{\alpha_j t} = f(t), \tag{6.3.6}$$

then (6.3.2) yields the inequality

$$\max_{a \leq t \leq a+d} |f(t)| \geq \left(\frac{d}{2e(a+d)}\right)^n |f(0)|. \tag{6.3.7}$$

Applications are discussed in Part II. Here we confine ourselves to indicating a few of them briefly. A passage to the limit gives an estimate similar to (6.3.7) for the general solution of homogeneous linear differential equations with constant coefficients. This does not depend on the coefficients of the equations at all [apart from the normalization (6.3.2), which is inessential] and hence offers various possible contributions to the *general* theory of differential equations. Furthermore, (6.3.7) in the case $\alpha_j = i\lambda_j$ ($\lambda_j \geq 0$ and an integer) gives a clue to several phenomena in the classical theory of gap series as we show in Part II. These and other applications seem to justify calling Theorem 6.1 the first *main* theorem.

6.4

It is natural to ask if $(n/2e(m+n))^n$ is the best constant in Theorem 6.1. According to the proof, if we succeed in determining the *exact* maximum of

$$\sum_{\nu} |c_\nu^{(3)}| \tag{6.4.1}$$

as a function of the z_j in $|z_j| \geq 1$, the lower bound in Theorem 6.1 will certainly be improved. If, for the extremal system $(z_1^*, \ldots, z_n^*)$, it turns out that we have

$$\sum_{\nu} |c_\nu^{(3)*}| = \left|\sum_{\nu} c_\nu^{(3)*}\right|$$

and also

$$z^*_{\mu_1} \neq z^*_{\mu_2} \quad \text{for} \quad \mu_1 \neq \mu_2,$$

then the b_j can be chosen so that

$$\sum_{j=1}^{n} b_j^* z_j^{*\nu} = 1, \qquad \nu = m+1, \ldots, m+n$$

and so

$$g^*(\nu) = \sum_{j=1}^{n} b_j^* z_j^{*\nu}$$

would be an *exact* extremal system. As we show, this is almost the case.

The extremal problem (6.4.1) (after Dancs's first improvement [2]) was solved independently by Makai [2] and deBruijn [1]. So, with the notation

$$\sum_{j=0}^{n-1} \binom{m+j}{j} 2^j = P_{m,n}, \tag{6.4.2}$$

the following theorem holds.

Theorem 6.2 (Makai–deBruijn). *With the normalization* (6.1.1) *the inequality*

$$\max_{\nu=m+1,\ldots,m+n} |g(\nu)| \geq P_{m,n}^{-1} |b_1 + \cdots + b_n| \tag{6.4.3}$$

holds. Inequality (6.4.3) *fails if we replace* $P_{m,n}^{-1}$ *by* $(P_{m,n} - \varepsilon)^{-1}$, $\varepsilon > 0$. *Equality in* (6.4.3) *can be attained for* $n \geq 2$, $m \geq 1$ *only if*

$$b_1 + b_2 + \cdots + b_n = 0.^* \tag{6.4.4}$$

Proof. We have only to investigate the maximum in (6.4.1). This will follow at once from the fact, discovered by Makai [3] and Dancs [2], that the quantities

$$(-1)^{m-1} c_\nu^{(3)} \tag{6.4.5}$$

regarded as polynomials in $(-1/z_j)$ have *nonnegative* coefficients. To show this [writing $f_2(z, m)$ instead of $f_2(z)$ in (6.2.4)] we form the generating function

$$B(t, z) = \sum_{m=0}^{\infty} (-1)^{m-1} f_2(z, m) t^m, \qquad |z| < 1, |t| < 1. \tag{6.4.6}$$

*If this is satisfied then the extremal systems $(z_1^*, \ldots, z_n^*)$ can be chosen in various ways, for example, $z_1^* = \cdots = z_n^*$ with $|z_j^*| \geq 1$.

Inserting the explicit form of $f_2(z)$ from (6.2.4) we obtain, due to (6.2.1)–(6.2.3), the equalities

$$B(t, z) = -\frac{1}{1+t} + f_1(z) \sum_{m=0}^{\infty} s_m\left(\frac{1}{f_1}\right)(-t)^m =$$

$$= -\frac{1}{1+t} + f_1(z) \sum_{m=0}^{\infty} (-t)^m \left\{1 + \sum_{\nu=1}^{m} c_\nu^{(2)} z^\nu\right\} =$$

$$= -\frac{1}{1+t} + f_1(z) \left\{\frac{1}{1+t} + \sum_{\nu=1}^{\infty} c_\nu^{(2)} z^\nu \sum_{m=\nu}^{\infty} (-t)^m\right\} =$$

$$= -\frac{1}{1+t} + f_1(z) \left\{\frac{1}{1+t} + \sum_{\nu=1}^{\infty} c_\nu^{(2)} \frac{(-zt)^\nu}{1+t}\right\} =$$

$$= \frac{1}{1+t}\left\{-1 + \frac{f_1(z)}{f_1(-zt)}\right\} = \frac{1}{1+t}\left\{-1 + \prod_{j=1}^{n} \frac{1 - z/z_j}{1 + zt/z_j}\right\}.$$

From the preceding we have the representation

$$B(t, z) = \frac{1}{1+t}\left\{-1 + \prod_{j=1}^{n}\left[1 - \frac{z/z_j}{1 + zt/z_j}(1+t)\right]\right\}. \tag{6.4.7}$$

Setting

$$-\frac{z/z_j}{1 + z_t/z_j} = x_j, \qquad j = 1, 2, \ldots, n \tag{6.4.8}$$

we have

$$x_j = \sum_{\nu=0}^{\infty} \left(-\frac{1}{z_j}\right)^{\nu+1} t^\nu z^{\nu+1}; \tag{6.4.9}$$

and from (6.4.7)

$$B(t, z) = \frac{1}{1+t}\left\{-1 + \prod_{j=1}^{n}\left[1 + x_j(1+t)\right]\right\}. \tag{6.4.10}$$

Since from (6.4.10) $B(t, z)$ appears as a polynomial in $(1 + t)$ and x_j with nonnegative coefficients, this, together with (6.4.9), establishes assertion (6.4.5).

Since

$$c_\nu^{(3)} = c_\nu^{(3)}\left(-\frac{1}{z_1}, \ldots, -\frac{1}{z_n}, m\right)$$

the maximum of $\left|c_\nu^{(3)}\right|$ for

$$|z_1| \geq 1, \ldots, |z_n| \geq 1$$

at *each* fixed ν and m is attained when

$$z_1 = \cdots = z_n = -1.$$

Thus

$$\text{(6.4.11)} \qquad \max_{z_j} \sum_{\nu=m+1}^{m+n} \left|c_\nu^{(3)}\right| = \sum_{\nu=m+1}^{m+n} \left|c_\nu^{(3)}(1,\ldots,1)\right| =$$

$$= \left| \sum_{\nu=m+1}^{m+n} c_\nu^{(3)}(1,1,\ldots,1) \right|.$$

The last expression is, however, the absolute value of $f_2(z)$ at $z = 1$ when all z_j are -1, that is, for

$$f_1(z) = (1 + z)^n.$$

Then we have

$$s_m\left(\frac{1}{f_1}\right) = \sum_{\nu=0}^{m} \binom{-n}{\nu} z^\nu,$$

and hence

$$\text{(6.4.12)} \qquad \max_{z_j} \sum_{\nu=m+1}^{m+n} \left|c_\nu^{(3)}\right| = |Q_{m,n}|$$

where

$$1 - 2^n \sum_{\nu=0}^{m} \binom{-n}{\nu} = Q_{m,n}.$$

To get an expression for $|Q_{m,n}|$ consisting of positive terms let

$$\sum_{m=0}^{\infty} Q_{m,n} t^m = \varphi(t)$$

for $|t| < 1$. From (6.4.12) we get

$$\varphi(t) = \frac{1 - 2^n(1+t)^{-n}}{1-t} = -\frac{1}{(1+t)^n} \frac{(1+t)^n - 2^n}{(1+t) - 2}$$

$$= -\sum_{\nu=0}^{n-1} 2^\nu (1+t)^{-\upsilon-1}$$

which yields

$$|Q_{m,n}| = \left|\sum_{\nu=0}^{n-1} 2^{\nu}\binom{-\nu-1}{m}\right| = \sum_{\nu=0}^{n-1} 2^{\nu}\binom{m+\nu}{\nu} = P_{m,n}.$$

6.5

To show that the quantity $P_{m,n}$ in Theorem 6.2 cannot be replaced by $P_{m,n} - \varepsilon$ (where $\varepsilon > 0$ is arbitrarily small) for *all* b_j coefficient systems, we note first that for continuity reasons there exists a $(z_1^*,\dots,z_n^*)$-system such that

$$(6.5.1) \qquad z_1^* = -1, \qquad |z_j^*| \geq 1 \qquad (j = 2,\dots,n)$$

$$z_\mu^* \neq z_\nu^* \quad \text{for} \quad \mu \neq \nu$$

and

$$(6.5.2) \qquad \left|\sum_{\nu=m+1}^{m+n} c_\nu^{(3)}(z_1^*,\dots,z_n^*)\right| > P_{m,n} - \varepsilon.$$

Fixing these z_j^* we can determine $b_1^*,\dots,b_n^*$ so that for

$$\sum_{j=1}^{n} b_j^* z_j^{*\nu} = g^*(\nu)$$

we have

$$(6.5.3) \qquad g^*(\nu) = 1, \qquad \nu = m+1,\dots,m+n.$$

Hence (6.2.7), (6.5.2), and (6.5.3) give

$$(6.5.4) \qquad \left|\sum_{j=1}^{n} b_j^*\right| = \left|\sum_{\nu=m+1}^{m+n} c_\nu^{(3)*}(z_1^*,\dots,z_n^*)\right| > P_{m,n} - \varepsilon =$$

$$= (P_{m,n} - \varepsilon) \max_{\nu=m+1,\dots,m+n} |g^*(\nu)|$$

which proves our assertion.

We need to investigate when equality can hold in (6.4.3). Suppose that

$$(6.5.5) \qquad b_1 + b_2 + \cdots + b_n \neq 0, \qquad n \geq 2, \qquad m \geq 1,$$

were *necessary* to achieve equality in (6.4.11). This could happen only if

$$(6.5.6) \qquad z_1 = \cdots = z_n \quad \text{and} \quad |z_j| = 1.$$

But then the validity of the equality sign would imply

$$\left|\sum_1^n b_j\right| = P_{m,n}^{-1}\left|\sum_{j=1}^n b_j\right|,$$

that is, for $n \geq 2$

$$\sum_{j=1}^n b_j = 0, \tag{6.5.7}$$

which contradicts (6.5.5). This completes the proof of Theorem 6.2. ■

6.6

It is curious that, whereas the important inequality (6.3.7) could be deduced from the inequality of Theorem 6.1 quite simply and intuitively, its deduction from the *sharper* inequality given by Theorem 6.2 was much less intuitive. The special case $m = 0$ is itself of interest.

Corollary. *For arbitrary complex* z_j *with* $\min_j|z_j| = 1$ *and complex* b_j *the inequality*

$$\max_{\nu=1,\ldots,n}\left|\sum_{j=1}^n b_j z_j^\nu\right| \geq \frac{1}{2^n - 1}|b_1 + \cdots + b_n| \tag{6.6.1}$$

holds and is the best possible result in the sense of Theorem 6.2.

For $b_1 = \cdots = b_n = 1$ the lower bound is only $n/(2^n - 1)$, which is very weak when compared to Theorem 3.1. This indicates how in the case of various *prescribed* b_j-systems, individual methods are needed to determine the *exact* solutions of the corresponding minimax problems.

6.7

For some applications, when we know $\Sigma_j 1/|z_j|$, it is better to have a lower bound in terms of $1/|z_j|$. Such a bound is furnished in the following result.

Theorem 6.3. *For* $\min_j|z_j| = 1$ *and positive integer m the inequality*

$$\max_{\nu=m+1,\ldots,m+n}|g(\nu)| \geq \tfrac{1}{3}|g(0)|\prod_{j=1}^n \frac{|z_j| - \exp(-1/m)}{|z_j| + 1} \tag{6.7.1}$$

holds.

Proof. This theorem follows easily from inequality (6.2.8), along with (6.2.9). Hence we have only to estimate $\Sigma|c_\nu^{(1)}|$ and $\Sigma|c_\nu^{(2)}|$ differently. For the first sum, (6.2.1) at once gives the upper bound

$$\prod_{j=1}^{n}\left(1+\frac{1}{|z_j|}\right). \tag{6.7.2}$$

For the other sum we have

$$\sum_{\nu=0}^{\infty}|c_\nu^{(2)}|r^\nu \le \prod_{j=1}^{n}\frac{1}{1-(r/|z_j|)} \qquad \text{for } 0<r<1$$

and hence

$$\sum_{\nu=0}^{m}|c_\nu^{(2)}| \le e\sum_{\nu=0}^{\infty}|c_\nu^{(2)}|e^{-\nu/m} \le e\prod_{j=1}^{n}\frac{1}{1-\dfrac{e^{-1/m}}{|z_j|}},$$

that is,

$$\max_{\nu=m+1,\dots,m+n}|g(\nu)| \ge \tfrac{1}{3}|g(0)|\prod_{j=1}^{n}\frac{|z_j|-e^{-1/m}}{|z_j|+1}.$$

This completes the proof. ■

Another simple but useful theorem of "simultaneous type" concerns two generalized power sums

$$g_1(\nu)=\sum_{j=1}^{n_1}b_j'\xi_j^\nu, \qquad g_2(\nu)=\sum_{j=1}^{n_2}b_j''\eta_j^\nu \tag{6.7.3}$$

where the ξ_j and η_j are complex numbers with

$$\min_j|\xi_j| = \min_j|\eta_j| = 1, \tag{6.7.4}$$

and b_j' and b_j'' are complex.

Theorem 6.4. *For arbitrary integer $m \ge 0$ there is an integer ν_0 with*

$$m+1 \le \nu_0 \le m+n_1n_2 \tag{6.7.5}$$

so that the inequalities

$$|g_1(\nu_0)| \ge \left(\frac{n_1n_2}{2e(m+n_1n_2)}\right)^{n_1n_2}\left\{\frac{\left|\sum_1^{n_1}b_j'\right|\left|\sum_1^{n_2}b_j''\right|}{\sum_1^{n_2}|b_j''||\eta_j|^{m+n_1n_2}}\right\} \tag{6.7.6}$$

and

(6.7.7) $$|g_2(\nu_0)| \geq \left(\frac{n_1 n_2}{2e(m+n_1 n_2)}\right)^{n_1 n_2} \left\{ \frac{\left|\sum_1^{n_1} b_j'\right| \left|\sum_1^{n_2} b_j''\right|}{\sum_1^{n_1} |b_j'| \, |\xi_j|^{m+n_1 n_2}} \right\}$$

hold simultaneously.

Proof. Observe that

$$g_3(\nu) = \sum_{j_1=1}^{n_1} \sum_{j_2=1}^{n_2} b_{j_1}' b_{j_2}'' (\xi_{j_1} \eta_{j_2})^{\nu}$$

is a generalized power sum with

$$n = n_1 n_2 \quad \text{and} \quad z_j = \xi_{j_1} \eta_{j_2}.$$

Applying Theorem 6.1 we find that for a ν_0 satisfying (6.7.5) we have

(6.7.8) $$|g_1(\nu_0) g_2(\nu_0)| \geq \left(\frac{n_1 n_2}{2e(m+n_1 n_2)}\right)^{n_1 n_2} \left|\sum b_j'\right| \left|\sum b_j''\right|.$$

From this the theorem follows at once by (6.7.4). ■

These inequalities are particularly simple when all ξ_j and η_j have absolute value 1. It is in this case they will be applied.

6.8

Some applications make it necessary, as mentioned in Section 2.5, to extend the results to "operator-type" expressions $G_1(\nu)$ as in (2.5.1). More precisely, let

(6.8.1) $$G_1(\nu) = \sum_{j=1}^{n} P_j(\nu) z_j^{\nu}$$

where the P_j are arbitrary complex polynomials in ν of degree $(k_j - 1)$ and

(6.8.2) $$\kappa = \sum_{j=1}^{n} k_j.$$

The first result in this direction was obtained by S. Dancs in his thesis [3], and was improved by van der Poorten [1] and Geysel [1]. We prove a result of Tijdeman next.

Theorem 6.5 (Tijdeman [2]). *For every integer* $m \geq 0$ *and* $\min_j |z_j| = 1$ *the inequality*

$$\max_{\substack{m+1 \leq \nu \leq m+\kappa \\ \nu \text{ integer}}} |G_1(\nu)| \geq \left\{ \sum_{j=0}^{\kappa-1} 2^j \binom{m+j}{j} \right\}^{-1} \left| \sum_{j=1}^{n} P_j(0) \right|.$$

holds.

The quantity in curly brackets is the best possible, as in Theorem 6.2.

This theorem can be deduced from Theorem 6.2 by a simple limiting process. Let

$$\gamma_j = \log z_j, \qquad j = 1, \ldots, n$$

with some branch of the logarithm. Then setting

$$F(x) = \sum_{j=1}^{n} P_j(x) e^{\gamma_j x} \tag{6.8.3}$$

let

$$z^\kappa + a_1 z^{\kappa-1} + \cdots + a_\kappa = 0 \tag{6.8.4}$$

be the equation for which γ_j is a root of multiplicity k_j. Then $F(x)$ is the solution of the equation

$$y^{(\kappa)} + a_1 y^{(\kappa-1)} + \cdots + a_\kappa y = 0 \tag{6.8.5}$$

with the initial conditions

$$y^{(j)}(0) = F^{(j)}(0) \qquad j = 0, 1, \ldots, \kappa - 1. \tag{6.8.6}$$

For arbitrarily small $\varepsilon > 0$ there exist complex numbers a_j' depending on ε ($j = 1, \ldots, \kappa$) so that the zeros $\gamma_1', \ldots, \gamma_\kappa'$ of

$$z^\kappa + a_1' z^{\kappa-1} + \cdots + a_\kappa' = 0 \tag{6.8.7}$$

are all simple and

$$|a_j' - a_j| \leq \varepsilon \qquad j = 1, 2, \ldots, \kappa. \tag{6.8.8}$$

Then the solution of the equation

$$Y^{(\kappa)} + a_1' Y^{(\kappa-1)} + \cdots + a_\kappa' Y = 0$$

with the initial conditions

$$Y^{(\nu)}(0) = F^{(\nu)}(0) \qquad \nu = 0, 1, \ldots, (\kappa - 1)$$

has the form

$$Y(x) = \sum_{j=1}^{\kappa} A_j e^{\gamma_j' x}. \tag{6.8.9}$$

We can now apply Theorem 6.2 to $Y(x)$. Hence

$$\max_{\substack{m+1 \le \nu \le m+\kappa \\ \nu \text{ integer}}} |Y(\nu)| \ge P_{m,\kappa}^{-1} |Y(0)| \tag{6.8.10}$$

$$= P_{m,\kappa}^{-1} |F(0)| = P_{m,\kappa}^{-1} \left| \sum_{j=1}^{n} P_j(0) \right|.$$

Since the right side of (6.8.10) is independent of ε, the passage to limit $\varepsilon \to 0$ completes the proof.

6.9

We have indicated in 6.3 how one can apply Theorem 6.1 to the theory of differential equations. To obtain an L_2-version of the inequality (6.3.7) one must investigate the quantity

$$\inf_{z_j} \frac{\sum_{\nu=m+1}^{m+n} |g(\nu)|^2}{\sum_{\nu=1}^{n} |g(\nu)|^2} \tag{6.9.1}$$

under the restriction $\min_j |z_j| = 1$. Using the identity (6.2.7) we have the following result.

Theorem 6.6. *Under the normalization* $\min_j |z_j| = 1$ *the inequality*

$$\frac{\sum_{\nu=m+1}^{m+n} |g(\nu)|^2}{\sum_{\nu=0}^{n} |g(\nu)|^2} \ge \frac{1}{n} \left[\frac{n}{2e(m+n+1)} \right]^{2n}$$

holds for $m \ge n$, *if* $\sum_{\nu=0}^{n} |g(\nu)|^2 \ne 0$.

Proof. First we observe that (6.2.7) gives, by Cauchy's inequality,

$$|g(0)|^2 \le \left\{ \sum_{\nu=m+1}^{m+n} |g(\nu)|^2 \right\} \left\{ \sum_{\nu=m+1}^{m+n} |c_\nu^{(3)}|^2 \right\}.$$

Since

$$\left|c_\nu^{(3)}\right|^2 \le \left(\sum_{\mu=0}^{n} \left|c_\mu^{(1)}\right|^2\right)\left(\sum_{j=0}^{m} \left|c_j^{(2)}\right|^2\right)$$

we get

$$|g(0)|^2 \le n\left(\sum_{\mu=0}^{n} \left|c_\mu^{(1)}\right|^2\right)\left(\sum_{j=0}^{m} \left|c_j^{(2)}\right|^2\right)\sum_{\nu=m+1}^{m+n} |g(\nu)|^2. \tag{6.9.2}$$

From $\left|c_\mu^{(1)}\right| \le \binom{n}{\mu}$,

$$\sum_{\mu=0}^{n} \left|c_\mu^{(1)}\right|^2 \le \binom{2n}{n} < 4^n$$

and from (6.2.11)

$$\sum_{\nu=0}^{m} \left|c_\nu^{(2)}\right|^2 \le \sum_{\nu=0}^{m} \binom{\nu+n-1}{n-1}^2 < \binom{m+n-1}{n-1}\sum_{\nu=0}^{m}\binom{\nu+n-1}{n-1}$$

$$= \binom{m+n-1}{n-1}\binom{m+n}{n} < e\left(e\frac{m+n}{n}\right)^{2n-1}.$$

Thus from (6.9.2)

$$|g(0)|^2 \le 2e\left(2e\frac{m+n}{n}\right)^{2n-1} n \sum_{\nu=m+1}^{m+n} |g(\nu)|^2. \tag{6.9.3}$$

For fixed $0 \le h \le n$ we apply (6.9.3) to $g(\nu + h)$ with $(m - h) \ge 0$; we then obtain

$$|g(h)|^2 \le 2e\left(2e\frac{m+n-h}{n}\right)^{2n-1} n \sum_{\nu=m+1}^{m+n} |g(\nu)|^2. \tag{6.9.4}$$

Summation with respect to h from 0 to n yields

$$\sum_{h=0}^{n} |g(h)|^2 \le \frac{(2e)^{2n}}{n^{2n-2}}\left(\sum_{\nu=m+1}^{m+n} |g(\nu)|^2\right)\sum_{\nu=m}^{m+n} \nu^{2n-1}$$

$$\le \frac{(2e)^{2n}}{n^{2n-1}}(m+n+1)^{2n}\sum_{\nu=m+1}^{m+n} |g(\nu)|^2,$$

which is another form of the assertion.

The inequality is a bit asymmetrical as it contains n terms in the numerator and $(n+1)$ terms in the denominator; this asymmetry could be made still larger. We show in Part II how smoothly we can come to L_2-forms from Theorem 6.6.

6.10

We conclude this section with a final application of the identity (6.2.7) (for $m = 0$).

Theorem 6.7. *For* $\min_j |z_j| = 1$ *and* $0 < r < 1$ *the inequality*

$$\sum_{\nu=1}^{n} |g(\nu)|^2 r^{2\nu} \geq \frac{|g(0)|^2 r^{2n}}{\sum_{\nu=0}^{n-1} \binom{n}{\nu}^2 r^{2\nu}} \tag{6.10.1}$$

holds.

For

$$b_j = \frac{1}{n}, \qquad z_j = e^{2\pi i j/n}$$

the left side of (6.10.1) is r^{2n} whereas the right side is

$$\frac{r^{2n}}{\sum_{\nu=0}^{n-1} \binom{n}{\nu}^2 r^{2\nu}},$$

and which for small r-values is asymptotically r^{2n}. In other words, Theorem 6.7 is, in this sense, the best possible result. It can be conjectured that the right side of (6.10.1) might be replaced by the quantity

$$|g(0)|^2 r^{2n} \tag{6.10.2}$$

for all fixed $0 \leq r < 1$ and $n > n_0(r)$. Applying the full identity (6.2.7) and also the Dancs–Makai observation (6.4.5) one could get an analogous lower bound for

$$\sum_{\nu=m+1}^{m+n} |g(\nu)|^2 r^{2\nu}. \tag{6.10.3}$$

Proof. We use the identity (6.2.7) for $m = 0$ when it takes the form [notation (6.2.1)]

$$\sum_{\nu=1}^{n} c_\nu^{(1)} g(\nu) = -\sum_{j=1}^{n} b_j.$$

Hence

$$\left|\sum_{j=1}^{n} b_j\right|^2 = \left|\sum_{\nu=1}^{n} [g(\nu) r^\nu]\left(\frac{c_\nu^{(1)}}{r^\nu}\right)\right|^2$$

$$\leq \left[\sum_{\nu=1}^{n} |g(\nu)|^2 r^{2\nu}\right]\left(\sum_{\nu=1}^{n} \frac{|c_\nu^{(1)}|^2}{r^{2\nu}}\right)$$

$$\leq \left[\sum_{\nu=1}^{n} |g(\nu)|^2 r^{2\nu}\right]\left[\sum_{\nu=1}^{n} \binom{n}{\nu}^2 \frac{1}{n^{2\nu}}\right]$$

$$= \left[\sum_{\nu=1}^{n} |g(\nu)|^2 r^{2\nu}\right]\frac{1}{r^{2n}}\left[\sum_{\nu=0}^{n-1} \binom{n}{\nu}^2 r^{2\nu}\right],$$

as required.

6.11

The papers of van der Poorten [1] and Tijdeman [2] contain interesting alternative proofs for Theorem 6.2. Their fundamental idea has some resemblance to an idea of Littlewood's [1] whose result, although specialized, was the initial step toward the first main theorem. Curiously enough, what Littlewood [1] actually needed was the corresponding *one-sided* theorem, and the only (slight) disadvantage of this streamlined new version, achieved after 30 years, is that *it* apparently cannot furnish one-sided theorems either.

7

CONTINUATION

7.1

We showed how the important inequality (6.3.7) followed from Theorem 6.1. However—as G. Halász observed—for $a \to +0$ the factor $[d/2e(a+d)]^n$ tends to $(2e)^{-n}$ and not to 1 or $[1+o(1)]^n$, so that for such cases (6.3.7) is weak. To deal with such cases he proved the following.

Theorem 7.1 (G. Halász, unpublished). *For any nonnegative integer* m, *any integer* $k \geq 2$, *and arbitrary complex numbers* b_j *and* z_j *with* $\min_j |z_j| = 1$ *the inequality*

$$\max_{\nu=m+1,\ldots,m+kn} |g(\nu)| \geq \frac{|g(0)|}{\sqrt{kn+1}} 2^{-m-1} \exp\left(-\frac{6n}{k}\right)$$

holds.

With the polynomial $\pi_k^*(z)$ of Lemma 5.8 let the first auxiliary polynomial be

$$F_1(z) = \prod_{j=1}^{n} \pi_k^*\left(\frac{z}{z_j}\right) = \sum_{\nu=0}^{kn} d_\nu^{(1)} z^\nu, \qquad d_0^{(1)} = 1. \tag{7.1.1}$$

We have

$$\|F_1\| \leq \sqrt{kn+1} \sqrt{\sum_{\nu=0}^{kn} \left|d_\nu^{(1)}\right|^2}$$

$$= \sqrt{kn+1} \left[\frac{1}{2\pi} \int_{-\pi}^{\pi} \left|F_1(e^{i\vartheta})\right|^2 d\vartheta\right]^{1/2}$$

$$\leq \sqrt{kn+1} \max_\vartheta \left|F_1(e^{i\vartheta})\right| \leq \sqrt{kn+1} \max_\vartheta \left|\pi_k^*(e^{i\vartheta})\right|^n,$$

that is,

$$\|F_1\| \le \sqrt{kn+1}\exp\frac{2n}{k} \tag{7.1.2}$$

due to part 4 of Lemma 5.8. Set

$$\frac{1}{F_1(z)} = \sum_{\nu=0}^{\infty} d_\nu^{(2)} z^\nu. \tag{7.1.3}$$

From (7.1.1) and part 1 of Lemma 5.8, $F_1(z)^{-1}$ is also regular for $|z| < 1$. Hence, from (5.8.1), we have

$$\max_{|z|=r}\frac{1}{|F_1(z)|} \le \max_{|z|=r}\frac{1}{|\pi_k^*(z)|^n} \le \exp\frac{4rn}{k(1-r)}.$$

From Cauchy's inequality

$$|d_\nu^{(2)}| \le \frac{1}{r^\nu}\exp\frac{4rn}{k(1-r)}$$

and thus

$$\sum_{\nu=0}^{m} |d_\nu^{(2)}| \le \frac{1}{r^m(1-r)}\exp\frac{4rn}{k(1-r)}.$$

For $r = \frac{1}{2}$

$$\sum_{\nu=0}^{m} |d_\nu^{(2)}| < 2^{m+1}\exp\left(\frac{4n}{k}\right). \tag{7.1.4}$$

Let

$$s_m\left(\frac{1}{F_1}\right) = \sum_{\nu=0}^{m} d_\nu^{(2)} z^\nu. \tag{7.1.5}$$

If we set

$$F_2(z) = 1 - F_1(z)s_m\left(\frac{1}{F_1}\right) \tag{7.1.6}$$

we obtain, as in Section 6.2, the equations

$$F_2(z) = \sum_{\nu=m+1}^{m+kn} d_\nu^{(3)} z^\nu \tag{7.1.7}$$

and

$$\sum_{\nu=m+1}^{m+kn} d_\nu^{(3)} z_j^\nu = 1 \qquad j = 1, 2, \ldots, n. \tag{7.1.8}$$

Hence we get the identity

$$\sum_{\nu=m+1}^{m+kn} d_\nu^{(3)} g(\nu) = g(0), \tag{7.1.9}$$

that is,

$$\max_{\nu=m+1,\ldots,m+kn} |g(\nu)| \geq \frac{|g(0)|}{\|F_2\|} \geq \frac{|g(0)|}{\|F_1\| \, \|s_m(1/F_1)\|}. \tag{7.1.10}$$

Theorem 7.1 is proved using (7.1.2) and (7.1.4). For $m = 0$, $k = n$ (≥ 2) there is an important consequence.

Corollary. *For* $\min_j |z_j| = 1$ *and arbitrary complex* b_j *the inequality*

$$\max_{\substack{1 \leq \nu \leq n^2 \\ \nu \text{ integer}}} |g(\nu)| > \frac{e^{-6}}{4} \frac{|g(0)|}{n} \tag{7.1.11}$$

holds.

It is surprising to compare (7.1.11) with (6.6.1). By applying the reasoning of Theorem 7.1 to the case $m = 0$ the constant $(4e^6)^{-1}$ can be improved.

7.2

In Theorem 6.5 we dealt with expressions of the form

$$G_1(\nu) = \sum_{j=1}^{n} \varphi_j(\nu) z_j^\nu$$

with polynomials $\varphi_j(\nu)$, which were first mentioned in (2.5.1). It is perhaps even more important to study the other operator-type expression

$$G_2(\nu) = \sum_{j=1}^{n} b_j \varphi(z_j) z_j^\nu \tag{7.2.1}$$

with a polynomial $\varphi(z)$, mentioned first in (2.5.2). Let

$$0 < \rho < 1 \tag{7.2.2}$$

and let

$$\varphi(z) = z^k + e_1 z^{k-1} + \cdots + e_k \tag{7.2.3}$$

be a polynomial with no zeros outside $|z| < \rho$ (so that $\varphi(z) \equiv 1$ is admissible). Then we assert the following.

Theorem 7.2. *With nonnegative integer m and*

$$\min_j |z_j| = 1 \tag{7.2.4}$$

and for the above $\varphi(z)$ the inequality

$$\max_{m+1 \le \nu \le m+n} |G_2(\nu)| \ge \left| \sum_{j=1}^{n} b_j \right| \frac{1}{2^n} \left(\frac{1+\rho}{2} \right)^m \left(\frac{1-\rho}{2} \right)^{n+k} \tag{7.2.5}$$

holds.

It is noteworthy that the lower bound is essentially independent of the coefficients of $\varphi(z)$ and that the length of S remains unchanged.

Applying Theorem 6.1 *directly* to $G_2(\nu)$ we would get a lower bound containing the factor $|\Sigma_{j=1}^{n} b_j \varphi(z_j)|$ instead of $|\Sigma_1^n b_j|$ which would obviously be useless. One could also raise the question (given Theorem 7.2) of the necessity of Theorem 6.1, since in this case $\varphi(z) \equiv 1$, $k = 0$; and, choosing

$$\rho = 1 - 2\frac{n}{m + 2n} \qquad \text{(where } m > 2n\text{)},$$

we have

$$\frac{1}{2^n} \left(\frac{1+\rho}{2} \right)^m \left(\frac{1-\rho}{2} \right)^{n+k} \ge \left[\frac{n}{2e(m+n)} \right]^n. \tag{7.2.6}$$

The answer lies in the fact that the case $\varphi(z) \equiv 1$ has the greatest number of applications, and its proof is much simpler than that of Theorem 7.2.

By the way, (7.2.6) shows in conjunction with Theorem 6.2 that (7.2.5) is essentially the best possible inequality.

7.3. PROOF OF THEOREM 7.2

We may suppose that the z_j are distinct. Let $f(z)$ be the polynomial of degree $\le n - 1$ defined by

$$f(z_j) = \frac{1}{z_j^{m+1} \varphi(z_j)} \qquad j = 1, 2, \ldots, n \tag{7.3.1}$$

(which makes sense due to (7.2.3) and (7.2.4)). Writing

$$f(z) = \sum_{\nu=0}^{n-1} c_\nu^{(1)} z^\nu \tag{7.3.2}$$

we obtain

$$\sum_{\nu=0}^{n-1} c_\nu^{(1)} z_j^{m+\nu+1} \varphi(z_j) = 1 \qquad j = 1,\dots,n. \tag{7.3.3}$$

Multiplying by b_j and summing with respect to j we get the identity

$$\sum_{\nu=0}^{n-1} c_\nu^{(1)} G_2(m+\nu+1) = \sum_{j=1}^{n} b_j. \tag{7.3.4}$$

Hence

$$\max_{\nu=m+1,\dots,m+n} |G_2(\nu)| \geq \frac{\left|\sum_{j=1}^{n} b_j\right|}{\|f\|}, \tag{7.3.5}$$

and we must give an upper bound for the denominator.

7.4

We write first $f(z)$ in the form

$$f(z) = c_0^{(0)} + c_1^{(0)}(z - z_1) + \cdots + c_{n-1}^{(0)}(z - z_1)\cdots(z - z_{n-1}). \tag{7.4.1}$$

Lemma 5.3 yields the representation

$$c_\nu^{(0)} = -\frac{1}{2\pi i}\int_{|\omega| = \frac{1+\rho}{2} = R} \frac{dw}{w^{m+1}\varphi(w)(w - z_1)\cdots(w - z_{\nu+1})}. \tag{7.4.2}$$

On the circle $|w| = R$ we have

$$|\varphi(w)| \geq (R - \rho)^k$$

and

$$|c_\nu^{(0)}| \leq \frac{1}{R^m (R-\rho)^k (|z_1| - R)\cdots(|z_{\nu+1}| - R)}$$

$$\leq \frac{1}{R^m(1-R)(R-\rho)^k(|z_1| - R)\cdots(|z_\nu| - R)}. \tag{7.4.3}$$

Writing

$$f(z) = \sum_{\nu=0}^{n-1} c_\nu^{(1)} z^\nu, \tag{7.4.4}$$

we get by comparing (7.4.4) and (7.4.1),

$$\|f\| \le \left|c_0^{(0)}\right| + \sum_{\nu=0}^{n-1} \left|c_\nu^{(0)}\right|(1 + |z_1|)\cdots(1 + |z_\nu|).$$

That is, from (7.4.3)

$$\|f\| \le \frac{1}{R^m(R-\rho)^k(1-R)} \sum_{\nu=0}^{n-1} \frac{1+|z_1|}{|z_1|-R}\cdot\frac{1+|z_2|}{|z_2|-R}\cdot\;\cdots\;\cdot\frac{1+|z_\nu|}{|z_\nu|-R}.$$

Since

$$\frac{1+x}{x-R} \le \frac{2}{1-R} = \frac{4}{1-\rho} \quad \text{for} \quad x \ge 1,$$

this gives

$$\|f\| < \left(\frac{2}{1+\rho}\right)^m \left(\frac{2}{1-\rho}\right)^{k+1} \sum_{\nu=0}^{n-1} \left(\frac{4}{1-\rho}\right)^\nu < \left(\frac{2}{1+\rho}\right)^m \left(\frac{2}{1-\rho}\right)^{n+k} 2^n.$$

This completes the proof of Theorem 7.2. ■

7.5

A general problem is finding, for a given domain D, systems of functions $f_0(z), f_1(z), \ldots$ such that for each $z_1, \ldots, z_n \in D$, and integer $m \ge 0$ and complex numbers $b_1, \ldots, b_n$ an inequality of type

$$\max_{\nu=m+1,\ldots,m+n} \left| \sum_{j=1}^{n} b_j f_\nu(z_j) \right| \ge c(m, n, D)\psi(b_1, \ldots, b_n, D) \tag{7.5.1}$$

holds with explicit positive $c(m, n, D)$ and ψ. Theorem 7.2 deals with such a case. The ingenious method of van der Poorten [1] permits construction of further interesting systems with this property. Since we do not know of any applications, we do not enter into their discussion here.

7.6

Let us return to Theorem 7.2 in the simplest case $\varphi(z) = z - a$, with a fixed and $|a| < 1$. Then

$$G_2^*(\nu) = \sum_{j=1}^{n} b_j z_j^{\nu}(z_j - a)$$

and the impossibility of

$$G_2^*(\nu) = 0 \qquad \nu = m+1,\ldots,m+n, \qquad G_2^*(\nu) \not\equiv 0$$

also follows from the fact that, if $z_j \neq a$ and $z_\mu \neq z_\nu$, then

$$\det\left|z_j^l(z_j - a)\right|_{\substack{j=1,\ldots,n\\ l=0,\ldots,n-1}} \neq 0 \quad \text{for} \quad \mu \neq \nu.$$

However, the situation changes if the coefficients of $\varphi(z)$ depend on ν.

The simplest such case is

$$\varphi(z) = (\nu + 1) - az \qquad (a \neq 0)$$

and $m = 0$. That is,

$$G^{**}(\nu) = \sum_{j=1}^{n} z_j^{\nu}(\nu + 1 - az_j) = (\nu+1)s_\nu - as_{\nu+1}. \tag{7.6.1}$$

The system

$$G^{**}(\nu) = 0 \qquad \nu = 1, 2, \ldots, n \tag{7.6.2}$$

is equivalent to the conditions

$$s_\nu = \frac{\nu!}{a^{\nu-1}} s_1 \qquad \nu = 2, 3, \ldots, n+1. \tag{7.6.3}$$

The Newton–Girard formulas

$$\begin{vmatrix} s_1 & 1 & 0 & \cdots & 0 \\ s_2 & s_1 & 2 & \cdots & 0 \\ \vdots & \vdots & \vdots & & \vdots \\ s_n & s_{n-1} & s_{n-2} & \cdots & n \\ s_{n+1} & s_n & s_{n-1} & \cdots & s_1 \end{vmatrix} = 0$$

and (7.6.3) imply that not only $z_1 = \cdots = z_n = 0$, but that there are (at most n) *nontrivial* systems satisfying (7.6.2). (These are not necessarily normalized by $\min|z_j| = 1$.) This shows, at any rate, that for polynomials $\varphi(z)$ with coefficients also depending on ν, no completely general theorem for the range $(m+1, m+n)$ can be expected. Because of their applicability it would be desirable to find such theorems for a longer ν-range.

7.7

S. Dancs called my attention to the fact that Theorem 6.1 was really proved for Banach spaces. So we formulate the following two corollaries of Theorem 6.1.

Corollary I. *Let $z_1,\ldots,z_n$ be complex numbers satisfying* (6.1.1) *and $b_1,\ldots,b_n$ be arbitrary elements of a Banach space. If $\|\cdot\|$ denotes the norm of the space then the inequality*

$$(7.7.1) \qquad \max_{\substack{m+1\le\nu\le m+n\\ \nu \text{ integer}}} \left\| \sum_{j=1}^{n} b_j z_j^{\nu} \right\| \ge \left[\frac{n}{2e(m+n)}\right]^n \left\| \sum_{j=1}^{n} b_j \right\|$$

holds.

Corollary II. *Let $b_1,\ldots,b_n$ be arbitrary elements and let $z_1,\ldots,z_n$ be invertible elements with*

$$\min_j \|z_j\| = 1$$

of a Banach algebra. Then the inequality (7.7.1) *holds.*

7.8

Relatively recently I realized the importance of an improvement of (7.1.11) even in the special case

$$(7.8.1) \qquad |z_1| = \cdots = |z_n| = 1, \qquad b_j > 0, \qquad j = 1,\ldots,n.$$

After the first significant result of the Dutch mathematician H. Leenman, the following stronger theorem was found independently by Cassels, D. J. Newman, and M. Szalay [2].

Theorem 7.3. *In the case* (7.8.1) *for arbitrary $c = c(n) \ge 1$, the inequality*

$$\max_{1\le\nu\le cn} |g(\nu)| \ge \sqrt{1 - \frac{n-1}{[cn]}} \cdot \frac{g(0)}{\sqrt{n}}$$

holds.

Hence if $c = 1 + \varepsilon$, $\varepsilon > 0$, then the right side is strictly larger than $c(\varepsilon) \cdot g(0)/\sqrt{n}$.

Proof. Let

$$k = [cn], \qquad a_\nu = 1 - \frac{|\nu|}{k+1} \tag{7.8.1}$$

and let

$$\sigma_k(t) = \sum_{\nu=-k}^{k} a_\nu e^{i\nu t} \tag{7.8.2}$$

be the kth Fejér kernel, which is nonnegative for all real t. We write

$$g(\nu) = \sum_{j=1}^{n} b_j e^{i\alpha_j \nu} \qquad (\alpha_j \text{ real}). \tag{7.8.3}$$

Then

$$\sum_{\nu=-k}^{k} a_\nu |g(\nu)|^2 = \sum_{j_1, j_2} b_{j_1} b_{j_2} \sigma_k(\alpha_{j_1} - \alpha_{j_2}) \geq \left(\sum_{j=1}^{n} b_j^2 \right) \sigma_k(0). \tag{7.8.4}$$

Denoting the $\max_{\nu=1,\ldots,k} |g(\nu)|$ by M, we obtain from (7.8.4),

$$|g(0)|^2 + M^2 \sum_{\substack{\nu=-k \\ \nu \neq 0}}^{k} a_\nu \geq (k+1) \sum_{j=1}^{n} b_j^2$$

$$\geq \frac{k+1}{n} \left(\sum_{j=1}^{n} b_j \right)^2 = \frac{k+1}{n} g(0)^2,$$

since $a_0 = 1$, that is,

$$|g(0)|^2 + M^2 k \geq \frac{k+1}{n} |g(0)|^2,$$

from which (7.8.4) follows. ■

7.9

It would be of great interest to increase the lower bound in Theorem 7.3 in case the interval is $[1, n^2]$ or, more generally, $[1, n^A]$ where A is a constant. The possible increases are surprisingly small. Erdős and Rényi [1] established by probabilistic methods for each $A \geq 2$ the existence of $(z_1^*, \ldots, z_n^*)$ with $|z_j^*| = 1$ $(j = 1, \ldots, n)$ so that

$$g^*(\nu) = \sum_{j=1}^{n} z_j^{*\nu} \qquad (\text{i.e. } g^*(0) = n)$$

and

$$\max_{1 \leq \nu \leq n^A} |g^*(\nu)| \leq c(A)\sqrt{\frac{\log n}{n}}\,|g^*(0)|.$$

We prefer to give an *explicit* example of this type due to Tijdeman (written communication).* Let n be sufficiently large and let $p_1 < p_2 < \cdots < p_\mu$ be the primes between $\sqrt{n \log n}$ and $3\sqrt{n \log n}$. Then it follows from the prime number theorem that

$$\sum_{j=1}^{\mu} p_j > n. \tag{7.9.1}$$

The index h is defined by

$$\sum_{j=1}^{h} p_j \leq n < \sum_{j=1}^{h+1} p_j \tag{7.9.2}$$

and

$$(0 \leq)\delta = n - \sum_{j=1}^{h} p_j < 3\sqrt{n \log n}\,. \tag{7.9.3}$$

If $\varepsilon_{\mu p}$ $(\mu = 1, 2, \ldots, p)$, stands for the pth roots of unity, let

$$g^*(\nu) = \sum_{j=1}^{h} \sum_{\mu=1}^{p_j} \varepsilon_{\mu p_j}^{\nu} + \delta \cdot 1^{\nu}. \tag{7.9.4}$$

Hence $g^*(0) = n$. Let $k = k(n)$, to be chosen later, and consider the interval

$$1 \leq \nu \leq k. \tag{7.9.5}$$

If we fix a ν in this interval, the inner sum in (7.9.4) is, for fixed j,

$$\begin{array}{ll} 0 & \text{for} \quad p_j \nmid \nu \\ p_j & \text{for} \quad p_j | \nu. \end{array}$$

*See also H. Leenman–R. Tijdeman [1].

This and (7.9.3) yields

$$|g^*(\nu)| \le 3\sqrt{n\log n} + \sum_{\substack{1\le j\le h\\ p_j|\nu}} p_j.$$

Denoting by κ the number of p_j-divisors of ν we have

$$\left(\sqrt{n\log n}\right)^{\kappa} \le \nu \le k; \qquad \kappa < \frac{2\log k}{\log n}$$

and hence

$$|g^*(\nu)| \le 3\sqrt{n\log n} + 3\sqrt{n\log n}\,\frac{2\log k}{\log n}.$$

For

$$k = n^A$$

this gives

$$\max_{1\le\nu\le n^A} |g^*(\nu)| \le (6A+3)\sqrt{n\log n} = (6A+3)\sqrt{\frac{\log n}{n}}\,|g^*(0)| \tag{7.9.6}$$

as required.

Another odd example due to H. Montgomery (written communication) shows that for infinitely many n the maximum

$$\max_{1\le\nu\le n^{2-\varepsilon}} \left|\sum_{j=1}^{n} z_j^{**\nu}\right| < 2\sqrt{n} \tag{7.9.7}$$

can occur with

$$|z_j^{**}| = 1 \qquad (j = 1,\dots,n)$$

(ε an arbitrarily small positive number). This follows easily from the identity

$$\left|\sum_{l=1}^{p-1} (\chi(l)\exp(2\pi il/p))^{\nu}\right| = \sqrt{p}, \quad \leftrightarrow \quad \text{for } p\nmid\nu, (p-1)\nmid\nu \tag{7.9.8}$$

$$= 1, \quad \leftrightarrow \quad \text{for } p\nmid\nu, (p-1)|\nu$$

$$= 0, \quad \leftrightarrow \quad \text{for } p|\nu, \ (p-1)\nmid\nu$$

$$= p-1, \quad \leftrightarrow \quad \text{for } p|\nu, \ (p-1)|\nu$$

where p is prime, χ is any nonprincipal character mod p with $\chi^{\nu} \ne \chi_0$ for $p-1\nmid\nu$ and $\nu = 1,2,\dots,p^2-p-1$.

8

THE CASE OF THE MAXIMUM NORM

Second Main Theorem

8.1

Throughout this section we write

$$g(\nu) = \sum_{j=1}^{n} b_j z_j^{\nu}$$

with

$$1 = |z_1| \geq |z_2| \geq \cdots \geq |z_n|. \tag{8.1.1}$$

Condition (8.1.1) implies an ordering on the b_j (which, when some $|z_j|$ are equal, is arbitrary but fixed). Then we assert the following.

Theorem 8.1. *For arbitrary integer $m \geq 0$ and complex b_j condition* (8.1.1) *implies the inequality*

$$\max_{\substack{m+1 \leq \nu \leq m+n \\ \nu \text{ integer}}} |g(\nu)| \geq 2\left(\frac{n}{8e(m+n)}\right)^{n} \min_{j=1,\ldots,n} |b_1 + \cdots + b_j|.$$

This gives, as before, for arbitrary complex η_j with

$$|\eta_1| \geq |\eta_2| \geq \cdots \geq |\eta_n|$$

the existence of an integer ν_0 in the interval

(8.1.2) $$m + 1 \leq \nu_0 \leq m + n$$

such that

(8.1.3) $$|b_1\eta_1^{\nu_0} + \cdots + b_n\eta_n^{\nu_0}| \geq 2\left(\frac{n}{8e(m + n)}\right)^n |\eta_1|^{\nu_0} \min_{1 \leq j \leq n} |b_1 + \cdots + b_j|.$$

Of course η_1 on the right-hand side of (8.1.3) can be replaced by any η_j with the same ν_0.

Inequality (8.1.3) is particularly simple when the b_j are in an angle with opening $\leq (\pi - \eta)$, $\eta > 0$. We note in particular the case $b_1 = b_2 = \cdots = b_n = 1$.

Corollary. *For integer $m \geq 0$ and complex η_j there is an integer ν_0 with*

(8.1.4) $$m + 1 \leq \nu_0 \leq m + n$$

so that the inequality

(8.1.5) $$|\eta_1^{\nu_0} + \cdots + \eta_n^{\nu_0}| \geq 2\left(\frac{n}{8e(m + n)}\right)^n \left(\max_j |\eta_j|\right)^{\nu_0}$$

holds.

Theorem 8.1 was first proved in my paper [5] with the constant e^{26} and in my book [12] with constant $24e^2$; in the paper Vera T. Sós–Turán [1] this was reduced to $2e^{(1+4/e)}$, and it was remarked by Uchiyama [5] that what we actually obtained was the constant $8e$. As we can show, through an ingenious example of Makai, the constant $8e$ in Theorem 8.1 cannot be pushed below $4e$. It would, however be of great importance to show in the corollary that the constant $8e$ can be replaced by $(1 + \varepsilon)$ for arbitrarily small $\varepsilon > 0$ provided that $m > m_0(n, \varepsilon)$.

The dependence of the lower bound on the z_j in Theorem 8.1 is surprisingly good compared to the result in first main theorem. The dependence upon the b_j is inconvenient in general. It is natural to ask whether or not it can be replaced *generally* by

(8.1.6) $$|b_1 + \cdots + b_n|$$

or by

(8.1.7) $$\min_j |b_j|.$$

The trivial examples

$$n \geq 2, \qquad z_1 = 1, \qquad z_2 = \cdots = z_n = 0$$

$$b_1 = 0, \qquad b_2 = \cdots = b_n = \frac{1}{n-1}$$

respectively,

$$n \geq 2, \qquad z_1 = \cdots = z_n = 1$$

$$b_1 = 1, \qquad b_2 = \cdots = b_{n-1} = -\frac{1}{n-1}$$

show that no such theorem can exist. Possibly a suitable geometric restriction, in addition to condition (8.1.1), can produce results of this type.

As we show in Part II, Theorem 8.1 has several applications, so it deserves to be called our second main theorem.

8.2. PROOF OF THEOREM 8.1

Suppose first that

$$z_1 = 1, \qquad z_\mu \neq z_\nu \quad \text{for} \quad \mu \neq \nu, \ |z_n| > 0. \tag{8.2.1}$$

Let

$$\delta = \begin{cases} \dfrac{m}{m+n} & \text{for} \quad m > 0 \\ 1 - \dfrac{1}{e} & \text{for} \quad m = 0 \end{cases} \tag{8.2.2}$$

and

$$f_0(z) = \prod_{j=1}^{n} (z - |z_j|). \tag{8.2.3}$$

Applying Lemma 5.1 to $f_0(z)$ and the interval $[\delta, 1]$ we obtain the existence of a ξ with

$$\delta \leq \xi \leq 1 \tag{8.2.4}$$

such that

$$|f_0(\xi)| \geq 2\left(\frac{1-\delta}{4}\right)^n. \tag{8.2.5}$$

But this means that on the circle $|z| = \xi$ we have, a fortiori,

$$\left|\prod_{j=1}^{n} (z - z_j)\right| \geq 2\left(\frac{1-\delta}{4}\right)^n \tag{8.2.6}$$

and, since on the circle $|z| = \xi$ we have

$$||z| - |z_j|| \leq 1,$$

it follows that *all* subproducts

$$\left|\prod_{j} (z - z_{i_j})\right|$$

of $|\prod_{j=1}^{n}(z - z_j)|$ have the lower bound

$$2\left(\frac{1-\delta}{4}\right)^n \quad \text{on} \quad |z| = \xi. \tag{8.2.7}$$

8.3

With δ as in (8.2.2) and ξ as in (8.2.4) we can split the z_j into two disjoint classes the "outer" (resp. "inner") z_j according to

$$(1 =)\ |z_1| \geq |z_2| \geq \cdots \geq |z_l| > \xi > |z_{l+1}| \geq \cdots \geq |z_n|. \tag{8.3.1}$$

The second class may be empty. The *strict* inequality $|z_l| > \xi$ (and ξ is larger than $|z_{l+1}|$ if the inner class is not empty) is ensured by (8.2.5). Let us consider first the auxiliary polynomial

$$f_1(z) = \prod_{j=l+1}^{n} (z - z_j) = \sum_{\nu=0}^{n-l} c_\nu^{(1)} z^\nu. \tag{8.3.2}$$

Due to the normalization (8.1.1) we have

$$\|f_1\| \leq 2^{n-l}. \tag{8.3.3}$$

8.4

A second auxiliary polynomial $f_2(z)$ of degree $\leq l - 1$ is defined by

$$f_2(z_j) = \frac{1}{f_1(z_j) z_j^{m+1}} \qquad j = 1, 2, \ldots, l. \tag{8.4.1}$$

We represent $f_2(z)$ in the form

$$f_2(z) = c_0^{(2)} + c_1^{(2)}(z - z_1) + c_2^{(2)}(z - z_1)(z - z_2) + \cdots + c_{l-1}^{(2)}(z - z_1) \cdots (z - z_{l-1}). \tag{8.4.2}$$

According to Lemma 5.3 we have the coefficient formula

$$c_\nu^{(2)} = \frac{-1}{2\pi i} \int_{|w| = \xi} \frac{dw}{w^{m+1} f_1(w)(w - z_1) \cdots (w - z_{\nu+1})},$$

hence

$$|c_\nu^{(2)}| \le \frac{1}{2\pi} \int_{|w| = \xi} \frac{|dw|}{\xi^{m+1} |f_1(w)(w - z_1) \cdots (w - z_{\nu+1})|}. \tag{8.4.3}$$

But by its construction the polynomial

$$f_1(w)(w - z_1) \cdots (w - z_{\nu+1})$$

is a subproduct of $\prod_{j=1}^n (w - z_j)$, and (8.2.7) is applicable. Hence, from (8.4.3) and (8.2.4), we get

$$|c_\nu^{(2)}| \le \tfrac{1}{2} \delta^{-m} \left(\frac{4}{1 - \delta} \right)^m \qquad \nu = 0, 1, \ldots, l - 1. \tag{8.4.4}$$

However we shall need $f_2(z)$ in the form

$$f_2(z) = c_0^{(3)} + c_1^{(3)} z + \cdots + c_{l-1}^{(3)} z^{l-1}. \tag{8.4.5}$$

Since

$$c_{l-1}^{(3)} = c_{l-1}^{(2)}$$

and we have

$$c_\nu^{(3)} = c_\nu^{(2)} - c_{\nu+1}^{(2)} \sum_{1 \le i_1 \le \nu+1} z_{i_1} + c_{\nu+2}^{(2)} \sum_{1 \le i_1 < i_2 \le \nu+2} z_{i_1} z_{i_2} - \cdots + (-1)^{l-\nu-1} c_{l-1}^{(2)} \sum_{1 \le i_1 < \cdots < i_{l-\nu-1} \le l-1} z_{i_1} z_{i_2} \cdots z_{i_{l-\nu-1}}$$

$\nu = 0, 1, \ldots, (l - 2)$

we get, using (8.4.4), the upper bound

$$|c_\nu^{(3)}| \leq \tfrac{1}{2}\delta^{-m}\left(\frac{4}{1-\delta}\right)^m \left\{1 + \binom{\nu+1}{1} + \binom{\nu+2}{2} + \cdots + \binom{l-1}{l-1-\nu}\right\}$$

$$= \frac{\delta^{-m}}{2}\left(\frac{4}{1-\delta}\right)^m \binom{l}{\nu+1}.$$

Thus we have

$$\|f_2\| \leq \frac{\delta^{-m}}{2}\left(\frac{4}{1-\delta}\right)^n 2^l. \tag{8.4.6}$$

8.5

We consider one more auxiliary polynomial

$$f_3(z) = z^{m+1} f_1(z) f_2(z) = \sum_{\nu=m+1}^{m+n} c_\nu^{(4)} z^\nu. \tag{8.5.1}$$

By construction we have

$$f_3(z_j) = \begin{cases} 1 & \text{for the outer} \quad z_j \\ 0 & \text{for the inner} \quad z_j, \end{cases}$$

that is,

$$\sum_{\nu=m+1}^{m+n} c_\nu^{(4)} z_j^\nu = \begin{cases} 1 & \text{for} \quad j = 1, 2, \ldots, l \\ 0 & \text{for} \quad j = (l+1), \ldots, n. \end{cases}$$

Multiplying by b_j and summing with respect to j, we get the fundamental identity

$$\sum_{\nu=m+1}^{m+n} c_\nu^{(4)} g(\nu) = b_1 + \cdots + b_l, \tag{8.5.2}$$

that is,

$$\max_{\substack{m+1 \leq \nu \leq m+n \\ \nu \text{ integer}}} |g(\nu)| \geq \frac{\min_j |b_1 + \cdots + b_j|}{\|f_3\|}. \tag{8.5.3}$$

Finally (8.5.1) implies

$$\|f_3\| \leq \|f_1\| \cdot \|f_2\| \leq \frac{\delta^{-m}}{2}\left(\frac{8}{1-\delta}\right)^n$$

by (8.3.3) and (8.4.6). Since we have

$$\delta^{-m} = \left(\frac{m+n}{m}\right)^m < e^n \quad \text{for} \quad m \geq 0,$$

Theorem 8.1 is proved under the restriction (8.2.1). But the general case follows by a passage to the limit. ■

8.6

The proof of the first main theorem was classical-algebraic. The foregoing proof of the second main theorem uses complex integration. The question arises as to whether an algebraic proof can be found. In the following we give such a proof found in collaboration with S. Dancs (Dancs–Turán [3]). It is presented in a somewhat stronger form (which could also have been given in the previous proof). This form is less elegant than that of Theorem 8.1 but more useful in application. The dependence of the lower bound on the b_j in Theorem 8.1 is rather inconvenient, useful in applications only when the real parts of the b_j are all nonnegative. The usefulness is obviously increased when the range of j in the factor

$$\min_{1 \leq j \leq n} |b_1 + \cdots + b_j| \tag{8.6.1}$$

is restricted (in particular from below). This may lead to circumstances when the quantity in (8.6.1) can be replaced essentially by the more manageable quantity $|b_1 + \cdots + b_n|$. The first theorem of this type was prepared from the preceding proof by S. Knapowski [3]; though it is not very elegant, the theorem showed itself to be very useful. Supposing that $m \geq 3n$ in addition to (8.1.1), he defined λ as any index satisfying

$$|z_\lambda| > \frac{4n}{m+n}$$

and then defined the index λ_1 as any satisfying

$$|z_{\lambda_1}| < |z_\lambda| - \frac{2n}{m+n}.$$

If the latter does not exist, set $\lambda_1 = n + 1$. Then his inequality establishes that

$$\max_{\nu = m+1, \ldots, m+n} |g(\nu)| \geq 2\left[\frac{n}{4(m+n)}\right]^n \left(\frac{|z_\lambda|}{2}\right)^m \min_{\lambda \leq j \leq \lambda_1} |b_1 + \cdots + b_j|.$$

Now let us assume, in addition to the normalization (8.1.1), that the integer m is nonnegative, and that δ_1 and δ_2 are restricted only by

$$\frac{m}{m+n} \leq \delta_2 < \delta_1 \leq 1. \tag{8.6.2}$$

Furthermore let l_1 and l_2 be indices for which

$$|z_{l_1}| \geq \delta_1, \qquad |z_{l_2}| \leq \delta_2. \tag{8.6.3}$$

If such an l_2 does not exist put

$$l_2 = n. \tag{8.6.4}$$

Then we assert the following relation.

Theorem 8.2. *With the preceding provisions,*

$$\max_{\substack{m+1 \leq \nu \leq m+n \\ \nu \text{ integer}}} |g(\nu)| \geq 2\left(\frac{\delta_1 - \delta_2}{8e}\right)^n \min_{l_1 \leq j \leq l_2} |b_1 + \cdots + b_j|.$$

8.7. PROOF

Let N be a fixed integer larger than m, let the z_j be distinct and $\neq 0$, and let R be a fixed number in the interval $(0, 1)$ restricted, at present, only by

$$R^N \neq z_j^N, \qquad j = 1, \ldots, n. \tag{8.7.1}$$

Further, let

$$f_n(z) = \prod_{j=1}^{n} (z - z_j) \tag{8.7.2}$$

and

$$F(z) = \sum_{j=1}^{n} \frac{f_n(z)}{f_n'(z_j)(z - z_j)} \left(\frac{z}{z_j}\right)^{m+1} \frac{z_j^N}{z_j^N - R^N}. \tag{8.7.3}$$

Then we have

$$F(z) = \sum_{\nu=m+1}^{m+n} d_\nu z^\nu \tag{8.7.4}$$

for coefficients $d_\nu = d_\nu(z_j, N, R)$.

Replacing z by z_j we have

$$\sum_{\nu=m+1}^{m+n} d_\nu z_j^\nu = \frac{z_j^N}{z_j^N - R^N} \qquad j = 1,\dots,n$$

and thus

$$\sum_{\nu=m+1}^{m+n} d_\nu g(\nu) = \sum_{j=1}^{n} b_j \frac{z_j^N}{z_j^N - R^N}. \tag{8.7.5}$$

Hence

$$\max_{\nu=m+1,\dots,m+n} |g(\nu)| \geq \frac{\left| \sum_{j=1}^{n} b_j \frac{z_j^N}{z_j^N - R^N} \right|}{\|F(z)\|} \tag{8.7.6}$$

where

$$\left\| \sum_{\nu=0}^{k} \mu_\nu z^\nu \right\| = \sum_{\nu} |\mu_\nu| \tag{8.7.7}$$

and we need to estimate the quantity $\|F(z)\|$ from above.

8.8

Let

$$f_\nu(z) = (z - z_1)(z - z_2)\cdots(z - z_\nu) \qquad \nu = 1,\dots,n \tag{8.8.1}$$

in agreement with (8.7.2), and let

$$f_0(z) \equiv 1. \tag{8.8.2}$$

The identity (5.2.5) gives

$$\frac{1}{z - z_j} = \sum_{l=1}^{j} \frac{f_{l-1}(z)}{f_l(z)} \qquad j = 1,\dots,n. \tag{8.8.3}$$

Substituting (8.8.3) in (8.7.3), we get

$$F(z) = \sum_{l=1}^{n} \frac{f_n(z)}{f_l(z)} z^{m+1} \sum_{j=l}^{n} \frac{f_{l-1}(z_j)}{f_n'(z_j)} \cdot \frac{z_j^{N-m-1}}{z_j^N - R^N}. \tag{8.8.4}$$

Since

$$\left\| \frac{f_n(z)}{f_l(z)} z^{m+1} \right\| \le 2^{n-l},$$

we get from (8.8.4), using the abbreviation

$$\sum_{j=l}^{n} \frac{f_{l-1}(z_j)}{f_n'(z_j)} \cdot \frac{z_j^{N-m-1}}{z_j^N - R^N} = U_l, \tag{8.8.5}$$

the inequality

$$\|F(z)\| \le \sum_{l=1}^{n} 2^{n-l} |U_l|. \tag{8.8.6}$$

8.9

For the estimation of U_l we use the partial-fraction representation

$$\frac{w^{N-m-1}}{w^N - R^N} = \sum_{\mu=1}^{N} \frac{1}{N(R\varepsilon^{\mu})^m} \cdot \frac{1}{w - R\varepsilon^{\mu}} \tag{8.9.1}$$

where

$$\varepsilon = \exp \frac{2\pi i}{N}. \tag{8.9.2}$$

Substituting (8.9.1) in (8.8.5) we get

$$U_l = -\frac{1}{NR^m} \sum_{\mu=1}^{N} \varepsilon^{-\mu m} \sum_{j=l}^{n} \frac{f_{l-1}(z_j)}{f_n'(z_j)} \cdot \frac{1}{R\varepsilon^{\mu} - z_j}. \tag{8.9.3}$$

We observe that

$$\sum_{j=l}^{n} \frac{f_{l-1}(z_j)}{f_n'(z_j)} \cdot \frac{1}{R\varepsilon^{\mu} - z_j} = \frac{f_{l-1}(R\varepsilon^{\mu})}{f_n(R\varepsilon^{\mu})}. \tag{8.9.4}$$

This and (8.9.3) yield

$$|U_l| \le \frac{1}{NR^m} \sum_{\mu=1}^{N} \frac{f_{l-1}(R\varepsilon^{\mu})}{f_n(R\varepsilon^{\mu})}.$$

That is, from (8.8.6),

$$\|F(z)\| \leq \frac{1}{NR^m} \sum_{l=1}^{n} 2^{n-l} \sum_{\mu=1}^{N} \left| \frac{f_{l-1}(R\varepsilon^{\mu})}{f_n(R\varepsilon^{\mu})} \right|. \tag{8.9.5}$$

Now we apply the reasoning of 8.2, but instead of the interval $(\delta, 1)$ we use the interval (δ_2, δ_1). This yields an R with

$$\delta_2 \leq R \leq \delta_1 \tag{8.9.6}$$

so that for $|z| = R$

$$\left| \frac{f_n(z)}{f_{l-1}(z)} \right| \geq 2\left(\frac{\delta_1 - \delta_2}{4} \right)^n \qquad l = 1, 2, \ldots, n. \tag{8.9.7}$$

This implies that $R \neq |z_j|$, that is, (8.7.1) is fulfilled. Let us also remark that R is independent of N, and therefore our choice of R satisfies (8.7.1) for all N. Inequalities (8.9.7), (8.9.6), and (8.9.5) give

$$\|F(z)\| \leq \frac{\delta_2^{-m}}{2} \left(\frac{8}{\delta_1 - \delta_2} \right)^n,$$

and, since from (8.6.2) we have

$$\delta_2^{-m} \leq \left(\frac{m+n}{m} \right)^m \leq e^n,$$

we get

$$\|F(z)\| \leq \frac{1}{2} \left(\frac{8e}{\delta_1 - \delta_2} \right)^n.$$

Thus from (8.7.6) we get

$$\max_{\nu = m+1, \ldots, m+n} |g(\nu)| \geq 2\left(\frac{\delta_1 - \delta_2}{8e} \right)^n \left| \sum_{j=1}^{n} b_j \frac{z_j^N}{z_j^N - R^N} \right|. \tag{8.9.8}$$

8.10

Let the index l be defined by the relationship

$$1 = |z_1| \geq \cdots \geq |z_l| > R > |z_{l+1}| \geq \cdots \geq |z_n|. \tag{8.10.1}$$

Since the maximizing ν on the left side of (8.9.8) can take only the values $m+1,\ldots$ or $(m+n)$ and R is independent of N, passage to the limit as $N \to \infty$ gives

$$\max_{\nu=m+1,\ldots,m+n} |g(\nu)| \geq 2\left(\frac{\delta_1 - \delta_2}{8e}\right)^n \left|\sum_{j=1}^{l} b_j\right|. \tag{8.10.2}$$

Finally, we have to limit l in terms of the prescribed δ_1 and δ_2 from above as well as from below. But according to (8.9.6) and the first part of (8.6.3) the sum

$$\left|\sum_{j=1}^{l} b_j\right|$$

is among the $|\sum_{j=1}^{\mu} b_j|$ quantities with

$$\mu \geq l_1;$$

and, by the second part of (8.6.3), among those sums with

$$\mu \leq l_2.$$

This, together with (8.10.2), completes the proof of Theorem 8.2. ■

8.11

In the applications of these theorems, notably of Theorem 8.2, it is better not to assume that m is an integer. Replacing m by $[m]$, supposing that δ_1 and δ_2 satisfy

$$\frac{m}{m+n} \leq \delta_2 < \delta_1 \leq 1$$

and that l_1 and l_2 are taken from (8.6.3)–(8.6.4), we obtain the inequality

$$\max_{\substack{m \leq \nu \leq m+n \\ \nu \text{ integer}}} |g(\nu)| \geq 2\left(\frac{\delta_1 - \delta_2}{8e}\right)^n \min_{l_1 \leq j \leq l_2} |b_1 + \cdots + b_j|. \tag{8.11.1}$$

In most applications we have only an upper bound N for n. To be able to work with N we only have to apply (8.11.1) to

$$g_1(\nu) = \sum_{j=1}^{[N]} b_j z_j^{\nu}$$

where

$$z_{n+1} = \cdots = z_{[N]} = 0, \qquad b_{n+1} = \cdots = b_{[N]} = 0.$$

These considerations yield the following.

Corollary to Theorem 8.2. *Let $m \geq 0$, $n \leq N$ and*

$$\frac{m}{m+N-1} \leq \delta_2 < \delta_1 \leq 1. \tag{8.11.2}$$

If the indices l_1 and l_2 are defined by

$$|z_{l_1}| \geq \delta_1, \qquad |z_{l_2}| \leq \delta_2 \tag{8.11.3}$$

then the inequality

$$\max_{\substack{m \leq \nu \leq m+N \\ \nu \text{ integer}}} |g(\nu)| \geq 2\left(\frac{\delta_1 - \delta_2}{8e}\right)^n \min_{l_1 \leq j \leq l_2} |b_1 + \cdots + b_j| \tag{8.11.4}$$

holds.

8.12

Another variant of the second main theorem is due to Makai [5]. Using, instead of the ordering (8.1.1), the ordering

$$0 = |1 - z_1| \leq |1 - z_2| \leq \cdots \leq |1 - z_n| \tag{8.12.1}$$

(and $\max_j |z_j| = 1$) his theorem runs as follows — after the successive improvements of Dancs [3], Tijdeman [2], and Geysel [1].

Theorem 8.3. *For a nonnegative integer m and the previously mentioned restrictions on the z_j the inequality*

$$\max_{\substack{m+1 \leq \nu \leq m+n \\ \nu \text{ integer}}} \left| \sum_{j=1}^{n} b_j z_j^\nu \right| \geq \frac{1}{4} \left[\frac{n-1}{8e(m+n)} \right]^{n-1} \min_j |b_1 + \cdots + b_j|$$

holds.

Though the lower bound in this theorem is somewhat better than that in Theorem 8.1, we see no possible application. So we confine ourselves to a reference to the original papers (see, however, Theorems 10.1 and 10.2).

8.13

We conclude this section with two examples that show that Theorem 8.1 is in a sense not far from best possible.

Theorem 8.4 (Uchiyama [5]). *For suitable z_j' in* (8.1.1), *complex b_j', and $m = 0$ the inequality*

$$(8.13.1) \qquad \max_{\substack{m+1 \le \nu \le m+n \\ \nu \text{ integer}}} \left| \sum_{j=1}^{n} b_j' z_j'^{\nu} \right| \le 3\sqrt{n} \left[\frac{n}{e(m+n)} \right]^n \min_j |b_1' + \cdots + b_j'|$$

holds.

Remark. The example of Makai–de Bruijn in (6.5.4) implies a fortiori for the maximum norm (8.1.1), the inequality

$$(8.13.2) \qquad \max_{\substack{m+1 \le \nu \le m+n \\ \nu \text{ integer}}} |g^*(\nu)| \le \frac{1}{P_{m,n} - \varepsilon} |b_1^* + \cdots + b_n^*|,$$

for all integers $m \ge 0$.

This is, in all but one respect, better than (8.13.1), the exception being that on the right of (8.13.2) we do not have the $\min_j |b_1^* + \cdots + b_j^*|$. It is quite possible that in the example of Makai–de Bruijn (8.13.2) the last factor can indeed be replaced by $\min_j |b_1^* + \cdots + b_j^*|$.

Proof of Theorem 8.4. With the notation (5.6.2)–(5.6.3) take

$$(8.13.3) \qquad z_j' = 1 - \frac{j-1}{n}, \qquad b_j' = \frac{1}{z_j' w'(z_j')}; \qquad j = 1, 2, \ldots, n.$$

Then (8.1.1) is fulfilled, and by (5.6.3) one has

$$\sum_{j=1}^{n} b_j' z_j'^{\nu} = \sum_{j=1}^{n} \frac{z_j'^{\nu-1}}{w'(z_j')} = \begin{cases} 0 & \text{for} \quad \nu = 1, 2, \ldots, n-1 \\ 1 & \text{for} \quad \nu = n \end{cases},$$

that is,

$$(8.13.4) \qquad \max_{\nu = 1, \ldots, n} \left| \sum_{1}^{n} b_j' z_j'^{\nu} \right| = 1.$$

Simple calculation shows that

$$b_l' = \frac{n^n}{n!} (-1)^{l-1} (l^n - 1),$$

that is,

$$b_1' + \cdots + b_j' = \frac{n^n}{n!}(-1)^{j-1}\binom{n-1}{j-1}, \qquad 1 \le j \le n,$$

and thus

$$\min_j |b_1' + \cdots + b_j'| = \frac{n^n}{n!} > \frac{e^n}{3\sqrt{n}}.$$

This and (8.13.4) complete the proof. ■

8.14

We provide a second example.

Theorem 8.5 (Makai [4]). *For* $0 < \varepsilon < 1$, *and for all* $m \ge (2/\varepsilon)n$ *there are numbers* b_j'', z_j'' *such that*

$$1 = |z_1''| \ge |z_2''| \ge \cdots \ge |z_n''|, \tag{8.14.1}$$

and the inequality

$$\max_{\substack{m+1 \le \nu \le m+n \\ \nu \text{ integer}}} \left| \sum_{j=1}^{n} b_j'' z_j''^{\nu} \right| \le 12(m+n)\left[\frac{1}{1-\varepsilon}\frac{n}{4e(m+n)}\right]^n \min_l |b_1'' + \cdots + b_l''|$$

holds.

Makai's original version holds only for an unspecified $m \ge m_0(\varepsilon, n)$. The present stronger form is due to M. Szalay; only minor changes in Makai's proof were required.

Proof. We start with the remark that from the system

$$\sum_{j=1}^{n} b_j z_j^{m+k} = (-1)^{m+k}, \qquad k = 1, 2, \ldots, n \tag{8.14.2}$$

where $z_\mu \ne z_\nu$ for $\mu \ne \nu$ and $z_1 z_2 \cdots z_n \ne 0$ we have

$$b_j = \frac{(-1)^{m+1}}{z_j^{m+1}} \prod_{\substack{l=1 \\ l \ne j}}^{n} \frac{l + z_l}{z_l - z_j} \qquad j = 1, 2, \ldots, n. \tag{8.14.3}$$

Now let

$$0 = x_1 < x_2 < \cdots < x_n \tag{8.14.4}$$

stand for the nonnegative zeros of the Hermitean polynomial $H_{2n-1}(x)$ [see (4.3.5)] and choose

$$z_j'' = 1 - \frac{x_j^2}{2m+2}, \qquad j = 1, 2, \ldots, n \tag{8.14.5}$$

with the b_j'' determined by (8.14.2). Choosing $m \geq 2n - 1$ and using $x_n^2 < 4n - 1$ from Section 5.6., we get from (8.14.5)

$$1 = z_1'' > z_2'' > \cdots > z_n'' > 0. \tag{8.14.6}$$

Hence from (8.14.3) we get

(8.14.7)

$$b_j'' = (-1)^{m+1}\left(1 - \frac{x_j^2}{2m+2}\right)^{-m-1} \prod_{\substack{l=1\\ l\neq j}}^{n} \left\{\frac{2 - [x_l^2/(2m+2)]}{(x^2 - x_l^2)/(2m+2)}\right\}$$

$$= (-1)^{m+1}\left(1 - \frac{x_j^2}{2m+2}\right)^{-m-1} \left\{\prod_{\substack{l=1\\ l\neq j}}^{n} (4m+4-x_l^2)\right\}\left\{\prod_{\substack{l=1\\ l\neq j}}^{n} (x_j^2 - x_l^2)\right\}^{-1}$$

$$j = 1, 2, \ldots, (n-1), n.$$

By the relation

$$H_{2n-1}(x) = 2^{2n-1} x \prod_{l=2}^{n} (x^2 - x_l^2)$$

the last factor in (8.14.7) is

$$2^{2n} H_{2n-1}'(x_j)^{-1} \quad \text{for} \quad j = 2, 3, \ldots, n$$

$$2^{2n-1} H_{2n-1}'(x_j)^{-1} \quad \text{for} \quad j = 1.$$

Hence for the b_j'' coefficients we have

$$b_j'' = (-1)^{m+1}\left(1 - \frac{x_j^2}{2m+2}\right)^{-m-1} \frac{1}{4m+4-x_j^2}$$

$$\times \left\{\prod_{l=1}^{n} (4m+4-x_l^2)\right\} 2^{2n} H_{2n-1}'(x_j)^{-1} \qquad j = 2, 3, \ldots, n,$$

that is,

$$(8.14.8)\quad b_j'' = (-1)^{m+1}2^{2n}\left\{\prod_{l=1}^{n}\left(4m+4-x_l^2\right)\right\}e^{-x_j^2/2}\left(1-\frac{x_j^2}{2m+2}\right)^{-m-1}$$

$$\cdot\frac{1}{4m+4-x_j^2}\left\{e^{-x_j^2/2}H_{2n-1}'(x_j)\right\}^{-1},$$

and this expression gives

$$(8.14.9)\qquad 2b_1'' \quad \text{for} \quad j=1.$$

For $0 \leq x < \sqrt{4n-1}$ let

$$f(x) = \left(1-\frac{x^2}{2m+2}\right)^{-m-1}e^{-x^2/2}.$$

Since

$$f'(x) = \frac{1}{2m+2}e^{-x^2/2}\left(1-\frac{x^2}{2m+2}\right)^{-m-2}x^3$$

the expression

$$e^{-x_j^2/2}\left(1-\frac{x_j^2}{2m+2}\right)^{-m-1}\frac{1}{4m+4-x_j^2}$$

increases monotonically with j.

Finally, using (5.6.1) we get for the last product in (8.14.8) the inequality

$$|b_1''| < \tfrac{1}{2}|b_2''| < \cdots < \tfrac{1}{2}|b_n''|.$$

Since the b_j'' are real and of alternating sign, we make the crucial observation that

$$(8.14.10)\qquad \min_{l=1,\ldots,n}|b_1''+\cdots+b_l''| = |b_1''|.$$

Since

$$H_{2n-1}'(0) = (-1)^{n-1}\frac{(2n)!}{n!}$$

we get

$$|b_1''| = 2^{2n-1}\prod_{l=2}^{n}\left(4m+4-x_l^2\right)\cdot\frac{n!}{(2n)!}.$$

This and (8.14.2) give, for $m \geq 2n-1$, the inequality

(8.14.11)

$$\max_{\substack{m+1 \leq \nu \leq m+n \\ \nu \text{ integer}}} \left| \sum_{j=1}^{n} b_j'' z_j''^{\nu} \right| = 1 = \frac{1}{|b_1''|} \min_l |b_1'' + \cdots + b_l''|$$

$$= \frac{(2n)!}{2^{2n-1} n! \prod_{l=2}^{n} (4m+4-x_l^2)} \min_l |b_1'' + \cdots + b_l''|$$

$$< 3\left(\frac{n}{e}\right)^n \left\{ \prod_{l=2}^{n} (4m+4-x_l^2)^{-1} \right\} \min_l |b_1'' + \cdots + b_l''|.$$

Since $0 \leq x_l < \sqrt{4n-1}$, we get for $m \geq (2/\varepsilon)n$ $(0 < \varepsilon < 1)$

$$\prod_{l=2}^{n} (4m+4-x_l^2) > [4(m-n)+5]^{n-1} > [4(m+n-2n)]^{n-1}$$

$$> [4(1-\varepsilon)(m+n)]^{n-1}.$$

This and (8.14.11) complete the proof of Theorem 8.5. ■

Hence, applying Theorem 8.5 with $\varepsilon = 10^{-r}$ (r a large integer) and $m = [(2/\varepsilon)n]$, we conclude that the best constant in Theorem 8.1 is between $4e$ and $8e$.

Uchiyama's result refers to $m = 0$, Makai's to $m \geq [(2/\varepsilon)n]$. As M. Szalay remarked, using the nonnegative zeros of the Chebychev-polynomial $T_{2n-1}(x)$ instead of the zeros of $H_{2n-1}(x)$, the inequality

(8.14.12)

$$\max_{\substack{m+1 \leq \nu \leq m+n \\ \nu \text{ integer}}} \left| \sum_{j=1}^{n} b_j z_j^{\nu} \right|$$

$$> 8(2n-1)\left(\frac{1+(m/n)}{4\sqrt{2}} \frac{n}{m+n}\right)^n \min_l |b_1 + \cdots + b_l|$$

cannot be generally true. For $0 \leq m \leq n$ this bound is better than Uchiyama's.

8.15

In view of the fact that all proofs up to this point—and also later—are based on suitable rational identities, it is perhaps not rash to assert that the entire theory could be built solely upon ideas of classical algebra (including the fundamental theorem of algebra). The proof of Theorem 6.1 could be presented in such a fashion that continuity is not used. The same holds for Theorem 8.2 since Chebychev's inequality could be proved, even in the sharper form

$$\max_{\substack{x=\cos(\nu\pi/n)\\ \nu=0,1,\ldots,n}} |x^n + \cdots + a_n| \geq 2^{1-n}$$

without using continuity. Nevertheless, we shall not press this point any further, although in principle it is interesting.

9

CONTINUATION

9.1

Again let $P_j(x)$ be polynomials of degree $(k_j - 1)$ $(j = 1,\dots,n)$ and $K = \sum_{j=1}^{n} k_j$. Performing a limiting process similar to the one in the proof of Theorem 6.5 we get the following results.

Theorem 9.1 (Tijdeman [2]). *For every integer* $m \geq 0$ *and with the ordering* (8.1.1) *the inequality*

$$\max_{\substack{m+1 \leq \nu \leq m+K \\ \nu \text{ integer}}} \left| \sum_{j=1}^{n} P_j(\nu) z_j^{\nu} \right| \geq \left[\frac{K}{8e(m+K)} \right]^{K} \min_{d} \left| \sum_{l=1}^{d} P_l(0) \right|$$

holds.

Theorem 9.2 (J. M. Geysel [1]). *For every integer* $m \geq 0$ *and with the ordering* (8.12.1) *the inequality*

$$\max_{\substack{m+1 \leq \nu \leq m+K \\ \nu \text{ integer}}} \left| \sum_{j=1}^{n} P_j(\nu) z_j^{\nu} \right| \geq \frac{1}{4} \left[\frac{K-1}{8e(m+K)} \right]^{K-1} \min_{d} \left| \sum_{l=1}^{d} P_l(0) \right|$$

holds.

Since these theorems are not used later we refer to the original papers for their proofs.

9.2

Next we prove an "operator-type" theorem (see Section 2.5) with the maximum norm. This theorem was designed for a very specific application which, however, has not yet been carried out. (We have a similar situation with the one-sided theorem in Section 13 where the originally intended application to the investigation of the sign changes of $\pi(x) - \operatorname{li} x$ could be carried out only later.) Let

$$1 = |z_1| \geq \cdots \geq |z_n| > 0 \tag{9.2.1}$$

and suppose that for

$$\varphi(z) = z^k + e_1 z^{k-1} + \cdots + e_k \tag{9.2.2}$$

where

$$0 < \rho < 1 \tag{9.2.3}$$

we have

$$\varphi(z) \neq 0 \quad \text{for} \quad |z| \geq \rho. \tag{9.2.4}$$

We refer to Case A (resp. Case B) if the nonnegative integer m satisfies

$$\frac{m}{m+n+k} \geq \rho \qquad \left(\text{resp. } \frac{m}{m+n+k} < \rho\right). \tag{9.2.5}$$

Theorem 9.3. *Under the foregoing provisions for*

$$G_2(\nu) = \sum_{j=1}^{n} b_j z_j^{\nu} \varphi(z_j) \tag{9.2.6}$$

we obtain in Case A:

(9.2.7)

$$\max_{\substack{m+1 \leq \nu \leq m+n \\ \nu \text{ integer}}} |G_2(\nu)| \geq \frac{1}{2^{n-1}} \left[\frac{n+k}{4e(m+n+k)}\right]^{n+k} \min_j |b_1 + \cdots + b_j|,$$

and in the Case B:

$$\max_{\substack{m+1 \leq \nu \leq m+n \\ \nu \text{ integer}}} |G_2(\nu)| \geq \frac{\rho^m}{2^{n-1}} \left(\frac{1-\rho}{4}\right)^{n+k} \min_j |b_1 + \cdots + b_j|. \tag{9.2.8}$$

It is remarkable that the dependence of the lower bound upon $\varphi(z)$ is restricted to k and ρ and that the range of ν is even entirely independent of these quantities.

We note that Theorem 8.1 is contained in Theorem 9.3 (with $\rho = 0$), and one might ask why we have proved them both. The reason is, once again, that several applications are based on Theorem 8.1 whose proof is much shorter than that of Theorem 9.3. In general, the applications are much more elegant if they can be deduced from a theorem with a *shorter* proof. (The proof of Theorem 9.3 has been shortened because we were able to incorporate some portions of the proof of Theorem 8.1. We will not go into details).

In Section 2.5 we indicated merely *one* aspect of the background of operator-type theorems. Another aspect can be formulated now that we have Theorem 8.1. We mentioned earlier that the applicability of this Theorem is in many cases hindered by the inconvenient dependence of the lower bound on the B_j-coefficients of $g(\nu) = \sum_{j=1}^{n} B_j z_j^{\nu}$. In Theorem 9.3 we have no such difficulty when $b_1 = \cdots = b_n$. Hence if we can determine a polynomial $\varphi(z)$ of the form (9.2.2) satisfying

$$\varphi(z_j) = B_j \qquad (j = 1, 2, \ldots, n), \tag{9.2.9}$$

then an application of Theorem 9.3 transforms the difficulty into finding circular domains containing all the zeros of $\varphi(z)$, that is, into determining $\varphi(z)$ so that (9.2.9) and (9.2.4) are satisfied.

9.3. PROOF OF THEOREM 9.3

Denote the zeros of $\varphi(z)$ by

$$\xi_1, \ldots, \xi_k \tag{9.3.1}$$

and denote the z_j and ξ_j by

$$w_1, w_2, \ldots, w_{n+k}, \tag{9.3.2}$$

which may be assumed nonzero and distinct. Define

$$\delta = \begin{cases} \dfrac{m}{m+n+k} & \text{if } \dfrac{m}{m+n+k} \geq \rho \\[2ex] \rho & \text{if } \dfrac{m}{m+n+k} < \rho \end{cases}. \tag{9.3.3}$$

Hence we have

$$|\xi_j| < \delta \qquad (j = 1, \ldots, k). \tag{9.3.4}$$

We shall apply Lemma 6.1 to the function

$$h(z) = \prod_{j=1}^{n+k} (z - |w_j|)$$

in the interval $[\delta, 1]$. We get, as in Section 8.2, that for a suitable R with

$$\delta \leq R \leq 1 \tag{9.3.5}$$

the inequality

$$\left| \prod_{\nu=1}^{d} (z - w_{i_\nu}) \right| \geq 2\left(\frac{1-\delta}{4}\right)^{n+k} \tag{9.3.6}$$

holds on the circle $|z| = R$ for all subproducts of $\prod_{\nu=1}^{n+k}(z - w_\nu)$
In particular, (9.3.6) holds for all subproducts of the form

$$\varphi(z)\prod(z - z_{i_\nu}). \tag{9.3.7}$$

Let us define the l-index by

$$1 = |w_1| \geq \cdots \geq |w_l| > R > |w_{l+1}| \geq \cdots \geq |w_{n+k}|;$$

and due to (9.3.4) this can be rephrased in the form

$$1 = |z_1| \geq |z_2| \geq \cdots \geq |z_l| > R > |w_{l+1}| \cdots \geq |w_{n+k}|. \tag{9.3.8}$$

9.4

Next let

$$h_1(z) = \sum_{j=l+1}^{n} (z - z_j) = \sum_{\nu=0}^{n-l} d_\nu^{(1)} z^\nu. \tag{9.4.1}$$

Then again (as in 8.3)

$$\|h_1\| \leq 2^{n-l}. \tag{9.4.2}$$

Define the polynomial $h_2(z)$ of degree $\leq l - 1$ by

$$h_2(z_j) = \frac{1}{z_j^{m+1} h_1(z_j)\varphi(z_j)}, \qquad j = 1, 2, \ldots, l. \tag{9.4.3}$$

Conditions (9.3.8) and (9.3.4) imply that this expression is meaningful. Again representing $h_2(z)$ in the form

$$h_2(z) = d_0^{(2)} + d_1^{(2)}(z - z_1) + \cdots + d_{l-1}^{(2)}(z - z_1)\cdots(z - z_{l-1}) \tag{9.4.4}$$

we have, from Lemma 5.3,

$$(9.4.5)\quad d_\nu^{(2)} = \frac{-1}{2\pi i}\int_{|w|=R}\frac{dw}{w^{m+1}h_1(w)\varphi(w)(w-z_1)\cdots(w-z_{\nu+1})}.$$

From (9.3.6) and (9.3.7) this yields

(9.4.6)

$$|d_\nu^{(2)}| \le \frac{1}{R^m}\frac{1}{2}\left(\frac{4}{1-\delta}\right)^{n+k} \le \frac{1}{2\delta^m}\left(\frac{4}{1-\delta}\right)^{n+k} \qquad (\nu = 0,1,\ldots,l-1).$$

Now writing $h_2(z)$ in the form

$$(9.4.7)\qquad h_2(z) = \sum_{\nu=0}^{l-1} d_\nu^{(3)} z^\nu$$

we proceed as in Section 8.4. Thus

$$(9.4.8)\qquad \|h_2\| \le \frac{2^{l-1}}{\delta^m}\left(\frac{4}{1-\delta}\right)^{n+k}.$$

Finally, setting

$$(9.4.9)\qquad h_3(z) = z^{m+1}h_1(z)h_2(z) = \sum_{\nu=m+1}^{m+n} d_\nu^{(4)} z^\nu$$

we have

$$h_3(z_j) = \begin{cases} \dfrac{1}{\varphi(z_j)} & \text{for } j = 1,2,\ldots,l \\ 0 & \text{for } j > l \end{cases}$$

and hence

$$\sum_{\nu=m+1}^{m+n} d_\nu^{(4)} z_j^\nu \varphi(z_j) = \begin{cases} 1 & \text{for } j = 1,\ldots,l \\ 0 & \text{for } j > l. \end{cases}$$

Multiplying by b_j and summing we get the identity

$$(9.4.10)\qquad \sum_{\nu=m+1}^{m+n} d_\nu^{(4)} G_2(\nu) = b_1 + \cdots + b_l,$$

that is,

$$(9.4.11)\qquad \max_{\substack{m+1\le\nu\le m+n \\ \nu\ \text{integer}}} |G_2(\nu)| \ge \frac{\min_j |b_1 + \cdots + b_j|}{\|h_3\|}.$$

Since (9.4.2) and (9.4.8) imply

$$\|h_3\| \le \|h_1\| \cdot \|h_2\| < \frac{2^{n-1}}{\delta^m}\left(\frac{4}{1-\delta}\right)^{n+k},$$

(9.4.11) and (9.3.3) complete the proof of Theorem 9.3. ■

9.5

This proof, together with an idea used in the proof of Theorem 8.2, gives a less elegant but more powerful theorem. Let m be a nonnegative integer and let $\varphi(z)$ in (9.2.2) satisfy (9.2.4). Let

$$\frac{m}{m+n+k} \le \delta_2 < \delta_1 \le 1 \tag{9.5.1}$$

and let the indices l_1 and l_2 be defined as in (8.6.3)–(8.6.4). Then we have the following result.

Theorem 9.4. *Under the foregoing conditions the last two factors in* (9.2.7) *and* (9.2.8) *can be replaced by*

$$\left(\frac{\delta_1-\delta_2}{4e}\right)^{n+k} \min_{l_1 \le j \le l_2} |b_1 + \cdots + b_j|.$$

9.6

The "operator" $e^{\alpha z}$, for fixed complex α, provides another important example of an operator-type theorem. In this case,

$$G_2(\nu) = \sum_{j=1}^{n} b_j z_j^{\nu} e^{\alpha z_j}. \tag{9.6.1}$$

For the following theorem, due to van der Poorten [1], assume

$$\Delta = |z_1| \ge |z_2| \ge \cdots \ge |z_n|. \tag{9.6.2}$$

Theorem 9.5. *With the foregoing notation the inequality*

$$\max_{\nu=0,1,\ldots,n-1} |G_2(\nu)| \ge \frac{1}{2n}\left|\sum_{j=1}^{n} b_j\right| \exp\{-(2\Delta+1)|\alpha|\}$$

holds.

Setting

$$F(z) = \sum_{j=1}^{n} b_j e^{z_j z} \tag{9.6.3}$$

this theorem implies

$$\max_{\nu=0,1,\ldots,n-1} |F^{(\nu)}(\alpha)| \geq \frac{1}{2n} |F(0)| \exp\{-(2\Delta + 1)|\alpha|\}.$$

Passage to the limit, as in Theorem 6.5, gives van der Poorten's result:

Theorem 9.6. *Set*

$$H(z) = \sum_{j=1}^{n} P_j(z) e^{z_j z}$$

where the $P_j(z)$ stand for polynomials of degree $\leq (k_j - 1)$, and let

$$K = \sum_{j=1}^{n} k_j.$$

Then the inequality

$$\max_{\nu=0,\ldots,K-1} |H^{(\nu)}(\alpha)| \geq \frac{1}{2K} \left| \sum_{j=1}^{n} P_j(0) \right| \exp\{-(2\Delta + 1)|\alpha|\} \tag{9.6.4}$$

holds for any complex α.

It is interesting to note that the right-hand side of (9.6.4) is also independent of the degrees of the polynomials $P_j(z)$. As the subsequent proof shows there are several alternative forms of this theorem. The original form of Theorems 9.5 and 9.6 is slightly stronger in that it does not require the factor $1/2n$. A better estimate than the crude one given in (9.7.5) would eliminate this factor.

9.7. PROOF OF THEOREM 9.5

With notation as in (5.6.4), let

$$l_\mu(z) = \sum_{\nu=0}^{n-1} d_{\nu\mu} z^\nu, \qquad \mu = 1,\ldots,n. \tag{9.7.1}$$

Then we get from (9.6.1),

$$\sum_{\nu=0}^{n-1} d_{\nu\mu} G_2(\nu) = b_\mu e^{\alpha z_\mu},$$

that is,

$$b_\mu = e^{-\alpha z_\mu} \sum_{\nu=0}^{n-1} d_{\nu\mu} G_2(\nu).$$

And,

$$(9.7.2) \qquad \left| \sum_{j=1}^{n} b_j \right| = \left| \sum_{\mu=1}^{n} e^{-\alpha z_\mu} \sum_{\nu=0}^{n-1} d_{\nu\mu} G_2(\nu) \right|$$

$$= \left| \sum_{\nu=0}^{n-1} G_2(\nu) \sum_{\mu=1}^{n} e^{-\alpha z_\mu} d_{\nu\mu} \right|$$

$$\leq \max_{\nu=0,1,\ldots,n-1} |G_2(\nu)| \sum_{\nu=0}^{n-1} \left| \sum_{\mu=1}^{n} e^{-\alpha z_\mu} d_{\nu\mu} \right|$$

$$= \|\pi_{n-1}(z)\| \max_{\nu=0,1,\ldots,n-1} |G_2(\nu)|$$

where $\pi_{n-1}(z)$ stands for the polynomial of degree $\leq n-1$ satisfying

$$(9.7.3) \qquad \pi_{n-1}(z_\mu) = e^{-\alpha z_\mu} \qquad \mu = 1,\ldots,n.$$

Writing (9.7.3) in the form

$$\pi_{n-1}(z) = e_0 + \sum_{\nu=0}^{n-1} e(z - z_1) \cdots (z - z_\nu)$$

and using Nörlund's representation (in Lemma 5.4) we get

(9.7.4)

$$e_\nu = \frac{1}{2\pi i} \int_{|w|=2\Delta+1} \frac{e^{-\alpha w}\, dw}{(w - z_1) \cdots (w - z_{\nu+1})} \qquad (\nu = 0,1,\ldots,n-1).$$

This immediately gives

$$(9.7.5) \qquad |e_\nu| \leq 2 \frac{e^{(2\Delta+1)|\alpha|}}{(\Delta+1)^\nu} \quad \text{for} \quad \nu = 0,1,\ldots,n-1.$$

So that

$$\|\pi_{n-1}(z)\| \leq |e_0| + \sum_{\nu=0}^{n-1} |e_\nu|(1+\Delta)^\nu \leq 2ne^{(2\Delta+1)|\alpha|},$$

which together with (9.7.2) completes the proof of Theorem 9.5. ■

The same proof also gives the following.

Theorem 9.7. *With the foregoing notations the inequality*

$$\max_{\nu=0,1,\ldots,n-1} \left| \sum_{j=1}^{n} b_j z_j^\nu h(z_j) \right| \geq \left| \sum_{j=1}^{n} b_j \right| \frac{1}{2n} \left\{ \frac{1}{2\pi} \int_{-\pi}^{\pi} \frac{d\vartheta}{|h((2\Delta+1)e^{i\vartheta})|} \right\}^{-1}$$

holds for any $h(z)$ that is regular and nonvanishing in $|z| \leq 1 + 2\Delta$.

10

TWO-SIDED THEOREMS WITH OTHER NORMS AND WITH OTHER *S*-SETS

10.1

We mentioned the inconvenience caused by the nature of the dependence upon the b_j in the applications of Theorem 8.1. Although this is essentially improved by Theorem 8.2, where the dependence is given by the factor

$$\min_{l_1 \leq j \leq l_2} |b_1 + \cdots + b_j|,$$

it seems to be desirable to have a like result for Theorem 8.1 where this factor has the form

$$\min_{L_1 \leq k \leq l \leq L_2} |b_k + b_{k+1} + \cdots + b_l| \tag{10.1.1}$$

Such a theorem was given first by Tijdeman [2].

Theorem 10.1 (Tijdeman). *Let* $z_1 = 1$ *and let* $z_2, \dots, z_n$ *be ordered so that*

$$0 = |1 - z_1| < |1 - z_2| \leq \cdots \leq |1 - z_n|. \tag{10.1.2}$$

If $m \geq -1$ *is an integer and* $n \geq 2$, *then the inequality*

$$\max_{\substack{m+1 \leq \nu \leq m+n \\ \nu \text{ integer}}} \left| \sum_{j=1}^{n} b_j z_j^{\nu} \right| \geq \frac{1}{8} \left[\frac{n-1}{8e(m+2n)} \right]^{n-1} \min_j |b_1 + \cdots + b_j|$$

holds.

The important aspect of this theorem is that we do not demand that $\max_j |z_j| = 1$. Hence the nature of the dependence of the lower bound on the b_j is "almost" of the required type. We prove, instead, a different theorem whose result is exactly of the required form since it also restricts the range of j in $|b_1 + \cdots + b_j|$. Specifically, let $m \geq 0$ be an integer and

$$0 < \delta_2 < \delta_1 < \frac{n}{m+n+1}. \tag{10.1.3}$$

Define [with the notation (10.1.2)] the index h as the maximal one satisfying

$$|1 - z_h| < \delta_2 \tag{10.1.4}$$

and the index l as the minimal one satisfying

$$|1 - z_{l+1}| > \delta_1 \tag{10.1.5}$$

if it exists; if not, let $l = n$.

Theorem 10.2. *With the foregoing notations the inequality*

$$\max_{\substack{m+1 \leq \nu \leq m+n \\ \nu \text{ integer}}} |g(\nu)| \geq 2\left(\frac{\delta_1 - \delta_2}{12e}\right)^n \min_{h \leq j \leq l} |b_1 + \cdots + b_j|$$

holds.

10.2

Let the z_j be distinct, and let

$$w_j = 1 - z_j. \tag{10.2.1}$$

Hence

$$0 = |w_1| < |w_2| \leq \cdots \leq |w_n|. \tag{10.2.2}$$

We apply (5.1.8) to the product

$$\psi(w) = \prod_{j=1}^{n} (w - |w_j|) \tag{10.2.3}$$

and to the interval $[\delta_2, \delta_1]$. Then there exists an R with

$$\delta_2 \leq R \leq \delta_1 \tag{10.2.4}$$

so that

$$(10.2.5)\quad |\psi(R)| \geq 2\left(\frac{\delta_1-\delta_2}{4}\right)^n \prod_{\left||w_j|-\frac{\delta_1+\delta_2}{2}\right| > \frac{\delta_1-\delta_2}{2}} \left|\frac{|w_j| - \frac{\delta_1+\delta_2}{2}}{\frac{1}{2}(\delta_1-\delta_2)}\right|.$$

In addition to the indices h and l we define the index (if it exists) by

$$(10.2.6)\quad 0 = |w_1| < |w_2| \leq \cdots \leq |w_h| < \delta_2$$
$$\leq |w_{h+1}| \leq \cdots \leq |w_k| < R < |w_{k+1}| \leq \cdots \leq |w_l|$$
$$\leq \delta_1 < |w_{l+1}| \leq \cdots \leq |w_n|, \qquad (h \leq k \leq l).$$

If such an index k does not exist, we set $k = n$. Then we have, a fortiori,

$$|\psi(R)| \geq 2\left(\frac{\delta_1-\delta_2}{4}\right)^n \prod_{j=l+1}^{n} \frac{|w_j| - [(\delta_1+\delta_2)/2]}{\frac{1}{2}(\delta_1-\delta_2)}$$

and with

$$\psi_1(w) = \prod_{j=1}^{l}(w - |w_j|) \prod_{j=l+1}^{n} \frac{w - |w_j|}{|w_j|},$$

we get from (10.2.6),

$$(10.2.7)\quad |\psi_1(R)| \geq 2\left(\frac{\delta_1-\delta_2}{4}\right)^n \prod_{j=l+1}^{n} \frac{2}{\delta_1-\delta_2}\left(1 - \frac{\delta_1+\delta_2}{2|w_j|}\right)$$
$$\geq 2\left(\frac{\delta_1-\delta_2}{4}\right)^n \left(\frac{1}{\delta_1}\right)^{n-l}.$$

Obviously, this also holds for all subproducts of $\psi_1(R)$ of the form

$$(10.2.8)\quad \frac{\psi_1(R)}{(R-|w_{i_1}|)\cdots(R-|w_{i_s}|)}, \qquad 1 \leq i_1 < \cdots < i_s \leq l.$$

Then we see, as in 8.8, that if

$$(10.2.9)\quad \psi_2(w) = \prod_{j=1}^{l}(w - w_j) \prod_{j=l+1}^{n} \frac{w - w_j}{w_j}$$

the inequalities

$$\left|\frac{\psi_2(w)}{(w-w_{i_1})\cdots(w-w_{i_s})}\right| \geq 2\left(\frac{\delta_1-\delta_2}{4}\right)^n\left(\frac{1}{\delta_1}\right)^{n-l} \tag{10.2.10}$$

hold simultaneously for all choices of the indices

$$1 \leq i_1 < \cdots < i_s \leq l \tag{10.2.11}$$

on the circle $|w| = R$.

10.3

Setting $w = 1 - z$ and using (10.2.1) we have

$$\psi_2(1-z) = f_0(z) = \prod_{j=1}^{l}(z_j - z)\prod_{j=l+1}^{n}\frac{z_j - z}{1 - z_j} \tag{10.3.1}$$

with

$$\begin{aligned} 0 = |1-z_1| &< |1-z_2| \leq \cdots \leq |1-z_h| < \delta_2 \\ &\leq |1-z_{h+1}| \leq \cdots \leq |1-z_k| < R < |1-z_{k+1}| \leq \cdots \\ &\leq |1-z_l| \leq \delta_1 < |1-z_{l+1}| \leq \cdots \leq |1-z_n|. \end{aligned} \tag{10.3.2}$$

On the circle $|1 - z| = R$ the inequalities

$$\left|\frac{f_0(z)}{(z-z_{i_1})\cdots(z-z_{i_s})}\right| \geq 2\left(\frac{\delta_1-\delta_2}{4}\right)^n\left(\frac{1}{\delta_1}\right)^{n-l} \tag{10.3.3}$$

hold simultaneously for all choices of the indices in (10.2.11).

Let

$$f_1(z) = \prod_{j=k+1}^{l}(z - z_j)\prod_{j=l+1}^{n}\frac{z - z_j}{1 - z_j} = \sum_{\nu=0}^{n-k} c_\nu^{(1)} z^\nu. \tag{10.3.4}$$

Then we have

$$\|f_1\| \leq \prod_{j=k+1}^{l}(1 + |z_j|)\prod_{j=l+1}^{n}\frac{1 + |z_j|}{|1 - z_j|}.$$

Since

$$1 + |z_j| \leq 2 + \delta_1 \quad \text{for} \quad j \leq l,$$

and

$$\frac{1+|z_j|}{|1-z_j|} = \frac{1+|1-(1-z_j)|}{|1-z_j|} \le 1 + \frac{2}{|1-z_j|} \le 1 + \frac{2}{\delta_1} \quad \text{for} \quad j > l,$$

we have

$$\|f_1\| \le \frac{(2+\delta_1)^{n-k}}{\delta_1^{n-l}}. \tag{10.3.5}$$

Let $f_2(z)$ be the polynomial of degree $\le k-1$ satisfying

$$f_2(z_j) = \frac{1}{z_j^{m+1} f_1(z_j)} \qquad j = 1, \ldots, k \tag{10.3.6}$$

[which is meaningful due to (10.3.2)].
Again letting

$$f_2(z) = \sum_{\nu=0}^{k-1} c_\nu^{(2)} (z-z_1) \cdots (z-z_\nu), \tag{10.3.7}$$

Lemma 5.4 gives

$$c_\nu^{(2)} = \frac{1}{2\pi i} \int_{|1-w|=R} \frac{dw}{w^{m+1} f_1(w)(w-z_1) \cdots (w-z_{\nu+1})}$$

since $z^{-m-1} f_1^{-1}(z)$ is regular in $|1-z| \le R$. And because

$$f_1(z)(z-z_1) \cdots (z-z_{\nu+1})$$

has the form of the left-hand side of (10.3.3),

$$|c_\nu^{(2)}| \le \frac{R}{(1-R)^{m+1}} \frac{1}{2} \left(\frac{4}{\delta_1 - \delta_2} \right)^n \delta_1^{n-l}. \tag{10.3.8}$$

Now write $f_2(z)$ in the form

$$f_2(z) = \sum_{\nu=0}^{k-1} c_\nu^{(3)} z^\nu \tag{10.3.9}$$

and we have

$$\|f_2\| \le \sum_{\nu=0}^{k-1} |c_\nu^{(2)}| (1+|z_1|) \cdots (1+|z_\nu|) \le \sum_{\nu=0}^{k-1} |c_\nu^{(2)}| (2+\delta_1)^\nu.$$

From (10.3.8), (10.1.3), and (10.2.4), in that order, we obtain

$$(10.3.10)\qquad \sum_{\nu=0}^{k-1} |c_\nu^{(3)}| \le \frac{R}{2(1-R)^{m+1}}\left(\frac{4}{\delta_1-\delta_2}\right)^n \delta_1^{n-l}\frac{(2+\delta_1)^k}{1+\delta_1}$$

$$\le \frac{1}{2}\left(\frac{4e}{\delta_1-\delta_2}\right)^n \delta_1^{n-l}\frac{(2+\delta_1)^k}{1+\delta_1}.$$

Finally let

$$(10.3.11)\qquad f_3(z) = z^{m+1}f_1(z)f_2(z) = \sum_{\nu=m+1}^{m+n} c_\nu^{(4)} z^\nu.$$

Again

$$\sum_{\nu=m+1}^{m+n} c_\nu^{(4)} z_j^\nu = \begin{cases} 1 & \text{for } j = 1,\ldots,k \\ 0 & \text{for } j > k. \end{cases}$$

Furthermore,

$$(10.3.12)\qquad \sum_{\nu=m+1}^{m+n} c_\nu^{(4)} g(\nu) = b_1 + \cdots + b_k$$

and

$$\max_{\nu=m+1,\ldots,m+n} |g(\nu)| \ge \frac{|b_1+\cdots+b_k|}{\|f_3\|}.$$

From (10.3.11), (10.3.5), and (10.3.10) we get

$$\|f_3\| \le \|f_1\|\cdot\|f_2\| \le \frac{1}{2}\left(\frac{4e(2+\delta_1)}{\delta_1-\delta_2}\right)^n \le \frac{1}{2}\left(\frac{12e}{\delta_1-\delta_2}\right)^n,$$

hence

$$\max_{m+1\le\nu\le m+n} |g(\nu)| \ge 2\left(\frac{\delta_1-\delta_2}{12e}\right)^n |b_1+\cdots+b_k|.$$

Since (10.3.2) implies that $h \le k \le l$, the proof is completed after passage to the limit. ■

10.4

Theorem 10.2 in effect compares $|g(\nu)|$ with $|b_k z_k^\nu|$ at a *prescribed* value of k for an appropriate integer ν in $[m+1, m+n]$. Here the ordering of the terms was determined by the distance of the z_js from z_k. In some applications it is more appropriate to use the ordering

$$(10.4.1) \qquad |z_1| \geq |z_2| \geq \cdots \geq |z_k| = 1 \geq \cdots \geq |z_n|.$$

We can in fact prove a theorem of this type. Let $m \geq 0$, k a fixed integer with $1 \leq k \leq n$, and let the indices k_1 and k_2 be determined by

$$(10.4.2) \quad |z_1| \geq \cdots \geq |z_{k_1}| > \frac{m+2n}{m+n} \geq |z_{k_1+1}| \geq \cdots$$
$$\geq |z_k| = 1 \geq \cdots \geq |z_{k_2-1}| > \frac{m}{m+n} \geq |z_{k_2}| \geq \cdots \geq |z_n|.$$

Then we assert the following.

Theorem 10.3 (Dancs–Turán [4]). *Under the foregoing circumstances there is an integer ν_1 with*

$$m + 1 \leq \nu_1 \leq m + n$$

so that the inequality

$$(10.4.3) \quad |g(\nu_1)| \geq \frac{1}{n}\left[\frac{n}{16e(m+n)}\right]^n \min_{k_1 < l_1 \leq k \leq l_2 < k_2} |b_{l_1} + \cdots + b_{l_2}|$$

holds.

The proof will be algebraic in the sense of Section 8. Using complex integration we could prove, for a suitable $m + 1 \leq \nu_2 \leq m + n$, the inequality

$$(10.4.4) \quad |g(\nu_2)| \geq \frac{1}{n}\left[\frac{n}{8e(m+2n)}\right]^n \min_{k_1 < l_1 \leq k \leq l_2 < k_2} |b_{l_1} + \cdots + b_{l_2}|.$$

10.5. PROOF OF THEOREM 10.3

We may suppose the z_j are distinct and nonzero. Let $m \geq 0$ and $N > m + 1$ be integers and

$$(10.5.1) \qquad \frac{n}{m+n} = \delta.$$

With this δ we have, by (10.4.2),

$$(10.5.2) \qquad |z_1| \geq |z_2| \geq \cdots \geq |z_{k_1}| > 1 + \delta \geq |z_{k_1+1}| \geq \cdots$$
$$\geq |z_k| = 1 \geq \cdots \geq |z_n|.$$

Let

$$f(z) = \prod_{j=1}^{k_1} \left(\frac{z}{z_j} - 1 \right) \prod_{j=k_1+1}^{n} (z - z_j) \tag{10.5.3}$$

and

$$q_n(x) = \prod_{j=1}^{k_1} \left(\frac{x}{|z_j|} - 1 \right) \prod_{j=k_1+1}^{n} (x - |z_j|). \tag{10.5.4}$$

We apply (5.1.8) to $q_n(x)$ in $[1 - \delta, 1]$. Then there exists an R_2 satisfying

$$1 - \delta \le R_2 < 1 \tag{10.5.5}$$

such that

$$|q_n(R_2)| \ge 2 \prod_{j=1}^{k_1} \frac{1}{|z_j|} \left(\frac{\delta}{4} \right)^n \prod{}' \frac{||z_j| - (1 - \delta/2)|}{\delta/2} \tag{10.5.6}$$

where the last product is extended to all z_j such that

$$\left| |z_j| - \left(1 - \frac{\delta}{2} \right) \right| \ge \frac{\delta}{2}.$$

In the last product of the right-hand side of (10.5.6) we keep only the z_j with $|z_j| > 1 + \delta$ and get

$$|q_n(R_2)| \ge 2 \left(\frac{\delta}{4} \right)^n \prod_{j=1}^{k_1} \frac{||z_j| - (1 - \delta/2)|}{(\delta/2)|z_j|}.$$

Hence, since

$$\frac{2}{\delta} \left(1 - \frac{1 - \delta/2}{z_j} \right) \ge \frac{3}{1 + \delta} \qquad (j = 1, 2, \ldots, k_1),$$

we get

$$|q_n(R_2)| \ge 2 \left(\frac{\delta}{4} \right)^n \left(\frac{3}{1 + \delta} \right)^{k_1}.$$

But this gives, as in Section 8.2, the inequality

$$|f(z)| \ge 2 \left(\frac{\delta}{4} \right)^n \left(\frac{3}{1 + \delta} \right)^{k_1} \tag{10.5.7}$$

on the whole circle $|z| = R_2$.

Next we apply (5.1.8) to $q_n(x)$ in $[1, 1+\delta]$. This gives an R_1 with

$$1 < R_1 \le 1 + \delta \tag{10.5.8}$$

such that

$$|q_n(R_1)| \ge 2\left(\frac{\delta}{4}\right)^n \prod_{\gamma=1}^{k_1} \frac{1}{|z_\gamma|} \prod{}' \frac{\left||z_j| - (1 + \delta/2)\right|}{\delta/2} \tag{10.5.9}$$

where the last product is extended to all z_j such that

$$\left||z_j| - \left(1 + \frac{\delta}{2}\right)\right| \ge \frac{\delta}{2}.$$

Again keep only those with $|z_j| > 1 + \delta$ in the last product in (10.5.9) and proceed as before to obtain

$$|f(z)| \ge 2\left(\frac{\delta}{4}\right)^n \left(\frac{1}{1+\delta}\right)^{k_1} \tag{10.5.10}$$

on the circle $|z| = R_1$. Obviously

$$R_1 \ne |z_j|, \qquad R_2 \ne |z_j| \qquad (j = 1, \ldots, n). \tag{10.5.11}$$

10.6

For the given R_1 and R_2 let

$$F(z) = \sum_{j=1}^{n} \frac{f(z)}{f'(z_j)(z - z_j)} \left(\frac{z}{z_j}\right)^{m+1} \frac{R_1^N}{R_1^N - z_j^N} \frac{z_j^N}{z_j^N - R_2^N}. \tag{10.6.1}$$

We have

$$F(z) = \sum_{\nu=m+1}^{m+n} d_\nu z^\nu. \tag{10.6.2}$$

This gives, as in Section 8.7, the identity

$$\sum_{\nu=m+1}^{m+n} d_\nu g(\nu) = \sum_{j=1}^{n} b_j \frac{R_1^N}{R_1^N - z_j^N} \frac{z_j^N}{z_j^N - R_2^N}, \tag{10.6.3}$$

that is,

$$\max_{\nu=m+1,\ldots,m+n} |g(\nu)| \ge \frac{\left|\sum_{j=1}^{n} b_j \frac{R_1^N}{R_1^N - z_j^N} \frac{z_j^N}{z_j^N - R_2^N}\right|}{\|F(x)\|} \tag{10.6.4}$$

with the notation of (8.7.7).

10.7

Next let

$$f_0(z) \equiv 1, \tag{10.7.1}$$

$$f_\nu(z) = \prod_{j=1}^{\nu}\left(\frac{z}{z_j} - 1\right) \quad \text{for} \quad \nu \le k_1 \tag{10.7.2}$$

and

$$f_\nu(z) = \prod_{j=1}^{k_1}\left(\frac{z}{z_j} - 1\right)\prod_{j=k_1+1}^{\nu}(z - z_j) \quad \text{for} \quad k_1 < \nu \le n. \tag{10.7.3}$$

We again use the identity (5.2.5). For $j \le k_1$ we get—introducing the $f_\nu(z)$—

$$\begin{aligned} \frac{1}{z - z_j} &= \frac{f_0(z_j)}{f_1(z)}\cdot\frac{1}{z_1} + \frac{f_1(z_j)}{f_2(z)}\cdot\frac{1}{z_2} \\ &\quad + \cdots + \frac{f_{j-1}(z_j)}{f_j(z)}\cdot\frac{1}{z_j} = \sum_{l=1}^{j}\frac{f_{l-1}(z_j)}{f_l(z)}\cdot\frac{1}{z_l} \end{aligned} \tag{10.7.4}$$

and for $j > k_1$

$$\begin{aligned} \frac{1}{z - z_j} &= \sum_{l=1}^{k_1}\frac{f_{l-1}(z_j)}{f_l(z)}\cdot\frac{1}{z_l} + \sum_{l=k_1+1}^{j}\frac{f_{l-1}(z_j)}{f_l(z)} \\ &= \sum_{l=1}^{k_1}\frac{f_{l-1}(z_j)}{f_l(z)}\cdot\frac{1}{z_l} + \sum_{l=k_1+1}^{n}\frac{f_{l-1}(z_j)}{f_l(z)}. \end{aligned} \tag{10.7.5}$$

Substituting these into (10.6.1) and changing the order of summation, as in Section 8.8, we get

$$\begin{aligned} F(z) &= \sum_{l=1}^{k_1}\frac{f_n(z)}{f_l(z)}\frac{z^{m+1}}{z_l}\sum_{j=l}^{n}\frac{f_{l-1}(z_j)}{f_n'(z_j)}\frac{R_1^N}{R_1^N - z_j^N}\cdot\frac{z_j^{N-m-1}}{z_j^N - R_2^N} \\ &\quad + \sum_{l=k_1+1}^{n}\frac{f_n(z)}{f_l(z)}z^{m+1}\sum_{j=l}^{n}\frac{f_{l-1}(z_j)}{f_n'(z_j)}\frac{R_1^N}{R_1^N - z_j^N}\cdot\frac{z_j^{N-m-1}}{z_j^N - R_2^N}. \end{aligned} \tag{10.7.6}$$

Since for $l \le k_1$ we have by (10.5.2)

$$\begin{aligned} \left\|\frac{f_n(z)}{f_l(z)}\frac{z^{m+1}}{z_l}\right\| &\le \prod_{j=l+1}^{k_1}\left(1 + \frac{1}{|z_j|}\right)\prod_{j=k_1+1}^{n}(1 + |z_j|) \\ &\le \frac{(2+\delta)^{n-l}}{(1+\delta)^{k_1-l}} < (2+\delta)^{n-l} \end{aligned}$$

and for $l > k_1$

$$\left\| \frac{f_n(z)}{f_l(z)} z^{m+1} \right\| \le \prod_{j=l+1}^{n} (1 + |z_j|) \le (2 + \delta)^{n-l},$$

(10.7.6) implies

$$\text{(10.7.7)} \quad \| F(z) \| \le \sum_{l=1}^{n} (2 + \delta)^{n-l} \left| \sum_{j=l}^{n} \frac{f_{l-1}(z_j)}{f_n'(z_j)} \frac{R_1^N}{R_1^N - z_j^N} \frac{z_j^{N-m-1}}{z_j^N - R_2^N} \right|.$$

10.8

Next we must estimate

$$\text{(10.8.1)} \qquad V_l = \sum_{j=l}^{n} \frac{f_{l-1}(z_j)}{f_n'(z_j)} \frac{R_1^N}{R_1^N - z_j^N} \frac{z_j^{N-m-1}}{z_j^N - R_2^N}.$$

Again using the notation (8.9.2) and the formula

$$\frac{\xi^{N-m-1}}{(R_1^N - \xi^N)(\xi^N - R_2^N)} = \frac{1}{N(R_1^N - R_2^N)} \sum_{\mu=1}^{N} \left(\frac{R_2^{-m} \varepsilon^{-\mu m}}{\xi - R_2 \varepsilon^{\mu}} - \frac{R_1^{-m} \varepsilon^{-\mu m}}{\zeta - R_1 \varepsilon^{\mu}} \right),$$

we obtain by reversing the order of summation

$$\text{(10.8.2)} \qquad V_l = \frac{R_1^N}{R_1^N - R_2^N} \{ R_2^{-m} V_l'' - R_1^{-m} V_l' \}$$

where

$$V_l'' = -\frac{1}{N} \sum_{\mu=1}^{N} \varepsilon^{-\mu m} \sum_{j=1}^{n} \frac{f_{l-1}(z_j)}{f_n'(z_j)} \cdot \frac{1}{R_2 \varepsilon^{\mu} - z_j}$$

and

$$V_l' = -\frac{1}{N} \sum_{\mu=1}^{N} \varepsilon^{-\mu m} \sum_{j=l}^{n} \frac{f_{l-1}(z_j)}{f_n'(z_j)} \cdot \frac{1}{R_1 \varepsilon^{\mu} - z_j}.$$

Now observe that the inner sums in V_l^* (resp. V_l') are

$$\frac{f_{l-1}(R_2 \varepsilon^{\mu})}{f_n(R_2 \varepsilon^{\mu})} \qquad \left(\text{resp. } \frac{f_{l-1}(R_1 \varepsilon^{\mu})}{f_n(R_1 \varepsilon^{\mu})} \right)$$

and hence

$$|V''_l| \le \max_{|z|=R_2} \left| \frac{f_{l-1}(z)}{f_n(z)} \right| \tag{10.8.3}$$

and

$$|V_l'| \le \max_{|z|=R_1} \left| \frac{f_{l-1}(z)}{f_n(z)} \right|. \tag{10.8.4}$$

10.9

We now estimate $|f_{l-1}(z)|$ on the circles $|z| = R_2$ and $|z| = R_1$. For $l \le k_1 + 1$ we have

$$\max_{|z|=R_2} |f_{l-1}(z)| \le \prod_{j=1}^{l-1} \left(1 + \frac{1}{|z_j|}\right) \le (2 + \delta)^{l-1}$$

and

$$\max_{|z|=R_1} |f_{l-1}(z)| \le \prod_{j=1}^{l-1} \left(1 + \frac{1+\delta}{|z_j|}\right) \le (2 + \delta)^{l-1}.$$

For $l > k_1 + 1$,

$$\max_{|z|=R_2} |f_{l-1}(z)| \le \prod_{j=1}^{k_1} \left(1 + \frac{1}{|z_j|}\right) \prod_{j=k_1+1}^{l-1} (1 + |z_j|) \le (2 + \delta)^{l-1}$$

and

$$\max_{|z|=R_1} |f_{l-1}(z)| \le \prod_{j=1}^{k_1} \left(1 + \frac{1+\delta}{|z_j|}\right) \prod_{j=k_1+1}^{l-1} (1 + \delta + |z_j|) \le (2 + 2\delta)^{l-1}.$$

These inequalities together with (10.5.7) and (10.5.10) give

$$|V_l''| \le \frac{1}{2}\left(\frac{4}{\delta}\right)^n \left(\frac{1+\delta}{3}\right)^{k_1} (2 + 2\delta)^{l-1} \tag{10.9.1}$$

and

$$|V_l'| \le \frac{1}{2}\left(\frac{4}{\delta}\right)^n (1+\delta)^{k_1} (2 + 2\delta)^{l-1}. \tag{10.9.2}$$

Hence by (10.8.2), (10.5.5), and (10.5.8), we get

$$|V_l| \le \frac{R_1^N}{R_1^N - R_2^N} \frac{1}{2}\left(\frac{4}{\delta}\right)^n (1+\delta)^{k_1} \left\{ \frac{(2+\delta)^{l-1}}{3^{k_1}} \frac{1}{(1-\delta)^m} + (2+2\delta)^{l-1} \right\}$$

$$< \frac{R_1^N}{R_1^N - R_2^N} \frac{1}{2}\left(\frac{4}{\delta}\right)^n (2+2\delta)^{l-1}(1+\delta)^{k_1} \left\{ \frac{1}{3^{k_1}(1-\delta)^m} + 1 \right\}$$

and thus from (10.7.7)

$$\|F(z)\| \le \frac{1}{2} \frac{nR_1^N}{R_1^N - R_2^N} \left(\frac{8}{\delta}\right)^n (1+\delta)^{n+k_1-1} \left\{ \frac{1}{3^{k_1}(1-\delta)^m} + 1 \right\}.$$

Since

$$\frac{(1+\delta)^{n+k_1-1}}{3^{k_1}(1-\delta)^m} < e^n (1+\delta)^{n-1} \left(\frac{1+\delta}{3}\right)^{k_1} < (2e)^n$$

and

$$(1+\delta)^{n+k_1-1} < 4^n < (2e)^n,$$

we have

$$\|F(z)\| \le n \frac{R_1^N}{R_1^N - R_2^N} \left(\frac{16e}{\delta}\right)^n.$$

Thus by (10.6.4),

$$\text{(10.9.3)} \qquad \max_{\nu = m+1, \dots, m+n} |g(\nu)|$$

$$\ge \frac{1}{n}\left[\frac{n}{16e(m+n)}\right]^n \frac{R_1^N - R_2^N}{R_1^N} \left| \sum_{j=1}^{n} b_j \frac{R_1^N}{R_1^N - z_j^N} \cdot \frac{z_j^N}{z_j^N - R_2^N} \right|.$$

10.10

Let the indices h_1 and h_2 be defined by

$$\text{(10.10.1)} \qquad |z_{h_1} - 1| > R_1 > |z_{h_1}|, \ |z_{h_2}| > R_2 > |z_{h_2+1}|$$

and pass to the limit $N \to \infty$. Since ν is always among $m+1, \dots, m+n$ and

R_1, R_2 are independent of N, (10.9.3) gives

$$\text{(10.10.2)} \qquad \max_{\nu=m+1,\ldots,m+n} |g(\nu)| \geq \frac{1}{n}\left[\frac{n}{16e(m+n)}\right]^n |b_{h_1} + \cdots + b_{h_2}|.$$

Since, from (10.5.2) and (10.10.1),

$$|z_{h_1}| < R_1 \leq 1 + \delta < |z_{k_1}|$$

respectively,

$$|z_{h_2}| > R_2 \geq 1 - \delta \geq |z_{k_2}|$$

we have

$$h_1 > k_1 \quad \text{and} \quad h_2 < k_2.$$

This, with (10.10.2), completes the proof of Theorem 10.3. ■

Passage to the limit in this theorem would give the corresponding theorem for $\sum_{j=1}^{n} P_j(\nu) z_j^{\nu}$; since no application of it is apparent, we do not formulate it explicitly.

11

CONTINUATION

11.1

To find lower bounds for

$$\max_{\nu=m+1,\ldots,m+n} \frac{|g(\nu)|}{\sum_{j=1}^{n} |b_j|\,|z_j|^\nu} \tag{11.1.1}$$

only in terms of m and n is in general impossible as shown by the example

$$z_1 = \cdots = z_n = 1$$

$$b_1 = 1, \qquad b_2 = \cdots = b_n = -\frac{1}{n-1}.$$

However, restricting the differences $|z_\mu - z_\nu|$ from below (for $\mu \neq \nu$), a useful lower bound was derived in Turán [12].

Theorem 11.1. *If m is a nonnegative integer and the z_j are restricted by*

$$\frac{\min\limits_{\mu\neq\nu}|z_\mu - z_\nu|}{\max\limits_{j}|z_j|} \geq \delta\ (>0), \qquad z_j \neq 0 \tag{11.1.2}$$

then the inequality

$$\max_{\nu=m+1,\ldots,m+n} \frac{|g(\nu)|}{\sum_{j=1}^{n} |b_j|\,|z_j|^\nu} \geq \frac{1}{n}\left(\frac{\delta}{2}\right)^{n-1}$$

holds.

It is interesting that the lower bound is independent of n.

Proof (simplified by Uchiyama). We may suppose for reasons of homogeneity that

$$\max_j |z_j| = 1 \tag{11.1.3}$$

and hence

$$\min_{\mu \neq \nu} |z_\mu - z_\nu| \geq \delta. \tag{11.1.4}$$

With the notation of Section 5.6, let μ be fixed and

$$K_\mu(z) = \left(\frac{z}{z_\mu}\right)^{m+1} l_\mu(z) = \sum_{\nu=m+1}^{m+n} a_{\nu\mu} z^\nu. \tag{11.1.5}$$

Then we have

$$\sum_{\nu=m+1}^{m+n} a_{\nu\mu} z_j^\nu = \begin{cases} 1 & \text{for} \quad j = \mu \\ 0 & \text{for} \quad j \neq \mu. \end{cases}$$

Multiplying by b_j and summing with respect to j we get

$$\sum_{\nu=m+1}^{m+n} a_{\nu\mu} g(\nu) = b_\mu. \tag{11.1.6}$$

If the integer ν_0 satisfies the equality

$$\max_{\nu=m+1,\ldots,m+n} |g(\nu)| = |g(\nu_0)| = M \tag{11.1.7}$$

we get from (11.1.6)

$$\begin{aligned} |b_\mu|\,|z_\mu|^{\nu_0} &\leq M|z_\mu|^{\nu_0} \sum_{\nu=m+1}^{m+n} |a_{\nu\mu}| \\ &\leq \frac{M}{|w'(z_\mu)|} \sum_{\nu=m+1}^{m+n} |a_{\nu\mu}|\,|w'(z_\mu)|\,|z_\mu|^{m+1}. \end{aligned} \tag{11.1.8}$$

Now we have from (11.1.5)

$$a_{\nu\mu} w'(z_\mu) z_\mu^{m+1} = \text{coefficient of } z^{\nu-m-1} \quad \text{in} \quad \frac{w(z)}{z - z_\mu},$$

$$\nu = m + 1, \ldots, m + n$$

that is, the last sum in (11.1.8) is

$$\leq \sum_{\nu=m+1}^{m+n} \left| \text{coeff } z^{\nu-m-1} \quad \text{in} \quad \frac{w(z)}{z - z_\mu} \right|$$

$$\leq \prod_{\substack{1 \leq \lambda \leq n \\ \lambda \neq \mu}} (1 + |z_\lambda|) \leq 2^{n-1},$$

from (11.1.3), and from (11.1.4)

$$|w'(z_\mu)| \geq \delta^{n-1}.$$

These, together with (11.1.8), give

$$|b_\mu|\, |z_\mu|^{\nu_0} \leq M\left(\frac{2}{\delta}\right)^{n-1}. \tag{11.1.9}$$

Summation with respect to μ completes the proof of Theorem 11.1, since ν_0 satisfies condition (11.1.7). ∎

The same proof gives a stronger, though less elegant result.

Corollary. *For integer* $m \geq 0$ *and* $\max_j |z_j| = 1$ *the inequality*

$$\max_{\nu=m+1,\ldots,m+n} \frac{|g(\nu)|}{\sum_{j=1}^{n} |b_j|\, |z_j|^\nu} \geq \frac{1}{n2^{n-1}} \min_j \prod_{\substack{l=1 \\ l \neq j}}^{n} |z_j - z_l| \tag{11.1.10}$$

holds independently of m.

11.2

How close are we (in 11.1.10) to the best possible result? An improvement to my original example was found by M. Szalay who noticed that if the factor $1/n2^{n-1}$ is replaced by $2/n^2$, (11.1.10) is no longer true in general. In fact, let

$$b_j^* = 1, \qquad z_j^* = \exp\frac{2\pi i j}{n+1}, \qquad j = 1,\ldots,n. \tag{11.2.1}$$

Then we have

$$\max_j |z_j^*| = 1$$

and

$$\sum_{j=1}^{n} |b_j^*|\,|z_j^*|^{\nu} = n. \tag{11.2.2}$$

Furthermore,

$$g^*(\nu) = -1 \quad \text{for} \quad n+1 \nmid \nu.$$

Thus choosing $m \equiv 0 \bmod (n+1)$, we get

$$\max_{\substack{m+1 \le \nu \le m+n \\ \nu \text{ integer}}} |g^*(\nu)| = 1, \tag{11.2.3}$$

and

$$\min_j \prod_{\substack{l=1 \\ l \ne j}}^{n} |z_j^* - z_l^*| = \min_j \frac{\prod_{t=1}^{n} \left|\exp\left(\frac{2\pi i t}{n+1}\right) - 1\right|}{\left|\exp\left(\frac{2\pi i j}{n+1}\right) - 1\right|}$$

$$\ge \frac{1}{2}\left|\prod_{t=1}^{n}\left(\exp\left(\frac{2\pi i t}{n+1}\right) - 1\right)\right| = \frac{n+1}{2} > \frac{n}{2}. \tag{11.2.4}$$

Conditions (11.2.3), (11.2.2), and (11.2.4) give

$$\max_{\substack{\nu = m+1, \dots, m+n \\ \nu \text{ integer}}} \frac{|g^*(\nu)|}{\sum_{j=1}^{n} |b_j^*|\,|z_j^*|^{\nu}} = \frac{1}{n} = \frac{2}{n^2} \cdot \frac{n}{2} < \frac{2}{n^2} \min_j \prod_{\substack{l=1 \\ l \ne j}}^{n} |z_j^* - z_l^*|$$

as required.

11.3

Here — in the notation (1.3.3) — the Bohr norm, the maximum norm, and the separation norm were directly involved. Indirectly, formula (11.1.9) gives a lower bound for M by the Cauchy norm too. An improved estimate of this type is due to Makai [5] in the following theorem. Denote by $\sigma_\nu(j)$ the νth elementary symmetric function of $z_1, z_2, \dots, z_{j-1}, z_{j+1}, \dots, z_n$.

Theorem 11.2. *For an arbitrary integer $m \ge -1$ the inequality*

$$\frac{\left(\sum_{\nu=m+1}^{m+n} |g(\nu)|^2\right)^{1/2}}{|b_j|\,|z_j|^{m+1}} \ge \frac{\prod_{\substack{l=1 \\ l \ne j}}^{n} |z_j - z_l|}{\left(1 + |\sigma_1(j)|^2 + \cdots + |\sigma_{n-1}(j)|^2\right)^{1/2}} \tag{11.3.1}$$

holds for $j = 1, \dots, n$, independently of m ($b_j, z_j \ne 0$).

Proof. Suppose, without loss of generality, that

$$j = n, \quad \begin{matrix} z_\mu \neq z_\nu & \text{for} & \mu \neq \nu, \\ b_n = 1 & \text{and} & m = -1. \end{matrix} \tag{11.3.2}$$

(taking $b_l z_l^{m+1}$ as b_l for $l = 1,\ldots,n$ if necessary). Set

$$A_{\mu\nu} = \sum_{l=0}^{n-1} z_\mu^l (\bar{z}_\nu)^l = \bar{A}_{\nu\mu} \quad \text{for} \quad 1 \leq \mu, \nu \leq n. \tag{11.3.3}$$

Then the left-hand side of (11.3.1) cannot sink below the square root of the minimum of

$$\sum_{1 \leq \mu, \nu \leq n} A_{\mu\nu} \xi_\mu \bar{\xi}_\nu \tag{11.3.4}$$

under the condition

$$\xi_n = 1. \tag{11.3.5}$$

This is

$$\left| \frac{D_n}{D_{n-1}} \right|^{1/2} \tag{11.3.6}$$

where

$$D_n = \det A_{\mu\nu} = \prod_{1 \leq \mu < \nu \leq n} |z_\mu - z_\nu|^2 \tag{11.3.7}$$

and D_{n-1} is the $(n-1)$th principal minor. By row-row composition we have

$$D_{n-1} = \det \left[\begin{pmatrix} 1 & z_1 & \cdots & z_1^{n-1} \\ \vdots & \vdots & & \vdots \\ 1 & z_{n-1} & \cdots & z_{n-1}^{n-1} \end{pmatrix} \begin{pmatrix} 1 & \bar{z}_1 & \cdots & \bar{z}_1^{n-1} \\ \vdots & \vdots & & \vdots \\ 1 & \bar{z}_{n-1} & \cdots & \bar{z}_{n-1}^{n-1} \end{pmatrix} \right], \tag{11.3.8}$$

which, according to the classical formula, is equal to the sum of the absolute values of the squares of all $(n-1) \times (n-1)$ minors of the first factor on the right. By a well-known formula,* the value of such a minor when the rth column is missing is

$$\pm \sigma_{n-r}(n) \prod_{1 \leq \mu < \nu \leq n-1} (z_\mu - z_\nu),$$

which completes the proof of Theorem 11.2. ■

*See G. Pólya–G. Szegő, "Aufgaben und Lehrsätze aus der Analysis," Vol. 2, Chapt. 7, Problem 10.

As the proof clearly shows, equality in (11.3.1) can be attained.

A natural extension of Theorem 11.1 would be an estimation of

$$\max \frac{\left|\sum_{j=1}^{n} P_j(\nu) z_j^{\nu}\right|}{\sum_{j=1}^{n} |z_j|^{\nu} \sum_{l=0}^{k_j-1} |a_{l_j}| \nu^l} \tag{11.3.9}$$

where

$$P_j(\nu) = \sum_{l=0}^{k_j-1} a_{l_j} \nu^l. \tag{11.3.10}$$

However, unlike the previous problems, this cannot be done by a simple passage to the limit from Theorem 11.1. When, however, the applicability of such theorems to value-distribution problems was discovered, such results were nevertheless derived in Dancs–Turán [1], Coates [1], and van der Poorten [1]. More recently Tijdeman has discovered that these applications can be obtained — even in stronger form — using only the first main theorem. We do not go into details.

11.4

As S. Dancs remarked, Uchiyama's idea leads to a theorem that gives lower bound by a "slightly isolated" term. Specifically, he proved (oral communication) the following useful estimate.

Theorem 11.3. *Let the integer m be nonnegative and μ be an index so that*

$$\frac{\min_{j \neq \mu} |z_j - z_\mu|}{|z_\mu|} \geq \delta. \tag{11.4.1}$$

Then there is an integer ν_0 such that

$$m + 1 \leq \nu_0 \leq m + n$$

and for which the inequality

$$|g(\nu_0)| \geq \left[\frac{\delta}{2(1 + \delta)}\right]^{n-1} |b_\mu z_\mu^{\nu_0}| \tag{11.4.2}$$

holds.

Proof. We suppose, without loss of generality, that

$$z_\mu = 1.$$

From the identity (11.1.6), we have only to give an upper bound for $\sum_{\nu=m+1}^{m+n} |a_{\nu\mu}|$. Writing $l_\mu(z)$ in the form

$$\prod_{\substack{|z_j|>1\\ |z_j-1|\geq\delta}} \frac{1-z/z_j}{1-1/z_j} \cdot \prod_{\substack{|z_j|\leq 1\\ |z_j-1|\geq\delta}} \frac{z-z_j}{1-z_j}$$

we have

$$\sum_{\nu=m+1}^{m+n} |a_{\nu\mu}| \leq \frac{2^{n-1}}{\prod_{\substack{|z_j|>1\\ |z_j-1|\geq\delta}} |1-1/z_j| \cdot \prod_{\substack{|z_j|\leq 1\\ |z_j-1|\geq\delta}} |1-z_j|}$$

and hence

$$\max_{\substack{m+1\leq\nu\leq m+n\\ \nu \text{ integer}}} |g(\nu)| \geq \frac{|b_\mu|}{2^{n-1}} \prod_{\substack{|z_j|>1\\ |z_j-1|\geq\delta}} \left|1-\frac{1}{z_j}\right| \cdot \prod_{\substack{|z_j|\leq 1\\ |z_j-1|\geq\delta}} |z_j-1|. \tag{11.4.3}$$

Denoting the number of factors in the products by J_1 and J_2 respectively, the second product in (11.4.3) is evidently

$$\geq \delta^{J_2} > \left(\frac{\delta}{1+\delta}\right)^{J_2}. \tag{11.4.4}$$

We see that for $|z_j| \geq 1+\delta$

$$\left|1-\frac{1}{z_j}\right| \geq \frac{\delta}{1+\delta}$$

and for $1 < |z_j| < 1+\delta$, $|z_j-1| \geq \delta$

$$\left|1-\frac{1}{z_j}\right| \geq \frac{\delta}{1+\delta}.$$

Hence the first product in (11.4.3) is bounded from below by

$$\left(\frac{\delta}{1+\delta}\right)^{J_1}. \tag{11.4.5}$$

Since $J_1 + J_2 = n-1$, (11.4.3), (11.4.4), and (11.4.5) prove Theorem 11.3.

11.5

So far we have been dealing with S-sets consisting of a single interval. Halász discovered that if S consists of two (disjoint) intervals then no normalizations are necessary. The intervals in question can be assumed to be

$$S^* = [-m_1 - n, -m_1 - 1] \cup [m_2 + 1, m_2 + n] \tag{11.5.1}$$

where m_1 and m_2 are nonnegative integers. We want to estimate $\max_{\nu \in S^*} |g(\nu)|$ by $|g(0)|$ and

$$m = \max(m_1, m_2). \tag{11.5.2}$$

Theorem 11.4 (Halász). *With the S^* in* (11.5.1) *the inequality*

$$M^* = \max_{\substack{\nu \in S^* \\ \nu \text{ integer}}} |g(\nu)| \geq \left(\frac{n}{8e(m+n)} \right)^n |g(0)|$$

holds if $z_1 \cdots z_n \neq 0$.

The proof given was found in collaboration with S. Dancs and differs from the (unpublished) proof of Halász.

11.6

We suppose without loss of generality that

$$z_\mu \neq z_\nu \quad \text{for} \quad \mu \neq \nu \tag{11.6.1}$$

and — replacing the z_μ's by z_μ^{-1} (changing m_1 and m_2 if necessary) —

$$\max_j |z_j| \geq 1. \tag{11.6.2}$$

Let

$$\frac{1}{z_j} = \xi_j \qquad j = 1, \ldots, n \tag{11.6.3}$$

and

$$\Delta = \frac{n}{m+n}. \tag{11.6.4}$$

If we find an index k with $1 \le k \le n$ and two polynomials

(11.6.5a) $$F(z) = \sum_{\nu=m_1+1}^{m_1+n} c_\nu z^\nu$$

(11.6.5b) $$f(z) = \sum_{\nu=m_2+1}^{m_2+n} d_\nu z^\nu$$

so that

(11.6.6a) $$\sum_{\nu=m_1+1}^{m_1+n} c_\nu \xi_j^\nu = \begin{cases} 0 & \text{for} \quad j \le k \\ 1 & \text{for} \quad j > k, \end{cases}$$

(11.6.6b) $$\sum_{\nu=m_2+1}^{m_2+n} d_\nu z_j^\nu = \begin{cases} 1 & \text{for} \quad j \le k \\ 0 & \text{for} \quad j > k \end{cases}$$

then we get (as before) the identity

(11.6.7) $$\sum_{\nu=m_1+1}^{m_1+n} c_\nu g(-\nu) + \sum_{\nu=m_2+1}^{m_2+n} d_\nu g(\nu) = g(0),$$

and hence

(11.6.8) $$M^* \ge \frac{|g(0)|}{\|F\| + \|f\|}.$$

11.7

Define the index l by

(11.7.1) $$|z_1| \ge \cdots \ge |z_l| \ge 1 > |z_{l+1}| \ge \cdots$$

if such an l exists; if not we can take $l = n$, by (11.6.2).

Let

(11.7.2a) $$\Phi(z) = \prod_{j=1}^{n} (1 - |z_j| z) \prod_{j=1}^{l} |\xi_j|$$

(11.7.2b) $$\varphi(z) = \prod_{j=1}^{n} (z - |z_j|) \prod_{j=1}^{l} |\xi_j|,$$

then

$$\Phi(z) = z^n \varphi\left(\frac{1}{z}\right). \tag{11.7.3}$$

We apply (5.1.8) to $\varphi(z)$ in $[1 - \Delta, 1]$. Then there exists an R with

$$1 - \Delta \le R \le 1 \tag{11.7.4}$$

so that

$$|\varphi(R)| \ge 2\left(\frac{\Delta}{4}\right)^n \prod_{j=1}^{l} \frac{2}{\Delta}\left(1 - \frac{1 - \Delta/2}{|z_j|}\right) \ge 2\left(\frac{\Delta}{4}\right)^n. \tag{11.7.5}$$

Furthermore, (11.7.3) implies

$$\left|\Phi\left(\frac{1}{R}\right)\right| = \frac{|\varphi(R)|}{R^n} \ge \frac{2}{R^n}\left(\frac{\Delta}{4}\right)^n. \tag{11.7.6}$$

With this R we define the index k, postulated in (11.6.6), by

$$|z_1| \ge \cdots \ge |z_l| \ge 1 > |z_{l+1}| \ge \cdots \ge |z_k| > R > |z_{k+1}| \ge \cdots$$

if such a k exists; if not we take $k = n$ if $|z_n| > R$ and $k = l$ if $R > |z_{l+1}|$. Write

$$\Phi(z) = \prod_{j=1}^{l} (|\xi_j| - z) \prod_{j=l+1}^{n} (1 - |z_j| z).$$

Since

$$\left|1 - \frac{|z_j|}{R}\right| = \frac{1}{R}|R - |z_j|\,| \le \frac{1}{R} \quad \text{for} \quad j > l$$

(11.7.6) shows that

$$\left|\prod_{j=1}^{l}\left(|\xi_j| - \frac{1}{R}\right) \prod_{j=l+1}^{\mu}\left(1 - |z_j|\frac{1}{R}\right)\right| \ge 2\left(\frac{\Delta}{4}\right)^n \frac{1}{R^{\mu}}$$

for (11.7.7) $l + 1 \le \mu \le n$.

But then it follows that on $|z| = 1/R$

$$\left|\prod_{j=1}^{l} (\xi_j - z) \prod_{j=l+1}^{\mu} (1 - z_j z)\right| \ge 2\left(\frac{\Delta}{4}\right)^n R^{-\mu} \tag{11.7.8a}$$

holds simultaneously for all μ such that $l + 1 \le \mu \le n$.

Let us write accordingly

$$\varphi(z) = \prod_{j=1}^{l} \left(\frac{z}{|z_j|} - 1 \right) \prod_{j=l+1}^{n} (z - |z_j|).$$

Since we have

$$\left| \frac{R}{|z_j|} - 1 \right| \leq 1, \quad \text{for} \quad 1 \leq j \leq l$$

and

$$|R - |z_j|\,| \leq 1, \quad \text{for} \quad j > l$$

we get that

$$\left| \prod_{1 \leq i_1 < \cdots < i_s \leq l} \left(\frac{R}{|z_{i_t}|} - 1 \right) \prod_{l+1 \leq j_1 < \cdots < j_r \leq n} (R - |z_{j_h}|) \right| \geq 2\left(\frac{\Delta}{4} \right)^n.$$

This shows again that on the circle $|z| = R$, and for all choices of the indices

$$1 \leq i_1 < \cdots < i_s \leq l$$

$$l + 1 \leq j_1 < \cdots < j_r \leq n$$

the inequalities

(11.7.8b) $$\left| \prod_{h=1}^{s} (\xi_{i_h} z - 1) \prod_{h=1}^{r} (z - z_{j_h}) \right| \geq 2\left(\frac{\Delta}{4} \right)^n$$

hold.

11.8

Let

(11.8.1a) $$F_1(z) = \prod_{j=1}^{k} (\xi_j - z)$$

(11.8.1b) $$f_1(z) = \prod_{j=k+1}^{n} (z - z_j).$$

Then we have

$$\|F_1\| \leq \prod_{j=1}^{k} (1 + |\xi_j|) = \prod_{j=1}^{l} \left(1 + \frac{1}{|z_j|} \right) \prod_{j=l+1}^{k} \left(1 + \frac{1}{|z_j|} \right) \leq 2^l \left(1 + \frac{1}{R} \right)^{k-l},$$

that is,

$$\|F_1\| \le \frac{2^k}{R^{k-l}}. \tag{11.8.2a}$$

Furthermore,

$$\|f_1\| \le \prod_{j=k+1}^{n} (1 + |z_j|) \le 2^{n-k}. \tag{11.8.2b}$$

11.9

Next we define the polynomials $F_2(z)$ [resp. $f_2(z)$] with the respective degrees at most $[n - (k - 1)]$ and $(k - 1)$ for which

$$F_2(\xi_j) = \frac{1}{\xi_j^{m_1+1} F_1(\xi_j)} \qquad j = k + 1, \ldots, n, \tag{11.9.1a}$$

respectively

$$f_2(z_j) = \frac{1}{z_j^{m_2+1} f_1(z_j)} \qquad j = 1, \ldots, k, \tag{11.9.1b}$$

hold. Write $F_2(z)$ first as

$$F_2(z) = e_0 + \sum_{\nu=0}^{n-k-1} e_\nu (z - \xi_{k+1}) \cdots (z - \xi_{k+\nu}). \tag{11.9.2}$$

Applying Nörlund's formula with $|w| = 1/R$ we have

$$e_\nu = -\frac{1}{2\pi i} \int_{|w|=1/R} \frac{dw}{w^{m_1+1} F_1(w)(w - \xi_{k+1}) \cdots (w - \xi_{k+\nu+1})}. \tag{11.9.3}$$

Inequality (11.7.8a) is not yet applicable to this expression. Let us introduce the e'_ν coefficients for $\nu = 0, 1, \ldots, n - k - 1$ by

$$e'_\nu = \frac{e_\nu}{z_{l+1} \cdots z_{k+l+1}}. \tag{11.9.4}$$

Then (11.9.3) can be rewritten as

$$e'_\nu = \frac{\pm 1}{2\pi i} \int_{|w|=1/R} \frac{dw}{w^{m_1+1} \prod_{j=1}^{l} (\xi_j - w) \prod_{j=l+1}^{k+\nu+1} (1 - z_j w)},$$

and now (11.7.8a) can be applied with $\mu = k + \nu + 1$. This gives

$$(11.9.5)\qquad |e'_\nu| \le \frac{R^{m_1}}{2}\left(\frac{4}{\Delta}\right)^n R^{k+\nu+1}.$$

Introducing the e'_ν coefficients into (11.9.2) we get

(11.9.6)

$$F_2(z) = (z_{l+1} \cdots z_k)\left(e'_0 z_{k+1} + \sum_{\nu=1}^{n-k-1} e'_\nu (zz_{k+1} - 1)\cdots(zz_{k+\nu} - 1)z_{k+\nu+1}\right).$$

11.10

The situation with $f_2(z)$ is a bit more difficult. We again write

$$(11.10.1)\qquad f_2(z) = h_0 + \sum_{\nu=1}^{k-1} h_\nu (z - z_1)\cdots(z - z_\nu).$$

Nörlund's formula with $|w| = R$ gives

$$(11.10.2)\quad h_\nu = -\frac{1}{2\pi i}\int_{|w|=R} \frac{dw}{w^{m_2+1}f_1(w)(w - z_1)\cdots(w - z_{\nu+1})}, \quad \nu = 0, 1, \ldots, k-1.$$

To apply (11.7.8b), we introduce

$$(11.10.3)\qquad \begin{aligned} h'_\nu &= h_\nu z_1 z_2 \cdots z_{\nu+1} && \text{for} \quad 0 \le \nu \le l - 1 \\ h'_\nu &= h_\nu z_1 z_2 \cdots z_l && \text{for} \quad l \le \nu \le k - 1. \end{aligned}$$

In the first case (11.10.2) takes the form

$$h'_\nu = -\frac{1}{2\pi i}\int_{|w|=R} \frac{dw}{w^{m_2+1}(\xi_1 w - 1)\cdots(\xi_{\nu+1}w - 1)(w - z_{k+1})\cdots(w - z_n)},$$

and then an application of (11.7.8b) gives

$$(11.10.4)\qquad |h'_\nu| \le \frac{1}{2R^{m_2}}\left(\frac{4}{\Delta}\right)^n.$$

In the second case the new form is

$$h'_\nu = \frac{-1}{2\pi i}\int_{|w|=R} \frac{dw}{w^{m_2+1}U(w)}$$

with

$$U(w) = (\xi_1 w - 1)\cdots(\xi_l w - 1)(w - z_{l+1})\cdots(w - z_{\nu+1})$$
$$\cdots(w - z_{k+1})\cdots(w - z_n)$$

for which (11.7.8b) is again applicable. This implies that the estimation (11.10.4) holds for

(11.10.5) $$0 \le \nu \le k - 1.$$

Introducing the h'_ν coefficients into (11.10.1) we get the representation

(11.10.6) $$f_2(z) = \frac{h'_0}{z_1} + \sum_{\nu=1}^{l-1} \frac{h'_\nu}{z_{\nu+1}}(\xi_1 z - 1)\cdots(\xi_\nu z - 1)$$
$$+ h'_l(\xi_1 z - 1)\cdots(\xi_l z - 1)$$
$$+ \sum_{\nu=l+1}^{k-1} h'_\nu(\xi_1 z - 1)\cdots(\xi_l z - 1)(z - z_{l+1})\cdots(z - z_\nu).$$

11.11

We need upper bounds for $\|F_2\|$ and $\|f_2\|$. Equations (11.9.6) and (11.9.5) yield the estimates

(11.11.1a) $$\|F_2\| \le |e'_0| + \sum_{\nu=1}^{n-k-1} |e'_\nu|(1 + |z_{k+1}|)\cdots(1 + |z_{k+\nu}|)$$
$$\le \frac{R^{m_1}}{2}\left(\frac{4}{\Delta}\right)^n R^{k+1} \sum_{\nu=0}^{n-k-1} 2^\nu < R^{m_1}\left(\frac{8}{\Delta}\right)^n \left(\frac{R}{2}\right)^{k+1}.$$

Furthermore, from (11.10.6) and (11.10.4),

(11.11.1b) $$\|f_2\| \le |h'_0| + \sum_{\nu=1}^{l-1} |h'_\nu|(1 + |\xi_1|)\cdots(1 + |\xi_\nu|)$$
$$+ |h'_l|(1 + |\xi_1|)\cdots(1 + |\xi_l|)$$
$$+ \sum_{\nu=l+1}^{k-1} |h'_\nu|(1 + |\xi_1|)\cdots(1 + |\xi_l|)(1 + |z_{l+1}|)$$
$$\cdots(1 + |z_\nu|) \le \frac{1}{2R^{m_2}}\left(\frac{4}{\Delta}\right)^n \left(1 + \sum_{\nu=1}^{k-1} 2^\nu\right) < \frac{2^{k-1}}{R^{m_2}}\left(\frac{4}{\Delta}\right)^n.$$

11.12

One can see immediately that the polynomials

$$z^{m_1+1}F_1 \cdot F_2, \qquad z^{m_2+1}f_1 \cdot f_2$$

satisfy the requirements in (11.6.6) and hence

$$F(z) = z^{m_1+1}F_1F_2, \qquad f(z) = z^{m_2+1}f_1f_2. \tag{11.12.1}$$

Hence from (11.8.2a) and (11.11.1a)

$$\|F\| \le \frac{R^{m_1+l+1}}{2}\left(\frac{8}{\Delta}\right)^n < \frac{1}{2R^m}\left(\frac{8}{\Delta}\right)^n$$

and from (11.8.2b) and (11.11.1b)

$$\|f\| \le \frac{1}{2R^{m_2}}\left(\frac{8}{\Delta}\right)^n < \frac{1}{2R^m}\left(\frac{8}{\Delta}\right)^n.$$

Since

$$\left(\frac{1}{R}\right)^m \le \left(\frac{1}{1-\Delta}\right)^m \le e^n$$

the proof is complete. ■

11.13

In these results the S-set consisted of two intervals of equal length (which could be shifted, of course), and our proof was based essentially on the idea of the proof of the second main theorem. It is of interest to note that the idea of the proof of the first main theorem, properly used, can also lead to a theorem of this type for the case of three intervals of equal length, giving a somewhat better lower bound. More specifically we state the following result.

Theorem 11.5. *For each integer $m \ge 0$ the inequality*

$$M^* = \max |g(\nu)| \ge \left(\frac{n}{6e(m+n)}\right)^n \frac{|g(0)|}{4}$$

holds for $z_1 \cdots z_n \ne 0$ where the max *refers to all integers ν that satisfy either*

$$m + 1 \le |\nu| \le m + n$$

or

$$2m + 1 \le \nu \le 2m + n.$$

We repeat that, just as in Theorem 11.4, no normalization of the z_j is required here. If $\min_j |z_j| \geq 1$, then Theorem 6.1 gives a stronger assertion. If $|z_j| < 1$ for all $j = 1,\ldots,n$, then $|z_j^{-1}| > 1$, and thus we are in the same situation since $|\nu|$ is in the range $m + 1 \leq |\nu| \leq m + n$. Hence we may suppose the existence of an index $k < n$ such that

$$|z_1| \geq \cdots \geq |z_k| \geq 1 > |z_{k+1}| \geq \cdots \geq |z_n|. \tag{11.13.1}$$

Denoting for convenience $z_{k+1},\ldots,z_n$ by $\xi_1,\ldots,\xi_l$ so that

$$k + l = n, \tag{11.13.2}$$

let

$$f_1(z) = \prod_{j=1}^{k}\left(1 - \frac{z}{z_j}\right) = \sum_{\nu=0}^{k} a_\nu^{(1)} z^\nu \tag{11.13.3}$$

$$f_2(z) = \prod_{j=1}^{l}\left(1 - \frac{\xi_j}{z}\right) = \sum_{\nu=0}^{l} \frac{a_\nu^{(2)}}{z^\nu}.$$

Condition (11.13.1) obviously gives

$$\|f_1(z)\| \leq 2^k, \qquad \left\|f_2\left(\frac{1}{z}\right)\right\| \leq 2^l. \tag{11.13.4}$$

Next we define the coefficients $c_\nu^{(1)}$ [resp. $c_\nu^{(2)}$] by

$$\frac{1}{f_1(z)} = \sum_{\nu=0}^{\infty} c_\nu^{(1)} z^\nu, \tag{11.13.5}$$

$$\frac{1}{f_2(z)} = \sum_{\nu=0}^{\infty} \frac{c_\nu^{(2)}}{z^\nu}$$

convergent for $|z| < 1$ (resp. $|z| > 1$). Then we get as in Section 6,

$$\left|c_\nu^{(1)}\right| \leq \binom{k + \nu - 1}{\nu} \tag{11.13.6}$$

$$\left|c_\nu^{(2)}\right| \leq \binom{l + \nu - 1}{\nu}.$$

Setting

$$s_{2m+l}\left(\frac{1}{f_1}\right) = \sum_{\nu=0}^{2m+l} c_\nu^{(1)} z^\nu \tag{11.13.7}$$

$$s_m\left(\frac{1}{f_2}\right) = \sum_{\nu=0}^{m} \frac{c_\nu^{(2)}}{z^\nu}$$

(11.13.6) together with (11.13.2) yields

(11.13.8)
$$\left\|s_{2m+l}\left(\frac{1}{f_1}\right)\right\| \le \binom{2m+n}{k}$$
$$\left\|s_m\left(\frac{1}{f_2}\right)\right\| \le \binom{l+m}{l}.$$

Set

(11.13.9)
$$F_1(z) = f_1(z)s_{2m+l}\left(\frac{1}{f_1}\right)$$
$$F_2(z) = f_2(z)s_m\left(\frac{1}{f_2}\right).$$

Then we have

(11.13.10)
$$F_1(z) = 1 - \sum_{\nu=m+l+1}^{2m+n} d_\nu^{(1)} z^\nu$$
$$F_2(z) = 1 - \sum_{\nu=m+1}^{m+l} \frac{d_\nu^{(2)}}{z^\nu}$$

and hence, from (11.13.4) and (11.13.8),

(11.13.11)
$$\|F_1\| \le \|f_1\| \cdot \left\|s_{2m+l}\left(\frac{1}{f_1}\right)\right\| \le 2^k\binom{2m+n}{k},$$
$$\|F_2\| \le 2^l\binom{l+m}{l}.$$

11.14

Finally, let

(11.14.1)
$$F(z) = 1 - F_1(z)F_2(z).$$

Using the representation (11.13.10) we get

(11.14.2)
$$F(z) = \sum_{\nu=2m+l+1}^{2m+n} d_\nu^{(1)} z^\nu + \sum_{\nu=m+1}^{m+l} \frac{d_\nu^{(2)}}{z^\nu} - \sum_{\nu=m+1}^{m+n-1} e_\nu z^\nu,$$

with complex e_ν coefficients. It follows from (11.14.1) and (11.13.9) that

$$F(z_j) = 1 \qquad j = 1,\ldots,n,$$

that is, for all j-indices

$$1 = \sum_{\nu=2m+l+1}^{2m+n} d_\nu^{(1)} z_j^\nu + \sum_{\nu=m+1}^{m+l} d_\nu^{(2)} z_j^{-\nu} - \sum_{\nu=m+1}^{m+n-1} e_\nu z_j^\nu.$$

From this we get the basic identity

$$g(0) = \sum_{\nu=2m+l+1}^{2m+n} d_\nu^{(1)} g(\nu) + \sum_{\nu=m+1}^{m+l} d_\nu^{(2)} g(-\nu) - \sum_{\nu=m+1}^{m+n-1} e_\nu g(\nu) \tag{11.14.3}$$

and hence

$$M^* \geq \frac{|g(0)|}{\|F_1\| + \|F_2\| + \|F_1\| \cdot \|F_2\|}.$$

Using (11.13.11) we obtain

$$M^* \geq \frac{|g(0)|}{\left[1 + 2^k\binom{2m+n}{k}\right]\left[1 + 2^l\binom{m+l}{l}\right]} \geq \frac{|g(0)|}{4} \frac{1}{2^n\binom{2m+n}{k}\binom{m+l}{l}}$$

$$\geq \frac{|g(0)|}{4} \frac{1}{2^n\binom{3m+n+l}{n}}$$

from which the assertion follows at once.

11.15

We return to a discussion of the Bohr norm. In (1.3.1) we have a *strong* inequality with *weak* localization (with specialized coefficients), whereas Theorem 11.1 gave a *weak* inequality with *strong* localization (with specialized vectors). An application of (1.3.1) was sketched in Section 1; applications of Theorem 11.1 are given in Part II. The application in Section 1 could obviously be substantially strengthened if the localization $[1, 5^n]$ could be replaced by $[1, c^n]$ with a "small" $c > 1$, without significantly changing the constant $\cos(2\pi/5)$ on the right-hand side. Strong limitations for an improvement of that sort were given by Erdős–Rényi [1] and Hlawka [1]. We state their result in the case $b_1 = \cdots = b_n$ and $|z_j| = 1$.

Theorem 11.6. *For each δ, $0 < \delta < 1$ there is a system $(x_1',\ldots,x_n')$ of real numbers so that for all integers ν with*

$$1 \le \nu \le \tfrac{1}{2} \cdot 2^{(\delta^2/4)n}$$

the inequality

$$\left| \sum_{j=1}^{n} \left(e^{2\pi i x_j'}\right)^{\nu} \right| \le \delta n$$

holds.

11.16. PROOF

We consider the integral

$$H = \int_0^1 \cdots \int_0^1 \left| e^{2\pi i x_1} + \cdots + e^{2\pi i x_n} \right|^{2P} dx_1 \cdots dx_n. \tag{11.16.1}$$

Parseval's formula gives — with the notation (5.9.1) —

$$H = J_n(P). \tag{11.16.2}$$

Denote by A_1 the set of all n-tuples $(x_1,\ldots,x_n)$ in the domain of integration with

$$\left| \sum_{j=1}^{n} e^{2\pi i x_j} \right| > \delta n \tag{11.16.3}$$

and by $m(A_1)$ its measure. Then

$$J_n(P) \ge m(A_1)(\delta n)^{2P}. \tag{11.16.4}$$

Since for all nonzero integers ν the value of the integral

$$\int_0^1 \cdots \int_0^1 \left| \sum_{j=1}^{n} e^{2\pi i x_j^{\nu}} \right|^{2P} dx_1 \cdots dx_n$$

is again H, then we get (with similar notation)

$$m(A_\nu) \le J_n(P)(\delta n)^{-2P} \tag{11.16.5}$$

and hence

$$\sum_{1 \le \nu \le (1/2)\cdot 2^{(\delta^2/4)n}} m(A_\nu) < \tfrac{1}{2}2^{(\delta^2/4)n}(\delta n)^{-2P} J_n(P) \le \tfrac{1}{2}2^{(\delta^2/4)n}\left(\frac{2P}{\delta^2 n}\right)^P$$

by Lemma 5.9. Choosing

$$P = \left[\frac{n\delta^2}{4}\right]$$

we get

$$\sum_{1 \le \nu \le (1/2)\cdot 2^{(\delta^2/4)n}} m(A_\nu) < 1$$

which completes the proof. ■

It would be interesting to find an *explicit* system $(x_1', \dots, x_n')$ that satisfies the theorem.

12

ONE-SIDED THEOREMS—MINIMUM NORM

12.1

Next we investigate extremal problems of the third and seventh types. As we demonstrate, the simple geometric restriction

$$\kappa \le |\operatorname{arc} z_j| \le \pi \qquad j = 1, \ldots, n \tag{12.1.1}$$

with

$$0 < \kappa \le \frac{\pi}{2} \tag{12.1.2}$$

will be sufficient to force the required type of theorem. Again, as in (9.2.2)–(9.2.6), let

$$G_2(\nu) = \sum_{j=1}^{n} b_j z_j^{\nu} \varphi(z_j) \tag{12.1.3}$$

and suppose that

$$\min_j |z_j| = 1. \tag{12.1.4}$$

Theorem 12.1. *We assume, along with* (12.1.4), *the argument condition* (12.1.1)–(12.1.2); *and for* ρ, $0 < \rho < 1$ *and real* a_ν,

$$\varphi(z) = z^k + a_1 z^{k-1} + \cdots + a_k \ne 0 \quad \text{for} \quad |z| > \rho. \tag{12.1.5}$$

Then for each integer $m \geq 0$ the inequalities

$$M_1 \doteq \max_{\nu} \operatorname{Re} G_2(\nu) \geq A \tag{12.1.6}$$

$$M_2 \doteq \min_{\nu} \operatorname{Re} G_2(\nu) \leq -A$$

hold, where

$$A = \frac{1}{6n \cdot 16^n}\left(\frac{1+\rho}{2}\right)^m \left(\frac{1-\rho}{2}\right)^{2n+k} \left|\operatorname{Re} \sum_{j=1}^{n} b_j\right|$$

and we take the max (*resp.* min) *over all integers satisfying*

$$m + 1 \leq \nu \leq m + n\left(3 + \left[\frac{\pi}{\kappa}\right]\right). \tag{12.1.7}$$

12.2. PROOF

Suppose that the z_j are pairwise distinct and

$$\operatorname{Re} \sum_{j=1}^{n} b_j > 0. \tag{12.2.1}$$

We define the numbers ξ_j by

$$\xi_j = z_j \qquad \text{for} \quad j = 1,\dots,n \tag{12.2.2}$$

$$\xi_j = \bar{z}_{j-n} \qquad \text{for} \quad j = n+1,\dots,2n$$

and let

$$\eta_1, \eta_2, \dots, \eta_l \tag{12.2.3}$$

be a maximal subset of distinct ξ_j. Thus

$$n \leq l \leq 2n. \tag{12.2.4}$$

Then we evidently have

$$\min_j |\eta_j| = 1, \qquad \kappa \leq |\operatorname{arc} \eta_j| \leq \pi \qquad j = 1,\dots,l. \tag{12.2.5}$$

12.3

Let the η_i satisfy

$$1 = |\eta_1| \leq |\eta_2| \leq \cdots \leq |\eta_l| \tag{12.3.1}$$

and let $f_1(z)$ be the polynomial of degree $\leq l-1$ defined by

$$(12.3.2) \qquad f_1(\eta_j) = \frac{1}{\eta_j^{m+1}\varphi(\eta_j)} \qquad j = 1,\dots,l.$$

If we set

$$(12.3.3) \qquad f_1(z) = \sum_{\nu=0}^{l-1} c_\nu^{(1)} z^\nu,$$

the coefficients $c_\nu^{(1)}$ are obviously real. Proceeding as in Section 7.4, we obtain

$$(12.3.4) \qquad \|f_1\| \leq 2^l\left(\frac{2}{1+\rho}\right)^m\left(\frac{2}{1-\rho}\right)^{l+k}.$$

12.4

Next consider the polynomial

$$(12.4.1) \qquad f_2(z) = \prod_{j=1}^{l}\left(1 - \frac{z}{\eta_j}\right).$$

The coefficients of $f_2(z)$ are real, and condition (12.2.5) implies that Lemmas 5.5 and 5.6 are applicable. Hence we have a polynomial

$$h(z) = 1 + \cdots$$

such that

$$(12.4.2) \qquad f_2(z)h(z) = 1 + \cdots$$

has nonnegative coefficients, is of degree

$$(12.4.3) \qquad \leq \frac{l}{2}\left(1 + \left[\frac{\pi}{\kappa}\right]\right),$$

and

$$(12.4.4) \qquad \|f_2 h\| \leq 2^l.$$

12.5

Next let the polynomial $f_3(z)$ be defined by

$$(12.5.1) \qquad f_3(z) = 2^{l+1}\left(\frac{2}{1+\rho}\right)^{l+k} f_2(z)h(z)(1 + z + \cdots + z^{l-1})$$

with $h(z)$ as in (12.4.2). The degree of $f_3(z)$ cannot exceed

$$\frac{l}{2}\left(3+\left[\frac{\pi}{\kappa}\right]\right)-1. \tag{12.5.2}$$

Its coefficients are obviously nonnegative and in fact, bounded from below. For, setting

$$f_3(z) = c_0^{(3)} + c_1^{(3)}z + \cdots \tag{12.5.3}$$

we have the bound

$$c_j^{(3)} \geq 2^{l+1}\left(\frac{2}{1+\rho}\right)^m\left(\frac{2}{1-\rho}\right)^{l+k} \quad \text{for } j = 0,1,\ldots,l-1 \tag{12.5.4}$$

by (12.4.2). Furthermore—using (12.5.1) and (12.4.4)— we obtain

$$\|f_3\| = \sum_j c_j^{(3)} \leq l \cdot 2^{2l+1}\left(\frac{2}{1+\rho}\right)^m\left(\frac{2}{1-\rho}\right)^{l+k}. \tag{12.5.5}$$

12.6

Our last two auxiliary polynomials $f_4(z)$ and $f_5(z)$ are defined by

$$f_4(z) = f_3(z) + f_1(z) = \sum_\nu c_\nu^{(4)} z^\nu \tag{12.6.1}$$

$$f_5(z) = f_3(z) - f_1(z) = \sum_\nu c_\nu^{(5)} z^\nu. \tag{12.6.2}$$

Their degrees are

$$\leq \frac{l}{2}\left(3+\left[\frac{\pi}{\kappa}\right]\right)-1. \tag{12.6.3}$$

From (12.5.4) and (12.3.4) we have

$$c_\nu^{(4)} \geq 0, \qquad c_\nu^{(5)} \geq 0, \quad \text{for } \nu = 0,1,\ldots,l-1, \tag{12.6.4}$$

and the same holds for the other coefficients $c_\nu^{(4)}$ and $c_\nu^{(5)}$ since all coefficients of $f_3(z)$ are nonnegative and the degree of $f_1(z)$ is $\leq l-1$. Since (12.5.1) gives

$$f_3(\eta_j) = 0 \qquad j = 1,\ldots,l$$

we get, from (12.3.2), (12.6.1), and (12.6.2),

$$\sum_\nu c_\nu^{(4)} \eta_j^{m+\nu+1}\varphi(\eta_j) = 1 \tag{12.6.5}$$

$$\sum_\nu c_\nu^{(5)} \eta_j^{m+\nu+1}\varphi(\eta_j) = -1, \quad \text{for all } \eta_j.$$

Hence with (12.6.5) we have for $j = 1,\ldots,n$

$$\sum_{\nu} c_{\nu}^{(4)} z_j^{m+\nu+1} \varphi(z_j) = 1$$

$$\sum_{\nu} c_{\nu}^{(5)} z_j^{m+\nu+1} \varphi(z_j) = -1,$$

and this shows, by (12.2.1),

$$\sum_{\nu} c_{\nu}^{(4)} \operatorname{Re} G_2(m+\nu+1) = \left|\operatorname{Re}\sum_{1}^{n} b_j\right| \tag{12.6.6}$$

$$\sum_{\nu} c_{\nu}^{(5)} \operatorname{Re} G_2(m+\nu+1) = -\left|\operatorname{Re}\sum_{1}^{n} b_j\right|.$$

Conditions (12.6.3) and (12.2.4) imply that the range of $(m+\nu+1)$ is contained in

$$\left[m+1,\, m+n\left(3+\left[\frac{\pi}{\kappa}\right]\right)\right].$$

With the notation (12.1.6) we get

$$M_1 \geq \frac{\left|\operatorname{Re}\sum_{1}^{n} b_j\right|}{\|f_4\|} \tag{12.6.7}$$

$$M_2 \leq -\frac{\left|\operatorname{Re}\sum_{1}^{n} b_j\right|}{\|f_5\|}.$$

Since, by (12.5.5) and (12.3.4),

$$\|f_4\| = \sum_{\nu} c_{\nu}^{(4)} = f_3(1) + f_1(1) \leq (l+1)2^{2l+1}\left(\frac{2}{1+\rho}\right)^{m}\left(\frac{2}{1-\rho}\right)^{l+k},$$

(12.2.4) yields

$$\|f_4\| \leq 6n \cdot 16^n \left(\frac{2}{1+\rho}\right)^{m}\left(\frac{2}{1-\rho}\right)^{2n+k}.$$

An analogous argument establishes the result for $\|f_5\|$. This completes the proof of Theorem 12.1. ■

12.7

The most interesting case of this theorem occurs when $\varphi(z) \equiv 1$. Then ρ can be chosen freely in the interval (0, 1). A better result can be obtained by following directly the steps of the previous proof in the case $\varphi(z) \equiv 1$. Since the proof—contrary to the situation in Section 9.2—presents no simplification, we formulate the result omitting details.

Theorem 12.2. *If m is a nonnegative integer, and if the normalization* (12.1.4) *and the argument condition* (12.1.1)–(12.1.2) *hold, then*

$$\overline{M}_1 \doteq \max_{\nu} \operatorname{Re} g(\nu) \geq B \tag{12.7.1}$$

$$\overline{M}_2 \doteq \min_{\nu} \operatorname{Re} g(\nu) \leq -B$$

where

$$B = \frac{1}{6n}\left|\operatorname{Re} \sum_{j=1}^{n} b_j\right| \min_{n \leq x \leq 2n} \left(\frac{x}{4e(m+x)}\right)^x.$$

The maximum (*minimum*) *ranges over all integers satisfying* (12.1.7).

Referring to Theorem 12.1, the requirement that the zeros ξ_j of $\varphi(z)$ be confined to the disk $|z| < \rho < 1$ can most probably replaced by the weaker condition that

$$\max \min_{\mu} |z_j - \xi_\mu| \tag{12.7.2}$$

be bounded from below.

Passage to the limit in Theorem 12.2 gives for

$$G_1(\nu) = \sum_{j=1}^{n} P_j(\nu) z_j^{\nu}$$

the following.

Theorem 12.3. *Setting*

$$B_1 = \frac{1}{6K}\left|\operatorname{Re} \sum_{j=1}^{n} P_j(0)\right| \left[\frac{K}{4e(m+2K)}\right]^{2K}$$

under the conditions of Theorem 12.2 *the inequalities*

$$\max_{\nu} \operatorname{Re} G_1(\nu) \geq B_1$$

$$\min_{\nu} \operatorname{Re} G_1(\nu) \leq -B_1$$

hold. Here the $P_j(x)$ *are polynomials of degree* $(k_j - 1)$ *and* $K = \sum_{j=1}^{n} k_j$ *and the range of* ν *is* $[m + 1, m + K(3 + [\pi/\kappa])]$.

For this theorem see also Dancs [4] and Tijdeman [2].

12.8

In addition to the normalizations (12.1.4) and the argument-condition (12.1.1)–(12.1.2) we make, for later applications, the following further restriction on the z_j:

(12.8.1) The system of z_j is symmetric to the real axis, with multiplicity.

Restricting ourselves to power sums $g(\nu) = \sum_{j=1}^{n} b_j z_j^{\nu}$ we also assume the following:

(12.8.2) The coefficients belonging to the complex conjugate $\bar{z}^{\nu}$ of z^{ν} are conjugate complex.

Then of course $g(\nu)$ is real. In this case the introduction of the ξ_j and the η_j and l in (12.2.2)–(12.2.3) is unnecessary, and the interval (12.1.7) is roughly halved. This gives, with some minor adjustments (see also Dancs–Turán [6]), the following.

Theorem 12.4. *If* $g(\nu)$ *satisfies, in addition to the normalization* (12.1.4) *and argument restriction* (12.1.1)–(12.1.2), *the two symmetry restrictions* (12.8.1)–(12.8.2), *then for an arbitrary integer* $m \geq 0$ *and for suitable integers* ν_1, ν_2 *with*

$$m + 1 \leq \nu_1, \nu_2 \leq m + \frac{n}{2}\left(3 + \frac{\pi}{\kappa}\right) \tag{12.8.3}$$

the inequalities

$$g(\nu_1) \geq \frac{1}{3n}\left[\frac{n}{8\sqrt{e}\,(m+n)}\right]^n |g(0)|$$

$$g(\nu_2) \leq -\frac{1}{3n}\left[\frac{n}{8\sqrt{e}\,(m+n)}\right]^n |g(0)|$$

hold.

13

ONE-SIDED THEOREMS–MAXIMUM NORM

Third Main Theorem

13.1

Next we turn to the proof of an operator type theorem which we call our third main theorem owing to the large number of applications, in particular when $\varphi(z) \equiv 1$. Let $m \geq 0$ be an integer, and

(13.1.1) $$\varphi(z) = z^k + d_1 z^{k-1} + \cdots + d_k$$

be a polynomial with real coefficients so that

(13.1.2) $$\varphi(z) \neq 0 \quad \text{for} \quad z \in \{|z| \geq \rho\} \cup \{|\text{arc}\, z| < \kappa\}$$

where $0 < \rho < 1$ and where $0 < \kappa \leq \pi/2$.

Suppose further that the quantities δ_1 and δ_2 are such that

(13.1.3) $$1 > \delta_1 > \delta_2 \geq \rho.$$

We let

$$G_2(\nu) = \sum_{j=1}^{n} b_j z_j^\nu \varphi(z_j)$$

and suppose that the argument-condition

(13.1.4) $$\kappa \leq |\text{arc}\, z_j| \leq \pi, \qquad j = 1, \ldots, n$$

and the normalization $\max_j |z_j| = 1$ are satisfied. Finally we define the indices h_1 and h_2 by

$$(13.1.5) \qquad 1 = |z_1| \geq \cdots \geq |z_{h_1}| \geq \delta_1 > |z_{h_1+1}| \geq \cdots$$
$$\geq |z_{h_2}| > \delta_2 \geq |z_{h_2+1}| \geq \cdots ; \qquad (h_1 \leq h_2).$$

(If some do not exist, replace them by n.) Let us denote

$$(13.1.6) \qquad I = \left[m+1, m+(n+k)\left(3+\frac{\pi}{\kappa}\right)\right]$$

$$M_1 = \max_{\substack{\nu \in I \\ \nu \text{ integer}}} \operatorname{Re} G_2(\nu)$$

$$M_2 = \min_{\substack{\nu \in I \\ \nu \text{ integer}}} \operatorname{Re} G_2(\nu).$$

Theorem 13.1. *We have the inequalities*:

$$M_1 \geq \frac{1}{3(n+k)}\left(\frac{\delta_1-\delta_2}{16}\right)^{2(n+k)} \cdot \delta_2^{m+(n+k)\left(1+\frac{\pi}{\kappa}\right)} \min_{h_1 \leq j \leq h_2} \left|\operatorname{Re}(b_1 + \cdots + b_j)\right|$$

and

$$M_2 \leq -\frac{1}{3(n+k)}\left(\frac{\delta_1-\delta_2}{16}\right)^{2(n+k)}$$
$$\cdot \delta_2^{m+(n+k)\left(1+\frac{\pi}{\kappa}\right)} \min_{h_1 \leq j \leq h_2} \left|\operatorname{Re}(b_1 + \cdots + b_j)\right|.$$

We notice that the length of I and the lower bound itself depend upon the operator $\varphi(z)$ only through the parameters k and ρ.

13.2. PROOF

We denote the zeros of $\varphi(z)$ by ξ and let w_j represent both the z_j and ξ_j. We may suppose, without loss of generality, that

$$(13.2.1) \qquad w_\mu \neq w_\nu \quad \text{for} \quad 1 \leq \mu \neq \nu \leq n+k.$$

Let

$$(13.2.2) \qquad \eta_1, \ldots, \eta_l$$

be a maximal subset of pairwise distinct elements among the

$$w_1, \ldots, w_{n+k}, \qquad \bar{w}_1, \ldots, \bar{w}_{n+k}.$$

That is, the system $\{\eta_i\}$ of (13.2.2) is symmetric with respect to the real axis and

$$n + k \le l \le 2(n+k). \tag{13.2.3}$$

Then we have (in addition to the condition $|\eta_j| \le 1$)

$$\kappa \le |\text{arc}\, \eta_j| \le \pi, \qquad j = 1, 2, \ldots, l. \tag{13.2.4}$$

13.3

Let

$$f_0(z) = \prod_{j=1}^{l} (z - \eta_j), \tag{13.3.1}$$

which obviously has real coefficients. By (13.2.4) we may apply Lemma 5.7. Hence we have

$$f_1(z) = f_0(z)\psi(z) = \sum_{j=1}^{N_1} c_j^{(1)} z^j \tag{13.3.2}$$

so that

$$c_{N_1}^{(1)} = 1 \tag{13.3.3}$$

$$N_1 \le \frac{l}{2}\left(1 + \left[\frac{\pi}{\kappa}\right]\right) \tag{13.3.4}$$

$$c_j^{(1)} \ge 0 \tag{13.3.5}$$

and

$$\sum_{j=0}^{N_1} c_j^{(1)} \le 2^l. \tag{13.3.6}$$

13.4

We apply Lemma 5.1 to $\prod_{j=1}^{l}(z - |\eta_j|)$ and $[\delta_2, \delta_1]$. Then, as before, there exists an R with

$$\delta_2 \le R \le \delta_1 \tag{13.4.1}$$

so that for $|z| = R$ the inequalities

$$\left|\prod_{1\le i_1<\cdots<i_s\le l}(z-\eta_{i_j})\right| \ge 2\left(\frac{\delta_1-\delta_2}{4}\right)^l \tag{13.4.2}$$

hold. Hence in particular we have

$$\left|\varphi(z)\prod_{1\le i_1<\cdots<i_s\le n}(z-z_{i_j})\right| \ge 2\left(\frac{\delta_1-\delta_2}{4}\right)^l$$

on the circle $|z| = R$. We define the indices r, r_1, and r_2 by

$$\begin{aligned} 1 = |\eta_1| \ge \cdots \ge |\eta_{r_1}| \ge \delta_1 > |\eta_{r_1+1}| \ge \cdots \\ \ge |\eta_r| > R > |\eta_{r+1}| \ge \cdots \ge |\eta_{r_2}| > \delta_2 \\ \ge |\eta_{r_2+1}| \ge \cdots \ge |\eta_l|, \qquad (r_1 \le r \le r_2). \end{aligned} \tag{13.4.3}$$

13.5

Next let $f_2(z)$ be defined by

$$f_2(z) = \prod_{\substack{j=r+1\\ \varphi(\eta_j)\ne 0}}^{l} (z-\eta_j). \tag{13.5.1}$$

All coefficients are real, and we have

$$\|f_2\| \le 2^{l-r}. \tag{13.5.2}$$

Let the polynomial $f_3(z)$ of degree $\le r-1$ be defined by

$$f_3(\eta_j) = \frac{1}{\eta_j^{m+N_1+1}\varphi(\eta_j)f_2(\eta_j)} \qquad j = 1,2,\ldots,r. \tag{13.5.3}$$

This expression is meaningful due to (13.4.3) and the fact that for these η_j

$$|\eta_j| \ge |\eta_r| > R \ge \delta_2 \ge \rho$$

so that $\varphi(\eta_j) \ne 0$. Moreover, writing $f_3(z)$ in the form

$$f_3(z) = \sum_{\nu=0}^{r-1} c_\nu^{(2)} z^\nu \tag{13.5.4}$$

all the coefficients of $f_3(z)$ are real. We write

$$f_3(z) = e_0 + \sum_{\nu=1}^{r-1} e_\nu (z - \eta_1) \cdots (z - \eta_\nu), \tag{13.5.5}$$

and again we have

$$c_\nu = -\frac{1}{2\pi i} \int_{|w| q R} \frac{dw}{w^{m+1+N_1} \varphi(w) f_2(w)(w - \eta_1) \cdots (w - \eta_{\nu+1})}.$$

We can apply (13.4.2); this gives

$$|e_\nu| \le \frac{1}{2R^{m+N_1}} \left(\frac{4}{\delta_1 - \delta_2} \right)^l. \tag{13.5.6}$$

Since from (13.5.5)

$$\|f_3\| \le |e_0| + \sum_{\nu=1}^{r-1} |e_\nu| (1 + |\eta_1|) \cdots (1 + |\eta_\nu|)$$

we get

$$\|f_3\| \le \frac{1}{2\delta_2^{m+N_1}} \left(\frac{4}{\delta_1 - \delta_2} \right)^l 2^r. \tag{13.5.7}$$

13.6

Let $f_4(z)$ be defined by

$$\begin{aligned} f_4(z) &= \left(\frac{8}{\delta_1 - \delta_2} \right)^l \delta_2^{-m - \frac{l}{2}\left(1 + \frac{\pi}{\kappa}\right)} f_1(z)(1 + z + \cdots + z^{l-1}) \\ &= \sum_{\nu=0}^{N_1 + l - 1} c_\nu^{(4)} z^\nu. \end{aligned} \tag{13.6.1}$$

From (13.3.5) we obtain

$$c_\nu^{(4)} \ge 0 \quad \text{for all } \nu \tag{13.6.2}$$

and, using (13.3.3),

$$c_\nu^{(4)} \ge \left(\frac{8}{\delta_1 - \delta_2} \right)^l \delta_2^{-m - \frac{l}{2}\left(1 + \frac{\pi}{\kappa}\right)} \tag{13.6.3}$$

for

(13.6.4) $$\nu = N_1, N_1 + 1, \ldots, N_1 + l - 1.$$

Finally let $f_5(z)$ and $f_6(z)$ be defined by

(13.6.5) $$f_5(z) = f_4(z) + z^{N_1} f_2(z) f_3(z) = \sum_{\nu=0}^{N_1+l-1} c_\nu^{(5)} z^\nu$$

and

(13.6.6) $$f_6(z) = f_4(z) - z^{N_1} f_2(z) f_3(z) = \sum_{\nu=0}^{N_1+l-1} c_\nu^{(6)} z^\nu.$$

By (13.3.4), their degrees cannot exceed

(13.6.7) $$\frac{l}{2}\left(3 + \frac{\pi}{\kappa}\right) - 1.$$

From (13.6.2) we have

(13.6.8) $$c_\nu^{(5)} \geq 0, \qquad c_\nu^{(6)} \geq 0 \quad \text{for} \quad \nu \leq N_1 - 1.$$

Since, from (13.5.2) and (13.5.7)

(13.6.9) $$\|f_2\| \|f_3\| \leq \frac{1}{2\delta_2^{m+N_1}} \left(\frac{8}{\delta_1 - \delta_2}\right)^l,$$

this and (13.6.3)–(13.6.4) implies that (13.6.8) holds for all ν. From the structure of the polynomials, we have

$$\sum_{\nu=0}^{N_1+l-1} c_\nu^{(5)} \eta_j^\nu = \begin{cases} \dfrac{1}{\eta_j^{m+1}\varphi(\eta_j)} & \text{for } j \leq r \\ 0 & \text{for } j > r, \qquad \varphi(\eta_j) \neq 0. \end{cases}$$

That is,

(13.6.10) $$\sum_{\nu=0}^{N_1+l-1} c_\nu^{(5)} \eta_j^{m+\nu+1} \varphi(\eta_j) = \begin{cases} 1 & \text{for } j \leq r \\ 0 & \text{for } j > r. \end{cases}$$

Define the index h by

(13.6.11) $$\begin{aligned} 1 = |z_1| \geq \cdots \geq |z_{h_1}| \geq \delta_1 > |z_{h_1+1}| \geq \cdots \\ \geq |z_h| > R > |z_{h+1}| \geq \cdots \geq |z_{h_2}| > \delta_2 \\ \geq |z_{h_2+1}| \geq \cdots \geq |z_n|, \qquad (h_1 \leq h \leq h_2). \end{aligned}$$

Expressed in terms of the z_j, condition (13.6.10) means that

$$\sum_{\nu=0}^{N_1+l-1} c_\nu^{(5)} z_j^{m+\nu+1}\varphi(z_j) = \begin{cases} 1 & \text{for } j \le h \\ 0 & \text{for } j > h \end{cases} \tag{13.6.12}$$

and

$$\sum_{\nu=0}^{N_1+l-1} c_\nu^{(6)} z_j^{m+\nu+1}\varphi(z_j) = \begin{cases} -1 & \text{for } j \le h \\ 0 & \text{for } j > h. \end{cases} \tag{13.6.13}$$

We may suppose without loss of generality that

$$\operatorname{Re}(b_1 + \cdots + b_h) \ge 0.$$

Then we get—as in Section 12.6—

$$\sum_{\nu=0}^{N_1+l-1} c_\nu^{(5)} \operatorname{Re} G_2(m+\nu+1) = |\operatorname{Re}(b_1 + \cdots + b_h)| \tag{13.6.14}$$

$$\sum_{\nu=0}^{N_1+l-1} c_\nu^{(6)} \operatorname{Re} G_2(m+\nu+1) = -|\operatorname{Re}(b_1 + \cdots + b_h)|.$$

13.7

Since the range of ν is contained in the interval $[m+1, m+(n+k)(3+\pi/\kappa)]$, (13.1.6) gives

$$M_1 \ge \frac{|\operatorname{Re}(b_1 + \cdots + b_h)|}{\sum_\nu c_\nu^{(5)}}, \tag{13.7.1}$$

$$M_2 \le -\frac{|\operatorname{Re}(b_1 + \cdots + b_h)|}{\sum_\nu c_\nu^{(6)}}.$$

But

$$\sum_\nu c_\nu^{(5)} = f_4(1) + f_2(1) f_3(1)$$

$$\le \left(\frac{8}{\delta_1 - \delta_2}\right)^l \delta_2^{-m-\frac{l}{2}\left(1+\frac{\pi}{\kappa}\right)} l f_1(1) + \|f_2\| \|f_3\|$$

and, using (13.6.9) and (13.3.6),

$$\le (l+1)\left(\frac{16}{\delta_1-\delta_2}\right)^{l}\delta_2^{-m-\frac{l}{2}\left(1+\frac{\pi}{\kappa}\right)}$$

$$\le 3(n+k)\left(\frac{16}{\delta_1-\delta_2}\right)^{2(n+k)}\left(\frac{1}{\delta_2}\right)^{m+(n+k)\left(1+\frac{\pi}{\kappa}\right)}.$$

The estimation of $\Sigma c_\nu^{(6)}$ is analogous. The proof of Theorem 13.1 is complete if we take into account the fact that

$$|\mathrm{Re}(b_1+\cdots+b_h)| \ge \min_{h_1\le j\le h_2}|\mathrm{Re}(b_1+\cdots+b_j)|. \quad \blacksquare$$

13.8

We now formulate the important special case of $\varphi(z)\equiv 1$.

Theorem 13.2. *Let m be nonnegative integer, and suppose the argument-condition* (13.1.4) *is satisfied with δ_1, δ_2 restricted only by*

$$0 \le \delta_2 < \delta_1 < 1.$$

If the indices h_1 and h_2 are defined by (13.1.5) *and*

$$I_1 = \left[m+1,\, m+n\left(3+\frac{\pi}{\kappa}\right)\right] \tag{13.8.1}$$

then, using the notation (13.1.6) *with I_1 instead of I, the inequalities*

$$M_1 \ge \frac{1}{3n}\left(\frac{\delta_1-\delta_2}{16}\right)^{2n}\delta_2^{m+n\left(1+\frac{\pi}{\kappa}\right)}\min_{h_1\le j\le h_2}|\mathrm{Re}(b_1+\cdots+b_j)|$$

$$M_2 \le -\frac{1}{3n}\left(\frac{\delta_1-\delta_2}{16}\right)^{2n}\delta_2^{m+n\left(1+\frac{\pi}{\kappa}\right)}\min_{h_1\le j\le h_2}|\mathrm{Re}(b_1+\cdots+b_j)|$$

hold.

In the important special case

$$\delta_1 = 1, \qquad \delta_2 = \frac{m+n\left(1+\frac{\pi}{\kappa}\right)}{m+n\left(3+\frac{\pi}{\kappa}\right)}$$

this gives the following consequence.

Corollary I. *With the foregoing notation there exist integers* ν_1 *and* ν_2 *satisfying*

$$m + 1 \leq \nu_1, \nu_2 \leq m + n\left(3 + \frac{\pi}{\kappa}\right)$$

for which the inequalities

$$\frac{\operatorname{Re} \sum_{j=1}^{n} b_j z_j^{\nu_1}}{\left(\max_j |z_j|\right)^{\nu_1}} \geq \frac{1}{3n} \min_{1 \leq j \leq n} \left|\operatorname{Re}(b_1 + \cdots + b_j)\right| \left\{ \frac{n}{8e\left(m + n\left(3 + \frac{\pi}{\kappa}\right)\right)} \right\}^{2n}$$

and

$$\frac{\operatorname{Re} \sum_{j=1}^{n} b_j z_j^{\nu_2}}{\left(\max_j |z_j|\right)^{\nu_2}} \leq -\frac{1}{3n} \min_{1 \leq j \leq n} \left|\operatorname{Re}(b_1 + \cdots + b_j)\right| \left\{ \frac{n}{8e\left(m + n\left(3 + \frac{\pi}{\kappa}\right)\right)} \right\}^{2n}$$

hold.

We obtain another important result when we choose $\delta_2 = \delta_1/2$.

Corollary II. *Suppose that*

$$1 = |z_1| \geq |z_2| \geq \cdots \geq |z_n|, \tag{13.8.2}$$

h_1 *is an index* $\leq n$, *and* h_2 *is any index with either*

$$|z_{h_2}| \leq \frac{|z_{h_1}|}{2} \tag{13.8.3}$$

or $h_2 = n$, *and that the argument condition* (13.1.4) *is satisfied. Then in the notation* (13.1.6) *with*

$$B \doteq \min_{h_1 \leq j \leq h_2} \left|\operatorname{Re}(b_1 + \cdots + b_j)\right| \tag{13.8.4}$$

we have, for $n \geq 4$

$$M_1 \geq B\left(\frac{|z_{h_1}|}{2}\right)^{m+n\left(12+\frac{\pi}{\kappa}\right)} \tag{13.8.5}$$

and

$$M_2 \leq -B\left(\frac{|z_{h_1}|}{2}\right)^{m+n\left(12+\frac{\pi}{\kappa}\right)}. \tag{13.8.6}$$

14

ONE-SIDED THEOREMS WITH OTHER NORMS

14.1

In the one-sided theorems of the last two sections we had in addition to the argument-condition

$$\kappa \leq |\operatorname{arc} z_j| \leq \pi \qquad \left(0 < \kappa \leq \frac{\pi}{2}\right) \tag{14.1.1}$$

the normalization $\max_j |z_j| = 1$ (resp. $\min_j |z_j| = 1$), and we obtained bounds in terms of

$$\min_{h \leq j \leq l} \left|\operatorname{Re}(b_1 + \cdots + b_j)\right|$$

(resp. $\left|\operatorname{Re}(b_1 + \cdots + b_n)\right|$). In some situations in which (14.1.1) is satisfied, it is more useful to assume the normalization $|z_k| = 1$ for a fixed k. This results in a lower bound depending on the quantity

$$\min \left|\operatorname{Re}(b_\mu + \cdots + b_\nu)\right|$$

in the sense of (10.1.1) and (10.1.2) or (10.4.1). We prove the corresponding theorems. Let m be an arbitrary nonnegative integer, and let the numbers z_j satisfy (14.1.1); it may be assumed that they are distinct. Let

$$|z_1| = 1 \qquad (\text{and of course } \kappa \leq |\operatorname{arc} z_1| \leq \pi) \tag{14.1.2}$$

and let the z_j be arranged so that,

$$0 = |z_1 - z_1| \leq |z_2 - z_1| \leq \cdots \leq |z_n - z_1|. \tag{14.1.3}$$

Let

$$(14.1.4) \qquad 0 < \delta_2 < \delta_1 \le \frac{2n}{m + n(3 + \pi/\kappa)}, \qquad z_1 = -1$$

and define the indices L_1 and L_2 by

$$(14.1.5) \quad \cdots \le |z_{L_2} - z_1| \le \delta_2 < |z_{L_2+1} - z_1| \le \cdots$$
$$\le |z_{L_1} - z_1| \le \delta_1 < |z_{L_1+1} - z_1| \le \cdots, \qquad (L_2 \le L_1),$$

when they exist; if not replace them by n.

Theorem 14.1 (Dancs–Turán, unpublished). *For*

$$g(\nu) = \sum_{j=1}^{n} b_j z_j^{\nu}$$

there are integers ν_1 and ν_2 with

$$m + 1 \le \nu_1, \nu_2 \le m + n\left(3 + \frac{\pi}{\kappa}\right)$$

so that

$$\operatorname{Re} g(\nu_1) \ge \frac{1}{5n}\left(\frac{\delta_1 - \delta_2}{24e}\right)^{2n} \min_{L_2 \le h \le L_1} \left|\operatorname{Re} \sum_{j=1}^{h} b_j\right|$$

and

$$\operatorname{Re} g(\nu_2) \le -\frac{1}{5n}\left(\frac{\delta_1 - \delta_2}{24e}\right)^{2n} \min_{L_2 \le h \le L_1} \left|\operatorname{Re} \sum_{j=1}^{h} b_j\right|.$$

The corresponding operator-type theorem could be obtained using ideas found in previous sections but we will not elaborate.

14.2

Among the z_j and their conjugates $\overline{z_j}$, let

$$(14.2.1) \qquad \eta_1, \ldots, \eta_l$$

be a maximal subset of pairwise distinct elements. Hence

$$(14.2.2) \qquad n \le l \le 2n.$$

Let $\eta_1 = z_1$ so that

$$|\eta_1| = 1, \qquad \kappa \le |\arc \eta_1| \le \pi \tag{14.2.3}$$

and setting

$$\eta_1 - \eta_j = \xi_j, \qquad j = 1,\dots,l \tag{14.2.4}$$

we impose an ordering so that

$$\begin{aligned} 0 = |\xi_1| \le \cdots \le |\xi_{l_2}| \le \delta_2 < |\xi_{l_2+1}| \cdots \\ \le |\xi_{l_1}| \le \delta_1 < |\xi_{l_1+1}| \le \cdots; \qquad (l_2 \le l_1). \end{aligned} \tag{14.2.5}$$

If either of the indices l_1, l_2 does not exist, replace those by l. We have

$$\kappa \le |\arc \eta_j| \le \pi, \tag{14.2.6}$$

and the η_j form a point system symmetric to the real axis.

14.3

Define

$$h_0(z) = \prod_{|\eta_j| \ge 1} \left(1 - \frac{z}{\eta_j}\right) \prod_{|\eta_j| < 1} (z - \eta_j) = B_1(z) B_2(z) \tag{14.3.1}$$

where the coefficients of B_1 and B_2 are real. Hence Lemma 5.6 (resp. 5.7) can be applied, with $R = 1$. If B_1 and B_2 have degrees J_1 and J_2 respectively, then these lemmas give polynomials $\psi_1(z)$ and $\psi_2(z)$ so that

$$B_1(z) = \psi_1(z) = \sum_{\nu=0}^{N_0} c_\nu z^\nu$$

with

$$c_0 = 1, \qquad c_\nu \ge 0, \qquad \sum_\nu c_\nu \le 2^{J_1}, \qquad N_0 \le \frac{J_1}{2}\left(1 + \frac{\pi}{\kappa}\right),$$

and

$$B_2(z)\psi_2(z) = \sum_{\nu=0}^{N} c_\nu^{(0)} z^\nu$$

with

$$c_{N_1}^{(0)} = 1, \qquad c_\nu^{(0)} \geq 0, \qquad \sum_\nu c_\nu^{(0)} \leq 2^{J_2}, \tag{14.3.2}$$

$$N_1 \leq \frac{J_2}{2}\left(1 + \frac{\pi}{\kappa}\right) \leq \frac{l}{2}\left(1 + \frac{\pi}{\kappa}\right).$$

Thus for

$$f_1(z) = h_0(z)\psi_1(z)\psi_2(z) = \sum_{\nu=0}^{N_2} c_\nu^{(1)} z^\nu \tag{14.3.3}$$

we have

$$c_{N_1}^{(1)} \geq 1, \qquad c_\nu^{(1)} \geq 0, \qquad \sum_\nu c_\nu^{(1)} \leq 2^l \tag{14.3.4}$$

and

$$N_2 \leq \frac{l}{2}\left(1 + \frac{\pi}{\kappa}\right). \tag{14.3.5}$$

Set

$$f_2(z) = \Omega f_1(z)(1 + z + \cdots + z^{l-1}) = \sum_{\nu=0}^{N_2+l-1} c_\nu^{(2)} z^\nu \tag{14.3.6}$$

where $\Omega > 0$ will be determined later, and we have

$$c_\nu^{(2)} \geq 0 \quad \text{for all } \nu; \tag{14.3.7}$$

moreover,

$$c_\nu^{(2)} \geq \Omega \quad \text{for} \quad \nu = N_1, N_1 + 1, \ldots, N_1 + l - 1, \tag{14.3.8}$$

$$\sum_\nu c_\nu^{(2)} \leq l 2^l \Omega, \tag{14.3.9}$$

and

$$f_2(\eta_j) = 0, \qquad j = 1, \ldots, l. \tag{14.3.10}$$

14.4

Let

$$f_3(x) = f_3(x, \eta_1) = \prod_{j=l_1+1}^{l}\left(1 - \frac{x}{|\xi_j|}\right)\prod_{j=1}^{l_1}(x - |\xi_j|). \tag{14.4.1}$$

Apply (5.1.8) to $f_3(x)$ and $[\delta_2, \delta_1]$ to get an x_1 with

$$\delta_2 \le x_1 \le \delta_1 \tag{14.4.2}$$

so that

$$|f_3(x_1)| \ge 2\left(\frac{\delta_1 - \delta_2}{4}\right)^l \left\{\prod{}'\left(\frac{2}{\delta_1 - \delta_2}\right)\left||\xi_j| - \frac{\delta_1 + \delta_2}{2}\right|\right\} \cdot \left\{\prod_{j=l_1+1}^{l} \frac{1}{|\xi_j|}\right\}. \tag{14.4.3}$$

The first product ranges over all j with

$$\left||\xi_j| - \frac{\delta_1 - \delta_2}{2}\right| > \frac{\delta_1 - \delta_2}{2}. \tag{14.4.4}$$

Since this is satisfied when $|\xi_j| > \delta_1$, that is, when $j > l_1$ we get from (14.4.3)

$$|f_3(x_1)| \ge 2\left(\frac{\delta_1 - \delta_2}{4}\right)^l \prod_{j=l_1+1}^{l} \frac{2}{\delta_1 - \delta_2}\left(1 - \frac{\delta_1 - \delta_2}{2|\xi_j|}\right)$$

$$\ge 2\left(\frac{\delta_1 - \delta_2}{4}\right)^l \left[\frac{2}{\delta_1 - \delta_2}\left(1 - \frac{\delta_1 + \delta_2}{2\delta_1}\right)\right]^{l-l_1}.$$

That is,

$$|f_3(x_1)| \ge 2\left(\frac{\delta_1 - \delta_2}{4}\right)^l \delta_1^{l_1 - l}, \tag{14.4.5}$$

Since

$$|x_1 - |\xi_j|| \le 1, \quad \text{for} \quad j \le l_1$$

we have from (14.4.5)

$$\left|\prod_{j=l_1+1}^{l}\left(1 - \frac{x_1}{|\xi_j|}\right) \prod_{1 \le i_1 < i_2 < \cdots \le l_1}^{\nu} (x_1 - |\xi_{i_\nu}|)\right| \ge 2\left(\frac{\delta_1 - \delta_2}{4}\right)^l \delta_1^{l_1 - l}. \tag{14.4.6}$$

This yields, as in Section 8.2, that on the circle $|z| = x_1$ the inequality

$$\left|\prod_{j=l_1+1}^{l}\left(1 - \frac{z}{\xi_j}\right) \prod_{1 \le i_1 < i_2 < \cdots \le l_1}^{\nu} (z - \xi_{i_\nu})\right| \ge 2\left(\frac{\delta_1 - \delta_2}{4}\right)^l \delta_1^{l_1 - l} \tag{14.4.7}$$

holds.

Define the index t by

$$(14.4.8)\quad 0 = |\xi_1| \le |\xi_2| \le \cdots \le |\xi_{l_2}| \le \delta_2 < |\xi_{l_2+1}| \le \cdots$$
$$\le |\xi_t| < x_1 < |\xi_{t+1}| \le \cdots \le |\xi_{l_1}| \le \delta_1 < |\xi_{l_1+1}| \le \cdots$$

where $(l_2 \le t \le l_1)$.

14.5

Let

$$(14.5.1)\qquad -z + \eta_1 = w$$

and

$$(14.5.2)\qquad f_4(z) = F(w) = \prod_{\nu=l_1+1}^{l} \left(1 - \frac{w}{\xi_\nu}\right) \prod_{\nu=t+1}^{l_1} (w - \xi_\nu)$$
$$= \prod_{\nu=l_1+1}^{l} \frac{z - \eta_\nu}{\xi_\nu} \prod_{\nu=t+1}^{l_1} (\eta_\nu - z).$$

Hence

$$(14.5.3)\qquad \|f_4(z)\| \le \prod_{\nu=l_1+1}^{l} \frac{1 + |\eta_\nu|}{|\xi_\nu|} \prod_{\nu=t+1}^{l_1} (1 + |\eta_\nu|).$$

Since by (14.2.5) for $\nu > l_1$

$$\frac{1 + |\eta_\nu|}{|\xi_\nu|} \le \frac{1 + |\eta_1| + |\xi_\nu|}{|\xi_\nu|} = 1 + \frac{2}{|\xi_\nu|} \le 1 + \frac{2}{\delta_1}$$

and for $\nu \le l_1$

$$1 + |\eta_\nu| \le 2 + |\xi_\nu| \le 2 + \delta_1,$$

(14.5.3) implies that

$$(14.5.4)\qquad \|f_4(z)\| \le (2 + \delta_1)^{l-t} \delta_1^{l_1 - l}.$$

14.6

Let $f_5(z) = \Phi(w)$ be the polynomial of degree $\le (t-1)$ defined by

$$(14.6.1)\qquad \Phi(\xi_s) = \frac{1}{(\eta_1 - \xi_s)^{m+N_1+1} F(\xi_s)}, \qquad s = 1,\ldots,t.$$

In other words,

$$(14.6.2)\qquad f_5(\eta_s) = \frac{1}{\eta_s^{m+N_1+1} f_4(\eta_s)}, \qquad s = 1,\ldots,t.$$

By (14.5.2) and (14.4.8) we have $f_4(\eta_s) \neq 0$ for these η, and

$$|\eta_s| = |\eta_1 - \xi_s| \geq 1 - |\xi_s| \geq 1 - |\xi_t| > 1 - x_1 \geq 1 - \delta_1 > 0$$

so (14.6.1) makes sense. Let us represent $\Phi(w)$ by

$$(14.6.3)\qquad \Phi(w) = e_0 + \sum_{\nu=1}^{t-1} e_\nu (w - \xi_1) \cdots (w - \xi_\nu).$$

Since the points ξ_s lie in the disk

$$|w| < x_1$$

and the function

$$\frac{1}{(\eta_1 - w)^{m+N_1+1} F(w)}$$

is regular here, Lemma 5.4 is applicable, and hence for $\nu = 0, 1, \ldots, t-1$

$$(14.6.4)\quad e_\nu = \frac{1}{2\pi i} \int_{|w|=x_1} \frac{dw}{(\eta_1 - w)^{m+N_1+1} F(w)(w - \xi_1) \cdots (w - \xi_{\nu+1})}.$$

Using (14.4.7) we have that on the circle $|w| = x_1$

$$(14.6.5)\qquad |F(w)(w - \xi_1) \cdots (w - \xi_{\nu+1})| \geq 2\left(\frac{\delta_1 - \delta_2}{4}\right)^l \delta_1^{l_1 - l},$$

and hence

$$(14.6.6)\qquad |e_\nu| \leq \frac{1}{2(1 - x_1)^{m+N_1+1}} \left(\frac{4}{\delta_1 - \delta_2}\right)^l \delta_1^{l - l_1}.$$

Since, from (14.6.3) and (14.5.1),

$$f_5(z) = e_0 + \sum_{\nu=1}^{t-1} e_\nu (\eta_1 - z) \cdots (\eta_\nu - z),$$

we have

$$(14.6.7)\qquad \|f_5(z)\| \le |e_0| + \sum_{\nu=1}^{t-1} |e_\nu|(1+|\eta_1|)\cdots(1+|\eta_\nu|)$$

$$\le \frac{\delta_1^{l-l_1}}{2(1-\delta_1)^{m+N_1+1}}\left(\frac{4}{\delta_1-\delta_2}\right)^l \sum_{\nu=0}^{t-1}(2+\delta_1)^\nu$$

$$\le \frac{\delta_1^{l-l_1}(2+\delta_1)^t}{2(1-\delta_1)^{m+N_1+1}}\left(\frac{4}{\delta_1-\delta_2}\right)^l .$$

14.7

Now let

$$(14.7.1)\qquad f_6(z) = z^{N_1} f_4(z) f_5(z) = \sum_{\nu=N_1}^{N_1+l-1} c_\nu^{(6)} z^\nu .$$

We can see that the coefficients of $f_6(z)$ are real. In fact, since $z_1 = -1$ and from the structure of $f_4(z)$ in (14.5.2) it follows that the coefficients of

$$(14.7.2)\qquad \left(\prod_{\nu=l_1+1}^{l} \xi_\nu\right) f_4(z)$$

are real. Write (14.6.2) in the form

$$\frac{f_5(\eta_s)}{\prod_{\nu=l_1+1}^{l} \xi_\nu} = \frac{1}{\eta_s^{m+N_1+1}\left\{\left(\prod_{\nu=l_1+1}^{l} \xi_\nu\right) f_4(\eta_s)\right\}};$$

since $z_1 = -1$ implies that the points $\eta_1, \dots, \eta_t$ form a point system symmetric to the real axis, it follows that the coefficients of

$$(14.7.3)\qquad \left(\prod_{\nu=l_1+1}^{l} \frac{1}{\xi_\nu}\right) f_5(z)$$

are real too. The assertion is proved by (14.7.2)–(14.7.3). Furthermore, (14.6.2) gives that

$$(14.7.4)\qquad \sum_{\nu=N_1}^{N_1+l-1} c_\nu^{(6)} \eta_j^\nu = \begin{cases} 0 & \text{for } j > t \\ \eta_j^{-m-1} & \text{for } j \le t, \end{cases}$$

and from (14.7.1), (14.5.4), and (14.6.7)

$$\|f_6(z)\| \le \|f_4(z)\|\|f_5(z)\| \le \frac{(2+\delta_1)^l}{2(1-\delta_1)^{m+N_1+1}}\left(\frac{4}{\delta_1-\delta_2}\right)^l. \tag{14.7.5}$$

Now define Ω in (14.3.6) by

$$\Omega = \frac{(2+\delta_1)^l}{(1-\delta_1)^{m+N_1+1}}\left(\frac{4}{\delta_1-\delta_2}\right)^l. \tag{14.7.6}$$

14.8

Finally we define $f_7(z)$ and $f_8(z)$ by

$$f_7(z) = f_2(z) + f_6(z) = \sum_{\nu=0}^{N_2+l-1} c_\nu^{(7)} z^\nu \tag{14.8.1}$$

$$f_8(z) = f_2(z) - f_6(z) = \sum_{\nu=0}^{N_2+l-1} c_\nu^{(8)} z^\nu. \tag{14.8.2}$$

Formulas (14.3.10) and (14.7.4) give

$$\sum_{\nu=0}^{N_2+l-1} c_\nu^{(7)} \eta_j^{m+1+\nu} = \begin{cases} 0 & \text{for } j > t \\ 1 & \text{for } j \le t \end{cases} \tag{14.8.3}$$

and

$$\sum_{\nu=0}^{N_2+l-1} c_\nu^{(8)} \eta_j^{m+1+\nu} = \begin{cases} 0 & \text{for } j > t \\ -1 & \text{for } j \le t. \end{cases} \tag{14.8.4}$$

We claim that

$$c_\nu^{(7)} \ge 0, \qquad c_\nu^{(8)} \ge 0, \quad \text{for all } \nu; \tag{14.8.5}$$

it is enough to show the first part. That they are real is shown by (14.3.7) and (14.7.2)–(14.7.3). The form of $f_0(z)$ in (14.7.1) and (14.3.7) holds for all ν unless

$$N_1 \le \nu \le N_1 + l - 1.$$

In this case, using the fact that the coefficients $c_\nu^{(6)}$ are real, (14.3.8), (14.7.5), and (14.7.6) show that

$$\begin{aligned} c_\nu^{(7)} &\ge \Omega - \left|\text{coeff } z^\nu \quad \text{in} \quad z^{N_1} f_4(z) f_5(z)\right| \\ &\ge \Omega - \|f_4(z)\|\|f_5(z)\| \ge \tfrac{1}{2}\Omega > 0. \end{aligned}$$

Define the index T by

$$(14.8.6)\quad \begin{aligned} 0 &= |z_1 - z_1| \le |z_2 - z_1| \le \cdots \le |z_{L_2} - z_1| \\ &\le \delta_2 < |z_{L_2+1} - z_1| \le \cdots \le |z_T - z_1| < x_1 \\ &< |z_{T+1} - z_1| \le \cdots \le |z_{L_1} - z_1| \le \delta_1 < |z_{L_1+1}| \le \cdots, \\ &\qquad (L_2 \le T \le L_1). \end{aligned}$$

Statements (14.8.3) and (14.8.4) give

$$(14.8.7)\qquad \sum_{\nu=0}^{N_2+l-1} c_\nu^{(7)} z_j^{m+\nu+1} = \begin{cases} 0 & \text{for } j > T \\ 1 & \text{for } j \le T, \end{cases}$$

respectively,

$$(14.8.8)\qquad \sum_{\nu=0}^{N_2+l-1} c_\nu^{(8)} z_j^{m+\nu+1} = \begin{cases} 0 & \text{for } j > T \\ -1 & \text{for } j \le T. \end{cases}$$

So,

$$\sum_{\nu=0}^{N_2+l-1} c_\nu^{(7)} \operatorname{Re} g(m+\nu+1) = \operatorname{Re} \sum_{j=1}^{T} b_j$$

and

$$\sum_{\nu=0}^{N_2+l-1} c_\nu^{(8)} \operatorname{Re} g(m+\nu+1) = -\operatorname{Re} \sum_{j=1}^{T} b_j.$$

Thus—by changing indices if it is necessary—we get the fundamental identities

$$(14.8.9)\qquad \begin{aligned} \sum_{\nu=0}^{N_2+l-1} c_\nu^{(7)} \operatorname{Re} g(m+\nu+1) &= \left|\operatorname{Re} \sum_{j=1}^{T} b_j\right| \\ \sum_{\nu=0}^{N_2+l-1} c_\nu^{(8)} \operatorname{Re} g(m+\nu+1) &= -\left|\operatorname{Re} \sum_{j=1}^{T} b_j\right|. \end{aligned}$$

From (14.8.5), (14.8.9), and from the inequality

$$(14.8.10)\qquad N_2 + l - 1 \le \frac{l}{2}\left(3 + \frac{\pi}{\kappa}\right) - 1 \le n\left(3 + \frac{\pi}{\kappa}\right) - 1,$$

which is a consequence of (14.3.5) and (14.2.2), we get

$$\max_{\substack{m+1\le\nu\le m+n(3+\pi/\kappa)\\ \nu\ \text{integer}}} \operatorname{Re} g(\nu) \ge \frac{\left|\operatorname{Re}\sum_{j=1}^{T} b_j\right|}{\sum_\nu c_\nu^{(7)}}, \tag{14.8.11}$$

respectively

$$\max_{\substack{m+1\le\nu\le m+n(3+\pi/\kappa)\\ \nu\ \text{integer}}} \operatorname{Re} g(\nu) \le -\frac{\left|\operatorname{Re}\sum_{j=1}^{T} b_j\right|}{\sum_\nu c_\nu^{(8)}}. \tag{14.8.12}$$

14.9

Now we can quickly complete the proof of Theorem 14.1. Since (14.7.6), (14.3.9), (14.7.5), and (14.3.2) yield

$$\sum_\nu c_\nu^{(7)} = \sum_\nu c_\nu^{(2)} + \sum_\nu c_\nu^{(6)} \le \sum_\nu c_\nu^{(2)} + \|f_6(z)\|$$

$$\le \Omega\left(l \cdot 2^l + \frac{1}{2}\right) \le (2n+1)\left(\frac{24}{\delta_1-\delta_2}\right)^{2n}\left(\frac{1}{1-\delta_1}\right)^{m+n(1+\pi/\kappa)+1},$$

and thus from (14.1.4),

$$\le 5n\left(\frac{24e}{\delta_1-\delta_2}\right)^{2n},$$

we get

$$\max_{\substack{m+1\le\nu\le m+n(3+\pi/\kappa)\\ \nu\ \text{integer}}} \operatorname{Re} g(\nu) \ge \frac{1}{5n}\left(\frac{\delta_1-\delta_2}{24e}\right)^{2n}\left|\operatorname{Re}\sum_{j=1}^{T} b_j\right|.$$

By the ordering (14.8.6), this completes the proof of the first part of Theorem 14.1; the second part follows similarly. ■

14.10

We prove a second theorem corresponding to the ordering

$$|z_1| \ge \cdots \ge |z_k| = 1 \ge \cdots \ge |z_n|. \tag{14.10.1}$$

Suppose again the argument-condition (14.1.1) holds; we may suppose that the z_j are distinct. Let m be an integer such that

$$m \geq 2n \tag{14.10.2}$$

and

$$\Delta = \frac{4n}{m + n\left(7 + \dfrac{\pi}{\kappa}\right)}. \tag{14.10.3}$$

Let

$$\begin{aligned} \cdots \geq |z_{l_1}| \geq 1 + \Delta > |z_{l_1+1}| \geq \cdots \geq |z_k| = 1 \geq \cdots \\ \geq |z_{l_2}| \geq 1 - \Delta > |z_{l_2+1}| \geq \cdots. \end{aligned} \tag{14.10.4}$$

Theorem 14.2. *Under the previous conditions and for suitable integers ν_1 and ν_2 with*

$$m + 1 \leq \nu_1, \nu_2 \leq m + n\left(3 + \frac{\pi}{\kappa}\right)$$

the inequalities

$$\operatorname{Re} g(\nu_1) \geq \frac{1}{4n}\left\{\frac{n}{3e\left[m + n\left(7 + \dfrac{\pi}{\kappa}\right)\right]}\right\}^{4n} \min\left|\operatorname{Re} \sum_{j=\mu_1}^{\mu_2} b_j\right|$$

and

$$\operatorname{Re} g(\nu_2) \leq -\frac{1}{4n}\left\{\frac{n}{3e\left[m + n\left(7 + \dfrac{\pi}{\kappa}\right)\right]}\right\}^{4n} \min\left|\operatorname{Re} \sum_{j=\mu_1}^{\mu_2} b_j\right|$$

hold. Here each min *is extended over all the pairs (μ_1, μ_2) such that*

$$l_1 + 1 \leq \mu_1 \leq \mu_2 \leq l_2.$$

14.11

The proof of this theorem contains several steps like those in the previous theorem; these are only sketched. Again let the numbers $\eta_1, \ldots, \eta_l$ be those that comprise a maximal set of distinct values among the quantities

$$z_1, z_2, \ldots, z_n, \bar{z}_1, \ldots, \bar{z}_n$$

and again suppose that

(14.11.1) $$n \le l \le 2n.$$

The indices L_1, L_2, K are defined by the inequality

(14.11.2) $$|\eta_1| \ge \cdots \ge |\eta_{L_1}| \ge 1 + \Delta > |\eta_{L_1+1}| \ge \cdots \ge |\eta_K| = 1$$
$$\ge \cdots \ge |\eta_{L_2}| \ge 1 - \Delta > |\eta_{L_2+1}| \ge \cdots \qquad (\eta_K = z_k).$$

The polynomials $h_0(z)$, $B_1(z)$, $B_2(z)$, $\psi_1(z)$, $\psi_2(z)$, $f_1(z)$, and $f_2(z)$ are defined as in 14.3, and only Ω is replaced by Ω_1, to be determined later. All inequalities in 14.3 are valid, as before. Let

(14.11.3) $$f_3(z) = \prod_{j=1}^{L_1} \left(1 - \frac{z}{|\eta_j|}\right) \prod_{j=L_1+1}^{L_2} (z - |\eta_j|).$$

We apply (5.1.8) to $f_3(z)$ in $[1, 1 + \Delta]$. This gives (as in 14.4) an x_1 satisfying

(14.11.4) $$1 < x_1 \le 1 + \Delta$$

so that

(14.11.5) $$f_3(x_1) \ge 2\left(\frac{\Delta}{4}\right)^{L_2} \frac{1}{(1 + \Delta)^{L_1}},$$

since

$$\left|\frac{2|\eta_j| - (2 + \Delta)}{\Delta}\right| = \frac{|\eta_j|}{\Delta}\left(2 - \frac{2 + \Delta}{|\eta_j|}\right) \ge \frac{|\eta_j|}{1 + \Delta} \ge 1 \quad \text{for} \quad 1 \le j \le L_1.$$

Define the index h_1 by

(14.11.6) $$|\eta_{h_1}| > x_1 > |\eta_{h_1+1}|$$

and let

(14.11.7) $$f_4(z) = \prod_{j=h_1+1}^{l} (z - |\eta_j|).$$

Applying (5.1.2) to $f_4(z)$ in $[1 - \Delta, 1]$, we obtain an x_2 satisfying

(14.11.8) $$1 - \Delta \le x_2 < 1$$

so that

(14.11.9) $$|f_4(x_2)| \ge 2\left(\frac{\Delta}{4}\right)^{l-h_1}.$$

Define the index h_2 by

$$|\eta_{h_2}| > x_2 > |\eta_{h_2+1}|. \tag{14.11.10}$$

14.12

We actually need certain subproducts of $f_3(x_1)$ and $f_4(x_2)$. Since we have

$$|x_2 - |\eta_j|| \le 1 + \Delta \quad \text{for} \quad h_1 < j \le h_2$$

and

$$|x_2 - |\eta_j|| \le 1 \quad \text{for} \quad j > h_2$$

inequality (14.11.9) shows that for each $\{i_k\}_{k=1,\ldots,s}$ with

$$h_1 + 1 \le i_1 < \cdots < i_s \le h_2, \tag{14.12.1}$$

$$\left| \prod_{\nu=1}^{s} (x_2 - |\eta_{i_\nu}|) \prod_{j=h_2+1}^{l} (x_2 - |\eta_j|) \right| \ge 2\left(\frac{\Delta}{4}\right)^{l-h_1} \frac{1}{(1+\Delta)^{h_2-h_1-s}}.$$

Then on the circle $|z| = x_2$ the inequalities

$$\left| \prod_{\nu=1}^{s} (z - \eta_{i_\nu}) \prod_{j=h_2+1}^{l} (z - \eta_j) \right| \ge 2\left(\frac{\Delta}{4}\right)^{l-h_1} (1+\Delta)^{s+h_1-h_2} \tag{14.12.2}$$

hold for all choices i_k in (14.12.1).

14.13

The corresponding inequalities for $f_3(x_1)$ need a little more attention. For

$$L_1 < j \le h_1$$

we have

$$\left| 1 - \frac{x_1}{|\eta_j|} \right| = \frac{|x_1 - |\eta_j||}{|\eta_j|} \ge \frac{|x_1 - |\eta_j||}{1+\Delta},$$

and for

$$h_1 + 1 \le j_1 < \cdots < j_s \le L_2 \tag{14.13.1}$$

we have

$$|x_1 - |\eta_{j_\nu}|| \le 1 + \Delta.$$

Hence for each system in (14.13.1), (14.11.5) yields the inequalities

$$\left|\prod_{j=1}^{h_1}\left(1-\frac{x_1}{|\eta_j|}\right)\prod_{\nu=1}^{s}\left(x_1-|\eta_{j_\nu}|\right)\right|$$
$$\geq 2\left(\frac{\Delta}{4}\right)^{L_2}\frac{1}{(1+\Delta)^{L_1}}\cdot\frac{1}{(1+\Delta)^{L_2-h_1-s}}\cdot\frac{1}{(1+\Delta)^{h_1-L_1}}.$$

Hence on the circle $|z| = x_1$ the inequalities

$$\left|\prod_{j=1}^{h_1}\left(1-\frac{z}{\eta_j}\right)\prod_{\nu=1}^{s}\left(z-\eta_{j_\nu}\right)\right| \geq 2\left(\frac{\Delta}{4}\right)^{L_2}\frac{1}{(1+\Delta)^{L_2-s}} \tag{14.13.2}$$

hold for all choices in (14.13.1).

14.14

We shall need an upper bound on the circle $|z| = x_2$ for the polynomial $f_5(z)$ of degree at most $(h_2 - h_1 - 1)$ defined by

$$f_5(\eta_s) = \prod_{j=1}^{h_1}\frac{1}{1-\dfrac{\eta_s}{\eta_j}}, \qquad s = h_1+1,\ldots,h_2. \tag{14.14.1}$$

The structure of the η and (14.11.10) show that the coefficients of $f_5(z)$ are real. Representing $f_5(z)$ in the form

$$f_5(z) = e_0 + \sum_{\nu=1}^{h_2-h_1-1} e_\nu(z-\eta_{h_1+1})\cdots(z-\eta_{h_1+\nu}) \tag{14.14.2}$$

Lemma 5.4 is applicable with $|z| = x_1$ as L-curve and

$$h(z) = \prod_{j=1}^{h_1}\frac{1}{1-\dfrac{z}{\eta_j}}.$$

Hence for $\nu = 0, 1, \ldots, (h_2 - h_1 - 1)$ we have

$$e_\nu = \frac{1}{2\pi i}\int_{|w|=x_1}\frac{dw}{\left\{\prod_{j=1}^{h_1}\left(1-\dfrac{w}{\eta_j}\right)\right\}(w-\eta_{h_1+1})\cdots(w-\eta_{h_1+\nu+1})}.$$

Then apply (14.13.2) and obtain

$$|e_\nu| \le \frac{1}{2}\left(\frac{4}{\Delta}\right)^{L_2}(1+\Delta)^{L_2-\nu-1}(1+\Delta). \tag{14.14.3}$$

Hence from (14.14.2) and for $|z| = x_2$ we have

$$\begin{aligned} |f_5(z)| &\le \frac{1}{2}\left(\frac{4}{\Delta}\right)^{L_2}(1+\Delta)^{L_2} \\ &\quad\cdot\left(1+\sum_{\nu=1}^{h_2-h_1-1}\frac{\left(x_2+|\eta_{h_1+1}|\right)\cdots\left(x_2+|\eta_{h_1}+\nu|\right)}{(1+\Delta)^\nu}\right) \\ &\le \frac{1}{2}\left(\frac{4}{\Delta}\right)^{L_2}(1+\Delta)^{L_2-h_2+h_1+1}(2+\Delta)^{h_2-h_1}. \end{aligned} \tag{14.14.4}$$

Next define $f_6(z)$ by

$$f_6(z) = \left\{\prod_{j=1}^{h_1}\left(1-\frac{z}{\eta_j}\right)\right\}\prod_{j=h_2+1}^{l}(z-\eta_j). \tag{14.14.5}$$

The coefficients of $f_6(z)$ are obviously real. Furthermore, we have

$$\|f_6(z)\| \le \left\{\prod_{j=1}^{h_1}\left(1+\frac{1}{|\eta_j|}\right)\right\}\prod_{j=h_2+1}^{l}(1+|\eta_j|) \le 2^{l-h_2+h_1}. \tag{14.14.6}$$

14.15

Now consider the polynomial $f_7(z)$ [of degree at most $(h_2 - h_1 - 1)$] defined by

$$f_7(\eta_s) = \frac{1}{\eta_s^{m+N_1+1}f_6(\eta_s)}, \qquad s = h_1+1,\ldots,h_2. \tag{14.15.1}$$

Obviously $f_7(z)$ has only real coefficients. Writing $f_7(z)$ in the form of a Newton-interpolation polynomial we observe that neither of the Nörlund formulas can be used for the estimation of its coefficients. However, we can rewrite $f_7(\eta_s)$ as

$$f_7(\eta_s) = \frac{f_5(\eta_s)}{\eta_s^{m+N_1+1}\prod_{j=h_2+1}^{l}(\eta_s-\eta_j)}, \qquad j = h_1+1,\ldots,h_2 \tag{14.15.2}$$

and then apply Lemma 5.3 with $|w| = x_2$ and

$$h(w) = \frac{f_5(w)}{w^{m+N_1+1} \prod_{j=h_2+1}^{l} (w - \eta_j)}.$$

Note that $h(w)$ is regular for $|w| \geq x_2$ and by (14.10.2) tends to zero as $w \to \infty$. Hence rewriting

$$\text{(14.15.3)} \quad f_7(z) = e_0^* + \sum_{\nu=1}^{h_2-h_1-1} e_\nu^*(z - \eta_{h_1+1}) \cdots (z - \eta_{h_1+\nu})$$

we have

(14.15.4)

$$e_\nu^* = -\frac{1}{2\pi i}$$

$$\cdot \int_{|w|=x_2} \frac{f_5(w)\, dw}{\left\{ w^{m+N_1+1} \prod_{j=h_2+1}^{l} (w - \eta_j) \right\} (w - \eta_{h_1+1}) \cdots (w - \eta_{h_1+\nu+1})}.$$

Now we apply inequalities (14.12.2), (14.14.4), and (14.11.8). They give

$$\text{(14.15.5)} \quad |e_\nu^*| \leq \frac{1}{4(1-\Delta)^{m+N_1}} \left(\frac{4}{\Delta} \right)^{l+L_2-h_1} (2+\Delta)^{h_2-h_1} (1+\Delta)^{L_2-\nu}.$$

Furthermore, we have

$$\|f_7(z)\| \leq |e_0^*| + \sum_{\nu=1}^{h_2-h_1-1} |e_\nu^*| \big(1 + |\eta_{h_1+1}|\big) \cdots \big(1 + |\eta_{h_1+\nu}|\big)$$

$$\leq \frac{1}{4(1-\Delta)^{m+N_1}} \left(\frac{4}{\Delta} \right)^{l+L_2-h_1} (2+\Delta)^{h_2-h_1} (1+\Delta)^{L_2}$$

$$\cdot \left(\sum_{\nu=0}^{h_2-h_1-1} \left(\frac{2+\Delta}{1+\Delta} \right)^{\nu} \right)$$

or

$$\text{(14.15.6)} \quad \|f_7(z)\| \leq \frac{1}{2(1-\Delta)^{m+N_1}} \left(\frac{4}{\Delta} \right)^{l+L_2-h_1} (2+\Delta)^{2(h_2-h_1)}$$

$$\cdot (1+\Delta)^{L_2-h_2+h_1}.$$

14.16

Let

$$f_8(z) = z^{N_1} f_6(z) f_7(z) = \sum_{\nu+N_1}^{N_1+l-1} c_\nu^{(8)} z^\nu. \tag{14.16.1}$$

Then the coefficients of $f_8(z)$ are real, and (14.15.1) implies that

$$\sum_{\nu=N_1}^{N_1+l-1} c_\nu^{(8)} \eta_j^\nu = \begin{cases} \eta_j^{-m-1} & \text{for} \quad h_1+1 \le j \le h_2 \\ 0 & \text{otherwise.} \end{cases} \tag{14.16.2}$$

Also, by (14.14.6) and (14.15.6),

$$\begin{aligned} \|f_8(z)\| &\le \|f_6(z)\| \|f_7(z)\| \\ &\le \frac{1}{2} \frac{2^{l-h_2+h_1}}{(1-\Delta)^{m+N_1}} \left(\frac{4}{\Delta}\right)^{l+L_2-h_1} (2+\Delta)^{2(h_2-h_1)} \\ &\quad \cdot (1+\Delta)^{L_2-h_2+h_1}. \end{aligned} \tag{14.16.3}$$

14.17

Finally set

$$\begin{aligned} f_9(z) &= f_2(z) + f_8(z) = \sum_{\nu=0}^{N_2+l-1} c_\nu^{(9)} z^\nu \\ f_{10}(z) &= f_2(z) - f_8(z) = \sum_{\nu=0}^{N_2+l-1} c_\nu^{(10)} z^\nu. \end{aligned} \tag{14.17.1}$$

Choosing

$$\Omega_1 = \frac{2^{l-h_2+h_1}}{(1-\Delta)^{m+N_1}} \left(\frac{4}{\Delta}\right)^{l+L_2-h_1} (2+\Delta)^{2(h_2-h_1)} (1+\Delta)^{L_2-h_2+h_1} \tag{14.17.2}$$

we can prove, as in 14.8, that

$$c^{(9)} \ge 0, \qquad c^{(10)} \ge 0. \tag{14.17.3}$$

Define the indices T_1 and T_2 by

$$|z_{T_1}| > x_1 > |z_{T_1+1}|$$

$$|z_{T_2}| > x_2 > |z_{T_2+1}|. \tag{14.17.4}$$

We get, as in 14.8, that

$$\sum_{\nu=0}^{N_2+l-1} c_\nu^{(9)} z_j^{m+1+\nu} = \begin{cases} 1 & \text{for } T_1+1 \le j \le T_2 \\ 0 & \text{otherwise} \end{cases}$$

$$\sum_{\nu=0}^{N_2+l-1} c_\nu^{(10)} z_j^{m+1+\nu} = \begin{cases} -1 & \text{for } T_1+1 \le j \le T_2 \\ 0 & \text{otherwise} \end{cases}$$

and hence

$$\sum_{\nu=0}^{N_2+l-1} c_\nu^{(9)} \operatorname{Re} g(m+\nu+1) = \operatorname{Re} \sum_{j=T_1+1}^{T_2} b_j \tag{14.17.5}$$

$$\sum_{\nu=0}^{N_2+l-1} c_\nu^{(10)} \operatorname{Re} g(m+\nu+1) = -\operatorname{Re} \sum_{j=T_1+1}^{T_2} b_j.$$

To complete the proof we need an upper bound for $\Sigma c_\nu^{(9)}$. By (14.3.9), (14.16.3), and (14.17.2) the quantity

$$2l \frac{2^{2l-h_2+h_1}}{(1-\Delta)^{m+N_1}} \left(\frac{4}{\Delta}\right)^{l+L_2-h_1} (2+\Delta)^{2(h_2-h_1)} (1+\Delta)^{L_2-h_2+h_1}$$

$$< \left[\frac{(2+\Delta)^2}{2(1+\Delta)}\right]^l 4n \frac{4^l(1-\Delta)^{2n}}{(1-\Delta)^{m+n(3+\frac{\pi}{\kappa})}} \left(\frac{16}{\Delta^2}\right)^l (1+\Delta)^l$$

$$< 4n \left[3e \frac{m + n\left(7 + \dfrac{\pi}{\kappa}\right)}{n}\right]^{4n}$$

is an upper bound. Then observe that

$$\left|\operatorname{Re} \sum_{j=T_1+1}^{T_2} b_j\right| \ge \min \left|\operatorname{Re} \sum_{j=\mu_1}^{\mu_2} b_j\right|$$

where the min ranges over all pairs (μ_1, μ_2) such that

$$l_1 + 1 \le \mu_1 \le \mu_2 \le l_2.$$

■

15

DIRECT CONNECTIONS WITH THE CLASSICAL THEORY OF DIOPHANTINE APPROXIMATIONS

15.1

Most of the two-sided problems we have investigated thus far have the following form. The coefficients b_j were fixed, the vectors z_j varied over a prescribed domain Z, the range S of the integers ν was chosen in advance, and then we sought the minimax $|g(\nu)|$. We deal now with a particular problem of this type. Let N and w be fixed positive integers, and assume

(15.1.1) $$0 < \eta < \frac{1}{2}$$

and

(15.1.2) $$n = (2w + 1)^N.$$

If $j_1, j_2, \ldots, j_N$ run independently through the integers in the interval $[-w, w]$, we order the $(j_1, \ldots, j_N)$ systems in some way and define the coefficients $b_1, \ldots, b_n$ by

(15.1.3) $$b_j = \prod_{l=1}^{N} \eta \left(\frac{\sin (j_l \pi \eta)}{j_l \pi \eta} \right)^2.$$

(If $j_l = 0$, the last factor $= 1$.) Setting

(15.1.4) $$[1/\eta^N] = K$$

we choose for the range S^* the interval

$$1 \le \nu \le K. \tag{15.1.5}$$

The domain Z^* of the vector $(z_1,\ldots,z_n)$ is given by

$$z_j = \exp\{2\pi i(j_1\vartheta_1 + \cdots + j_N\vartheta_N)\}, \qquad j = 1,\ldots,n; \tag{15.1.6}$$

$(\vartheta_1,\ldots,\vartheta_N$ are real). We want to get a lower bound for

$$\min_{(z_1,\ldots,z_n)\in Z^*} \max_{\nu\in S^*} |g(\nu)|$$

in terms of our parameters η, N, w.

15.2

With this objective in mind consider the expression

$$V = \sum_{\substack{\nu=-K \\ \nu\neq 0}}^{K} \left(1 - \frac{|\nu|}{K+1}\right) g(\nu). \tag{15.2.1}$$

Using the abbreviation

$$\eta\left(\frac{\sin(l\pi\eta)}{l\pi\eta}\right)^2 = a_l, \tag{15.2.2}$$

V can be written as

(15.2.3)

$$V = \sum_{j_1=-w}^{w} \cdots \sum_{j_N=-w}^{w} a_{j_1} \cdots a_{j_N} \sum_{\substack{\nu=-K \\ \nu\neq 0}}^{K} \left(1 - \frac{|\nu|}{K+1}\right) \exp\{2\pi i\nu(j_1\vartheta_1 + \cdots + j_N\vartheta_N$$

The contribution of the term corresponding to the $j_1 = \cdots = j_N = 0$ is obviously equal to

$$a_0^N K.$$

The contribution from the other $(j_1,\ldots,j_N)$ systems is

$$\ge a_0^N - \sum_{j_1=-w}^{w} \cdots \sum_{j_N=-w}^{w} a_{j_1} \cdots a_{j_N} = a_0^N - \left(\sum_{j=-w}^{w} a_j\right)^N,$$

by the following classical inequality of Fejér:

$$\sum_{\substack{j=-k \\ j\neq 0}}^{k} \left(1 - \frac{|j|}{k+1}\right) e^{2\pi i j x} \geq -1 \qquad \text{(for real } x\text{)}$$

and the fact that the a_j are nonnegative. Hence

$$V \geq (K+1)a_0^N - \left(\sum_{j=-\infty}^{\infty} a_j\right)^N. \tag{15.2.4}$$

Using the fact that for

$$F_0(x) = \begin{cases} 0 & \text{for} \quad \eta \leq |x| \leq \frac{1}{2} \\ 1 - \dfrac{|x|}{\eta} & \text{for} \quad |x| \leq \eta \end{cases} \tag{15.2.5}$$

which can be continued periodically, the representation

$$F_0(x) = \sum_{j=-\infty}^{\infty} a_j \exp(2\pi i j x) \tag{15.2.6}$$

holds, and we get immediately from (15.2.4) that

$$V \geq (K+1)\eta^N - 1 \tag{15.2.7}$$

independent of w. Hence from (15.2.1) we get that

$$[1/\eta^N] \max_{1 \leq |\nu| \leq K} |g(\nu)| \geq (1 + [1/\eta^N])\eta^N - 1$$

or rather

$$[1/\eta^N] \max_{1 \leq \nu \leq K} |g(\nu)| \geq (1 + [1/\eta^N])\eta^N - 1. \tag{15.2.8}$$

15.3

Hence we have a *positive* lower bound for $\max_{\nu \in S^*} |g(\nu)|$ independent not only of the ϑ_j but also of w. The range S^* does not depend on w either but the power sum $g(\nu)$ does; so we write (15.2.8) in the form

$$\max_{\substack{\nu \in S^* \\ \nu \text{ integer}}} |g(\nu, w)| \geq c = c(N, \eta), \tag{15.3.1}$$

where the right side is positive. From the point of view of the applications, the estimate (15.2.8) does what is required of it, nevertheless, consider what happens as $w \to \infty$. Since ν can take on only finitely many values, there is a fixed integer ν_0 with

$$1 \leq \nu_0 \leq K \tag{15.3.2}$$

so that for a suitable sequence

$$w_1 < w_2 < \cdots \to \infty$$

all inequalities

$$|g(\nu_0, w_j)| \geq c(N, \eta) \qquad j = 1, 2, \ldots \tag{15.3.3}$$

hold. It follows at once from (15.2.5), (15.2.6), and (15.3.3) that

$$F_0(\nu_0\vartheta_1)F_0(\nu_0\vartheta_2)\cdots F_0(\nu_0\vartheta_N) \geq c(N, \eta) > 0. \tag{15.3.4}$$

But by (15.2.5) this implies that the inequalities

$$\{\vartheta_j\nu_0\} < \eta$$

hold* simultaneously for $j = 1, 2, \ldots, N$ if ν_0 is a suitable integer with $1 \leq \nu_0 \leq [1/\eta^N]$. For $\eta = 1/q$ (q an integer ≥ 3) this is Dirichlet's theorem, but this result holds more generally for all $0 < \eta < \frac{1}{2}$.

15.4

Next let

$$\alpha_1 = 0, \qquad \alpha_2, \ldots, \alpha_n \quad \text{be real and fixed } (n \geq 2) \tag{15.4.1}$$

and let

$$b_j = e^{-2\pi i \alpha_j} \qquad j = 1, \ldots, n. \tag{15.4.2}$$

Let $\eta(x)$ be monotonically decreasing and tending to 0 for $x \to +\infty$. Let $(\lambda_2, \ldots, \lambda_n)$ run over all systems for which the inequality

$$\{d_2\lambda_2 + \cdots + d_n\lambda_n\} \geq \eta(x), \qquad x > n \tag{15.4.3}$$

holds for all integral d_ν satisfying

$$|d_\nu| \leq x, \qquad \nu = 2, 3, \ldots, n, \tag{15.4.4}$$

*The expression $\{x\}$ means the distance of x from the nearest integer.

but excluding $d_2 = d_3 = \cdots = d_n = 0$. The domain Z^{**} now consists of the systems $(z_1,\ldots,z_n)$ defined by

$$(z_1,\ldots,z_n) = \left(1, e^{2\pi i\lambda_2},\ldots,e^{2\pi i\lambda_n}\right). \tag{15.4.5}$$

Let ε ($\le \frac{1}{2}$) be an arbitrarily small, fixed positive number and let $k_0 = k_0(n, \varepsilon)$ be the smallest positive integer such that the inequality

$$\frac{n^2 \log 2x}{2x} \le 4\varepsilon^2 \tag{15.4.6}$$

holds for $x \ge k_0$. Obviously

$$k_0 > n. \tag{15.4.7}$$

The range S^{**} is the interval

$$0 \le \nu \le \frac{(2k_0)^n}{\eta(k_0)} = w. \tag{15.4.8}$$

Then we must investigate

$$\inf_{Z^{**}} \max_{\substack{\nu \in S^{**} \\ \nu \text{ integer}}} |g(\nu)|. \tag{15.4.9}$$

15.5

To do this, we investigate

$$U = \frac{1}{w} \sum_{\nu=0}^{w-1} |g(\nu)|^{2k_0}. \tag{15.5.1}$$

On one hand,

$$U \le \max_{\substack{\nu \in S^{**} \\ \nu \text{ integer}}} |g(\nu)|^{2k_0}. \tag{15.5.2}$$

On the other, with the notation

$$c_j = \frac{k_0!}{j_1! \cdots j_n!} e^{-2\pi i(j_2\alpha_2 + \cdots + j_n\alpha_n)}, \qquad j_2\lambda_2 + \cdots + j_n\lambda_n = \mu_j \tag{15.5.3}$$

we have

$$g(\nu)^{k_0} = \sum_{j_1 + \cdots + j_n = k_0} c_j e^{2\pi i \nu \mu_j}, \tag{15.5.4}$$

that is,

$$U = \sum_{\substack{j_1 + \cdots + j_n = k_0 \\ j_1' + \cdots + j_n' = k_0}} c_j \bar{c}_{j'} \frac{1}{w} \sum_{\nu=0}^{w-1} e^{2\pi i \nu(\mu_j - \mu_{j'})}. \tag{15.5.5}$$

Hence

$$U \geq \sum_{j_1 + \cdots + j_n = k_0} |c_j|^2 - \frac{1}{w} \sum_{j \neq j'}\sum |c_j c_{j'}| \frac{1}{|\sin \pi(\mu_j - \mu_{j'})|}.$$

The last factor is, by (15.4.3), (15.4.4), and (15.4.7),

$$\leq \frac{1}{2\{\mu_j - \mu_{j'}\}} \leq \frac{1}{2\eta(k_0)}.$$

Taking into account that

$$\sum_{j_1 + \cdots + j_n = k_0} |c_j| = n^{k_0}$$

we get

$$U \geq \sum_{j_1 + \cdots + j_n = k_0} |c_j|^2 - \frac{n^{2k_0}}{2w\eta(k_0)}. \tag{15.5.6}$$

Apply Cauchy's inequality to obtain

$$\sum_{j_1 + \cdots + j_n = k_0} |c_j|^2 \geq \frac{1}{\binom{k_0 + n - 1}{n - 1}} n^{2k_0} > n^{2k_0}(k_0 + n)^{-n+1} > 2\frac{n^{2k_0}}{(2k_0)^n},$$

and we get, from (15.5.2), (15.5.6), (15.4.8), and (15.4.6),

$$\max_{\substack{\nu \in S^{**} \\ \nu \text{ integer}}} |g(\nu)| \geq n(2k_0)^{-n/2k_0} > n - n^2 \frac{\log 2k_0}{2k_0} \geq n - 4\varepsilon^2. \tag{15.5.7}$$

15.6

This inequality already produces a lower bound for our inf max in (15.4.9). But we use it to draw another conclusion. Equation (15.5.7) implies that for a suitable integer ν_0 with

$$\nu_0 \in S^{**}$$

the inequality

$$\left|1 + \sum_{j=2}^{n} e^{2\pi i(\lambda_j \nu_0 - \alpha_j)}\right| \geq n - 4\varepsilon^2 \tag{15.6.1}$$

holds.

But this gives, for $2 \leq j \leq n$,

$$|1 + e^{2\pi i(\lambda_j \nu_0 - \alpha_j)}| + (n - 2) \geq n - 4\varepsilon^2$$

$$|\cos \pi(\lambda_j \nu_0 - \alpha_j)| \geq 1 - 2\varepsilon^2$$

$$\cos^2 \pi(\lambda_j \nu_0 - \alpha_j) \geq 1 - 4\varepsilon^2$$

$$2\{\lambda_j \nu_0 - \alpha_j\} \leq |\sin \pi(\lambda_j \nu_0 - \alpha_j)| \leq 2\varepsilon.$$

So for each $j = 2, \ldots, n$

$$\{\lambda_j \nu_0 - \alpha_j\} \leq \varepsilon. \tag{15.6.2}$$

Let

$$1, \lambda_2, \ldots, \lambda_n \tag{15.6.3}$$

be linearly independent over the rationals. Then one can determine a suitable $\eta(x)$; in this way we obtain Kronecker's classical theorem.

15.7

Thus the approach that motivates this book also yields proofs for the two best known theorems of the classical theory of diophantine approximations; these proofs appear here as limiting cases. These are, in our opinion, not inferior to the arithmetical proofs in elegance and clarity. One is due to H. L. Montgomery (see [1] which is somewhat stronger than a previous proof of Cassels in Cassels [1]), the other to H. Bohr [3] except for some slight adjustments. Bohr's proof for Minkowski's theorem, as well as the proofs of Siegel [1] and [2] and of Mordell [1], shows that the idea to formulate and treat problems of diophantine approximations in terms of exponential sums is by no means new. This point of view was already present in H. Weyl's theory on uniform distribution mod 1 as early as 1913 (which was "finalized" in our paper [1] with Erdős). The essential novelty of this method in the theory of diophantine approximations seems to be the systematic investigation of problems concerning exponential sums, treating them not just as tools but as part of the theory itself.

16

SOME OPEN PROBLEMS IN CONNECTION WITH PART I

This collection of problems was arranged in accordance with the arrangement of Part I. Sometimes we state as a separate problem a special case of another; this means that the more general problem seems to be very difficult but that the special case would also be useful. In contrast, several of the problems are really a complex of problems.

PROBLEM 1

For a finite set S of integers and two generalized power sums

$$g_1(\nu) = \sum_{j=1}^{n} b_j' z_j'^{\nu}$$

$$g_2(\nu) = \sum_{j=1}^{n} b_j'' z_j''^{\nu}$$

with fixed coefficients, what is

$$\inf_{z_j} \max_{\nu \in S} \left(\frac{|g_1(\nu)|}{M_2(\nu)}, \frac{|g_2(\nu)|}{M_2(\nu)} \right)$$

and what the extremal systems?

PROBLEM 2

If S_1 and S_2 are finite sets of integers and

$$h(\mu, \nu) = \sum_{j=1}^{n} b_j z_j^{\nu} w_j^{\mu}$$

then for fixed b_j, what is

$$\inf_{z_j, w_j} \max_{\substack{\nu \in S_1 \\ \mu \in S_2}} \frac{h(\mu, \nu)}{\max_j |z_j|^\nu |w_j|^\mu}$$

and what are the extremal systems?

PROBLEM 3

With the notation (3.1.2) what is

$$\inf_{z_j} \max_{\substack{1 \le \nu \le n \\ \nu \text{ integer}}} |s_\nu|$$

when in addition to (3.1.3) $s_1 = 0$ is also required?

PROBLEM 4

What is

$$\inf_{z_j} \max_{\substack{1 \le \nu \le n \\ \nu \text{ integer}}} |s_\nu|$$

if $\max_j |z_j| = 1$ and $s_1 = 0$?

PROBLEM 5 (P. ERDŐS).

Under the minimum norm what is

$$\inf_{z_j} \max_{\substack{2 \le \nu \le n+1 \\ \nu \text{ integer}}} |s_\nu| ?$$

PROBLEM 6

What is the *exact* value of

$$\inf_{z_j} \max_{\substack{2 \le \nu \le n+1 \\ \nu \text{ integer}}} |s_\nu|$$

if $\max_j |z_j| = 1$?

PROBLEM 7

Under the assumption $\max_j |z_j| = 1$ is

$$\inf_{z_j} \max_{\substack{2 \le \nu \le n+2 \\ \nu \text{ integer}}} |s_\nu|$$

exponentially small?

PROBLEM 8

Give an upper bound for

$$\inf_{z_j} \max_{\substack{4 \le \nu \le n+3 \\ \nu \text{ integer}}} |s_\nu|$$

based on Uchiyama's theorem in 4.6 (with $\max_j |z_j| = 1$, of course).

PROBLEM 9

Improve M. Szalay's upper bound (3.7.2).

PROBLEM 10

If $\min_j |z_j| = 1$ then what is

$$\inf_{z_j} \max_{\substack{1 \le \nu \le n^2 \\ \nu \text{ integer}}} |s_\nu|?$$

PROBLEM 11

Does $\max_j |z_j| = 1$ imply that $\sum_{\nu=1}^{n} |s_\nu|^2 > cn$ with a constant c independent of n?*

PROBLEM 12

Find the best possible constant in Theorem 3.4, by Atkinson, to replace the constant $\frac{1}{6}$.

PROBLEM 13

What is

$$\inf_{z_j} \max_{\substack{1 \le \nu \le 2n-1 \\ \nu \text{ integer}}} |s_\nu|$$

if

$$\min_j |z_j| = 1.$$

PROBLEM 14

What is the sharp form of (3.4.6)?

*This conjecture has been disproved by M. Szalay [2]. For $\max_j |z_j| = 1$, $n > n_0$ he proved that

$$\tfrac{1}{5}\log^2 n < \min_{z_j} \sum_{\nu=1}^{n} |s_\nu|^2 < \log^2 n.$$

PROBLEM 15

Does there exist a positive Θ such that $\Theta < 1$ and

$$\inf_{z_j} \max_{\substack{1 \le \nu \le n \\ \nu \text{ integer}}} |s_\nu| \le \Theta$$

if $\max_j |z_j| = 1$?

PROBLEM 16

What is

$$\inf_{z_j} \max_{\substack{1 \le \nu \le n \\ \nu \text{ integer}}} \left(\frac{|s_\nu|}{n} \right)^{1/\nu}$$

if not only $\max_j |z_j| = 1$ but $s_1 = 0$ also is required?

PROBLEM 17

Show that for arbitrarily small $\varepsilon > 0$ there is an $m_0(\varepsilon, n)$ such that for every integer $m > m_0(\varepsilon, n)$ there exists a system $(z_1^*, \ldots, z_n^*)$ with $\max_j |z_j^*| = 1$ for which the inequality

$$\max_{\substack{m+1 \le \nu \le m+n \\ \nu \text{ integer}}} |s_\nu^*| \le \varepsilon^n$$

holds.

PROBLEM 18

Does the assertion of the previous problem hold if $\max_j |z_j^*| = 1$ is replaced by

$$\min_j |z_j^*| = 1?$$

PROBLEM 19

Is it true—generalizing Ławrynowicz's theorem in (3.12.3)—that for fixed $m > 1$ and $\max_j |z_j| = 1$ the extremum

$$\min_{z_j} \max_{\substack{m+1 \le \nu \le m+n \\ \nu \text{ integer}}} |s_\nu|$$

is attained when

$$|s_{m+1}^*| = |s_{m+2}^*| = \cdots = |s_{m+n}^*|?$$

PROBLEM 20

Prove, under some argument condition, an inequality of type (6.1.6).

PROBLEM 21

Can one deduce from a positive solution of the previous problem an analogous result for the "simultaneous" Theorem 6.4?

PROBLEM 22

Improve the "simultaneous" Theorem 6.4.

PROBLEM 23

Dancs–Makai's observation (6.4.5) amounts to the fact that with notation (6.2.1)–(6.2.3) the maximum of

$$\frac{1}{2\pi}\int_{|z|=1}\left|1 - f_1(z)s_m\left(\frac{1}{f_1}\right)\right|^2 |dz|$$

under the normalization $\min_j|z_j| = 1$ and $m \geq 1$ is attained only for $z_1 = \cdots = z_n = \xi$, $|\xi| = 1$. What is the maximum if the z_j are restricted not just by $\min_j|z_j| = 1$ but also by

$$z_1 = \bar{z}_2 = e^{i\alpha} \qquad 0 < \alpha < \pi?$$

PROBLEM 24

Improve Theorem 6.3.

PROBLEM 25

Improve Theorem 7.3 for an appropriate $c(n)$.

PROBLEM 26

In a context more general than that of Theorem 7.2, find systems $f_0(z)$, $f_1(z)$,... of entire functions such that for each system $(z_1,\ldots,z_n)$ with $\min_j|z_j| = 1$, integer $m \geq 0$, and complex constant b_j the inequality

$$\max_{m+1\leq\nu\leq m+n}\left|\sum_{j=1}^{n} b_j f_\nu(z_j)\right| \geq c(m, n)\psi(b_1,\ldots,b_n)$$

holds with an explicit $c(m, n)$ and nonnegative ψ.

PROBLEM 27

Do there exist theorems of the type

$$\max_{\substack{m+1\leq\nu\leq m+n \\ \nu \text{ integer}}} |g(\nu)| \geq \psi(m, n, \delta)\min_j|b_j|$$

provided that

$$\max_j |z_j| = 1$$

and

$$\max_{\mu,\nu} \left| \operatorname{arc} \frac{z_\mu}{z_\nu} \right| \geq \delta$$

where the last max is taken for all pairs (μ, ν) such that

$$1 - \frac{n}{2(m+n)} \leq |z_\mu|, |z_\nu| \leq 1?$$

PROBLEM 28

Can the operator-type Theorem 7.2 with minimum norm be proved replacing the restriction (7.2.3) by a lower bound for

$$\max_j \min_h |z_j - \zeta_h|$$

[where $\varphi(\zeta_h) = 0$]?

PROBLEM 29

Find the best constant in Theorem 8.1 to replace the constant $8e$.

PROBLEM 30

Can the constant $8e$ in Theorem 8.1 in the case $b_1 = \cdots = b_n = 1$ be replaced by $(1 + \varepsilon)$ for arbitrarily small $\varepsilon > 0$ and $m > m_0(\varepsilon, n)$?

PROBLEM 31

Can a geometric restriction imply theorems of type (8.1.6) or (8.1.7)?

PROBLEM 32

Can the factor $|b_1^* + \cdots + b_n^*|$ in the example (8.13.2) of Makai–de Bruijn be replaced by $\min_j |b_1^* + \cdots + b_j^*|$?

PROBLEM 33

Can the operator-type Theorem 9.3 with max norm be proved replacing the restriction (9.2.4) by a lower bound for

$$\max_j \min_h |z_j - \zeta_h|?$$

[ζ_h stand for the zeros of $\varphi(z)$.]

PROBLEM 34

Does

$$\max_{\substack{m+1 \le \nu \le m+n \\ \nu \text{ integer}}} |z_1^\nu + \cdots + z_n^\nu| > e^{-o(n)}$$

hold if we have, for $j = 1, \ldots, n$, $\max_j |z_j| = 1$ and $\kappa \le |\text{arc } z_j| \le \pi$?

PROBLEM 35

Let b_j be arbitrary complex numbers and let the z_j be such that each angle of opening π/M $(M \ge 2)$ from the origin contains at most N $(\le n)$ of the z_j. Is it true that

$$(16.1) \qquad \min_{\substack{m+1 \le \nu \le m+(2M/\log 2)\log n \\ \nu \text{ integer}}} \frac{|g(\nu)|}{\left(\sum_{j=1}^{n} |b_j|^2 |z_j|^{2\nu} \right)^{1/2}} \le \sqrt{N+2}\,?$$

[With n replacing $N + 2$ this holds trivially for a fixed ν. On the other hand, (16.1) cannot hold without some geometric restrictions. For $\max_j |z_j| = 1$ the slightly weaker inequality

$$\min |g(\nu)| \le 3\sqrt{N + N\frac{\log M}{\log n}} \left(\sum_{j=1}^{n} |b_j|^2 |z_j|^{2m+2} \right)^{1/2}$$

where the min refers to the range in (16.1), can be proved. The problem obviously belongs to the extremal problems of the fifth type, the "dual conditional problems" of 2.4.]

PROBLEM 36

Does there exist a theorem with max norm for $m + 1 \le \nu \le m + kn$ (analogous to Theorem 7.1 of Halász)? In particular in the case $b_1 = \cdots = b_n = 1$?

PROBLEM 37

Does there exist a reasonable theorem for $m + 1 \le |\nu| \le m + kn$ (analogous to Theorem 11.4 of Halász) with max norm?

PROBLEM 38

Does Theorem 12.1 hold if (12.1.5) is replaced by the requirement that

$$\max_j \min_\mu |z_j - \zeta_\mu|$$

is bounded from below [where $\varphi(\zeta_\mu) = 0$]?

PROBLEM 39

Does "the onesided operator-type theorem with maximum norm" 13.1 hold if (13.1.2) is replaced by a lower bound for

$$\max_j \min_h |z_j - \zeta_h|?$$

[Again $\varphi(\zeta_h) = 0$.]

PROBLEM 40

Can Theorem 14.2 be proved without the restriction $m \geq 2n$?

PROBLEM 41

How can one find an *explicit* $(x_1', \ldots, x_n')$ system of real numbers with the property in Theorem 11.6?

PROBLEM 42

Find the "continuous" analogue of the problems dealt with in the first part, that is, find lower bounds for

$$\max_{a \leq x \leq b} \frac{|G(x)|}{|G(0)|}$$

where

$$G(z) = \int_{(l)} g(\zeta) e^{\zeta z} \, d\zeta$$

(l is a Jordan arc).

PROBLEM 43

Do there exist $(1/n)\psi(n)$ values of the integer ν, at least for $\psi(n) \geq n$ in $[m + 1, m + \psi(n)]$ for which the inequalities

$$|g(\nu)| \geq \left[\frac{n}{3e(m+n)}\right]^n |g(0)|$$

hold simultaneously, assuming only that $\min_j |z_j| = 1$?

PROBLEM 44

If $\varphi(z) = z^k + a_1 z^{k-1} + \cdots + a_k$ is an arbitrary polynomial of degree $k < n$ with fixed coefficients (independent of ν) and $m \geq 0$ is an integer, does there exist an inequality of the type

$$\max_{\substack{m+1 \le \nu \le m+n+k \\ \nu \text{ integer}}} \left| \sum_{j=1}^{n} z_j^\nu \varphi(z_j) \right| \ge c(m, n, k)$$

without any restriction on the zeros of $\varphi(z)$ provided that $\max_j |z_j| = 1$?

PROBLEM 45

Let $\Phi(z) = z^k + a_1 z^{k-1} + \cdots + a_k$ be an arbitrary polynomial whose coefficients are polynomials in ν of degree at most l, and let $m \ge 0$ be an integer. Do there exist $U = U(m, n, k, l)$ and $V = V(m, n, k, l)$ so that $\max_j |z_j| = 1$ implies

$$\max_{\substack{m+1 \le \nu \le m+U \\ \nu \text{ integer}}} \left| \sum_{j=1}^{n} z_j^\nu \Phi(z_j) \right| \ge V?$$

PROBLEM 46

Are there one-sided theorems having the "sparseness-condition" of Problem 35 instead of the argument-condition $\kappa \le |\text{arc } z_j| \le \pi$?

PROBLEM 47. (P. ERDŐS)

Does there exist for each integer $1 \le l \le n-1$ a $c = c(l)$ so that

$$\max_{\substack{1 \le \nu \le c(l)n \\ \nu \text{ integer}}} \frac{|z_1^\nu + \cdots + z_n^\nu|}{M_1(\nu)} \ge l?$$

PROBLEM 48

Are there *arithmetical equivalents* of the main theorems?

PROBLEM 49

Can the first main theorem be improved if the coefficients b_j are all positive? Can it be improved in particular for $m = 0$?

PROBLEM 50

Does there exist an inequality of the type

$$\max_{1 \le \nu \le n^2} \text{Re} \sum_{j=1}^{n} b_j z_j^\nu > c(n) \sum_{j=1}^{n} b_j$$

if all the b_j are positive?

PROBLEM 51

If $H_\nu(x)$ stands, as in (4.3.5), for the νth Hermitean polynomial and

$$U_\nu = \sum_{j=1}^{n} H_\nu(z_j),$$

then what can be said about

$$\inf_{z_j} \max_{\substack{1 \le \nu \le n \\ \nu \text{ integer}}} |U_\nu| = \psi(n, \alpha)$$

when $z_1, \ldots, z_n$ run over all systems such that

$$\operatorname{Re} z_1 \le \operatorname{Re} z_2 \le \cdots \le \operatorname{Re} z_n = \alpha?$$

PROBLEM 52

What can be said about $\psi(n, \alpha)$ of the previous problem if the z_j satisfy

$$|\operatorname{Re} z_1| \le |\operatorname{Re} z_2| \le \cdots \le |\operatorname{Re} z_n| = \alpha?$$

PROBLEM 53

Let $m = m(n) \ge n$ be fixed. What can be said about

$$\min_{z_j} \max_{1 \le \nu \le m} \left| \frac{s_\nu}{n} \right|^{1/\nu}$$

where $s_\nu = z_1^\nu + \cdots + z_n^\nu$ and $\max_j |z_j| = 1$?

PROBLEM 54

Does there exist an $w(x) \nearrow \infty$ such that

$$b_j > 0, \qquad |z_j| = 1, \qquad j = 1, \ldots, n$$

implies for $g(\nu) = \sum_{j=1}^{n} b_j z_j^\nu$ the inequality

$$\max_{1 \le \nu \le n^{100}} |g(\nu)| > \frac{w(n)}{\sqrt{n}} |g(0)|$$

(as an improvement of Theorem 7.3)?

II

APPLICATIONS

17

INTRODUCTION

In the following material we show some applications of the theorems in Part I. We write "some" instead of "the" applications because only a fraction of all published or unpublished existing applications can be discussed in this book owing to limitations of space. For others, we refer to the bibliography at the end. Our applications cover a rather wide spectrum, but it is almost impossible to determine the best selection or the best arrangement.

In our selection all *groups* of applications are represented, and all results are proved using our methods. These methods were novel and important in their time although some were later superceded by other methods. (It seems respect is now given only to the record holder.) We also demonstrate the strength of our methods by obtaining stronger versions of previously known results. There were several alternatives for arranging these results: collecting all applications of the same "power sum theorem" together, ordering according to the degree of sophistication of the reduction to a power sum theorem, or to simply arrange according to the topics of application. We essentially chose the last principle.

Our aim is, of course, to exhibit the various ways in which the principles of Part I can be applied. Thus after sketching the proof, we refer to the original paper concerning details that do not require our methods.

18

SOME BASIC INEQUALITIES

18.1

As a first indication of the applicability of our methods we gave a short proof for inequality (6.3.7). Because of its importance, we formulate it (Turán [2]) as a theorem.

Theorem 18.1. *If*

$$f(t) = \sum_{j=1}^{n} b_j e^{i\alpha_j t} \tag{18.1.1}$$

with

$$\min_j \operatorname{Re} \alpha_j = 0, \tag{18.1.2}$$

and

$$a > 0, \qquad d > 0, \tag{18.1.3}$$

then the inequality

$$\max_{a \le t \le a+d} |f(t)| \ge \left[\frac{d}{2e(a+d)}\right]^n |f(0)| \tag{18.1.4}$$

holds.

Theorem 18.1 has the following consequence.

Corollary I. *The inequality* (18.1.4) *holds for all solutions* $y = f(t)$ *of the differential equation*

$$y^{(n)}(t) + a_1 y^{(n+1)}(t) + \cdots + a_n y(t) = 0, \tag{18.1.5}$$

where the a_ν are constants, provided that the zeros $\alpha_1, \ldots, \alpha_n$ of the characteristic equation

$$x^n + a_1 x^{n-1} + \cdots + a_n = 0 \tag{18.1.6}$$

satisfy the normalization (18.1.2).

Of course the normalization (18.1.2) implies that for arbitrary real numbers A, B, C with

$$A \le B < C, \tag{18.1.7}$$

the inequality

$$|y(A)| \le \left(2e\frac{C-A}{C-B}\right)^n \max_{B \le t \le C} |y(t)| \tag{18.1.8}$$

holds for all solutions of Equation (18.1.5). With the notation

$$A < B < C < D, \tag{18.1.9}$$

this gives the following.

Corollary II. *With the foregoing conventions, the inequality*

$$\max_{A \le t \le B} |y(t)| \le \left(2e\frac{D-A}{D-C}\right)^n \max_{C \le t \le D} |y(t)|$$

holds. (*Valid in fact if $A \le B \le C < D$.*)

18.2

Next we suppose that

$$\alpha_j = i\lambda_j, \qquad \operatorname{Im} \lambda_j = 0 \qquad j = 1, \ldots, n, \tag{18.2.1}$$

and with the conventions (18.1.1) and (18.1.9) we want to estimate

$$M_1 = \max_{A \le t \le D} |f(t)|$$

from above by

$$M_2 = \max_{B \le t \le C} |f(t)|.$$

Without loss of generality we may suppose that M_1 is attained either for

$$A \leq t \leq B$$

or for

$$C \leq t \leq B.$$

In the first case, apply Corollary II to $f(t)$ and get

$$M_1 \leq \left(2e\frac{C-A}{C-B}\right)^n M_2. \tag{18.2.2}$$

In the second case, (18.2.1) permits an application of Corollary II to

$$f_1(t) = f(-t) = \sum_{j=1}^{n} b_j e^{-i\lambda_j t}$$

for the intervals $[-D, -C]$, $[-C, -B]$.

Hence in this case

$$\begin{aligned} M_1 &= \max_{C \leq t \leq D} |f(t)| = \max_{-D \leq t \leq -C} |f_1(t)| \\ &\leq \left(2e\frac{D-B}{C-B}\right)^n \max_{-C \leq t \leq -B} |g_1(t)| = \left(2e\frac{D-B}{C-B}\right)^n \max_{B \leq t \leq C} |f(t)|. \end{aligned} \tag{18.2.3}$$

(18.2.2) and (18.2.3) yield another consequence.

Corollary III. *For*

$$k(t) = \sum_{j=1}^{n} b_j e^{i\lambda_j t} \quad (\textit{the } \lambda_j \textit{ are real}) \tag{18.2.4}$$

with the notation (18.1.9), *the inequality*

$$\max_{A \leq t \leq D} |k(t)| \leq \left[\frac{2e}{C-B}\max(C-A, D-B)\right]^n \max_{B \leq t \leq C} |k(t)| \tag{18.2.5}$$

and the slightly weaker inequality

$$\max_{A \leq t \leq D} |k(t)| \leq \left(2e\frac{D-A}{C-B}\right)^n \max_{B \leq t \leq C} |k(t)| \tag{18.2.6}$$

hold.

18.3

The case when the λ_j are rational integers m_j is particularly important. Then with $A = 0$ and $D = 2\pi$ we have the following corollary.

Corollary IV. *For the rational polynomial*

$$F(z) = \sum_{j=1}^{n} b_j z^{m_j} \qquad (0 \leq m_1 < m_2 < \cdots < m_n), \tag{18.3.1}$$

with $0 \leq \alpha < \alpha + \delta \leq 2\pi$, *the inequality*

$$\max_{|z|=1} |F(z)| \leq \left(\frac{4e\pi}{\delta}\right)^n \max_{\substack{|z|=1 \\ \alpha \leq \operatorname{arc} z \leq \alpha+\delta}} |F(z)| \tag{18.3.2}$$

holds.

The main point of Corollary IV is that the exponent on the right of (18.3.2) is the number of terms and *not* the degree of $F(z)$. An inequality of type

$$\max_{|z|=1} |F(z)| \leq C(\delta)^{m_n} \max_{\substack{|z|=1 \\ \alpha \leq \operatorname{arc} z \leq \alpha+\delta}} |F(z)|$$

where $C(\delta)$ depends only on δ could be obtained by a simple direct argument but has no applications to the theory of lacunary power series, whereas Corollary IV is effective in such problems.

As Gaier remarked (Gaier [2]), no inequality of type (18.3.2) holds if the unit circle with its subarc is replaced by a rather general type of Jordan curve together with its suitable subarc in $|z| \leq 1$.

18.4

Consider now a special case of Theorem 18.1 when

$$\alpha_\mu = i\lambda_\mu \qquad (\mu = 0,\ldots,n-1),\ n = 2N+1, \tag{18.4.1}$$

$$\lambda_0 = 0, \qquad \lambda_{N+j} = -\lambda_j = \text{real}$$

$$j = 1,\ldots,N,$$

and with real A_j, B_j,

$$b_0 = A_0, \qquad b_{N+j} = \bar{b}_j = \tfrac{1}{2}(A_j + iB_j), \qquad j = 1,\ldots,N. \tag{18.4.2}$$

This gives the following.

Corollary V. *For $a > 0$, $d > 0$ the inequality*

$$\max_{a \le t \le a+d} \left| A_0 + \sum_{j=1}^{N} \left(A_j \cos \lambda_j t + B_j \sin \lambda_j t \right) \right| \ge \left[\frac{d}{2e(a+d)} \right]^{2N+1} \left| \sum_{j=0}^{N} A_j \right|$$

holds.

In Littlewood's paper of 1937 (Littlewood [1]) he proves the following two results. Let

$$a_j > 0, \qquad \lambda_j > 0 \quad \text{for} \quad 1 \le j \le N$$

and

$$f(x) = 1 + \sum_{j=1}^{N} a_j \cos \lambda_j x.$$

With the notation

$$\max_{\xi - 1 \le x \le \xi + 1} |f(x)| = U = U(f, \xi)$$

and

$$\max_{\xi - \xi^{3/4} \le x \le \xi + \xi^{3/4}} |f(x)| = U^* = U^*(f, \xi),$$

the following statements are true: If $\xi > 1$ and for $j = 1, \ldots, N$, $\lambda_j \le 1$, then the inequality

$$U \ge (A_1 N \xi)^{-2N-1}$$

holds. Dropping all restrictions on the λ_j but requiring

$$N < \exp \xi^{1/3}, \qquad a_j < \exp \xi^{1/3}, \qquad j = 1, \ldots, N$$

the inequality

$$U^* > (A_2 N \xi)^{-2N-1}$$

holds if $\xi > A_3$. Here A_1, A_2, and A_3 stand for positive constants.

These results follow at once from Corollary V. They can be considered as forerunners of our power sum theorems (in "continuous" form). As we mentioned at the end of Section 6, Littlewood's method was quite different from ours.

We also need in the sequel the following corollary which is obtained from (18.2.6) by using the substitution (18.4.1)–(18.4.2).

Corollary VI. *Denote the trigonometric polynomial in Corollary V by* $h(t)$. *Then the inequality*

$$\max_{0 \le t \le 2\pi} |h(t)| \le \left(\frac{4e\pi}{\delta}\right)^{2N+1} \max_{\alpha \le t \le \alpha+\delta} |h(t)|$$

holds for $0 \le \alpha < \alpha + \delta \le 2\pi$.

18.5

Theorem 18.1 gives as an upper bound for $|f(A)|$ (A is real) the absolute maximum of f on a segment of the real axis to the right of A, subject to the normalization (18.1.2). The same holds, of course, also on every line of the complex plane parallel to the real axis. For some applications we need analogous estimates on an arbitrary straight line in the complex plane. Therefore we state a form of Theorem 18.1 without the condition (18.1.2). If

$$\min_j \operatorname{Re} \alpha_j = \Lambda, \tag{18.5.1}$$

then by Theorem 18.1 (for $f(t)e^{-\Lambda t}$) we obtain for $a > 0$ and $d > 0$ a t^* with

$$a \le t^* \le a + d \tag{18.5.2}$$

so that

$$|f(0)| \le \left(2e\frac{a+d}{d}\right)^n |f(t^*)| e^{-\Lambda t^*}.$$

A fortiori, setting

$$\max_j |\alpha_j| = K \tag{18.5.3}$$

the inequality

$$|f(0)| \le \left(2e\frac{a+d}{d}\right)^n e^{K(a+d)} \max_{a \le t \le a+d} |f(t)| \tag{18.5.4}$$

holds. Now if

$$f(z) = \sum_{j=1}^{n} b_j e^{\alpha_j z} \tag{18.5.5}$$

and the segment I is given by

$$I = \{ z : \operatorname{arc} z = \varphi, a \leq |z| \leq a + d \} \tag{18.5.6}$$

where a, d, and φ are fixed, $a > 0$, $d > 0$, then, by (18.5.4), we get

$$\max_{z \in I} |f(z)| = \max_{a \leq t \leq a+d} \left| \sum_{j=1}^{n} b_j e^{(\alpha_j e^{i\varphi})t} \right| \geq \left[\frac{d}{2e(a+d)} \right]^n e^{-K(a+d)} |f(0)|.$$

From this we get, as in (18.1.8), the following.

Corollary VII. *Let l be an arbitrary straight line in the complex plane and A, B, C three different points on l (in this order). Then, for $f(z)$ as in* (18.5.5), *with the notation* (18.5.3) *and $J = \overline{BC}$, the inequality*

$$|f(A)| \leq \left(2e \left| \frac{C-A}{C-B} \right| \right)^n e^{K|C-A|} \max_{z \in J} |f(z)| \tag{18.5.7}$$

holds.

Passage to the limit gives the same inequality for all solutions of (18.1.5).

The form of the first main theorem which has the shortest proof, and hence is most elegant in applications, is the form with $[n/2e(m+n)]^n$. There is, however, one application in Section 24 where using the best constant $\left[\sum_{h=0}^{n-1} 2^h \binom{m+h}{h} \right]^{-1}$ is important. By a remark of Tijdeman we have

$$\sum_{h=0}^{n-1} 2^h \binom{m+h}{h} \leq 2^{n-1} \sum_{h=0}^{n-1} \binom{m+h}{h} = 2^{n-1} \binom{m+n}{n-1} < \frac{[2(m+n)]^{n-1}}{(n-1)!}$$

and for $n \geq 2$

$$(n-1)! > \left(\frac{n}{e} \right)^{n-1},$$

and hence for $n \geq 2$ (18.1.4) can be replaced by

$$\max_{a \leq t \leq a+d} |f(t)| \geq \left[\frac{d}{2e(a+d)} \right]^{n-1} |f(0)|. \tag{18.5.8}$$

Thus in Corollary VII (18.5.7) can be replaced by the inequality

$$|f(A)| \leq \left(2e \left| \frac{C-A}{C-B} \right| \right)^{n-1} e^{K|C-A|} \max_{z \in J} |f(z)|. \tag{18.5.9}$$

18.6

The previously given bounds for solutions of the differential equation (18.1.5) are not bad if the zeros of the characteristic equation (18.1.6) are in the right halfplane "not too far" from the imaginary axis. When some zeros are too far away, we get better lower bounds from Theorem 6.3. So we formulate only one result, which can be verified using the reasoning of 6.3.

Theorem 18.2. *Under the restriction* (18.1.2) *on the zeros of the characteristic equation* (18.1.6) *we have for all solutions of equation* (18.1.5) *and for* $a, d > 0$, $d \le a^*$ *the inequality*

$$\max_{a \le t \le a+d} |y(t)| \ge \tfrac{1}{3}|y(0)| \prod_{j=1}^{n} \frac{\exp([d/n]\operatorname{Re}\alpha_j) - \exp(-d/an)}{\exp([d/n]\operatorname{Re}\alpha_j) + 1}.$$

18.7

Next we consider Theorem 11.1. Drop the restriction that m is an integer and choose m and the z_j as before:

$$m = a\frac{n}{d}, \qquad z_j = e^{w_j(d/n)}.$$

Theorem 18.3. *Let* $a > 0$, $d > 0$, *and* w_ν *be arbitrary complex numbers and* $\delta > 0$ *such that*

$$\frac{\min\limits_{\mu \ne \nu} |e^{w_\mu \cdot d/n} - e^{w_\nu \cdot d/n}|}{\max\limits_{j} |e^{w_j \cdot d/n}|} \ge \delta. \tag{18.7.1}$$

Then, for a suitable x_0 *with*

$$a \le x_0 \le a + d, \tag{18.7.2}$$

the inequality

$$\left|\sum_{j=1}^{n} b_j e^{w_j x_0}\right| \ge \frac{1}{n}\left(\frac{\delta}{2}\right)^{n-1} \sum_{j=1}^{n} |b_j|\,|e^{w_j x_0}| \tag{18.7.3}$$

holds.

In particular, if $w_\mu = i\lambda_\mu$ $(\mu = 1,\dots,n)$ and the λ_μ are real, then we have the following consequence.

*The condition $d \le a$ can be omitted because, for arbitrary $m_1 > 0$, the proof of Theorem 6.3 yields (6.7.1) for $m = 0$ with $\exp(-1/m_1)$ instead of $\exp(-1/m)$.

Corollary. *If $a > 0$, $d > 0$, and*

$$\min_{\mu \neq \nu} \left| \sin \frac{\lambda_\mu - \lambda_\nu}{2n} d \right| \geq \frac{\delta}{2}, \tag{18.7.4}$$

then the inequality

$$\sum_{j=1}^{n} |b_j| \leq n \left(\frac{2}{\delta} \right)^{n-1} \max_{a \leq t \leq a+d} \left| \sum_{j=1}^{n} b_j e^{i\lambda_j t} \right| \tag{18.7.5}$$

holds.

Hence the field of application includes, in principle, coefficient estimates. The proof of Theorem 11.1 in the case of $\alpha_j = i\lambda_j$ (λ_j are real), if followed with care, shows that inequality (18.7.5) could be improved substantially.

18.8

Let

$$F(z) = \sum_{j=1}^{n} b_j e^{z_j z}.$$

Then Theorem 9.5 of van der Poorten can be rewritten as

$$\max_{\nu = 0, 1, \ldots, n-1} \left| F^{(\nu)}(z) \right| \geq \frac{|F(0)|}{2n} e^{-(2\Delta + 1)|z|}.$$

Now standard arguments lead easily to the following generalization.

Theorem 18.4 (van der Poorten). *For*

$$H(z) = \sum_{j=1}^{n} P_j(z) e^{\alpha_j z} \tag{18.8.1}$$

with polynomials $P_j(z)$ and

$$\max_j |\alpha_j| = \Delta$$

and

$$\sigma = \sum_{j=1}^{n} \{ \operatorname{gr} P_j(z) + 1 \}, \tag{18.8.2}$$

the inequality

$$\max_{\nu=0,1,\ldots,\sigma-1} |H^{(\nu)}(z)| \geq \frac{1}{2\sigma}|H(0)|e^{-(2\Delta+1)|z|} \tag{18.8.3}$$

holds.

The proof in Tijdeman's thesis [2] gives a somewhat stronger conclusion: the factor $1/2\sigma$ in (18.8.3) is eliminated. He also gave a good lower bound for

$$\max_{\nu=m,\ldots,m+\sigma} |H^{(\nu)}(z)|.$$

For details see Tijdeman [2].

Some applications of Theorem 12.2 and 12.3 are based on the following form of Theorem 12.3.

Theorem 18.5. *With the notation* (18.8.1) *and* (18.8.2) *and* $a > 0$, $d > 0$, $0 \leq \kappa \leq \pi/2$ *we suppose*

$$\min_j \operatorname{Re} \alpha_j = 0 \tag{18.8.4}$$

$$\kappa \leq \operatorname{Im}\left(\frac{d}{\sigma}\alpha_j\right) \leq 2\pi - \kappa \bmod[0, 2\pi) \tag{18.8.5}$$

for $j = 1,\ldots,n$. *Then, the inequalities*

$$\max_{a \leq x \leq a+d\left(3+\frac{\pi}{\kappa}\right)} \operatorname{Re} H(x) \geq \frac{1}{6\sigma}|\operatorname{Re} H(0)| \cdot \left(\frac{d}{4e(a+2d)}\right)^{2\sigma} \tag{18.8.6}$$

and

$$\min_{a \leq x \leq a+d\left(3+\frac{\pi}{\kappa}\right)} \operatorname{Re} H(x) \leq -\frac{1}{6\sigma}|\operatorname{Re} H(0)|\left[\frac{d}{4e(a+2d)}\right]^{2\sigma} \tag{18.8.7}$$

hold.

The proof is the same as in Section 6.3 so we do not repeat it.

18.9

The situation with Theorem 18.6 is slightly different. Here

$$f(t) = \sum_{j=1}^{n} b_j e^{\alpha_j t} \tag{18.9.1}$$

where, in addition to the normalization (18.1.2) and the condition

$$\min_j |\operatorname{Im} \alpha_j| \geq \lambda > 0, \tag{18.9.2}$$

the following conditions are satisfied:

(18.9.3)

1. $f(t)$ is real for real t.
2. The α_j are distinct.
3. $\prod_{j=1}^{n} (z - \alpha_j)$ has only real coefficients.

Since $\Sigma_{j=1}^{n} b_j \alpha_j^{\nu} = f^{(\nu)}(0)$, the e^{α_j} and b_j satisfy the symmetry restrictions (12.8.1)–(12.8.2), and hence Theorem 12.4 is applicable. Let $m = 0$, let d be so small that

$$0 \leq \frac{d}{n} \max_j |\operatorname{Im} \alpha_j| < \frac{\pi}{2}, \tag{18.9.4}$$

and let

$$z_j = e^{\alpha_j (d/n)}. \tag{18.9.5}$$

Then one can choose

$$K = \frac{d\lambda}{n}. \tag{18.9.6}$$

From (12.8.4) obtain the inequalities

$$f\left(\frac{d\nu_1}{n}\right) \geq \frac{1}{3n} 16^{-n} |f(0)| \tag{18.9.7}$$

$$f\left(\frac{d\nu_2}{n}\right) \leq -\frac{1}{3n} 16^{-n} |f(0)|$$

for suitable integers ν_1 and ν_2 satisfying the inequalities

$$0 < \frac{d\nu_1}{n}, \frac{d\nu_2}{n} \leq \frac{d}{2}\left(3 + \frac{\pi n}{d\lambda}\right) = \frac{\pi n}{2\lambda} + \frac{3}{2} d.$$

Hence

$$\max_{0 \leq t \leq \pi n/2\lambda + (3/2)d} f(t) \geq \frac{1}{3n} 16^{-n} |f(0)|.$$

Letting $d \to +0$ we get

$$\max_{0 \le t \le \pi n/2\lambda} f(t) \ge \frac{1}{3n} 16^{-n}|f(0)| \tag{18.9.8}$$

and analogously

$$\max_{0 \le t \le \pi n/2\lambda} f(t) \le -\frac{1}{3n} 16^{-n}|f(0)|. \tag{19.9.9}$$

Applying this to $f(t+\gamma)$ with an arbitrary real γ and a trivial passage to the limit yields a result of Dancs and Turán.

Theorem 18.6. *Let*

$$y^{(n)}(t) + a_{n-1}y^{(n-)}(t) + \cdots + a_0 y(t) = 0$$

an arbitrary linear differential equation with real constant coefficients so that the zeros of the equation

$$z^n + a_{n-1}z^{n-1} + \cdots + a_0 = 0$$

are all outside the strip $|\mathrm{Im}\, z| < \lambda$.* *Then every real solution* $f(t)$ *changes sign in all real intervals*

$$\left[\gamma, \gamma + \frac{n\pi}{2\lambda}\right], \quad \text{if} \quad f(\gamma) \ne 0. \tag{18.9.10}$$

As the example $y'' + \lambda^2 y = 0$ ($\lambda > 0$) shows, with $n = 2$ and $y = \sin \lambda t$, the theorem cannot be improved in general. For the proof, see Dancs–Turán [6].

*If $\min_j \mathrm{Re}\, \alpha_j = \delta \ne 0$ consider instead $f(t)e^{-\delta t}$.

19

CONTINUATION

19.1

We want to find the L_2-analogue of Theorem 18.1, or rather of Corollary II. Tijdeman [5] proved the following result.

Theorem 19.1. *Under the normalization* (18.1.2) *for the zeros of the characteristic equation, for all solutions* $y(t)$ *of the equation* (18.1.5) *and for* $a \geq d > 0$ *we have the inequality*

$$\int_a^{a+d} |y(t)|^2 \, dt \geq \frac{e^{-4}}{n} \left[\frac{d}{2e(a+d)} \right]^{2n} \int_0^d |y(t)|^2 \, dt.$$

For $0 < a < D$, *we get the same inequality with* $4e$ *instead of* $2e$.

The first theorem of this type is contained in Turán [34] with the exponent n^2 instead of $2n$.

19.2. PROOF

Apply Theorem 6.6 with

$$(19.2.1) \qquad z_j = \exp\left(\frac{d}{n+1} \alpha_j \right), \qquad b_j = a_j \exp(\alpha_j t_0) \qquad j = 1, \ldots, n$$

t_0 real, a_j arbitrary complex numbers and $\min_j \operatorname{Re} \alpha_j = 0$. This substitution with $m \geq n$ and with the notation

$$(19.2.2) \qquad r(t) = \sum_{j=1}^{n} a_j e^{\alpha_j t},$$

yields the inequality

$$\sum_{\nu=m+1}^{m+n}\left|r\left(\frac{d\nu}{n+1}+t_0\right)\right|^2 \geq \frac{1}{n}\left[\frac{n}{2e(m+n+1)}\right]^{2n}\sum_{\nu=0}^{n}\left|r\left(\frac{d\nu}{n+1}+t_0\right)\right|^2.$$

Integrating with respect to t_0 over the interval $[0, d/(n+1)]$ we get

$$(19.2.3)\qquad \int_{d\frac{m+1}{n+1}}^{d\frac{m+n+1}{n+1}}|r(t)|^2\,dt \geq \frac{1}{n}\left[\frac{n}{2e(m+n+1)}\right]^{2n}\int_0^d|r(t)|^2\,dt.$$

Now let m_1 be an arbitrary real number in $[n, \infty)$ and apply (19.2.3) with $m = [m_1]$. Then

$$\int_{d\frac{m_1}{n+1}}^{d\frac{m_1+n+1}{n+1}}|r(t)|^2\,dt \geq \frac{1}{n}\left[\frac{n}{2e(m_1+n+1)}\right]^{2n}\int_0^d|r(t)|^2\,dt.$$

Choosing

$$m_1 = a\frac{n+1}{d} \geq n$$

this gives

$$\int_a^{a+d}|r(t)|^2\,dt \geq \frac{1}{n}\left(\frac{1}{1+1/n}\right)^{2n}\left[\frac{d}{2e(a+d)}\right]^{2n}\int_0^d|r(t)|^2\,dt \quad \text{for} \quad a \geq d > 0.$$

For $0 < a' < d'$ we apply the foregoing inequality for $r(t + a' - d')$ and $a = d = d' > 0$ then add to it $\int_{a'}^{a'+d'}|r(t)|^2\,dt$. From this one obtains the corresponding inequality for $0 < a' < d'$ with $4e$, instead of $2e$, and a factor $\frac{1}{2}$.

This completes the proof of Theorem 19.1 after the usual passage to the limit. ■

19.3

Theorems 18.1 and 19.1 are useful if a is not too small. If however $a \to +0$, then one would expect a factor on the right-hand side of (18.1.4) that tends "essentially" to 1. As indicated in Section 7, such an improvement was achieved by Halász's Theorem 7.1. Suppose

$$(19.3.1)\qquad 0 < a \leq \frac{2}{3\log 2}d.$$

Theorem 19.2 (Halász). *For all solutions* $y(t)$ *of Equation* (18.1.5), *with the normalization* (18.1.2) *and* a, d *satisfying* (19.3.1), *the inequality*

$$\max_{a \le t \le a+d} |y(t)| \ge \frac{|y(0)|}{5\sqrt{n}} \left(\frac{a}{d}\right)^{1/4} \exp\left(-6n\sqrt{\frac{a}{d}}\right)$$

holds.

19.4. PROOF

Suppose that $y(t)$ has the form (18.1.1). As in 6.3, for arbitrary nonnegative m_1, integral $k \ge 2$ and $\min_j |z_j| = 1$, Theorem 7.1 yields the inequality

$$(19.4.1) \qquad \max_{\substack{m_1 \le \nu \le m_1 + kn \\ \nu \text{ integer}}} |g(\nu)| \ge \frac{|g(0)|}{\sqrt{kn+1}} 2^{-m_1-1} \exp\left(-\frac{6n}{k}\right).$$

Again we choose

$$m_1 = a\frac{n}{d}, \qquad z_j = \exp\left(\frac{d}{n}\alpha_j\right), \qquad j = 1, \dots, n.$$

This gives

$$\max_{a \le d\nu/n \le a+kd} \left| \sum_{j=1}^{n} b_j e^{\alpha_j (d\nu/n)} \right| \ge \frac{1}{2\sqrt{kn+1}} \left| \sum_{j=1}^{n} b_j \right| \exp\left\{-n\left(\frac{a}{d}\log 2 + \frac{6}{k}\right)\right\}$$

and, replacing d by d/k,

$$(19.4.2) \qquad \max_{a \le x \le a+d} \left| \sum_{j=1}^{n} b_j e^{\alpha_j x} \right| \ge \frac{\left| \sum_{j=1}^{n} b_j \right|}{2\sqrt{kn+1}} \exp\left\{-\left(\frac{a}{d}\log 2 \cdot k + \frac{6}{k}\right) n\right\}.$$

Finally we must choose the integer $k \ge 2$. By (19.3.1), we may choose

$$k = \left[\sqrt{\frac{6d}{a \log 2}}\right],$$

that is,

$$\sqrt{\frac{6d}{a\log 2}} \ge k \ge \sqrt{\frac{6d}{a\log 2}} - 1 \ge \frac{2}{3}\sqrt{\frac{6d}{a\log 2}} \ge 2.$$

With (19.4.2) this completes the proof after taking the limit. ■

Setting

$$a = \frac{d}{n^2}$$

(19.3.1) is fulfilled for $n \geq 2$, and hence Theorem 19.2 implies the following, say for $d = 1$.

Corollary. *For all solutions $y(t)$ of the equation* (18.1.5) *with normalization* (18.1.2) *the inequality*

$$|y(0)| \leq 5e^6 n \max_{1/n^2 \leq t \leq (1+1/n^2)} |y(t)|$$

holds for $n \geq 2$.

G. Halász has also proved the inequality

$$|y(0)| \leq cn^5 \int_0^1 |y(t)|dt$$

where c is a numerical constant (oral communication).

19.5

All previous results on solutions of Equation (18.1.5) depended upon the normalization (18.1.2). Halász has found some other types of results that do not need any normalization. Let a_1, a_2, and d be positive numbers and

(19.5.1) $$\max(a_1, a_2) = a.$$

Set

(19.5.2) $$S = [-a_1 - d, -a_1] \cup [a_2, a_2 + d].$$

Theorem 19.3 (Halász). *For all solutions $y(t)$ of all equations of the form* (18.1.5), *the inequality*

$$\max_{x \in S} |y(t)| \geq \left[\frac{d}{8e(a+d)}\right]^n |y(0)|$$

holds.

19.6. PROOF

We use Theorem 11.4. If m_1' and m_2' are positive, then apply this theorem with

$$m_1 = [m_1'], \qquad m_2 = [m_2'] \quad \text{for} \quad z_1, \ldots, z_n \neq 0$$

and put

$$S' = [-m_1' - n, -m_1'] \cup [m_2', m_2' + n] \tag{19.6.1}$$

$$m' = \max(m_1', m_2').$$

We have $m' \geq m$ and

$$\max_{\substack{\nu \in S' \\ \nu \text{ integer}}} |g(\nu)| \geq \left[\frac{n}{8e(m' + n)}\right]^n |g(0)|. \tag{19.6.2}$$

If the b_j and β_j are arbitrary complex numbers then, setting

$$m_j' = a_j \frac{n}{d} \qquad (j = 1, 2)$$

$$z_j = \exp\left(\frac{d}{n}\beta_j\right) \qquad (j = 1, \ldots, n)$$

(19.6.2) implies that

$$\max_{\substack{(d\nu/n) \in S'' \\ \nu \text{ integer}}} \left|\sum_{j=1}^{n} b_j e^{(d\nu/n)\beta_j}\right| \geq \left[\frac{d}{8e(a + d)}\right]^n \left|\sum_{j=1}^{n} b_j\right|$$

where

$$S'' = [-a_1 - d, -a_1] \cup [a_2, a_2 + d].$$

Theorem 19.3 is proven after passage to the limit. ■

Next we turn to applications in complex function theory.

20

APPLICATIONS TO FUNCTION THEORY I.

The Gap Theorem of Fabry

20.1

The original gap theorem of Fabry applies to power series. We prove the following more general form for Dirichlet series, due essentially to Carlson–Landau [1] and O. Szász [1]. Let $s = \sigma + it$ and

$$\lambda_1 < \lambda_2 < \cdots \to \infty. \tag{20.1.1}$$

Theorem 20.1. *If*

$$f(s) = \sum_{n=1}^{\infty} a_n e^{-\lambda_n s}$$

converges for $\sigma > 0$ *and*

$$\lim_{n \to \infty} \frac{n}{\lambda_n} = 0, \tag{20.1.2}$$

then either all points of $\sigma = 0$, *or none, belong to the domain of regularity of* $f(s)$.

20.2. PROOF (SEE TURÁN [12])

It is enough to suppose that $f(s)$ is regular at $s = 0$ to deduce that $f(s)$ is regular in a vertical strip containing the line $\sigma = 0$ in its interior. We may suppose without loss of generality that $f(s)$ is regular for

(20.2.1) $$\sigma \geq -\eta, \qquad |t| \leq 2$$

with an η such that

$$0 < \eta < \tfrac{1}{25}.$$

Consider the polynomials $Y_n(s)$ (defined in Appendix B) and the expression

(20.2.2) $$Y_{N+1}(\sigma_0 + it) - Y_N(\sigma_0 + it) \doteq U_N(t)$$

with a fixed

(20.2.3) $$-\frac{\eta}{2} \leq \sigma_0 \leq \frac{\eta}{2}.$$

On one hand the representation (B. 2.4) gives

(20.2.4) $$U_N(t) = \sum_{j=1}^{N} \left\{ (\beta_{N+1,j} - \beta_{N,j}) a_j e^{-\lambda_j \sigma_0} \right\} e^{-\lambda_j it}$$

$$+ \beta_{N+1,N+1} a_{N+1} e^{-\lambda_{N+1}\sigma_0} \cdot e^{-\lambda_{N+1} it},$$

which has the form treated in Corollary III of Theorem 18.1. Applying this corollary for $t_0 > 1$ with

$$n = N + 1, \qquad A = B = -1, \qquad C = 1, \qquad D = t_0$$

we get from (18.2.6),

(20.2.5) $$\max_{-1 \leq t \leq t_0} |U_N(t)| \leq \{e(t_0 + 1)\}^{N+1} \max_{-1 \leq t \leq 1} |U_N(t)|.$$

On the other hand, by Lemma II.1 for $-1 \leq t \leq 1$:

(20.2.6)

$$|U_N(t)| = |[Y_{N+1}(\sigma_0 + it) - f(\sigma_0 + it)] - [Y_N(\sigma_0 + it) - f(\sigma_0 + it)]|$$

$$\leq |Y_{N+1}(\sigma_0 + it) - f(\sigma_0 + it)| + |Y_N(\sigma_0 + it) - f(\sigma_0 + it)| \leq$$

$$\leq 2ce^{-(\lambda_N/4)\eta}.$$

Substituting this into (20.2.5) we get

$$\max_{-1 \leq t \leq t_0} |U_N(t)| \leq \{e(t_0 + 1)\}^{N+1} \cdot 2ce^{-(\lambda_N/4)\eta}$$

$$= \left\{ (2c)^{1/\lambda_N} [e(t_0 + 1)]^{(N+1)/\lambda_N} \cdot e^{-\eta/4} \right\}^{\lambda_N}.$$

But by the gap condition (20.1.2), for all sufficiently large N

$$(2c)^{1/\lambda_N}\left[e(t_0+1)\right]^{(N+1)/\lambda_N} < e^{\eta/8},$$

that is, for these N,

$$\max_{-1\le t\le t_0} |U_N(t)| \le e^{-(\eta/8)\lambda_N} \le e^{-(\eta/8)N}$$

uniformly for $-\eta/2 \le \sigma_0 \le \eta/2$. The reasoning is similar when $t_0 \le -1$. Thus the polynomials $Y_N(s)$ converge uniformly in the parallelogram

$$-\frac{\eta}{2} \le \sigma \le \frac{\eta}{2}, \qquad -t_0 \le t \le t_0.$$

That is, they converge to a function analytic in every finite part of the strip $-\eta/2 \le \sigma \le \eta/2$. But for the parallelogram

$$-\frac{\eta}{2} \le \sigma \le \frac{\eta}{2}, \qquad |t| \le 1$$

they converge to $f(s)$. This completes the proof. ■

20.3

If $f(z) = \sum_{n=1}^{\infty} a_n z^{\lambda_n}$ is regular for $|z| < 1$ and the *integers* λ_ν satisfy the Fabry gap-condition (20.1.2), then $f(z)$ has at least one singular point on $|z| = 1$; alternatively, $f(e^{-s})$ has at least one singular point on the line $\sigma = 0$. Thus the first alternative in Theorem 20.1 does not hold. This is Fabry's classical theorem. Such a proof for this theorem was postulated in 1934 by N. Wiener (Paley–Wiener [1] p. 125.) who settled the special case

$$\lambda_{n+1} - \lambda_n \to \infty \quad \text{for} \quad n \to \infty. \tag{20.3.1}$$

Considering only integers λ_ν, the proof of Theorem 20.1 also shows that if the exponents of the series

$$\sum_{\nu=1}^{\infty} (a_\nu \cos \lambda_\nu x + b_\nu \sin \lambda_\nu x) \tag{20.3.2}$$

satisfy the gap-condition (20.1.2) and the series is say, Abel summable in a subinterval of $[0, 2\pi]$ to a real, analytic function here, then the series is the Fourier series of a function analytic in $[0, 2\pi]$. Wiener's proof shows that if the series (20.3.2) satisfies the gap-condition (20.3.1) and is Abel summable in a subinterval of $[0, 2\pi]$ to a function in $L_2[0, 2\pi]$ then the series is also the Fourier series of a function in $L_2[0, 2\pi]$.

The preceding results are specific examples of the general problem of "real propagation," first formulated by Mandelbrojt. Suppose that the series in (20.3.2) is summable, in a subinterval of $[0, 2\pi]$, to a function having a prescribed property A. Then what sort of gap-condition (depending on A) must be imposed on the exponents to guarantee that the series is a Fourier series of a function that also satisfies A on $[0, 2\pi]$? There are some surprising results in this area. For example, Erdös and Rényi [2] showed that the L_2-theorem does not generalize to L_q $(q > 2)$ under the gap-conditions (20.3.1). In Turán [38] it is shown that no such result for $q > 6$ can hold even under the stronger gap-condition

$$\lambda_{n+1} - \lambda_n > \lambda_n^{1/(q+6)}. \tag{20.3.3}$$

On the other hand, the problem has not even been considered for L_p $(1 < p < 2)$. The methods in Part I of this book offer some possibilities for dealing with such problems.

20.4

The proof of Theorem 20.1 works mutatis mutandis if the abscissa of convergence is replaced by some other abscissae of summability. The same is true for other results of this type, using other gap-conditions. But these generalizations are meaningless when the summability, together with the gap-condition, implies convergence. Such Tauber-type theorems were called "high indices theorems" by Hardy and Littlewood. They proved such results for Abel summation; we return to the case of Borel summability in Section 22.

21

APPLICATIONS TO FUNCTION THEORY II

The Pólya Theory of Entire Functions with Gaps

21.1

In his classic paper [1] Pólya made a broad attack on Hadamard's problem: to investigate the influence of the coefficients on the properties of the function. He laid the foundations of a theory of entire functions depending exclusively on the properties of the (positive integer) exponents λ_n occurring in the series

$$f(z) = \sum_{n=1}^{\infty} a_n z^{\lambda_n}. \tag{21.1.1}$$

His paper is filled with surprising theorems and interesting open problems. Pólya himself emphasized two conclusions, both under the condition

$$\lim_{n\to\infty} \frac{n}{\lambda_n} = 0. \tag{21.1.2}$$

Denote, as usual,

$$\max_{\varphi} |f(re^{i\varphi})| = M(r, f) = M(r),$$

$$\max_{\alpha\le\varphi\le\beta} |f(re^{i\varphi})| = M(r, \alpha, \beta, f) = M(r, \alpha, \beta), \tag{21.1.3}$$

$$\min_{\varphi} |f(re^{i\varphi})| = m(r, f) = m(r).$$

Pólya's results are as follows.

1. If $f(z)$ satisfies the gap-condition (21.1.2) and is of finite order, then its order with respect to an arbitrarily small angle is the same as that with respect to the whole plane.

2. If $f(z)$ satisfies condition (21.1.2) and is of finite order, then for any continuous path tending to ∞ we have

$$\varlimsup_{z\to\infty \text{ on } \Gamma} \frac{|\log f(z)|}{M(|z|, f)} = 1.$$

The results and problems of this paper provoked a number of further important investigations; from the literature (prior 1956) we mention only those of V. Bernstein, S. Mandelbrojt, and Laurent Schwartz. In spite of this, a number of problems raised in Pólya's paper (or inspired quite naturally by it) remained inaccessible; solutions for some of these will be discussed in the sequel. An up-to-date survey of this theory cannot be given here; we wish only to illustrate how the methods of Part I can be applied here.

21.2

We begin with a theorem on harmonic entire functions

$$h(r, \varphi) = \sum_{n=1}^{\infty} r^{\lambda_n}(a_n \cos \lambda_n\varphi + b_n \sin \lambda_n\varphi) \tag{21.2.1}$$

which converge on the whole plane and satisfy the gap-condition (21.1.2). We use the notation (21.1.3) with h instead of f.

Theorem 21.1 (Turán [11]). *If $h(r, \varphi)$ is the harmonic entire function just described, then for all $0 < \varepsilon \le \frac{1}{2}$ and $0 \le \alpha < \beta \le 2\pi$ the inequality*

$$M(r, h) \le M(2r, h)^{\varepsilon} M(r, \alpha, \beta, h) \tag{21.2.2}$$

holds for $r > r_0(h, \varepsilon, \beta - \alpha)$.

For arbitrarily small fixed $\lambda > 0$ and $r > r_1(h, \varepsilon, \lambda, \beta - \alpha)$ the proof actually yields the inequality

$$M(r, h) \le M[(1 + \lambda)r, h]^{\varepsilon} M(r, \alpha, \beta, h),$$

but we want to avoid introducing an extra parameter.

21.3. PROOF

Let

$$s_k(re^{i\varphi}) = \sum_{\nu=1}^{k} \left(a_\nu r^{\lambda_\nu}\cos \lambda_\nu\varphi + b_\nu r^{\lambda_\nu}\sin \lambda_\nu\varphi\right). \tag{21.3.1}$$

Apply Corollary VI from Section 18.4. This gives

$$\max_{\varphi}\left|s_k(re^{i\varphi})\right| \le \left(\frac{4e\pi}{\beta-\alpha}\right)^{2k+1} \max_{\alpha\le\varphi\le\beta}\left|s_k(re^{i\varphi})\right|. \tag{21.3.2}$$

Let $\max_\varphi |h(re^{i\varphi})|$ be attained for $\varphi = \varphi_0$. Then

$$\begin{aligned}\max_{\varphi}\left|s_k(re^{i\varphi})\right| &\ge \left|s_k(re^{i\varphi_0})\right| \\ &= \left|h(r, \varphi_0) - \sum_{\nu=k+1}^{\infty} r^{\lambda_\nu}(a_\nu\cos \lambda_\nu\varphi_0 + b_\nu\sin \lambda_\nu\varphi_0)\right| \\ &\ge M(r, h) - \sum_{\nu=k+1}^{\infty} r^{\lambda_\nu}(|a_\nu| + |b_\nu|).\end{aligned} \tag{21.3.3}$$

Using the (crude) coefficient estimation

$$|a_\nu| \le 2\frac{M(2r, h)}{(2r)^{\lambda_\nu}}, \qquad |b_\nu| \le 2\frac{M(2r, h)}{(2r)^{\lambda_\nu}} \tag{21.3.4}$$

we have

$$\max_{\varphi}\left|s_k(re^{i\varphi})\right| \ge M(r, h) - 8\frac{M(2r, h)}{2^{\lambda_k}}. \tag{21.3.5}$$

If $\max_{\alpha\le\varphi\le\beta}\left|s_k(re^{i\varphi})\right|$ is attained for $\alpha \le \varphi = \varphi_1 \le \beta$, then we have

$$\begin{aligned}\max_{\alpha\le\varphi\le\beta}\left|s_k(re^{i\varphi})\right| &= \left|s_k(re^{i\varphi_1})\right| \\ &= \left|h(re^{i\varphi_1}) - \sum_{\nu=k+1}^{\infty} r^{\lambda_\nu}(a_\nu\cos \lambda_\nu\varphi_1 + b_\nu\sin \lambda_\nu\varphi_1)\right| \\ &\le M(r, \alpha, \beta, h) + \frac{8M(2r, h)}{2^{\lambda_k}}.\end{aligned} \tag{21.3.6}$$

For all $r > 0$ and an integer $k \geq 1$ (21.3.2), (21.3.5), and (21.3.6) yield the inequality

(21.3.7)

$$M(r, h) \leq \left(\frac{4\pi e}{\beta - \alpha}\right)^{2k+1} M(r, \alpha, \beta, h) + \frac{16}{2^{\lambda_k}}\left(\frac{4\pi e}{\beta - \alpha}\right)^{2k+1} M(2r, h).$$

21.4

The gap condition (21.1.2) implies that for a suitable $w(k)$ tending monotonically to $+\infty$ we have

$$\lambda_k \geq kw(k). \tag{21.4.1}$$

If we succeed in establishing the inequality

$$\frac{16}{2^{\lambda_k}}\left(\frac{4\pi e}{\beta - \alpha}\right)^{2k+1} M(2r, h) \leq \tfrac{1}{2} M(r, h), \tag{21.4.2}$$

then (21.3.7) will take the form

$$M(r, h) \leq 2\left(\frac{4\pi e}{\beta - \alpha}\right)^{2k+1} M(r, \alpha, \beta, h). \tag{21.4.3}$$

To ensure (21.4.2) by a proper choice of the integer $k = k(r)$ we write it first in the form

$$\left\{\frac{1}{2}\left(\frac{4\pi e}{\beta - \alpha}\right)^{(2k+1)/\lambda_k}\right\}^{\lambda_k} \leq \frac{1}{32}\,\frac{M(r, h)}{M(2r, h)}. \tag{21.4.4}$$

If $k(r) \to \infty$ as $r \to \infty$, then

$$\left(\frac{4\pi e}{\beta - \alpha}\right)^{3/w(k)} \leq \frac{3}{2} \tag{21.4.5}$$

will hold, provided that

$$r > r_0(h, \beta - \alpha). \tag{21.4.6}$$

Equation (21.4.4) is now certainly satisfied if

$$\left(\frac{4}{3}\right)^{\lambda_k} \geq 32\frac{M(2r, h)}{M(r, h)},$$

and by (21.4.1) this is certainly true if

$$\left(\frac{4}{3}\right)^{kw(k)} \geq 32\frac{M(2r, h)}{M(r, h)}. \tag{21.4.7}$$

If in addition we succeed in determining k so that

$$\left(\frac{8\pi e}{\beta - \alpha}\right)^{2k+1} \leq M(2r, h)^{\varepsilon}, \tag{21.4.8}$$

then by (21.4.3), the proof of Theorem 21.1 will be complete. But the crucial requirements (21.4.7) and (21.4.8) are both satisfied with the choice

$$k = k_0 = \left[\frac{\varepsilon \log M(2r, h)}{3\log(8\pi e/(\beta - \alpha))}\right] \qquad (\to +\infty). \tag{21.4.9}$$

For (21.4.8) this is evident; for $r > r_1(\varepsilon, h, \beta - \alpha)$ we have

$$k_0 w(k_0)\log\frac{4}{3} > \frac{\varepsilon}{2}\,\frac{\log M(2r, h)}{3\log(8\pi e/(\beta - \alpha))}\cdot w(k_0)$$

$$> \log M(2r, h) > \log\left(32\,\frac{M(2r, h)}{M(r, h)}\right)$$

so by the choice (21.4.9) the requirement (21.4.7) is also fulfilled. ■

21.5

The harmonic entire function $h(r, \varphi)$ is of (finite) order d if

$$d = \overline{\lim_{r\to\infty}}\,\frac{\log\log M(r, h)}{\log r} < \infty. \tag{21.5.1}$$

We define for such functions the type τ by

$$\tau = \overline{\lim_{r\to\infty}}\,\frac{\log M(r, h)}{r^d}. \tag{21.5.2}$$

We first deduce the following corollary of Theorem 21.1.

Corollary. *If $h(r, \varphi)$ is a harmonic entire function of finite order with the gap-condition* (21.1.2), *then its order and type with respect to an arbitrarily small angle are the same as those with respect to the whole plane.*

For the sake of simplicity we restrict ourselves to the case $d > 0$ and normal type $0 < \tau < \infty$ (as Pólya has done). Then for an arbitrarily small $0 < \eta < \tau/4$, and for all $r > r_2(h, \eta)$,

$$M(r, h) \leq \exp\{(\tau + \eta)r^d\}, \tag{21.5.3}$$

and for a suitable sequence

$$r_1 < r_2 < \cdots \to +\infty$$

also

$$M(r_\nu, h) \geq \exp\{(\tau - \eta)r_\nu^d\}. \tag{21.5.4}$$

Applying (21.2.2) with $r = r_\nu$ sufficiently large and using (21.5.3)–(21.5.4) we get

$$\exp\{(\tau - \eta)r_\nu^d\} \leq M(r_\nu, \alpha, \beta, h)\exp\{\varepsilon(\tau + \eta)(2r_\nu)^d\}$$

which proves the corollary, provided ε is chosen sufficiently small.

Next we deduce Pólya's theorem I from this corollary. Let the order (resp. type) of $f(z)$ with respect to the whole plane be K (resp. T) so that for all $r > r_3(f, \eta)$, $\eta > 0$

$$M(r, f) \leq \exp\{(T + \eta)r^K\}, \tag{21.5.5}$$

and for suitable

$$R_1 < R_2 < \cdots \to \infty$$

we have

$$M(R_\nu, f) \geq \exp\{(T - \eta)R_\nu^K\}. \tag{21.5.6}$$

(Again we deal only with the case $T > 0$.) Take

$$R'_\nu = R_\nu\left(1 + \frac{\eta}{100KT}\right) \tag{21.5.7}$$

and apply (21.2.2) with $r = R'_\nu$ (ν sufficiently large) and

$$h(r, \varphi) = \operatorname{Re} f(re^{i\varphi}).$$

This gives

$$M(R'_\nu, \alpha, \beta, f) \geq M(R'_\nu, \alpha, \beta, h) \geq M(2R'_\nu, f)^{-\varepsilon} M(R'_\nu, h). \tag{21.5.8}$$

Using the Poisson–Schwartz formula

$$f(R_\nu e^{i\varphi}) = \frac{1}{2\pi}\int_{-\pi}^{\pi} h(R'_\nu e^{i\vartheta})\frac{R'_\nu e^{i\vartheta} + R_\nu e^{i\varphi}}{R'_\nu e^{i\vartheta} - R_\nu e^{i\varphi}}\,d\vartheta$$

and also (21.5.7), we obtain

$$M(R_\nu, f) \le \frac{200KT + \eta}{\eta} M(R'_\nu, h).$$

Substituting this into (21.5.8) we obtain, if ε is suitably small,

$$M(R'_\nu, \alpha, \beta, f) \ge M(R_\nu, f)\exp\{-2\varepsilon(T + \eta)(2R'_\nu)^K\}$$

$$\ge \exp\{(T - 2\eta)R_\nu^K\} = \exp\left\{R_\nu'^K \frac{T - 2\eta}{(1 + \eta/100KT)^K}\right\}$$

$$> \exp\{(T - 3\eta)R_\nu'^K\}$$

for all sufficiently large ν and $\eta < T/4$.

21.6

An immediate extension to functions of infinite order is difficult since order and type practically lose their meaning. Formerly this difficulty was avoided by estimating $M(r, f)$ for all sufficiently large r by

(21.6.1)
$$\max_{\substack{(1-\varepsilon)r \le \rho \le (1+\varepsilon)r \\ \alpha \le \varphi \le \beta}} |f(\rho e^{i\varphi})|$$

respectively by

$$\max_{\alpha \le \varphi \le \beta} |f(\mathrm{Re}^{i\varphi})|$$

with an $R > r$. Recently it turned out to be useful to estimate $M(r, f)$ from above by $M(r, \alpha, \beta)^{1+\varepsilon}$ which can be done, however, only with the exception of a "small" r-set. This modification of the problem makes it meaningful to ask (even in the case of entire functions of finite order) on how big an r-set the inequality

(21.6.2)
$$M(r, f) \le M(r, \alpha, \beta, f)^{1+\varepsilon}$$

can fail? If E is any measurable set on the positive axis, and $E(a, b) = E \cap [a, b]$, one defines the upper logarithmic density of E by

$$\overline{\lim_{w \to \infty}} \frac{1}{\log w} \int_{E(1, w)} \frac{dr}{r}.$$

Kővári proved [1] the following.

Theorem 21.2. *If $h(r, \varphi)$ is a harmonic entire function of finite order satisfying the Fabry gap-condition* (21.1.2), *then for arbitrarily small $\eta > 0$ and $\delta > 0$ the inequality*

$$M(r)^{1-\eta} \leq \min_{\alpha} M(r, \alpha, \alpha + \delta)$$

holds with the exception (perhaps) of a set E of r-values with zero logarithmic density.

To deduce this theorem from Theorem 21.1 (with $\varepsilon = \eta^2$), it is enough to show that the set of r-values on which

$$M(2r, h) > M(r, h)^{1/\eta} \tag{21.6.3}$$

is of zero logarithmic density. Since $h(r, \varphi)$ is of finite order we have for some $K > 0$ and $r > r_2(f)$

$$M(r, h) \leq \exp r^K. \tag{21.6.4}$$

Set $x > \log \max\{r_0(h, \eta^2, \beta - \alpha), r_2\} = c(\eta)$,

$$f(x) = \log \log M(e^x, h). \tag{21.6.5}$$

Then $f(x)$ is monotonically increasing, continuous, tending with x to infinity, and (21.6.3) takes the form

$$f(x + \log 2) \geq f(x) + \log \frac{1}{\eta}, \tag{21.6.6}$$

whereas (21.6.4) gives

$$f(x) \leq Kx, \quad \text{for} \quad x > c. \tag{21.6.7}$$

It is enough to consider those values $x > c$ where $f(x)$ is positive.

Let x_0 be the smallest $x \geq c(\eta)$ (if any) satisfying (21.6.6) and

$$\xi_0 = x_0 + \log 2.$$

Also let x_1 be the smallest $x \geq \xi_0$ satisfying (21.6.6) (if any) and

$$\xi_1 = x_1 + \log 2,$$

and so forth. The $[x_\nu, \xi_\nu]$ intervals cover the set U of x-values for which (21.6.6) is satisfied. Thus for large X

$$KX \geq f(X) \geq \sum_{\xi_\nu \leq X} [f(\xi_\nu) - f(x_\nu)] \geq \log\frac{1}{\eta} \sum_{\xi_\nu \leq X} 1.$$

Hence the measure of $U(0, X)$ is

$$\leq c(\eta) + \log 2 + \frac{K}{\log(1/\eta)} X. \tag{21.6.8}$$

Returning to the r-axis, the set E of r-values satisfying (21.6.3) is covered by the intervals (e^{x_ν}, e^{ξ_ν}), and hence from (21.6.8) $(Y = e^X)$

$$\overline{\lim_{Y}} \frac{1}{\log Y} \int_{E(1, Y)} \frac{dr}{r} = \overline{\lim_{X}} \frac{1}{X} \int_{U(1, X)} dx \leq \frac{2K}{\log(1/\eta)}.$$

Since η is arbitrarily small, Theorem 21.2 is proved. ■

It is easy to deduce from Theorem 21.2 the corresponding theorem for entire functions of finite order; we do not go into details.

21.7

The foregoing reasoning made it possible to prove, more than 30 years after its publication, a conjecture formulated at the very end of Pólya's paper [1]. This conjecture, which is stronger than Pólya's aforementioned Theorem II, asserted [with the notation (21.1.3)] that if $f(z)$ is an entire function of finite order with Fabry gaps, then for arbitrarily small $\varepsilon > 0$ there are radii

$$R_1 < R_2 < \cdots \to \infty$$

so that

$$m(R_\nu, f) \geq M(R_\nu, f)^{1-\varepsilon}. \tag{21.7.1}$$

Let

$$m(r, f) = |f(re^{i\varphi_0})| = |f(\zeta)|.$$

Then, given arbitrarily small $\delta > 0$, $\varepsilon > 0$, Theorem 21.2 shows—even on "almost all" circles $|z| = r$—that on the arc

$$|\operatorname{arc} z - \varphi_0| \leq \delta, \qquad |z| = r \tag{21.7.2}$$

there is a z^* so that

$$\log|f(z^*)| > (1-\varepsilon)\log M(r,f). \tag{21.7.3}$$

Hence if one could prove that apart from an r-set of zero logarithmic density we have

$$\big|\log|f(z_1)| - \log|f(z_2)|\big| \leq \frac{\varepsilon}{2}\log M(r,f)$$

for each z_1, z_2 with

$$|z_1| = |z_2| = r, \qquad \left|\operatorname{arc}\frac{z_1}{z_2}\right| \leq \delta,$$

then the conjecture (21.7.1) is proved. In other words, one has only to show that apart from an r-set of "small" logarithmic density the integral

$$r\int_{\alpha_1}^{\alpha_2}\left|\frac{f'}{f}(re^{i\varphi})\right|d\varphi$$

is "small compared with $\log M(r, f)$." W. H. J. Fuchs in his paper [1] in 1963 succeeded in doing this; namely, he proved that for each entire function $f(z)$ of order ρ, the inequality

$$r\int_{\alpha_1}^{\alpha_2}\left|\frac{f'}{f}(re^{i\varphi})\right|d\varphi \leq c(\rho,\eta)\,\delta\log\frac{1}{\delta}\log M(r,f) \tag{21.7.5}$$

holds for all α_1, α_2 with

$$|\alpha_1 - \alpha_2| \leq \delta \tag{21.7.6}$$

with the exception (perhaps) of an r-set of logarithmic density η. Since his proof for (21.7.5) is not short and does not use our methods, we refer to his original paper for details.

21.8

The corresponding situation for entire functions of infinite order is different. This was foreshadowed by the theorem of Erdős–Macintyre (Macintyre [1]) which disproved the analogue of Pólya's conjecture for functions of infinite order in all cases with

$$\sum_n \frac{1}{\lambda_n} = \infty. \tag{21.8.1}$$

Possibly, in the general case, the gap-condition

$$\sum_n \frac{1}{\lambda_n} < \infty \tag{21.8.2}$$

of Fejér will be the proper one; in most problems we are still far from it. The analogue of inequality (21.6.2) was proved in my paper (Turán [21]) under the gap-condition

$$\lambda_n \geq n^{1+\varepsilon} \quad (n \geq n_0,\ \varepsilon \text{ arbitrarily small, positive})$$

for a suitable sequence $r_1 < r_2 < \cdots \to \infty$. We prove a stronger result.

Theorem 21.3 (Kővári [2]). *Let*

$$f(z) = \sum_{n=1}^{\infty} d_n z^{\lambda_n} \tag{21.8.3}$$

be an arbitrary entire function satisfying the gap-condition

$$\lambda_n \geq n \log^{2+\eta} n, \qquad \eta > 0. \tag{21.8.4}$$

*If γ is arbitrary and positive, and ε is an arbitrarily small positive number, then the inequality**

$$M(r)^{1-\varepsilon} \leq \min_{\alpha \text{ real}} M(r, \alpha, \alpha + \log^{-\gamma} M(r)) \tag{21.8.5}$$

holds with the exception (perhaps) of an r-set of finite logarithmic measure.

A set of r-values is of finite logarithmic measure if it can be covered by intervals $[\alpha_\nu, \beta_\nu]$ so that

$$\sum_\nu \log \frac{\beta_\nu}{\alpha_\nu} < \infty. \tag{21.8.6}$$

*Occasionally we write $M(r)$ instead of $M(r, f)$ or $M(r, h)$ when the context makes this unambiguous.

The same inequality can also be proved for harmonic entire functions. The proof is performed in this form [with the notation (21.3.1)].

21.9. PROOF

We again use (21.3.2) (i.e., Corollary VI of the first main theorem) but this time with

$$k = \left[5 \log M(r)(\log\log M(r))^{-1-\eta/2}\right] \tag{21.9.1}$$

and

$$\beta - \alpha = \log^{-\gamma} M(r). \tag{21.9.2}$$

Then we get—with the notation (21.3.1)—

(21.9.3)

$$\max_{\varphi} |s_k(re^{i\varphi})| \leq \exp\{3k \log[40 \log^{\gamma} M(r)]\} \max_{\alpha \leq \varphi \leq \alpha + \log^{-\gamma} M(r)} |s_k(re^{i\varphi})|$$

and using (21.9.1)

$$\max_{\varphi} |s_k(re^{i\varphi})| \leq \exp\left\{c(\gamma) \log M(r)[\log\log M(r)]^{-\eta/2}\right\} \tag{21.9.4}$$

$$\max_{\alpha \leq \varphi \leq \alpha + \log^{-\gamma}(r)} |s_k(re^{i\varphi})|.$$

Now instead of the coefficient estimates (21.3.4) we use

$$|a_n| \leq 2\frac{M(\rho)}{\rho^n}, \qquad |b_n| \leq 2\frac{M(\rho)}{\rho^n},$$

but this time with

$$\rho = r\left(1 + \frac{1}{[\log\log M(r)]^{1+\eta/2}}\right). \tag{21.9.5}$$

Then we get

$$\left|\sum_{n>k} r^{\lambda_n}(a_n \cos \lambda_n \varphi + b_n \sin \lambda_n \varphi)\right|$$

$$\leq 4M(\rho)[\log\log M(r)]^{1+\eta/2} \exp\left\{-\frac{\lambda_k}{2[\log\log M(r)]^{1+\eta/2}}\right\}$$

$$< \frac{M(\rho)}{[M(r)]^2}.$$

This last inequality is shown by (21.9.1) and (21.8.4). Thus we get, as in 21.3,

(21.9.6)

$$M(r) < \exp\left\{\frac{\log M(r)}{[\log\log M(r)]^{\eta/3}}\right\}\left\{M(r, \alpha, \alpha + \log^{-\gamma} M(r)) + \frac{M(\rho)}{[M(r)]^2}\right\}$$

for $r > r_1(\gamma)$.

Applying (C.1.4) from Appendix C the proof is complete. ■

21.10

We have already mentioned that the analogue of the theorem of Pólya–Fuchs for functions of infinite order does not even hold for sequences λ_n that satisfy the gap-condition (21.8.2). Kővári found [2] that in the case of the gap-condition (21.8.4) the answer is positive. To prove this he combined Theorem 21.3 with a suitably modified form of Fuchs's inequality (21.7.5)–(21.7.6). If $f(z)$ is an entire function then the inequality

(21.10.1)
$$r\left|\frac{f'}{f}(re^{i\varphi})\right| \le c(f)\log^4 M(r, f)$$

holds except perhaps on an r-set, of finite logarithmic measure. So he proved that if the entire function $f(z)$ satisfies the gap-condition (21.8.4), then for arbitrarily small $\varepsilon > 0$ the inequality

(21.10.2)
$$m(r, f) > M(r, f)^{1-\varepsilon}$$

holds except perhaps on an r-set of finite logarithmic measure. For the proof of (21.10.1) we again refer to the original paper.

Using a more refined argument. Hayman [1] proved inequality (21.10.2) for all entire functions satisfying the weaker gap-condition

(21.10.3)
$$\lim_{x\to\infty}\left(\frac{x}{\log x(\log\log x)^{\alpha}}\right)^{-1}\sum_{\lambda_\nu \le x} 1 = 0, \qquad \alpha > 2$$

with the exception perhaps of an r-set of zero logarithmic density. Neither of these last two theorems contains the other; in Section 22 we give an example where Kővári's theorem is more advantageous. Comparison of these theorems indicates that further weakening of the gap-condition, perhaps reaching the gap-condition (21.8.2) of Fejér, can only be expected by finding the "proper" Φ-density of the exceptional r-set instead of the logarithmic one.

21.11

To indicate how these results lead to the value distribution properties of gap-series we first remark that Pólya's Theorem II implies that no entire functions of finite order with Fabry gap can have an asymptotic path with finite limit. Therefore, from a general theorem, such a function takes every finite a-value infinitely often. That is, it has no Picard exceptional value.* Pólya raised the question whether or not the theorem holds for functions of infinite order. Hayman's theorem implies that this is certainly the case when the gap-condition (21.10.3) holds. But much more can be said. Let $f(z)$ be an arbitrary entire function with $f(0) = 1$. If $f(z) \neq 0$ on the segment

$$0 \leq |z| \leq r, \qquad \operatorname{arc} z = \Theta,$$

then starting from $\operatorname{arc} f(0) = 0$ let $v(t, \Theta)$ for $0 \leq t \leq r$ be defined as the value of $\operatorname{arc}[f(te^{i\Theta})]$ by continuous variation of the argument along $\operatorname{arc} z = \Theta$ from 0 to t. Denote by $n(r, \Theta_1, \Theta_2)$ the number of zeros in the open sector

$$0 < |z| < r, \qquad 0 \leq \Theta_1 < \operatorname{arc} z < \Theta_2 \leq 2\pi.$$

Hayman (in [1]) deduced from the argument principle the general inequality

$$\left| \int_0^r \frac{n(t, \Theta_1, \Theta_2)}{t} dt - \frac{1}{2\pi} \int_{\Theta_1}^{\Theta_2} \log|f(re^{i\vartheta})| \, d\vartheta \right| \leq 4 \left\{ \log M(r) \frac{1}{2\pi^2} \int_0^{2\pi} d\Theta \int_0^r \frac{|v(t, \Theta)|}{t} dt \right\}^{1/2}. \tag{21.11.1}$$

He then needed to show that apart from an r-set of zero logarithmic density the expression in curly bracket is $o(\log M(r))$ (which also required several new ideas). So he proved that if $f(z)$ is an entire function of finite order then not only is every radius a Julia line (even if f has no exceptional value) but also that $f(z)$ does not have any Borel exceptional value relative to an arbitrarily small angle.

*See Pólya [1]; page 640.

22

APPLICATIONS TO FUNCTION THEORY III

The High-Indices Theorem for Borel Summability

22.1

The high-indices theorem of Abel summability, due to Hardy and Littlewood (1926), mentioned at the end of Section 20, asserts that if

$$\lim_{x \to 1-0} \sum_{n=1}^{\infty} a_n x^{\lambda_n} = A \qquad (\lambda_n \text{ integers})$$

and

$$\frac{\lambda_{n+1}}{\lambda_n} \geq q > 1$$

then thc series $\Sigma_n a_n$ is convergent and the gap condition is the "best possible." For the other classic "function theoretical" method of summation, Borel summability, the question was raised by Hardy and Littlewood but solved only in 1965 by Gaier [1]. For a series $\Sigma_n a_n$ with partial sums s_ν, the B (and B') summability methods are defined generally by

$$B[(s_n)] = \lim_{x \to +\infty} e^{-x} \sum_{\nu=0}^{\infty} \frac{s_\nu}{\nu!} x^\nu \tag{22.1.1}$$

and

$$(22.1.2)\qquad B'[(s_n)] = \lim_{x\to+\infty}\int_0^x e^{-t}\left(\sum_{\nu=0}^{\infty}\frac{a_\nu t^\nu}{\nu!}\right)dt.$$

As remarked by Erdős in 1952, if $\{\varepsilon_n\}$ is an arbitrary sequence tending to 0 one can construct a series $\Sigma_n a_n^*$ so that

$$a_n^* = 0 \quad \text{for} \quad n \neq \lambda_k; \qquad \lambda_{k+1} - \lambda_k > \varepsilon_k\sqrt{\lambda_k}\,.$$

Furthermore, $B(s_n^*)$ exists and is finite, but $\Sigma_n a_n^*$ is divergent. Gaier's theorem runs as follows.

Theorem 22.1. *If the series $\Sigma_n a_n$ satisfies—with an arbitrary positive constant ϑ —the cap-condition*

$$(22.1.3)\qquad a_n = 0 \quad \textit{for} \quad n \neq \lambda_k, \qquad \lambda_{k+1} - \lambda_k \geq \vartheta\sqrt{\lambda_k}$$

and is B (or B') summable, then it is convergent.

The original proof of Gaier can be shortened quite substantially using Kővári's inequality (21.10.1). (Gaier's original proof used the first main theorem directly.) Because of the known relations*

$$B(s_n) \to s \quad \text{implies} \quad B'(s_n) \to s$$

and

$$B(s_n) = B(a_n) + B'(s_n)$$

it follows from the hypothesis that

$$(22.1.4)\qquad f_0(x) = \sum_{n=1}^{\infty}\frac{a_n}{n!}x^n = o(e^x)$$

if $x \to \infty$ along the positive axis. Since the gap-condition (21.8.4) is amply satisfied by (22.1.3), (21.10.2) is applicable. So inequality

$$(22.1.5)\qquad M(r, f_0) < m(r, f_0)^{1/1-\varepsilon} \leq f_0(r)^{1/(1-\varepsilon)} = o(e^{r/(1-\varepsilon)})$$

holds for all positive r values except perhaps for a set covered by intervals (a_k, b_k) with

$$(22.1.6)\qquad \sum_k \log\frac{b_k}{a_k} < \infty.$$

*G. H. Hardy, Divergent series, p. 182.

But then for $k \to \infty$ we have

$$b_k = a_k[1 + o(1)] \tag{22.1.7}$$

and hence $a_k < r < b_k$

$$M(r, f_0) \le M(b_k, f_0) = o(e^{b_k}) \le e^{r[1+o(1)]}, \tag{22.1.8}$$

that is, (22.1.8) holds for *all* positive r. Then the Cauchy coefficient estimate can be applied with $r = n$, which gives

$$\left|\frac{a_n}{n!}\right| \le \frac{e^{n(1+\varepsilon)}}{n^n}$$

and so

$$a_n = O[(1 + \varepsilon)^n] \tag{22.1.9}$$

for arbitrarily small $\varepsilon > 0$. For the case when (22.1.9) in addition to B-summability and the gap-condition (22.1.3) is assumed, convergence was already proved by Meyer–König and Zeller.* This completes the proof. ■

* *Math. Z.* Bd. **66**, p. 203.

23

APPLICATIONS TO FUNCTION THEORY IV

Quasi-Analytic Function Classes

23.1

A class of functions defined on a set S is called "a quasi-analytic class in sense A" if the class is closed under sum and difference, and if two members are "close to each other in a neighborhood of a point B of S in the sense A," it follows that these members are "globally equal" on S (i.e., equal everywhere or almost everywhere). In the classical sense, for a set of functions f infinitely differentiable on $[-1,1]$, "sense A of closeness at $x = x_0$" means that

$$f_1^{(n)}(x_0) = f_2^{(n)}(x_0), \qquad n = 0, 1, \ldots, \tag{23.1.1}$$

and "globally equal" means "equal on $[-1,1]$." Suppose a class of functions is defined by the inequalities

$$|f^{(n)}(x)| \le k^n M_n$$

with a positive $k = k(f)$ for $[-1,1]$ and $n = 0, 1, \ldots$. The classic theorem of Denjoy–Carleman says this class is quasi-analytic in the sense of (23.1.1) if and only if the least monotonically decreasing majorant of the series $\Sigma(1/M_\nu^{1/\nu})$ diverges. Another important "sense A" is due to S. Bernstein; it says the functions f_1 and f_2 are equal in an arbitrarily small subinterval of $[0, 2\pi]$ almost everywhere, and "globally equal" means equality almost everywhere in $[0, 2\pi]$. Such a class is formed, for example, by the radial limits of functions regular and bounded in $|z| < 1$. Still another "sense A" is the sense of

Mandelbrojt, which—at a fixed $\alpha > 0$—requires that for a real x_0

$$\lim_{h \to +0} e^{1/h^\alpha} \int_{x_0 - h}^{x_0} |f_1(x) - f_2(x)|\, dx < \infty. \tag{23.1.2}$$

One can also define the local closeness of two functions f_1 and f_2 at $x = x_0$ by requiring

$$\lim_{h \to +0} \frac{1}{\varphi(h)} \sup_{x \in (x_0 - h,\, x_0)} |f_1(x) - f_2(x)| < \infty. \tag{23.1.3}$$

Here $\varphi(x)$ in an interval $[0, c]$ is continuous and strictly monotonically increasing with

$$\varphi(0) = 0 \tag{23.1.4}$$

so that the function

$$H(h) = \frac{\log \varphi(h)}{\log h} \tag{23.1.5}$$

is strictly increasing and tends to $+\infty$ for $h \to +0$. Let the inverse of

$$w = \tfrac{2}{3} H(x) \qquad 0 \le x \le c$$

be

$$x = k(w) \qquad w \ge H(c).$$

Then we have the following.

Theorem 23.1. *Consider the class U of functions*

$$f(x) = \sum_{\nu=1}^{\infty} a_\nu e^{i\lambda_\nu x} \tag{23.1.6}$$

where the λ_ν are real and the coefficients satisfy the condition

$$\overline{\lim_{w \to \infty}} \frac{1}{\varphi[k(w)]} \sum_{\nu > w} |a_\nu| < \infty. \tag{23.1.7}$$

Then the class U is a quasi-analytic class in the sense of (23.1.3).

23.2

In the case

$$\varphi(x) = \exp\left(-\frac{1}{x^\alpha}\right) \tag{23.2.1}$$

with a fixed positive α we get easily, for large enough w,

$$k(w) \geq \left(\frac{2}{\alpha} w \log w\right)^{-1/\alpha}$$

Thus (23.1.7) certainly holds if

$$\overline{\lim_{w \to \infty}} \, e^{(2/\alpha) w \log w} \sum_{\nu > w} |a_\nu| < \infty. \tag{23.2.2}$$

In this case, the example

$$f_0(x) = \exp\left(-\frac{1}{\sin^\beta x}\right) \qquad \beta = 2\left[\frac{1}{\varepsilon}\right] \tag{23.2.3}$$

shows that the theorem is no longer true if in (23.2.2)—in the exponent—the factor $w \log w$ is replaced by $w^{1-\varepsilon}$.

The functions in this class are continuous, but if the exponents are chosen to increase rapidly enough, the functions can be nowhere differentiable, or worse.

For the sake of orientation we introduce the class U_1 of functions of the form

$$f(x) \sim \sum_n (a_n \cos m_n x + b_n \sin m_n x) \tag{23.2.4}$$

where m_n are integers chosen so that the exponent of convergence of $\Sigma_n(1/m_n^\sigma)$ is $\delta < 1$. Mandelbrojt proved [1] that U_1 is a quasi-analytical class in the sense of (23.2.1)–(23.1.3) if

$$\delta < \frac{\alpha}{1+\alpha}$$

and this is no longer true for

$$\delta = \frac{\alpha}{1+\alpha}.$$

This theorem characterizes a class exclusively by the exponents, whereas our theorem is a "coefficient-theorem." Also, Levinson [1] proved that the class of functions

$$f(x) = \frac{1}{\sqrt{2\pi}} \int_{-\infty}^{\infty} g(u) e^{ixu} \, du$$

is a quasi-analytical class in Bernstein's sense if $g \in L(-\infty, +\infty)$ and for $u \to +\infty$

$$g(u) = e^{-\vartheta(u)} \tag{23.2.5}$$

where

$$\int_1^\infty \frac{\vartheta(u)}{u^2}\,du = \infty. \tag{23.2.6}$$

The following uniqueness theorem follows at once from Theorem 23.1.

Theorem 23.2. *If the coefficients of*

$$f(s) = \sum_{n=1}^{\infty} a_n e^{-\lambda_n s}$$

satisfy the inequality

$$\varlimsup_{w\to\infty} e^{(2/\alpha)w\log w} \sum_{\nu>w} |a_\nu| < \infty$$

and for some real t_0

$$\varliminf_{h\to+0} e^{1/h^\alpha} \max_{t_0-h\le t\le t_0} |f(t)| < \infty,$$

then $f(s) \equiv 0$.

23.3. PROOF

We must show that if for an $f(x)$ of the form (23.1.6), for a real x_0, and for a suitable sequence

$$h_1 > h_2 > \cdots \to 0 \tag{23.3.1}$$

the inequality

$$\max_{x_0-h_\nu\le x\le x_0} |f(x)| \le c_1\varphi(h_\nu) \tag{23.3.2}$$

holds, then (23.1.7) implies that

$$f(x) \equiv 0.$$

If not, there exists a real x_1 so that $f(x_1) \neq 0$. Without loss of generality we may assume that

$$f(x_1) = 1 \tag{23.3.3}$$

and also

$$x_1 = 0, \qquad x_0 = 1, \qquad h_\nu < \tfrac{1}{2}. \tag{23.3.4}$$

Fix ν and let

$$N = N(h_\nu) \to \infty$$

to be determined later. We write

$$f(x) = \sum_{\nu \le N} a_\nu e^{i\lambda_\nu x} + \sum_{\nu > N} a_\nu e^{i\lambda_\nu x}. \tag{23.3.5}$$

For all x-values in $[1 - h_\nu, 1]$ and for all ν (23.3.2) and (23.1.7) imply

$$c_1\varphi(h_\nu) \ge \left| \sum_{\nu \le N} a_\nu e^{i\lambda_\nu x} \right| - c_2\varphi(k(N)),$$

that is,

$$\max_{1 - h_\nu \le x \le 1} \left| \sum_{\nu \le N} a_\nu e^{i\lambda_\nu x} \right| \le c_1\varphi(h_\nu) + c_2\varphi[k(N)]. \tag{23.3.6}$$

Apply the first main theorem in the form (18.1.4) with $a = 1 - h_\nu$, $d = h_\nu$. This and (23.3.6) give

$$c_1\varphi(h_\nu) + c_2\varphi[k(N)] \ge \left(\frac{h_\nu}{6}\right)^N \left| \sum_{\nu \le N} a_\nu \right|,$$

that is, recalling (23.3.3) and (23.3.4),

$$c_1\varphi(h_\nu) + c_2\varphi[k(N)] > \tfrac{1}{2} h_\nu^{(5/4)N} > h_\nu^{(4/3)N}, \tag{23.3.7}$$

for all sufficiently large ν. Choosing

$$N = \tfrac{2}{3} H(h_\nu) \tag{23.3.8}$$

we have

$$\varphi[k(N)] = \varphi(h_\nu)$$

and thus (23.3.7) yields

$$c_3\varphi(h_\nu) > h_\nu^{\frac{4}{3}\cdot\frac{2}{3}H(h_\nu)} = h_\nu^{\frac{8}{9}\cdot\frac{\log \varphi(h_\nu)}{\log h_\nu}} = \varphi(h_\nu)^{8/9},$$

which is a contradiction if ν is large.

24

APPLICATIONS TO FUNCTION THEORY V

Value Distribution of Entire Functions Satisfying a Linear Differential Equation (with a Digression to Transcendental Numbers)

24.1

The simplest linear differential equation $Y^{(n+1)}(t) \equiv 0$ has as solutions all polynomials of degree $\leq n$. By "the fundamental theorem of algebra" *all* solutions of the equation take *all* values in *all* disks at most n-times. One then asks what can be said about the value distribution of solutions of the equation

$$y^{(n)} + a_1 y^{(n-1)} + \cdots + a_n = 0 \qquad (a_\nu \text{ constant}). \tag{24.1.1}$$

These solutions are the functions

$$y(z) = \sum_{\nu=1}^{k} P_\nu(z) e^{\alpha_\nu z} \tag{24.1.2}$$

where $\alpha_1, \ldots, \alpha_k$ are the distinct zeros of the characteristic equation

$$t^n + a_1 t^{n-1} + \cdots + a_n = 0, \tag{24.1.3}$$

and $P_\nu(z)$ are polynomials so that

$$\sum_{\nu=1}^{k} \operatorname{degr} P_\nu(z) = n. \tag{24.1.4}$$

The question has—from a different origin—a long history, dating back at least to Poincaré. A typical result of these investigations is due to Pólya [1] who proved, for the number $n(r)$ of solutions of $y = 0$ in $|z| \le r$, the asymptotic formula

$$n(r) = Ar + O(\log r)$$

where A does not depend on r. The error term was improved to $O(1)$ by G. Dickson [1], however, this depends heavily on the coefficients of the P_ν. Let

$$Y(z) = \sum_{j=1}^{n} b_j e^{\alpha_j z}. \tag{24.1.5}$$

Theorem 24.1 (Turán [26]). *Let M stand for the radius of the smallest disk containing all the α_ν; let*

$$\min_{\mu \ne \nu} |\alpha_\mu - \alpha_\nu| = D \tag{24.1.6}$$

and suppose ϑ is an arbitrary real number with $0 < \vartheta < 1$. Then the number of solutions of $Y(z) = 0$ in an arbitrary disk $|z - z_0| \le R$ (including multiplicity) cannot exceed

$$(e + 5\vartheta)MR + (n+1) + n\log\left(1 + \frac{n^3}{2\vartheta RD}\right). \tag{24.1.7}$$

The point of this theorem is, of course, the fact that the upper bound does not depend on the b_j, or on the choice of z_0, and depends on the exponents α_j rather loosely, by way of M and D only. Observing that M cannot exceed

$$\frac{\Delta}{\sqrt{3}} = \frac{1}{\sqrt{3}} \max_{\mu, \nu} |\alpha_\mu - \alpha_\nu|, \tag{24.1.8}$$

the upper bound depends on the exponents—more symmetrically—only Δ and D. Easy examples show that the dependence upon M, R, and n is of the right size.

Let β be an arbitrary complex number and let us try to get an upper bound for the number of β-places of $Y(z)$ in $|z - z_0| \le R$. Since

$$Y(z) - \beta = \sum_{\nu=1}^{n+1} b_j e^{\alpha_j z}$$

with $b_{n+1} = -\beta$ and $\alpha_{n+1} = 0$, Theorem 24.1 has the following immediate consequence.

Corollary. *If M_1 denotes the radius of the smallest disk containing*

$$\alpha_1, \ldots, \alpha_n, \alpha_{n+1}, \; D_1 = \min_{\substack{\mu, \nu = 1, \ldots, n+1 \\ \mu \neq \nu}} |\alpha_\mu - \alpha_\nu|,$$

then the number of β places of $Y(z)$ in $|z - z_0| \leq R$ cannot exceed

$$(24.1.9) \qquad (e + 5\vartheta) M_1 R + (n+2) + (n+1) \log\left(1 + \frac{(n+1)^3}{2\vartheta R D_1}\right).$$

Thus the upper bound is independent of β.

24.2

To prove Theorem 24.1 using Theorem 18.3 we first notice that since the zeros of $Y(z)$ and $Y(z)e^{cz}$ are the same, we may suppose that

$$(24.2.1) \qquad \max_{j=1,\ldots,n} |\alpha_j| \leq M$$

in (24.1.5), and second

$$(24.2.2) \qquad z_0 = 0.$$

Consider the vectors

$$(24.2.3) \qquad \boldsymbol{\delta}_{\mu\nu} = \boldsymbol{\alpha}_\mu - \boldsymbol{\alpha}_\nu = |\boldsymbol{\alpha}_\mu - \boldsymbol{\alpha}_\nu| \mathbf{e}^{\mathbf{i}\varphi_{\mu\nu}}, \qquad 1 \leq \mu \neq \nu \leq n.$$

There is an open angle measuring $2\pi/n(n-1)$ with vertex at the origin so that neither the angle nor its opposite contains any of the $\boldsymbol{\delta}_{\mu\nu}$. Hence there is a γ so that for each $\mu \neq \nu$ either

$$(24.2.4) \qquad -\frac{\pi}{2} + \frac{\pi}{n(n-1)} \leq \varphi_{\mu\nu} - \gamma \leq \frac{\pi}{2} - \frac{\pi}{n(n-1)}$$

or

$$(24.2.5) \qquad \frac{\pi}{2} + \frac{\pi}{n(n-1)} \leq \varphi_{\mu\nu} - \gamma \leq \frac{3\pi}{2} - \frac{\pi}{n(n-1)}$$

holds. Let

$$(24.2.6) \qquad 0 < \vartheta < 1$$

be arbitrary; apply Theorem 18.3 with

$$a = 0, \qquad d = \vartheta R, \qquad w_j = \alpha_j e^{-i\gamma}. \tag{24.2.7}$$

In the case (24.2.4) we have

$$\begin{aligned}|e^{w_\mu \cdot d/n} - e^{w_\nu \cdot d/n}| &= \left|\exp\left(\vartheta \frac{R}{n}\alpha_\nu e^{-i\gamma}\right)\right| \cdot \left|-1 + \exp\left(\vartheta \frac{R}{n}|\alpha_\mu - \alpha_\nu| e^{i(\varphi_{\mu\nu}-\gamma)}\right)\right| \\ &\geq \left[-1 + \exp\left(\vartheta \frac{R}{n} D \sin \frac{\pi}{n(n-1)}\right)\right] \min_\nu \left|\exp\left(\vartheta \frac{R}{n}\alpha_\nu e^{-i\gamma}\right)\right| \\ &\geq \frac{2\vartheta RD}{n^3} \cdot \min_\nu \left|\exp\left(\vartheta \frac{R}{n}\alpha_\nu e^{-i\gamma}\right)\right|.\end{aligned}$$

In the case (24.2.5),

$$\begin{aligned}|e^{w_\mu \cdot d/n} - e^{w_\nu \cdot d/n}| &\geq \left\{1 - \exp\left[-\vartheta RD \frac{2}{n^2(n-1)}\right]\right\} \\ &\quad \cdot \min_\nu \left|\exp\left(\vartheta \frac{R}{n}\alpha_\nu e^{-i\gamma}\right)\right| \\ &\geq \frac{2\vartheta RD}{n^3 + 2\vartheta RD} \min_\nu \left|\exp\left(\vartheta \frac{R}{n}\alpha_\nu e^{-i\gamma}\right)\right|.\end{aligned}$$

Hence, for all pairs $\mu \neq \nu$ and with (24.2.1), we get

$$\frac{\min_{\mu \neq \nu} |e^{w_\mu \cdot d/n} - e^{w_\nu \cdot d/n}|}{\max_\nu |e^{w_\nu \cdot d/n}|} \geq \frac{2\vartheta RD}{n^3 + 2\vartheta RD} \exp\left(-2\vartheta \frac{R}{n} M\right). \tag{24.2.8}$$

Hence condition (18.7.1) is fulfilled with

$$\delta = \frac{2\vartheta RD}{n^3 + 2\vartheta RD} \exp\left(-\frac{2\vartheta R}{n} M\right), \tag{24.2.9}$$

and thus (18.7.2)–(18.7.3) is applicable. We obtain an x^* with

$$0 \leq x^* \leq \vartheta R$$

so that

$$\begin{aligned}\frac{\sum_{j=1}^{n} |b_j| \left|\exp\left(\alpha_j x^* e^{-i\gamma}\right)\right|}{\left|\sum_{j=1}^{n} b_j \exp\left(\alpha_j x^* e^{-i\gamma}\right)\right|} &\leq n \left\{2 \frac{n^3 + 2\vartheta DR}{2\vartheta RD} \cdot \exp\left(\frac{2\vartheta RM}{n}\right)\right\}^n \\ &< e^{n+1}\left(1 + \frac{n^3}{2\vartheta RD}\right)^n \exp(2\vartheta RM).\end{aligned} \tag{24.2.10}$$

This means for the $Y(z)$ in (24.1.5) that for a suitable z^* the inequality

$$\frac{\sum_{j=1}^{n} |b_j||e^{\alpha_j z^*}|}{|Y(z^*)|} < e^{n+1}\left(1 + \frac{n^3}{2\vartheta RD}\right)^n \exp(2\vartheta RM)$$

holds on the segment

$$(24.2.11) \qquad 0 \le |z| \le \vartheta R, \qquad \operatorname{arc} z = -\gamma.$$

But then, using (24.2.1), on the circle $|z - z^*| = \rho$

$$\max_{|z-z^*|=\rho} |Y(z)| = \max_{|z-z^*|=\rho} \left| \sum_{j=1}^{n} b_j e^{\alpha_j z^*} e^{\alpha_j (z-z^*)} \right|$$
$$\le \left(\sum_{j=1}^{n} |b_j||e^{\alpha_j z^*}| \right) \exp(M\rho).$$

The last two inequalities give for all $\rho > 0$

(24.2.12)

$$\log \frac{\max_{|z-z^*|=\rho} |Y(z)|}{|Y(z^*)|} \le n + 1 + M\rho + 2\vartheta RM + n \log\left(1 + \frac{n^3}{2\vartheta RD}\right).$$

Since $n(R)$ cannot be greater than the number of zeros of Y in the disk $|z - z^*| \le (1 + \vartheta)R$, we can apply Jensen's inequality and (24.2.12) with $\rho = e(1 + \vartheta)R$. This completes the proof. ∎

24.3

The dependence of the upper bound on D prohibited immediate extensions of the foregoing results to an arbitrary solution of *any* equation (24.1.1). This was subsequently accomplished in our papers with S. Dancs [1–2] but our lower bound depended again on D. Van der Poorten [1–3], and Tijdeman [2–3] succeeded in removing this last blemish.

Theorem 24.2 (Tijdeman). *Denote by M the radius of the smallest disk containing all zeros of* (24.1.3). *The number $n(R)$ of zeros (with multiplicity) of any given solution of* (24.1.1) *in an arbitrary disk $|z - z_0| \le R$ cannot exceed*

$$MR \frac{\lambda_1 + \lambda_2}{\log \dfrac{\lambda_2^2 + \lambda_1}{\lambda_2(1 + \lambda_1)}} + n \frac{1 + \log(1 + \lambda_2/\lambda_1)}{\log \dfrac{\lambda_2^2 + \lambda_1}{\lambda_2(1 + \lambda_1)}},$$

where λ_1 and λ_2 are positive constants subject only to

(24.3.1) $$\lambda_2 > 1, \qquad \lambda_2 > \lambda_1 > 0.$$

If we set

(24.3.2) $$\lambda_1 = 0.1072, \qquad \lambda_2 = 4.6949$$

we get the very elegant upper bound

(24.3.3) $$\frac{10}{3}(MR + n).$$

This, however, is not always better than (24.1.7).

24.4. PROOF

Suppose without loss of generality

(24.4.1) $$z_0 = 0,$$

(24.4.2) $$f(z) = \sum_{j=1}^{n} b_j e^{\alpha_j z},$$

and

(24.4.3) $$\max_j |\alpha_j| = M.$$

With the notation (24.3.1), let

$$R_1 = \lambda_1 R, \qquad R_2 = \lambda_2 R,$$

that is,

(24.4.4) $$R < R_2, \qquad R_1 < R_2.$$

Let $z^* = R_2 e^{i\varphi_0}$ be such that

(24.4.5) $$|f(z^*)| = \max_{|z| = R_2} |f(z)| = M(R_2)$$

and apply Corollary VII of the first main theorem in Section (18.5.7) with

$$A = z^*, \qquad B = R_1 e^{i\varphi_0}, \qquad C = -R_1 e^{i\varphi_0}.$$

Since the maximum of $|f(z)|$ on the segment $\overline{BC}$ cannot exceed $M(R_1)$, this corollary implies that

$$\text{(24.4.6)} \qquad \frac{M(R_2)}{M(R_1)} \leq \left[e\left(\frac{R_2+R_1}{R_1}\right)\right]^n \cdot e^{M(R_1+R_2)}.$$

Applying Lemma D.1 (see Appendix D), this shows that the number $n(R)$ of zeros in $|z| \leq R$ satisfies the assertion of the theorem.

24.5

Using form (18.5.9) of the first main theorem, the previous argument gives the inequality

$$\text{(24.5.1)} \quad n(R) \leq MR\frac{\lambda_1+\lambda_2}{\log\frac{\lambda_2^2+\lambda_1}{\lambda_2(1+\lambda_1)}} + (n-1)\frac{1+\log(1+\lambda_2/\lambda_1)}{\log\frac{\lambda_2^2+\lambda_1}{\lambda_2(1+\lambda_1)}}.$$

Choosing λ_2, at fixed $\lambda_1 = 1$, so large that

$$\text{(24.5.2)} \qquad \frac{1+\log(1+\lambda_2)}{\log\frac{\lambda_2^2+1}{2\lambda_2}} = \frac{2n-1}{2n-2},$$

then choosing R so small that

$$\text{(24.5.3)} \qquad RM\frac{1+\lambda_2}{\log\frac{\lambda_2^2+1}{2\lambda_2}} \leq \frac{1}{4},$$

we obtain $n(R) \leq n - \frac{1}{4}$, that is,

$$\text{(24.5.4)} \qquad n(R) \leq n-1.$$

There is no sharper inequality. ■

The choice (24.3.2) automatically gives the general inequality

$$\text{(24.5.4)} \qquad n(R) \leq \frac{10}{3}(MR+n-1).$$

Reasoning as in 24.1 at once gives the following.

Corollary. *Denote by β an arbitrary nonzero complex number and by M_1 the radius of the smallest disk containing all zeros of* (24.1.3) *including* 0. *Then*

$n(R, z_0, \beta)$, the number of β-places of an arbitrary solution of (24.1.1) *in an arbitrary disk $|z - z_0| \leq R$, satisfies the inequality*

$$n(R, z_0, \beta) \leq \frac{10}{3}(MR + n),$$

independent of the solution, of β and of z_0.

24.6

One may also use Theorem 18.4 of van der Poorten to prove a similar type of result. With the notation (24.1.1)–(24.1.2) and with M denoting the radius of the least disk containing all α_j's, we may again suppose that

$$\max_j |\alpha_j| = M.$$

Theorem 18.4 gives, for all complex z_0, the inequality

$$\max_{\nu=0,1,\dots,n} \left|y^{(\nu)}(z_0)\right| \leq \frac{|y(0)|}{2n} e^{-(2M+1)|z_0|}. \tag{24.6.1}$$

Applying it to $y(z_0 - z)$, this yields for all solutions of (24.1.1) the inequality

$$\left|y(z_0)\right| \leq 2ne^{(2M+1)|z_0|} \max_{\nu=0,1,\dots,n} \left|y^{(\nu)}(0)\right|. \tag{24.6.2}$$

Choosing z_0 on $|z| = R_2 > R$ so that

$$\left|y(z_0)\right| = \max_{|z|=R_2} \left|y(z)\right| = M(R_2)$$

and applying Cauchy's estimate with a $0 < R_1 < R_2$,

$$\left|y^{(\nu)}(0)\right| \leq \nu! \frac{M(R_1)}{R_1^\nu}.$$

(24.6.2) implies

$$\frac{M(R_2)}{M(R_1)} \leq 2ne^{(2M+1)R_2} \max_{\nu=0,1,\dots,n-1} \frac{\nu!}{R_1^\nu}. \tag{24.6.3}$$

Applying Lemma D.1 we get the following inequality.

Theorem 24.3 (van der Poorten). *With the restriction* (24.3.1) *the inequality*

$$n(R) \leq \frac{\lambda_2}{\log \dfrac{\lambda_2^2 + \lambda_1}{\lambda_2(1 + \lambda_1)}} R(2M + 1) + \frac{\log 2n + \log \max\limits_{\nu=0,1,\ldots,n-1} \nu/(\lambda_1 R)^{\nu}}{\log \dfrac{\lambda_2^2 + \lambda_1}{\lambda_2(1 + \lambda_1)}}$$

holds.

24.7

At a colloquium in Imperial College London in September 1964, speaking on my joint results with S. Dancs contained in papers [1–2], I made two conjectures. The first asserted that our bounds can be freed from dependence on D [in (24.1.6)]. The second asserted that our results can be inverted. In other words, if all y-solutions of an equation

$$y^{(n)} + f_1(z)y^{(n-1)} + \cdots + f_n(z)y = 0 \qquad (24.7.1)$$

with entire functions $f_1(z)$ have the property that *all* values, in *all* disks with radius R, are taken at most

$$c_1(y)R + c_2(y)$$

times, then the $f_\nu(z)$ are constants. The first conjecture was proved by van der Poorten and Tijdeman. The second was solved by H. Wittich [1] in a stronger form. Namely, the same conclusion can be drawn on the $f_\nu(z)$ from the weaker assumption that *a single* R_0 value exists so that *all* y-solutions of (24.7.1) take, in *all* disks $|z - z_0| \leq R_0$, *one* $\beta = \beta(y)$ value at most c-times where c is independent of z_0 (but can depend on y). Thus for the first time the possibility that value distribution properties can *characterize* the type of differential equation was confirmed.

24.8

We conclude this section by briefly explaining the effect of the first main theorem on the theory of transcendental numbers through Tijdeman's Theorem 24.2. Gelfond began, in 1949, the investigation of algebraic independence of certain groups of classical numbers [apart from the earlier corollary of Lindeman's theorem that for rationally independent algebraic $\gamma_1,\ldots,\gamma_n$ the numbers e^{γ_ν} $(\nu = 1,\ldots,n)$ are algebraically independent]. What he calls his first fundamental theorem (Gelfond [1]) asserts that if α_0, α_1, α_2 are linearly independent over the field Q of rationals (i.e., rationally independent), and the same holds for η_0, η_1, η_2, and moreover

$$|x_0\eta_0 + x_1\eta_1 + x_2\eta_2| > e^{-cx\log x} \quad \text{for all integers } x_j \text{ satisfying}$$

$$\sum_0^2 |x_j| > 0, \qquad |x_j| \leq x, \tag{24.8.1}$$

then at least two of the numbers

$$\alpha_0, \alpha_1, \alpha_2, e^{\alpha_j\eta_\mu} \qquad j = 1,2,3, \qquad \mu = 0,1,2 \tag{24.8.2}$$

are algebraically independent over Q. (And they are transcendental, of course.) Tijdeman proved [4] the same *without* the requirement (24.8.1). His proof runs parallel to Gelfond's except at a crucial point [the only point where (24.8.1) is used by Gelfond]. From the assumptions on the η_j and α_j [without (24.8.1), of course] and the Gelfond–Schneider theorem, he deduces that there is a transcendental number w among the numbers

$$\alpha_0, \alpha_1, e^{\alpha_0\eta_0}, e^{\alpha_1\eta_0}.$$

If the theorem were false, each of the numbers in (24.8.2) would lie in a field $Q(w, w_1)$ where w_1 is algebraic over $Q(w)$. Then he constructs an exponential polynomial

$$f(z) = \sum_{k_0=0}^{K-1} \sum_{k_1=0}^{K-1} \sum_{k_2=0}^{K-1} A_{k_0k_1k_2} \exp\{(k_0\alpha_0 + k_1\alpha_1 + k_2\alpha_2)z\}, \tag{24.8.3}$$

with

$$A_{k_0k_1k_2} = \sum_{k=0}^{P-1} C_{k_0k_1k_2k} w^k$$

where $K = K(N)$, $P = P(N)$, N are large, and $C_{k_0k_1k_2k}$ rational integers so that

$$f^{(\nu)}(\eta_0x_0 + \eta_1x_1 + \eta_2x_2) = 0 \tag{24.8.4}$$

for

$$0 \leq \nu < S = S(N), \qquad 0 \leq x_j < N, \qquad j = 0,1,2. \tag{24.8.5}$$

This can be easily done by considering an appropriate homogeneous linear system with more than twice as many variables than equations. For an appropriate integer T of the field $Q(w, w_1)$, the quantities

$$T\alpha_k, \qquad Te^{\alpha_k\eta_j}$$

are integers of $Q(w, w_1)$, and the quantities

$$T^{\mu} f^{(\nu)}(\eta_0 x_0 + \eta_1 x_1 + \eta_2 x_2) \tag{24.8.6}$$

in the range

$$0 \leq \nu < S^* = S^*(N) = cS \tag{24.8.7}$$
$$0 \leq x_j < N, \qquad j = 0, 1, 2$$

are all integers of this field for a suitable rational integer $\mu = \mu(N)$. Here c is a constant larger than 1, depending only on the degree of w_1 with respect to $Q(w)$. An ingenious use of Cauchy's integral representation, combined with (24.8.4), makes it possible—as in Gelfond's proof—to all make all quantities in (24.8.6)–(24.8.7) "very small" depending on N. Now a vital point of Gelfond's argument was to show that there is at least one element of the set (24.8.6)–(24.8.7) that is not zero.

For this he devised an ingenious analytical argument dealing *directly* with the values $(\eta_0 x_0 + \eta_1 x_1 + \eta_2 x_2)$, but then he could control the situation only by requiring (24.8.1). However Tijdeman showed, using inequality (24.3.3), that the total number of zeros of $f(z)$ in a domain containing all points $(\eta_0 x_0 + \eta_1 x_1 + \eta_2 x_2)$ in (24.8.7) is less than $S^* N^3$. This implies the existence of the crucial nonzero element without using (24.8.1). The rest of the argument is Gelfond's. The nonzero element can be rewritten as a

$$\text{"very small"} \quad |P_N(w)| > 0$$

where $P_N(x)$ is a polynomial with rational integer coefficients whose degree and height are under good control. Choosing $K(N)$, $P(N)$, $S(N)$, $S^*(N)$ properly, one can apply Gelfond's last lemma to conclude that w is algebraic; this is a contradiction.

A similar argument was given in a paper of Waldschmidt [1] where van der Poorten's Theorem 24.3 was used instead of Theorem 24.2. In this paper it was shown that if $x_1, \ldots, x_N$ (resp. $y_1, y_2, \ldots, y_M$) are rationally independent and $MN \geq 2(M + N)$, then at least two of the numbers $e^{x_\nu y_\mu}$ are algebraically independent. His conjecture that here one can take $N = 2$ and that at least $(M - 1)$ among the $e^{x_\nu y_\mu}$ quantities are algebraically independent seems, to me, not hopeless; more generally the effect of Theorems 24.2 and 24.3 on the theory of transcendental numbers seems far from being exhausted.

25

APPLICATIONS TO FUNCTION THEORY VI

Linear Combinations of Certain Entire Functions

25.1

It is a long-standing and vague statement of Borel's that if

$$(25.1.1)\qquad f_1(x) = \sum_{j=1}^{n_1} b_j' e^{i\nu_j' x}$$

$$f_2(x) = \sum_{j=1}^{n_2} b_j'' e^{i\nu_j'' x} \qquad \nu_j', \nu_j'' \text{ real}$$

are "close" to each other on a "large" interval then in some sense

$$(25.1.2)\qquad b_j' \sim b_j'', \qquad \nu_j' \sim \nu_j''.$$

The exact meaning of this is still unknown. In this Section we supply a meaning to it when closeness in a disk is required of the functions

$$(25.1.3)\qquad \Phi_1(z) = \sum_{j=1}^{n_1} b_j' e^{\lambda_j' z}$$

$$\Phi_2(z) = \sum_{j=1}^{n_2} b_j'' e^{\lambda_j'' z} \qquad \lambda_j', \lambda_j'' \text{ complex.}$$

Combining the exponents λ_j', λ_j'' in one sequence λ_j and setting

$$\Phi_1(z) - \Phi_2(z) = F(z) = \sum_{j=1}^{n} b_j e^{\lambda_j z},$$

we ask when we can conclude that the condition

$$|F(z)| \le \varepsilon \quad \text{in} \quad |z| \le R \tag{25.1.4}$$

implies that

$$\sum_j |b_j|^2 \le \varepsilon^2. \tag{25.1.5}$$

This investigation has an interesting consequence. It is easy to see that if $f(z)$ is regular for $|z| \le R$ and $\varepsilon > 0$ is arbitrarily small, then

$$\left| f(z) - \sum_{j=1}^{n_1} d_j' e^{\mu_j' z} \right| \le \varepsilon \tag{25.1.6}$$

for suitable complex d_j', μ_j' numbers. Suppose now that in addition to (25.1.6) the inequality

$$\left| f(z) - \sum_{j=1}^{n_2} d_j'' e^{\mu_j'' z} \right| \le \varepsilon \tag{25.1.7}$$

holds as well. Then after uniting the sequences μ_j', μ_j'' we have

$$\left| \sum_{j=1}^{n} \left(d_j' - d_j'' \right) e^{\mu_j z} \right| \le 2\varepsilon, \quad \text{in} \quad |z| \le R. \tag{25.1.8}$$

Applying our result concerning (25.1.4)–(25.1.5), we conclude that

$$\sum_j \left| d_j' - d_j'' \right|^2 \le \varepsilon^2,$$

that is, $f(z)$ can be "well" approximated by an exponential polynomial in a "large" disk "essentially uniquely."

25.2

To investigate the problem in a more general context let

$$h(z) = \sum_{\nu=0}^{\infty} a_\nu z^\nu \tag{25.2.1}$$

be a prescribed entire function. Gelfond [2] investigated the conditions on $h(z)$

necessary for every $f(z)$ (regular in $|z| \le R$) to be approximated here so that

(25.2.2) $$\left| f(z) - \sum_{j=1}^{n} d_j h(\lambda_j z) \right| \le \varepsilon$$

for arbitrarily small $\varepsilon > 0$ and suitable complex d_j and λ_j. He found that the necessary and sufficient condition for this is

(25.2.3) $$a_\nu \ne 0, \qquad \nu = 0, 1, \ldots$$

—necessity being trivial. So instead of (25.1.4)–(25.1.5) we investigate when the validity of

(25.2.4) $$\left| \sum_{j=1}^{n} b_j h(\lambda_j z) \right| \le \varepsilon$$

for $|z| \le R$ implies

(25.2.5) $$\sum_{j=1}^{n} |b_j|^2 \le \varepsilon^2.$$

Instead of (25.1.6)–(25.1.7) we investigate the uniqueness of approximation by

(25.2.6) $$F(z) = \sum_{j=1}^{n} d_j h(\lambda_j z) \quad \text{for} \quad |z| \le R.$$

25.3

The functions of (25.2.6) are obviously entire functions; if $h(z)$ is of finite order so is $F(z)$, and its order cannot exceed that of $h(z)$. Can it be smaller? If

(25.3.1) $$h_0(z) = \sum_{\nu=0}^{\infty} \frac{z^{5\nu+1}}{(5\nu+1)!},$$

and choosing

(25.3.2) $$n = 5, \qquad d_1 = \cdots = d_5 = 1, \qquad \lambda_j = e^{2\pi i j/5},$$

the corresponding $F(z)$ is identically 0. So we must impose a restriction on $h(z)$. We say that $h(z)$ in (25.2.1) satisfies the $E_n(w)$-condition, with a positive w, if for every positive integer g the inequality

(25.3.3) $$\min_{g \le \mu, \nu \le g+n-1} \left| \frac{a_\mu}{a_\nu} \right| \ge \frac{A}{(g+n)^{nw}}$$

holds; A can depend on n and w but not on g. This is somewhat stronger than (25.2.3). The function $h(z) = e^z$ obviously satisfies the $E_n(1)$-condition for all natural numbers n. We first establish the following.

Theorem 25.1. *Suppose the entire function $h(z)$ of finite order satisfies the $E_n(w)$-condition for an n and $w > 0$. Then the function*

$$F(z) = \sum_{j=1}^{n} d_j h(\lambda_j z) \tag{25.3.4}$$

—for complex d_j's and distinct λ_j—can be of smaller order than $h(z)$ if and only if

$$\sum_{j=1}^{n} |d_j|^2 = 0. \tag{25.3.5}$$

An obvious reformulation of the theorem states that the $E_n(w)$ functions are "asymptotically linearly independent over the complex number field."

We use the convention

$$|\lambda_1| \geq |\lambda_2| \geq \cdots \geq |\lambda_n|. \tag{25.3.6}$$

When we are interested in the order *and type* of $F(z)$ we assume the type cannot exceed that of $h(\lambda_1 z)$.

Theorem 25.2. *If the entire function $h(z)$ of finite order and type satisfies the $E_n(w)$ condition for an n and $w > 0$, then the function $F(z)$ in* (25.3.4) *has the order and type of $h(\lambda_1 z)$, except in the case* (25.3.5).

The conclusion of this last theorem is stronger than that of Theorem 25.1; still we discuss both since the inequality leading to Theorem 25.1 is in one respect stronger than the corresponding statement in the proof of Theorem 25.2, and this will be of importance in the uniqueness question.

25.4

The proofs of these two theorems run parallel, up to a certain point. Since

$$F(z) = \sum_{\nu=0}^{\infty} z^\nu a_\nu \left(\sum_{j=1}^{n} d_j \lambda_j^\nu \right), \tag{25.4.1}$$

Parseval's formula gives

$$M_2(R, F) = \frac{1}{2\pi}\int_0^{2\pi}|F(\mathrm{R}e^{i\varphi})|^2\,d\varphi \tag{25.4.2}$$

$$= \sum_{\nu=0}^{\infty}|a_\nu|^2\left|\sum_{j=1}^{n} d_j\lambda_j^\nu\right|^2 R^{2\nu}$$

$$= \sum_{l=0}^{\infty}\sum_{\nu=\mathrm{ln}}^{(l+1)n-1}|a_\nu|^2\left|\sum_{j=1}^{n} d_j\lambda_j^\nu\right|^2 R^{2\nu}.$$

Denote

$$\frac{\min\limits_{\mu\neq\nu}|\lambda_\mu-\lambda_\nu|}{\max\limits_j|\lambda_j|} = \delta \tag{25.4.3}$$

and apply Theorem 11.1 with $m = ln - 1$, l fixed. This gives an integer ν_l satisfying

$$ln \leq \nu_l \leq (l+1)n - 1 \tag{25.4.4}$$

so that

$$\left|\sum_{j=1}^{n} d_j\lambda_j^{\nu_l}\right| \geq \frac{1}{2}\left(\frac{\delta}{2}\right)^n\sum_{j=1}^{n}|d_j|\,|\lambda_j|^{\nu_l}.$$

We also have

$$\left|\sum_{j=1}^{n} d_j\lambda_j^{\nu_l}\right|^2 R^{2\nu_l} \geq \frac{1}{4n^2}\left(\frac{\delta}{2}\right)^{2n} R^{2\nu_l}\cdot\sum_{j=1}^{n}|d_j|^2|\lambda_j|^{2\nu_l}$$

and from (25.4.2)

$$M_2(R, F) \geq \frac{1}{4n^2}\left(\frac{\delta}{2}\right)^{2n}\sum_{j=1}^{n}|d_j|^2\left(\sum_{l=0}^{\infty}|a_{\nu_l}|^2(R|\lambda_j|)^{2\nu_l}\right). \tag{25.4.5}$$

Now we consider for

$$\rho \geq 1 \tag{25.4.6}$$

$$U(\rho) = \sum_{l=0}^{\infty}|a_{\nu_l}|^2\rho^{2\nu_l}. \tag{25.4.7}$$

Apply the $E_n(w)$-condition. For an arbitrary integer k with

$$ln \le k \le (l+1)n - 1$$

we obtain

$$\frac{|a_{\nu_l}|}{|a_k|} \ge \frac{A}{(ln+n)^{nw}} \ge \frac{A}{(k+n)^{nw}},$$

that is, by (25.4.6)

$$|a_{\nu_l}|^2 \rho^{2\nu_l} \ge \frac{A^2}{(k+n)^{2nw}} \frac{|a_k|^2 \rho^{2k}}{\rho^{2n}}$$

and also

$$|a_{\nu_l}|^2 \rho^{2\nu_l} \ge \frac{A^2}{n\rho^{2n}} \sum_{k=ln}^{(l+1)n-1} \frac{|a_k|^2 \rho^{2k}}{(k+n)^{2nw}}. \tag{25.4.8}$$

If

$$R > \frac{1}{|\lambda_n|}, \tag{25.4.9}$$

then (25.4.6) is fulfilled with $\rho = R|\lambda_j|$ and (25.4.8) yields

$$\begin{aligned} M_2(R, F) \ge \frac{A^2}{4n^3}\left(\frac{\delta}{2}\right)^{2n} \frac{1}{R^{2n}} \sum_{j=1}^{n} \frac{|d_j|^2}{|\lambda_j|^{2n}} \\ \cdot \left(\sum_{k=0}^{\infty} \frac{|a_k|^2}{(k+n)^{2nw}} (R|\lambda_j|)^{2k} \right). \end{aligned} \tag{25.4.10}$$

Using the well known formula

$$\frac{1}{(k+n)^c} = \frac{1}{\Gamma(c)} \int_0^\infty e^{-(k+n)t} t^{c-1}\, dt,$$

we get

$$\begin{aligned} \sum_{k=0}^{\infty} \frac{a_k^2}{(k+n)^{2nw}} \rho^{2k} &= \frac{1}{\Gamma(2nw)} \int_0^\infty \left\{ e^{-nt} t^{2nw-1} \cdot \sum_{k=C}^{\infty} |a_k|^2 (\rho e^{-t/2})^{2k} \right\} dt \\ &= \frac{1}{\Gamma(2nw)} \int_0^\infty e^{-nt} t^{2nw-1} M_2(\rho e^{-t/2}, h)\, dt \\ &= \frac{2}{\Gamma(2nw)} \int_0^\rho \left(\frac{y}{\rho}\right)^{2n} \left(2 \log \frac{\rho}{y}\right)^{2nw-1} M_2(y, h) \frac{dy}{y}. \end{aligned}$$

with the abbreviation

$$\frac{A^2}{4n^3}\cdot\left(\frac{\delta}{2}\right)^{2n}\frac{2^{2nw}}{\Gamma(2nw)} = A_1 \tag{25.4.11}$$

we have obtained

$$M_2(R, F) \geq \frac{A_1}{R^{2n}}\sum_{j=1}^{n}\frac{|d_j|^2}{\lambda_j^{2n}}\int_0^{R|\lambda_j|}\left(\frac{y}{R|\lambda_j|}\right)^{2n} \cdot\left(\log\frac{R|\lambda_j|}{y}\right)^{2nw-1} M_2(y, h)\frac{dy}{y}. \tag{25.4.12}$$

Since the last integral is larger than

$$\int_{(R|\lambda_j|/4)}^{(1/e)R|\lambda_j|} > \frac{1}{4^{2n}}M_2\left(\frac{R|\lambda_j|}{4}, h\right)\log\frac{4}{e},$$

then for $R > 1/|\lambda_n|$ the inequality

$$M_2(R, F) \geq \frac{A_1}{2^{4n+1}R^{2n}}\sum_{j=1}^{n}\frac{|d_j|^2}{|\lambda_j|^{2n}}M_2\left(\frac{R|\lambda_j|}{4}, h\right) \tag{25.4.13}$$

holds.

25.5

The proofs for both Theorems 25.1 and 25.2 follow from (25.4.12), respectively (25.4.13). This last gives

$$M_2(R, F) \geq \frac{A_1 M_2[(R|\lambda_n|/4), h]}{|\lambda_1|^{2n}2^{4n+1}R^{2n}}\sum_{j=1}^{n}|d_j|^2, \tag{25.5.1}$$

and, in view of the easy inequalities

$$M(R, F)^2 \geq M_2(R, F)$$

$$M_2(\rho, h) \geq \tfrac{1}{2}M\left(\frac{\rho}{2}, h\right)^2,$$

we also have the inequality

$$M(R, F) \geq \frac{\sqrt{A_1}\left(\sum_{j=1}^{n}|d_j|^2\right)^{1/2}}{|\lambda_1|^n 2^{2n+1}R^n}M\left(\frac{R|\lambda_n|}{8}, h\right), \quad \text{for} \quad R > \frac{1}{|\lambda_n|}. \tag{25.5.2}$$

This immediately proves Theorem 25.1. From (25.4.12) [see (25.3.6)] it follows that for $R > 1/|\lambda_n|$

$$M_2(R, F) \geq \frac{A_1}{R^{2n}} \frac{|d_1|^2}{|\lambda_1|^{2n}} \int_0^{R|\lambda_1|} \left(\frac{y}{R|\lambda_1|}\right)^{2n} \left(\log\frac{R|\lambda_1|}{y}\right)^{2nw-1} M_2(y, h)\frac{dy}{y}.$$

Thus replacing the integral by

$$\int_{R|\lambda_1|(1-\varepsilon)}^{R|\lambda_1|(1-\varepsilon/2)}$$

we get

$$M_2(R, F) \geq \frac{A_1}{R^{2n}} \frac{|d_1|^2}{|\lambda_1|^{2n}} (1-\varepsilon)^{2n} \cdot \log^{2nw-1}\frac{1}{1-\varepsilon/2} \tag{25.5.3}$$

$$\cdot M_2(R|\lambda_1|(1-\varepsilon), h) \cdot \log\frac{1-\varepsilon/2}{1-\varepsilon},$$

from which Theorem 25.2 follows as before.

25.6

Suppose now that for $|z| \leq R$

$$\left|f(z) - \sum_{j=1}^{n} d_j' h(\lambda_j z)\right| \leq \varepsilon \tag{25.6.1}$$

$$\left|f(z) - \sum_{j=1}^{n} d_j'' h(\lambda_j z)\right| \leq \varepsilon$$

and $h(z)$ satisfies the $E_n(w)$-condition in (25.3.3). Then

$$\left|\sum_{j=1}^{n} (d_j' - d_j'') h(\lambda_j z)\right| \leq 2\varepsilon$$

and inequality (25.5.2) can be applied. For $R > 1/|\lambda_n|$ this gives

$$\left(\sum_{j=1}^{n} |d_j' - d_j''|^2\right)^{1/2} \leq \varepsilon \frac{4^{n+1}|\lambda_1|^n}{\sqrt{A_1}} \frac{R^n}{M(R|\lambda_n|/8, h)}.$$

Theorem 25.3. *If $h(z)$ satisfies the $E_n(w)$-condition, R is larger than $1/|\lambda_n|$ and*

$$\frac{4^{n+1}|\lambda_1|^n}{\sqrt{A_1}} R^n \leq \max_{|z|=R|\lambda_n|/8} |h(z)|$$

then (25.6.1) *implies*

$$\sum_{j=1}^{n} \left|d_j' - d_j''\right|^2 \leq \varepsilon^2.$$

The case $h(z) = e^z$ is important enough to receive a direct formulation.

Corollary. *If R is larger than $1/|\lambda_n|$ and*

$$3\sqrt{(2n+2)!}\left(\frac{4|\lambda_1|}{\delta}\right)^n R^n \leq \exp\left(\frac{R|\lambda_n|}{8}\right), \tag{25.6.2}$$

then the inequalities (25.1.6)–(25.1.8) *imply*

$$\sum_{j=1}^{n} \left|d_j' - d_j''\right|^2 \leq \varepsilon^2.$$

Condition (25.6.2) could be improved.

25.7

Using, instead of Theorem 11.1, the first (resp. second) main theorems, we could get different inequalities estimating $M_2(R, F)$ from below by $M_2(R|\lambda_n|/8, h)$. These have an advantage over (25.5.2) in that they do not depend on δ (in A_1), but their dependence upon the d_j coefficients is less favorable than in (25.5.2). Since the proofs are similar we confine ourselves to the formulation of the following.

Theorem 25.4. *If $h(z)$ satisfies the $E_n(w)$-condition for some n, then for $R > 1/|\lambda_n|$ the inequality*

$$\max_{|z|=R}\left|\sum_{j=1}^{n} d_j h(\lambda_j z)\right| \geq \frac{c(n, w, A)}{R^{n(w+2)}}\left|\sum_{j=1}^{n} d_j\right| \max_{|z|=R|\lambda_n|/8} |h(z)|$$

holds with an explicit $c(n, w, A)$.

26

APPLICATIONS TO FUNCTION THEORY VII

Varia

A. REMARKS ON RATIONAL FUNCTIONS

26.1

Let

$$R_n(z) = \frac{u(z)}{v(z)}$$

be a rational function where $u(z)$ and $v(z)$ are polynomials without common zeros of degrees less than n, and n, respectively. Let

$$I(r, R_n) = \frac{1}{2\pi}\int_{-\pi}^{\pi} \left|R_n(re^{i\varphi})\right|^2 d\varphi. \tag{26.1.1}$$

We reproduce here some special cases of results presented in the joint paper with Q. I. Rahman [1]. There we gave nontrivial lower bounds for this integral mean in terms of very limited information about R_n, provided that all its poles are in $|z| > r$.

26.2

Theorem 26.1 (Rahman–Turán [1]). *Suppose that all the poles of R_n are in the annulus*

$$r < |z| < 1$$

and it is normalized by $R_n(0) = 1$. *Then*

$$I(r, R_n) \geq 1 + \frac{r^{2n}}{\sum_{\nu=0}^{n-1} \binom{n}{\nu}^2 r^{2\nu}}.$$

Note that the estimate does not depend on the coefficients of the numerator and denominator of R_n. For r tending to zero the inequality is sharp, as shown by the example $R_n(z) = 1/(1 - z^n)$ where

$$I\left(r, \frac{1}{1 - z^n}\right) = \frac{1}{1 - r^{2n}} = 1 + [1 + o(1)] r^{2n} \qquad (r \to 0).$$

The logarithmic derivative of a polynomial is an important special case.

Corollary (Rahman–Turán [1]). *Let* $p_n(z)$ *be a polynomial of degree n with all its zeros in the annulus* $r < |z| < 1$. *Then, with the notation* (26.1.1),

$$I\left(r, \frac{p_n'}{p_n}\right) \geq \frac{n^2 r^{2(n-1)}}{\sum_{\nu=0}^{n-1} \binom{n}{\nu}^2 r^{2\nu}}.$$

This case is special in that the bound does not even depend on $\frac{p_n'}{p_n}(0)$. For $r \to 0$ the estimate is again accurate since

$$I\left(r, \frac{(1 - z^n)'}{1 - z^n}\right) = n^2 r^{2(2n-1)} \cdot \sum_{k=0}^{\infty} (k + 1)^2 r^{2nk} \tag{26.2.1}$$

$$= n^2 r^{2(n-1)} [1 + o(1)] \qquad (r \to 0).$$

26.3

We have so far restricted the *largest* pole of $R_n(z)$. There is obviously a deeper problem if we only restrict the smallest. To give a bound in terms of a few simple quantities is then possible only for the previous special case.

Theorem 26.2 (Rahman–Turán [1]). *Suppose that* $p_n(z)$ *is a polynomial of degree n having its smallest zero on* $|z| = 1$. *Then—with the notation* (26.1.1)—

$$I\left(\rho, \frac{p_n'}{p_n}\right) \geq \frac{1}{(1 + e)^2 \sum_{\nu=1}^{n} [1/\nu^2 \rho^{2(\nu-1)}]} \qquad (\rho < 1).$$

As (26.2.1) shows, this is essentially the best possible result for small ρ, as the right-hand side here is

$$(1+e)^{-2} n^2 \rho^{2(n-1)}[1+o(1)] \qquad (\rho \to 0).$$

26.4

To prove Theorem 26.1, observe that its hypotheses imply the representation

$$R_n(z) = \sum_{i=1}^{n} \frac{b_i}{1 - z_i z} = \sum_{\nu=0}^{\infty} g(\nu) z^{\nu} \qquad (|z| = r)$$

where

$$g(\nu) = \sum_{i=1}^{n} b_i z_i^{\nu}, \qquad g(0) = 1, \qquad |z_i| \geq 1.$$

The conclusion follows from Parseval's inequality and Theorem 6.7:

$$I(r, R_n) \geq \sum_{\nu=0}^{n} |g(\nu)|^2 r^{2\nu} \geq 1 + \frac{r^{2n}}{\sum_{\nu=0}^{n-1} \binom{n}{\nu}^2 r^{2\nu}}.$$

26.5

To deduce the corollary, we use the preceding inequality with

$$R_n(z) = 1 - \frac{1}{n} z \frac{p_n'(z)}{p_n(z)}$$

satisfying all the assumptions. By Parseval's formula we have

$$I(r, R_n) = 1 + I\left(r, \frac{1}{n} z \frac{p_n'(z)}{p_n(z)}\right),$$

and the corollary follows.

26.6

To prove Theorem 26.2 we use the representation

$$\frac{p_n'(z)}{p_n(z)} = -\sum_{i=1}^{n} \frac{z_i}{1 - z z_i} = -\sum_{\nu=1}^{\infty} s_{\nu} z^{\nu-1} \qquad (|z| < 1)$$

where

$$s_\nu = \sum_{i=1}^{n} z_i^\nu, \qquad \max_{i=1,\ldots,n} |z_i| = 1.$$

By Parseval's inequality

$$I\left(\rho, \frac{p_n'}{p_n}\right) \geq \sum_{\nu=1}^{n} |s_\nu|^2 \rho^{2(\nu-1)}. \tag{26.6.1}$$

Now use inequality (3.16.4) in the proof of Buchholtz's Theorem 3.6 with $r = e^{1/n}$. By Cauchy's inequality,

$$1 \leq (1+e) \sum_{\nu=1}^{n} \cdot \frac{|s_\nu|}{\nu}$$

$$\leq (1+e)\sqrt{\sum_{\nu=1}^{n} |s_\nu|^2 \rho^{2(\nu-1)} \sum_{\nu=1}^{n} \left[1/\nu^2 \rho^{2(\nu-1)}\right]}. \tag{26.6.2}$$

Combining (26.6.1) and (26.6.2) completes the proof. ■

B. COEFFICIENT ESTIMATIONS

(No clue has been left to the contents under this heading which was obviously intended to collect a number of interesting, though less important, consequences of power sum inequalities—Remark of the editors.)

27

APPLICATIONS TO DIFFERENTIAL EQUATIONS I

Growth of the Solutions of Ordinary Equations in L_2-Norm, Remarks on Stability

27.1

Let $z_1, \ldots, z_n$ be complex, let

$$X(t) = \begin{pmatrix} x_1(t) \\ \vdots \\ x_n(t) \end{pmatrix}, \qquad Z = \begin{pmatrix} z_1 \\ \vdots \\ z_n \end{pmatrix} \tag{27.1.1}$$

be $n \times 1$ column vectors, similarly $Y(t)$, $W(Z, t), \ldots,$ and set

$$|X(t)|^2 = \sum_{j=1}^{n} |x_j(t)|^2. \tag{27.1.2}$$

Let $\mathbf{A} = (a_{jk})$ be an $n \times n$ matrix with complex entries, $\lambda_1, \ldots, \lambda_n$ its eigenvalues, not necessarily distinct, with

$$\min_j \operatorname{Re} \lambda_j = -\Lambda, \tag{27.1.3}$$

and the norm $|\mathbf{A}|$ of $\mathbf{A}$ defined by

$$|\mathbf{A}|^2 = \sum_{j,k} |a_{jk}|^2. \tag{27.1.4}$$

We also use the norm

$$\overline{|\mathbf{A}|} = \max_{j,k} |a_{jk}|. \tag{27.1.5}$$

27.2

We consider the equation in the usual normal form

$$\frac{dX(t)}{dt} = \mathbf{A}X(t) + W[X(t), t] \tag{27.2.1}$$

and let $X(t)$ be a solution that exists for $t \geq 0$. We suppose that for all vectors Z and all $t \geq 0$ the inequality

$$|W(Z, t)| \leq k|Z| \tag{27.2.2}$$

holds with a positive constant k; furthermore, assume that every component of $W(Z, t)$ is continuous in the half-space $t \geq 0$ of the $(2n + 1)$-dimensional $(\xi_1, \eta_1, \ldots, \xi_n, \eta_n, t)$-space, where $z_j = \xi_j + i\eta_j$ with real ξ_j and η_j.

If $\gamma > 1$ is a constant and $\gamma t = \tau$, $X(t) = X_1(\tau)$, then (27.2.1) takes the form

$$\frac{dX_1(\tau)}{d\tau} = \mathbf{A}_1 X_1(\tau) + W_1[X_1(\tau), \tau] \tag{27.2.3}$$

where

$$\mathbf{A}_1 = \frac{1}{\gamma}\mathbf{A}$$

and

$$W_1(Z, \tau) = \frac{1}{\gamma} W\left(Z, \frac{\tau}{\gamma}\right),$$

and thus

$$|W_1(Z, \tau)| \leq \frac{k}{\gamma}|Z|.$$

Hence k in the constraint (27.2.2) can be normalized. To get a possibly simple inequality, supposing $\mathbf{A} \neq \mathbf{0}$, we take

$$k = \sqrt{\frac{\overline{|\mathbf{A}|}}{d}} \exp\left(-n\left[6 + 4d\overline{|\mathbf{A}|}\right]\right) \tag{27.2.4}$$

with a $d \geq 1$.

Theorem 27.1. *Under the constraint* (27.2.2)–(27.2.4) *all solutions of* (27.2.1) *satisfy, for all* $\alpha \geq 0$, *the (explicit) inequality*

$$\int_{\alpha}^{\alpha+2d} |X(t)|^2 e^{2\Lambda t}\, dt \geq e^{-11n}\left(\int_0^d |X(t)|^2 e^{2\Lambda t}\, dt\right) e^{-(11n/d)\alpha} \tag{27.2.5}$$

if $n \geq 2$.

This inequality is slightly better than the one in Turán [39]; the improvement is due to the use of Theorem 19.1 of Tijdeman instead of our analogous inequality with n^2 in the exponent instead of $2n$. Still, no effort is made here to derive the best constants the method can yield, though for some applications it could be significant.

If we have a solution with

$$\lim_{t \to +\infty} |X(t)| = 0, \tag{27.2.6}$$

then it is enough to require (27.2.2)–(27.2.4) only for

$$|Z| \leq c, \qquad t \geq 0. \tag{27.2.7}$$

Then (27.2.5) holds for

$$\alpha \geq 0$$

if the inequality

$$|X(t)| \leq c$$

holds for $t \geq 0$.

27.3

A classic case arises when $W(Z, t)$ satisfies the inequality

$$|W(Z, t)| \leq \varepsilon |Z| \tag{27.3.1}$$

with an arbitrary small $\varepsilon > 0$ if only

$$|Z| \leq \delta(\varepsilon), \qquad t > t_0(\varepsilon). \tag{27.3.2}$$

Then in the case (27.2.6) for all $\alpha > t_1(\varepsilon, X)$, d in (27.2.4) can be chosen arbitrarily large, and the *proof* of Theorem 27.1 gives the following consequence.

Corollary. *For all solutions $X(t)$ with* (27.2.6), *and for arbitrarily small $\eta > 0$, each interval $[\alpha, \alpha + 2n/\eta]$ contains a $t^* = t^*(X)$ satisfying the inequality*

$$|X(t^*)| \geq c(\eta, X)e^{-(\Lambda + 7\eta)\alpha} \tag{27.3.3}$$

if (27.3.1)–(27.3.2) *is fulfilled, $\alpha \geq \alpha_0(\eta, X)$ (and $|X(\alpha_0)| \neq 0$).*

Perron proved, under mild additional requirements on $W(Z, t)$, that one *always* has a solution $X_0(t)$ with

$$\overline{\lim_{t \to \infty}} \frac{\log|X_0(t)|}{t} = -\Lambda. \tag{27.3.4}$$

Hence, for each member of this large subclass of equations, (27.3.3) is "essentially" the best possible result and yields much better information on $X_0(t)$ than does (27.3.4).

27.4

The first main theorem also has ramifications on the theory of asymptotical stability. In this theory the normal form of the equation is (27.2.1) with

$$W(0, t) \equiv 0,$$

and the stability of the solution $X(t) \equiv 0$ is investigated.

Theorem 27.2. *Suppose*

$$\mathbf{A} \neq \mathbf{0}, \quad \Lambda > 0 \quad [\text{see } (27.1.3)] \tag{27.4.1}$$

and let $\Delta > 1$ be such that

$$[2e(\Delta + 1)]^n e^{-\Lambda\Delta} \leq 1, \qquad 2\sqrt{n} \leq e^{\Lambda\Delta}. \tag{27.4.2}$$

Suppose that for all Z and $t \geq 0$

$$|W(Z, t)| \leq K_1|Z| \quad [K_1 \text{ in } (27.7.5)] \tag{27.4.3}$$

and (27.2.6) *is fulfilled. Then, for every* $\alpha \geq 0$, *there is a* τ *with*

$$\alpha \leq \tau \leq \alpha + \Delta^2 + \Delta \tag{27.4.4}$$

so that the inequality

$$|X(\tau)| \geq |X(0)| e^{-(1+3/\Delta)\Lambda\tau} \tag{27.4.5}$$

holds.

As before, the inequality in (27.4.5) is "essentially" the best possible for a large class of equations. Easy modifications of the proof would give an analogous assertion for all sufficiently large α if (27.4.3) is required only for $|Z| \leq a_1$, $t \geq a_2$. Some significance can be attached to the fact that the "nearly best possible" inequality (27.4.5) can be asserted in explicitly given finite intervals instead of the $\overline{\lim}$-form of Perron.

27.5. PROOF OF THEOREM 27.1

For the sake of convenience, suppose first that

$$\lambda_\mu \neq \lambda_\nu \quad \text{for} \quad \mu \neq \nu, \tag{27.5.1}$$

and let $b \geq 0$ be fixed. Let $Y(t)$ be the solution of

$$\frac{dY}{dt} = \mathbf{A}Y \tag{27.5.2}$$

with

$$Y(b) = X(b). \tag{27.5.3}$$

Then

$$Y(t) = \mathbf{C}\begin{pmatrix} e^{\lambda_1 t} \\ \vdots \\ e^{\lambda_n t} \end{pmatrix} \tag{27.5.4}$$

for some $\mathbf{C}$. Applying Theorem 19.1 to each component of $Y(t+b)e^{\Lambda(t+b)}$ we get, for $a - b \geq d$, the inequality

$$\int_b^{b+d} |Y(t)|^2 e^{2\Lambda t}\, dt \leq e^4 n \left(2e\frac{a-b+d}{d}\right)^{2n} \cdot \int_a^{a+d} |Y(t)|^2 e^{2\Lambda t}\, dt. \tag{27.5.5}$$

For $a = b + d$, this gives

$$\int_b^{b+d} |Y(t)|^2 e^{2\Lambda t}\, dt \le ne^{6n+4} \int_{b+d}^{b+2d} |Y(t)|^2 e^{2\Lambda t}\, dt. \tag{27.5.6}$$

From the inequalities

$$|Y(t)|^2 \le 2|X(t)|^2 + 2|Y(t) - X(t)|^2$$

$$|Y(t)|^2 \ge \tfrac{1}{2}|X(t)|^2 - |Y(t) - X(t)|^2$$

and from (27.5.6) we get

$$\int_b^{b+d} |X(t)|^2 e^{2\Lambda t}\, dt \le 4ne^{6n+4} \int_{b+d}^{b+2d} |X(t)|^2 e^{2\Lambda t}\, dt + (2 + 4ne^{6n+4}) \int_b^{b+2d} |X(t) - Y(t)|^2 e^{2\Lambda t}\, dt. \tag{27.5.7}$$

27.6

Now consider

$$J = \int_b^{b+2d} |X(t) - Y(t)|^2 e^{2\Lambda t}\, dt. \tag{27.6.1}$$

If $\mathbf{H}(t)$ is the $n \times n$ matrix consisting of n solutions of the equation (27.5.2) so that

$$\mathbf{H}(b) = \mathbf{E} \quad (\mathbf{E} \text{ is the } n \times n \text{ unit matrix}) \tag{27.6.2}$$

then Liouville's formula gives (for $t \ge b$)

$$X(t) - Y(t) = \int_b^t \mathbf{H}(t + b - \tau) W(X(\tau), \tau)\, d\tau.$$

Hence, with the notation (27.1.4),

$$|X(t) - Y(t)|^2 \le \left(\int_b^t |\mathbf{H}(t + b - \tau)|^2\, d\tau \right) \cdot \left(\int_b^t |W(X(\tau), \tau)|^2\, d\tau \right). \tag{27.6.3}$$

Applying the classical lemma of Ljapunov and Perron to each column and summing over all columns we get,

$$|\mathbf{H}(r)|^2 \le n \exp(2n \overline{|\mathbf{A}|}(r - b)), \quad \text{for} \quad r \ge b.$$

Thus

$$(27.6.4)\qquad \int_b^t |\mathbf{H}(t+b-\tau)|^2\,d\tau \le \frac{1}{2|\overline{\mathbf{A}}|}\exp[2n|\overline{\mathbf{A}}|(t-b)].$$

Using the abbreviations

$$(27.6.5)\qquad \sqrt{\frac{|\overline{\mathbf{A}}|}{2dn}}\,e^{-2n-3-4nd|\overline{\mathbf{A}}|} = K,$$

we have, by (27.2.2)–(27.2.4),

$$\int_b^t |W(X(\tau),\tau)|^2\,d\tau \le K^2\cdot\int_b^t |X(\tau)|^2\,d\tau.$$

From (27.6.3) and (27.6.4), for $b \le t \le b+2d$,

$$|X(t)-Y(t)|^2 \le \frac{1}{2|\overline{\mathbf{A}}|}K^2\exp(4n|\overline{\mathbf{A}}|d)\int_b^t |X(\tau)|^2\,d\tau.$$

Hence, from (27.6.1), we have

$$J \le \frac{K^2}{2|\overline{\mathbf{A}}|}\exp(4n|\overline{\mathbf{A}}|d)\int_b^{b+2d}|X(\tau)|^2 e^{2\Lambda\tau}\left[\int_\tau^{b+2d} e^{2\Lambda(t-\tau)}\,dt\right]d\tau$$

and, since the inner integral cannot exceed

$$2d\cdot\max(1,e^{4\Lambda d}) \le 2d\cdot\exp[2d(\Lambda+|\Lambda|)] \le 2d\cdot\exp(4nd|\overline{\mathbf{A}}|),$$

also

$$(27.6.6)\qquad J \le \frac{K^2 d}{|\overline{\mathbf{A}}|}\exp(8nd|\overline{\mathbf{A}}|)\left\{\int_b^{b+d} + \int_{b+d}^{b+2d}\right\}|X(\tau)|^2 e^{2\Lambda\tau}\,d\tau.$$

Thus (27.5.7) takes the form

$$\int_b^{b+d}|X(\tau)|^2 e^{2\Lambda\tau}\,d\tau \le ne^{6n+6}\int_{b+d}^{b+2d}|X(\tau)|^2 e^{2\Lambda\tau}\,d\tau$$

$$+ne^{6n+6+8nd|\overline{\mathbf{A}}|}\cdot\frac{K^2 d}{|\overline{\mathbf{A}}|}\left\{\int_b^{b+d}|X(\tau)|^2 e^{2\Lambda t}\,dt\right.$$

$$\left.+\int_{b+d}^{b+2d}|X(t)|^2 e^{2\Lambda T}\,dt\right\},$$

and, from (27.6.5), we get

(27.6.7)

$$\int_b^{b+d} |X(\tau)|^2 e^{2\Lambda\tau}\, d\tau \le 2\left(ne^{6n+6} + \frac{1}{2}\right)\int_{b+d}^{b+2d} |X(\tau)|^2 e^{2\Lambda\tau}\, d\tau$$

$$< e^{7n+7}\int_{b+d}^{b+2d} |X(\tau)|^2 e^{2\Lambda\tau}\, d\tau.$$

Since this inequality is independent of the configuration of the λ_j, passage to the limit gives (27.6.7) without the restriction (27.5.1).

Replace b by $b + d, b + 2d, \ldots, b + \nu d$ in (27.6.7) to obtain

$$\int_b^{b+d} |X(\tau)|^2 e^{2\Lambda t}\, d\tau \le e^{\nu(7n+7)}\int_{b+\nu d}^{b+(\nu+1)d} |X(\tau)|^2 e^{2\Lambda\tau}\, d\tau. \tag{27.6.8}$$

Choosing $b = 0$ and ν as the smallest integer such that

$$\nu d > \alpha,$$

we have

$$(\nu + 1)d \le (\alpha + d) + d = \alpha + 2d.$$

Thus $[\nu d, (\nu + 1)d]$ is contained in $[\alpha, \alpha + 2d]$. Since

$$\nu = \left[\frac{\alpha}{d}\right] + 1 \le \frac{\alpha}{d} + 1$$

the proof is complete. ∎

27.7. PROOF OF THEOREM 27.2

We may again suppose (27.5.1), and with $b \ge 0$, consider $Y(t)$ in (27.5.2)–(27.5.3). Then, for a suitable ν $(1 \le \nu \le n)$, we have

$$|Y_\nu(b)| \ge \frac{1}{\sqrt{n}}|Y(b)| = \frac{1}{\sqrt{n}}|X(b)|. \tag{27.7.1}$$

Applying the first main theorem to $Y_\nu(t + b)e^{\Lambda t}$ we get, for a suitable t^* such that

$$b + \Delta^2 \le t^* \le b + \Delta^2 + \Delta, \tag{27.7.2}$$

the inequality

$$|Y(t^*)|e^{\Lambda(t^*-b)} \ge \frac{1}{\sqrt{n}}|X(b)|\left[\frac{1}{2e(\Delta + 1)}\right]^n. \tag{27.7.3}$$

This implies that

(27.7.4)

$$\frac{|X(b)|^2}{n}[2e(\Delta + 1)]^{-2n} \le 2e^{2\Lambda(t^* - b)}\left[|X(t^*)|^2 + |Y(t^*) - X(t^*)|^2\right].$$

Estimating $|X(t) - Y(t)|^2$ for $b + \Delta^2 \le t \le b + \Delta^2 + \Delta$ by the aid of Liouville's formula as in 27.6 and using the abbreviation

$$\frac{1}{2}\sqrt{\frac{\overline{|A|}}{(\Delta^2 + \Delta)}}\; e^{-3n\overline{|\mathbf{A}|}(\Delta^2 + 2\Delta)} = K_1, \tag{27.7.5}$$

we have

$$|X(t) - Y(t)|^2 \le \frac{K_1^2}{2\overline{|\mathbf{A}|}} e^{2n\overline{|\mathbf{A}|}(\Delta^2 + \Delta)} \int_b^t |X(\tau)|^2\, d\tau$$

for $b \le t \le b + \Delta^2 + \Delta$. Hence

(27.7.6)

$$|X(t) - Y(t)|^2 \le \frac{K_1^2}{\overline{|\mathbf{A}|}} e^{2n\overline{|\mathbf{A}|}(\Delta^2 + \Delta)}\left\{\int_b^t |Y(\tau)|^2\, d\tau + \int_b^t |X(\tau) - Y(\tau)|^2\, d\tau\right\}.$$

The first integral in the curly bracket is, by the inequality of Ljapunov and Perron,

$$\le (\Delta^2 + \Delta)|Y(b)|^2 e^{2n\overline{|\mathbf{A}|}(\Delta^2 + \Delta)} = |X(b)|^2 e^{2n\overline{|\mathbf{A}|}(\Delta^2 + \Delta)}(\Delta^2 + \Delta) \tag{27.7.7}$$

for $b \le t \le b + \Delta^2 + \delta$.

If $|X(\tau) - Y(\tau)|$ takes its maximum M with respect to $[b, b + \Delta^2 + \Delta]$ at $\tau = t^{**}$, then, using (27.7.6) with $t = t^{**}$,

$$M^2 \le \frac{K_1^2}{\overline{|\mathbf{A}|}} e^{2n\overline{|\mathbf{A}|}(\Delta^2 + \Delta)}\left\{|X(b)|^2 e^{2n\overline{|\mathbf{A}|}(\Delta^2 + \Delta)} \cdot (\Delta^2 + \Delta) + M^2(\Delta^2 + \Delta)\right\}.$$

Thus by (27.7.5)

$$|Y(t^*) - X(t^*)|^2 \le K_1^2 e^{4n\overline{|\mathbf{A}|}(\Delta^2 + \Delta)} \frac{2}{\overline{|\mathbf{A}|}} |X(b)|^2 (\Delta^2 + \Delta),$$

and by (27.7.4)

$$2e^{2\Lambda(\Delta^2+\Delta)}|X(t^*)|^2$$

$$\geq \frac{|X(b)|^2}{n}\left\{[2e(\Delta+1)]^{-2n} - 2e^{2\Lambda(\Delta^2+\Delta)}\left(\frac{2nK_1^2(\Delta^2+\Delta)}{|\mathbf{A}|}\right)e^{4n|\mathbf{A}|(\Delta^2+\Delta)}\right\}.$$

From (27.4.2) and (27.7.5) we get

$$|X(t^*)| \geq |X(b)|e^{-\Lambda(\Delta^2+3\Delta)}. \tag{27.7.8}$$

Setting $b = t_0$, $t^* = t_1$, we have

$$|X(t_1)| \geq e^{-\Lambda(\Delta^2+3\Delta)}|X(t_0)| = U|X(t_0)|$$

for a suitable t_1 satisfying the restriction

$$\Delta^2 \leq t_0 + \Delta^2 \leq t_1 \leq t_0 + (\Delta^2 + \Delta).$$

Then the restriction concerning the simplicity of the λ_j can be dropped as before.

Choosing t_1 as b, there exists a t_2 with

$$t_1 + \Delta^2 \leq t_2 \leq t_1 + (\Delta^2 + \Delta)$$

such that

$$|X(t_2)| \geq U^2|X(t_0)|.$$

Continuing this process we get a sequence $t_0, t_1, \ldots$ such that

$$t_\nu + \Delta^2 \leq t_{\nu+1} \leq t_\nu + \Delta^2 + \Delta$$

and

$$|X(t_\nu)| \geq U^\nu|X(t_0)|.$$

Hence for $\alpha \geq t_0$ the interval

$$\alpha \leq t \leq \alpha + \Delta^2 + \Delta$$

contains some t_ν, and since

$$t_0 + \nu\Delta^2 \leq t_\nu \leq t_0 + \nu(\Delta^2 + \Delta), \qquad \nu(\Delta^2 + 3\Delta) \leq (t_\nu - t_0)\left(1 + \frac{3}{\Delta}\right),$$

choosing $t_0 = 0$, the proof is finished. ∎

28

APPLICATIONS TO DIFFERENTIAL EQUATIONS II

Oscillation Theorem

28.1

Consider the equation

$$(28.1.1)\qquad y^{(n)}(t)+[a_{n-1}+\varphi_{n-1}(t)]\,y^{(n-1)}(t)+\cdots+[a_0+\varphi_0(t)]\,y(t)=0$$

with real constants a_ν and real functions $\varphi_\nu(t)$ which are assumed to be ν times continuously differentiable for $t \geq t_0$. Suppose that the zeros $\lambda_1,\ldots,\lambda_n$ of the equation

$$(28.1.2)\qquad w(\lambda)=\lambda^n+a_{n-1}\lambda^{n-1}+\cdots+a_0=0$$

satisfy the restriction

$$(28.1.3)\qquad \lambda=\min_{j=1,\ldots,n}|\operatorname{Im}\lambda_j|>0$$

and that the λ_j are simple. Let

$$\max(1,|a_0|,\ldots,|a_{n-1}|)=B.$$

Theorem 28.1. *There is an (explicitly calculable)*

(28.1.4)

$$H = H(B, n, \lambda) \qquad \left(= \frac{1}{7} \frac{(8e)^{-n}}{(B+3)^n} \exp\left\{ -(B+1)\frac{\pi}{2\lambda}(n^2+2n) \right\}, e.g. \right)$$

so that the inequality

$$(28.1.5.) \quad |\varphi_\nu^{(l)}(t)| \le H \qquad \nu = 0, 1, \ldots, n-1; \qquad l = 0, 1, \ldots, \nu;$$

$$t \ge t_0$$

implies that all real solutions $y(t)$ of Equation (28.1.1) [*with* (28.1.3)] *change sign in every interval*

$$(28.1.6) \qquad \left[\gamma, \gamma + \frac{n\pi}{2\lambda}\right] \quad \text{if} \quad \gamma \ge t_0, y(\gamma) \ne 0 \quad \text{and}$$

$$|y(t)| \le |y(\gamma)| e^{Bn(t-\gamma)} \quad \text{for} \quad \gamma \le t \le \gamma + \frac{n\pi}{2\lambda}.$$

28.2

The zeros of $w(\lambda)$ are all simple. Let $f(t) = \sum_{j=1}^{n} b_j e^{\lambda_j t}$ be real for $t \ge \gamma$. We can apply the reasoning of Sections 12.2–12.6 (with $l \le n$ since the a_ν's are real) and Section 6.3 to the function

$$f(t+\gamma)e^{-\Lambda t} \qquad \left(\Lambda = \min_j \operatorname{Re} \lambda_j\right)$$

with $a = 0$, $\gamma \ge t_0$, and $d > 0$ so small that $(d/n) \max |\operatorname{Im} \lambda_j| < \pi/2$. Then κ might be chosen as $d\lambda/n$, and the interval $[a, a + (d/2)(3 + \pi/\kappa)]$ becomes $[0, 3d/2 + \pi n/2\lambda]$. Since the right-hand side is independent of d we can let $d \to +0$. Thus, for suitable t_1, t_2 with

$$(28.2.1) \qquad \gamma \le t_1, t_2 \le \gamma + \frac{n\pi}{2\lambda},$$

the inequalities

$$(28.2.2) \quad f(t_1)e^{-\Lambda(t_1-\gamma)} = \max_{0 \le t \le n\pi/2\lambda} f(t+\gamma)e^{-\Lambda t} \ge |f(\gamma)| \frac{1}{6n}(8e)^{-n}$$

and

$$(28.2.3)\quad f(t_2)e^{-\Lambda(t_2-\gamma)} = \min_{0\le t\le n\pi/2\lambda} f(t+\gamma)e^{-\Lambda t} \le -|f(\gamma)|\frac{1}{6n}(8e)^{-n}$$

hold.* Hence $f(t)$ has a sign change in the interval $[\gamma, \gamma + n\pi/2\lambda]$ if $f(\gamma) \neq 0$.

28.3

If we suppose that $h(t)$ is continuous for $t \ge t_0$, then all solutions of the equation

$$z^{(n)}(t) + a_{n-1}z^{(n-1)}(t) + \cdots + a_0 z(t) = h(t)$$

have, with an arbitrary but fixed $\gamma \ge t_0$, the form

$$(28.3.1)\qquad z(t) = \sum_{j=1}^{n} c_j e^{\lambda_j t} + \int_\gamma^t h(\tau)F(t-\tau)\,d\tau \quad \text{for} \quad t \ge \gamma$$

where

$$(28.3.2)\qquad F(u) = \sum_{j=1}^{n} \frac{e^{\lambda_j u}}{w'(\lambda_j)}.$$

[In our case $F(u)$ is real for real u.]

If $h(t)$ and $z(t)$ are real for $t \ge \gamma$, then $\sum_{j=1}^{n} c_j e^{\lambda_j t}$ in (28.3.1) is real for $t \ge \gamma$.

Write Equation (28.1.1) in the form

$$y^{(n)} + a_{n-1}y^{(n-1)} + \cdots + a_0 y = -\sum_{j=1}^{n-1} \varphi_j(t)y^{(j)}(t).$$

Setting

$$(28.3.3)\qquad \begin{cases} -\sum_{j=0}^{n-1} \varphi_j(t)y^{(j)}(t) = h(t) \\ \sum_{j=1}^{n} c_j e^{\lambda_j t} = U(t) \end{cases}$$

(28.3.1) gives

$$(28.3.4)\qquad y(t) = U(t) - \sum_{\nu=0}^{n-1} I_\nu \quad \text{for} \quad t \ge \gamma$$

*Some constants could be improved (cf. Sections 12.8, 18.9).

where

$$I_\nu = \int_\gamma^t F(t-\tau)\varphi_\nu(\tau)y^{(\nu)}(\tau)\,d\tau. \tag{28.3.5}$$

Since

$$F(0) = F'(0) = \cdots = F^{(n-2)}(0) = 0$$
$$F^{(n-1)}(0) = 1,$$

partial integration gives

$$I_\nu = S_1 + S_2 \tag{28.3.6}$$

where

$$\begin{cases} S_1 = \sum_{j=0}^{\nu-1} (-1)^{\nu-j} y^{(j)}(\gamma) \dfrac{d^{(\nu-j-1)}}{d\tau^{\nu-j-1}}[\varphi_\nu(\tau)F(t-\tau)]_{\tau=\gamma} \\ S_2 = (-1)^\nu \int_\gamma^t y(\tau)\dfrac{d^\nu}{d\tau^\nu}[\varphi_\nu(\tau)F(t-\tau)]\,d\tau. \end{cases} \tag{28.3.7}$$

Observe that S_1 has the form

$$\sum_{j=1}^{n} c_j' e^{\lambda_j t},$$

and hence (28.3.4) gives with a suitable (real)

$$U_1(t) = \sum_{j=1}^{n} c_j'' e^{\lambda_j t}$$

the formula

$$U_1(t) - y(t) = \sum_{\nu=0}^{n-1} (-1)^\nu \int_\gamma^t y(\tau)\frac{d^\nu}{d\tau^\nu}[\varphi_\nu(\tau)F(t-\tau)]\,d\tau.$$

Or

$$U_1(t) - y(t) = \int_\gamma^t y(\tau)\Phi_0(t,\tau)\,d\tau \tag{28.3.8}$$

with

$$\Phi_0(t,\tau) = \sum_{j=0}^{n-1} \frac{d^j F(t-\tau)}{d\tau^j} \sum_{\nu=0}^{n-1} (-1)^\nu \binom{\nu}{j} \varphi_\nu^{(\nu-j)}(\tau). \tag{28.3.9}$$

28.4

We use the representation

$$F(u) = \frac{1}{2\pi i}\int_l \frac{e^{zu}}{w(z)}\,dz \tag{28.4.1}$$

where l is the circle

$$|z| = B + 2. \tag{28.4.2}$$

(The disk $|z| \le B + 1$ contains all the λ_j.) Then we obtain the final form of the "transition formula"

$$U_1(t) - y(t) = \int_\gamma^t y(\tau)\Phi(t, \tau)\,d\tau \qquad (t \ge \gamma) \tag{28.4.3}$$

where

$$\Phi(t, \tau) = \frac{1}{2\pi i}\int_l \frac{e^{z(t-\tau)}}{w(z)} \sum_{j=0}^{n-1} (-z)^j \sum_{\nu=j}^{n-1} (-1)^\nu \binom{\nu}{j} \varphi_\nu^{(\nu-j)}(\tau)\,dz. \tag{28.4.4}$$

Notice, that from (28.4.3),

$$U_1(\gamma) = y(\gamma). \tag{28.4.5}$$

28.5. PROOF OF THEOREM 28.1

Apply (28.2.2)–(28.2.3) with

$$f(t) = U_1(t) \qquad (t \ge \gamma).$$

By (28.4.5), with t_1, t_2 in (28.2.1), the inequalities

$$U_1(t_1)e^{-\Lambda(t_1-\gamma)} \ge |y(\gamma)|\frac{1}{6n}(8e)^{-n} \tag{28.5.1}$$

$$U_1(t_2)e^{-\Lambda(t_2-\gamma)} \le -|y(\gamma)|\frac{1}{6n}(8e)^{-n}$$

hold.

Since

$$y(t_1) = U_1(t_1) - [U_1(t_1) - y(t_1)] \ge U_1(t_1) - |U_1(t_1) - y(t_1)|$$

we get, from (28.5.1),

(28.5.2)

$$y(t_1)e^{-\Lambda(t_1-\gamma)} \geq |y(\gamma)|\frac{1}{6n}(8e)^{-n} - |U_1(t_1) - y(t_1)|e^{-\Lambda(t_1-\gamma)}$$

and analogously

(28.5.3)

$$y(t_2)e^{-\Lambda(t_2-\gamma)} \leq -|y(\gamma)|\frac{1}{6n}(8e)^{-n} + |U_1(t_2) - y(t_2)|e^{-\Lambda(t_2-\gamma)}.$$

28.6

We need an upper bound for

(28.6.1) $$M = \max_{\gamma \leq t \leq \gamma + n\pi/2\lambda} |U_1(t) - y(t)|.$$

Using the transition formula (28.4.3) we get

(28.6.2) $$|U_1(t) - y(t)| \leq \int_\gamma^t |y(\tau)|\,|\Phi(t,\tau)|\,d\tau \quad \text{for} \quad t \geq \gamma.$$

Thus, for $\gamma \leq t \leq \gamma + n\pi/2\lambda$,

(28.6.3) $$|U_1(t) - y(t)| \leq \int_\gamma^{\gamma+n\pi/2\lambda} |y(\tau)|\,|\Phi(t,\tau)|\,d\tau.$$

Hence from (28.6.1) we get a suitable t with $\gamma \leq t \leq \gamma + n\pi/2\lambda$ and

(28.6.4) $$M \leq \int_\gamma^{\gamma+n\pi/2\lambda} |y(\tau)|\,|\Phi(t,\tau)|\,d\tau.$$

28.7

We need an upper bound for $|\Phi(t,\tau)|$ if $t, \tau \in [\gamma, \gamma + n\pi/\lambda]$. If

(28.7.1) $$\max_{0 \leq j \leq \mu \leq n-1} \; \max_{\gamma \leq \tau \leq \gamma + n\pi/2\lambda} \left|\varphi_\mu^{(j)}(\tau)\right| \leq H,$$

(28.4.4) implies that

$$(28.7.2)\qquad |\Phi(t,\tau)| \le \exp\left[(B+2)\frac{n\pi}{2\lambda}\right]H\sum_{j=0}^{n-1}(B+2)^{j+1}\binom{n}{j+1}$$

$$\le H\exp\left[(B+2)\frac{n\pi}{2\lambda}\right](B+3)^n.$$

So if H satisfies the inequality

$$(28.7.3)\qquad H\exp\left[(B+2)\frac{n\pi}{2\lambda}\right](B+3)^n \le 1,$$

then $|\Phi(t,\tau)| \le 1$.

From (28.6.4),

$$(28.7.4)\qquad M \le H\exp\left[(B+2)\frac{n\pi}{2\lambda}\right](B+3)^n\int_\gamma^{\gamma+n\pi/2\lambda}|y(\tau)|\,d\tau.$$

And from (28.1.6), for $\gamma \le \tau \le \gamma + n\pi/2\lambda$,

$$|y(\tau)| \le |y(\gamma)|e^{Bn(\tau-\gamma)},$$

and hence, from (28.7.4),

$$(28.7.5)\qquad M \le |y(\gamma)|H\exp\left[(B+2)\frac{n\pi}{2\lambda}\right](B+3)^n\int_\gamma^{\gamma+n\pi/2\lambda}e^{Bn(\tau-\gamma)}\,d\tau$$

$$\le |y(\gamma)|\frac{1}{Bn}H\exp\left[(B+2)\frac{n\pi}{2\lambda}\right](B+3)^n\exp\left(B\frac{n^2\pi}{2\lambda}\right).$$

For $\nu = 1, 2$—also using the inequality $|\Lambda| \le B + 1$—

$$(28.7.6)\qquad |U_1(t_\nu) - y(t_\nu)|e^{-\Lambda(t_\nu-\gamma)}$$

$$\le |y(\gamma)|\frac{1}{Bn}H(B+3)^n\exp\left[(2B+2)\frac{n\pi}{2\lambda} + (B+1)\frac{n^2\pi}{2\lambda}\right].$$

This together with (28.5.2) implies

$$y(t_1)e^{-\Lambda(t_1-\gamma)} \ge \frac{1}{n}|y(\gamma)|\left\{\tfrac{1}{6}(8e)^{-n}\right.$$

$$\left. - \frac{1}{B}H(B+3)^n\exp\left[(B+1)\frac{\pi}{2\lambda}(n^2+2n)\right]\right\} > 0$$

if (28.1.6) holds, $y(\gamma) \neq 0$, and

$$H < \frac{B}{6} \frac{(8e)^{-n}}{(B+3)^n} \exp\left[-(B+1)\frac{\pi}{2\lambda}(n^2+2n)\right]. \tag{28.7.7}$$

Analogously,

$$y(t_2)e^{-\Lambda(t_2-\gamma)} < 0.$$

Therefore $y(t)$ changes sign in $[\gamma, \gamma + n\pi/2\lambda]$ if (28.1.6) holds and $y(\gamma) \neq 0$.
Equations (28.7.3) and (28.7.7) are clearly satisfied if we choose, for example,

$$H = \frac{1}{7} \frac{(8e)^{-n}}{(B+3)^n} \exp\left[-(B+1)\frac{\pi}{2\lambda}(n^2+2n)\right]. \quad \blacksquare$$

29

APPLICATIONS TO NUMERICAL ALGEBRA I

Eigenvalues of Matrices and Latent Roots of λ-Matrices

29.1

Let A be an $n \times n$ matrix with complex entries and eigenvalues $\lambda_1,\ldots,\lambda_n$ arranged so that

$$(29.1.1) \qquad |\lambda_1| \leq |\lambda_2| \leq \cdots \leq |\lambda_n|.$$

Since the eigenvalues of A^ν are λ_j^ν $(j = 1,\ldots,n)$, formula (3.1.5) proves the following.

Theorem 29.1. *An arbitrary $n \times n$ matrix A has at least one eigenvalue in the disk*

$$|z| \leq \max_{\nu=1,\ldots,n} |\operatorname{tr} A^\nu|^{1/\nu}.$$

where $\operatorname{tr} A^\nu$ *denotes the sum of the entries in the main diagonal of A^ν. This cannot be improved if and only if the eigenvalues form n vertices of a regular $n + 1$-gon with center the origin.*

29.2

Consider the more general λ-matrix

$$(29.2.1) \qquad K(\lambda) = \lambda^r E_n - \lambda^{r-1} A_1 - \cdots - A_r$$

where the A_ν are $n \times n$ matrices with complex entries and E_n is the $n \times n$ unit matrix. $K(\lambda)$ is an $n \times n$ matrix where the entries are polynomials of degree at most r. The roots of the equation

$$\det K(\lambda) = 0 \tag{29.2.2}$$

play a role in the theory of aerodynamic stability (see Franklin [1]) and in the theory of electricity (see P. Lancaster [1]). In the case $r = 1$ these roots are eigenvalues of A_1, but not for $r \geq 2$*; in this case the roots of (29.2.2) are called latent roots and we denote them $\mu_1, \mu_2, \ldots, \mu_{rn}$, numbered so that

$$|\mu_1| \leq |\mu_2| \leq \cdots \leq |\mu_{rn}|. \tag{29.2.3}$$

Define the matrices S_ν recursively so that

$$S_\nu = S_{\nu-1}A_1 + S_{\nu-2}A_2 + \cdots + S_1 A_{\nu-1} + \nu A_\nu \tag{29.2.4}$$

for $\nu \leq r$ and

$$S_\nu = S_{\nu-1}A_1 + S_{\nu-2}A_2 + \cdots + S_{\nu-r}A_r \tag{29.2.5}$$

for $\nu > r$. Franklin [1] proved that the relation

$$\operatorname{tr} S_\nu = \sum_{j=1}^{rn} \mu_j^\nu \tag{29.2.6}$$

holds. Application of (3.1.5) with matrices of size $n \times n$ only yields the following.

Theorem 29.2. *The disk*

$$|z| \leq \max_{\nu=1,2,\ldots,rn} |\operatorname{tr} S_\nu|^{1/\nu} \tag{29.2.7}$$

contains a root of (29.2.2) *if the* S_ν *are given by* (29.2.4)–(29.2.5). *This disk cannot be diminished if and only if* $\mu_1, \ldots, \mu_{rn}$ *are the rn vertices of a regular* $(rn + 1)$*-gon with the center the origin.*

*They are, however, the eigenvalues of the $rn \times rn$ matrix

$$\begin{bmatrix} A_1 & A_2 & \cdots & A_{r-1} & A_r \\ E_n & 0 & \cdots & 0 & 0 \\ 0 & E_n & \cdots & 0 & 0 \\ \vdots & \vdots & & \vdots & \vdots \\ 0 & 0 & \cdots & E_n & 0 \end{bmatrix}$$

29.2

So far we have estimates for the absolutely smallest eigenvalue or latent root. Can one obtain analogous upper bounds for the maximal eigenvalue, that is, for all of them? The most elegant upper bound can be obtained from (3.4.5).

Theorem 29.3. *All eigenvalues of an arbitrary $n \times n$ matrix A are in the disk*

$$|z| \leq \max_{\nu=1,\ldots,2n-1} |\operatorname{tr} A^{\nu}|^{1/\nu}.$$

The estimate is sharp, for example, for

$$A = \begin{pmatrix} 1 & 0 & \cdots & 0 \\ 0 & 0 & \cdots & 0 \\ \vdots & \vdots & & \vdots \\ 0 & 0 & \cdots & 0 \end{pmatrix}$$

but also for several other matrices.

Theorem 29.4. *All roots of Equation* (29.2.2) *are in the disk*

$$|z| \leq \max_{\nu=1,2,\ldots,2rn-1} |\operatorname{tr} S_{\nu}|^{1/\nu}.$$

The need to lessen our computational work suggests that we should try to reduce the interval for ν. Then (3.9.2) and (3.15.2) can be applied.

Theorem 29.5. *All eigenvalues of an arbitrary $n \times n$ matrix A are in the disk*

$$|z| \leq 5 \max_{\nu=1,\ldots,n} \left(\frac{1}{n}|\operatorname{tr} A^{\nu}|\right)^{1/\nu}.$$

30

APPLICATIONS TO NUMERICAL ALGEBRA II

Modifications of the Graeffe Method

30.1

Consider the equation

$$f_0(z) = 0 \tag{30.1.1}$$

where

$$f_0(z) = a_0 + a_1 z + \cdots + a_n z^n, \qquad a_n = 1; \tag{30.1.2}$$

let $z_1, \ldots, z_n$ be its zeros. The theory of approximative solution of the equation $f_0(z) = 0$ started with Newton and D. Bernoulli. The method of Bernoulli–Graeffe–Lobatshevsky consists in forming the "Graeffe transforms" $f_\nu(z)$ defined recursively by

$$f_{\nu+1}(z) = (-1)^n f_\nu(\sqrt{z}\,) f_\nu(-\sqrt{z}\,), \qquad \nu = 0, 1, \ldots. \tag{30.1.3}$$

Set

$$f_\nu(z) = a_{0\nu} + a_{1\nu} z + \cdots + a_{n\nu} z^n, \qquad \text{with } a_{n\nu} = 1. \tag{30.1.4}$$

Assuming

$$|z_1| < |z_2| < \cdots < |z_n|, \tag{30.1.5}$$

it was easy to show, for $j = 1,\ldots,n$, the relations

$$z_j = \lim_{\nu \to \infty} \left| \frac{a_{j-1,\nu}}{a_{j,\nu}} \right|^{2^{-\nu}}. \tag{30.1.6}$$

There are many variants of this method; for example, for $j = n$ it is enough to suppose that

$$|z_1| \leq \cdots \leq |z_{n-1}| < |z_n|. \tag{30.1.7}$$

There are also many deficiencies in this method. There are no known *necessary and sufficient* conditions even for (30.1.7); that the assertion (30.1.6) without (30.1.7) is false for $j = n$ can be shown by the simple example $f_0(z) = z^3 - 1$. But even if (30.1.7) holds, the lack of error estimate makes (30.1.6) practically useless. So, *from a mathematical point of view*, this algorithm is very bad. *From the computer's point of view* the algorithm can be easily performed, although "overflowing" was formerly thought to cause difficulties. The same can be said for eigenvalues of matrices. The first real progress was made by A. Ostrowski [2], where all aspects of the problem are thoroughly analyzed and essentially theoretically solved. His basic new tool was the use of Newton diagrams, which is certainly appropriate from the theoretical point of view. Practical difficulties may have induced D. H. Lehmer [1] to write in 1961: "...A recent survey of available methods indicates that no one method is desirable for automatic computers... ." G. Birkhoff [1] also mentioned as the first problem in the new numerical algebra: "Finding the roots of polynomial equations of degrees up to 100." These difficulties increase quite substantially for eigenvalues of matrices.

30.2

Our aim in this section is to give a number of rules for the approximative *determination* of $|z_1|$ and $|z_n|$ (resp. $|\lambda_1|$ and $|\lambda_n|$) where $z_1,\ldots,z_n$ are roots of the equation $f_0(z) = 0$ with

$$|z_1| \leq |z_2| \leq \cdots \leq |z_n| \tag{30.2.1}$$

(resp. $\lambda_1,\ldots,\lambda_n$ are the eigenvalues of an $n \times n$ matrix A with

$$|\lambda_1| \leq |\lambda_2| \leq \cdots \leq |\lambda_n|). \tag{30.2.2}$$

Actually, almost all the theorems of Section 3 give rise to such rules. More generally, let

$$M_n(m) = \min \max_{\substack{1 \leq \nu \leq m \\ \nu \text{ integer}}} |\xi_1^\nu + \cdots + \xi_n^\nu| \quad \text{for} \quad m = m(n) \geq n, \tag{30.2.3}$$

where the minimum ranges over all $(\xi_1, \xi_2, \ldots, \xi_n)$-systems with

$$\max_{j=1,\ldots,n} |\xi_j| = 1. \tag{30.2.4}$$

Let $M_n^*(m)$ be any number $\le M_n(m)$.

Theorem 30.1. *The inequality*

$$\frac{1}{n} \le \frac{|z_n|}{\max\limits_{\substack{1 \le \nu \le m \\ \nu \text{ integer}}} |z_1^\nu + \cdots + z_n^\nu|^{1/\nu}} \le \frac{1}{M_n^*(m)}$$

holds whenever (30.2.1) *holds.*

We use the abbreviation

$$z_1^\nu + \cdots + z_n^\nu = s_\nu. \tag{30.2.5}$$

Assume that $\max_{1 \le \nu \le m} |s_\nu|^{1/\nu}$ is attained for $\nu = \nu_0$. Since evidently

$$|s_{\nu_0}|^{1/\nu_0} \le n^{1/\nu_0}|z_n| \le n|z_n|$$

we already have the lower bound.

Next apply (30.2.3)–(30.2.4) with

$$\xi_j = \frac{z_j}{z_n} \qquad j = 1, \ldots, n.$$

Then for a suitable index ν_1 with $1 \le \nu_1 \le m$ we have

$$\frac{|s_{\nu_1}|}{|z_n|^{\nu_1}} \ge M_n^*(m). \tag{30.2.6}$$

Since $(1, 0, \ldots, 0)$ satisfies (30.2.4), we have $(0 <)\ M^*(m) \le 1$, and thus, from (30.2.5),

$$\frac{|s_{\nu_1}|}{|z_n|^{\nu_1}} \ge M_n^*(m)^{\nu_1}$$

or

$$|z_n| \le \frac{1}{M_n^*(m)} \cdot |s_{\nu_1}|^{1/\nu_1} \le \frac{1}{M_n^*(m)} \max_{\substack{1 \le \nu \le m \\ \nu \text{ integer}}} |s_\nu|^{1/\nu},$$

which completes the proof of Theorem 30.1. ■

We apply this theorem to $z_j^{2^k}$ $(j = 1,\ldots,n)$ where k is an arbitrary positive integer.

Corollary. *For all equations* $f_0(z) = 0$ *and* $m = m(n) \geq n$ *the inequality*

$$\left(\frac{1}{n}\right)^{1/2^k} \leq \frac{|z_n|}{\left\{\max\limits_{\substack{1\leq\nu\leq m\\ \nu\,\text{integer}}} |z_1^{\nu 2^k} + \cdots + z_n^{\nu 2^k}|^{1/\nu}\right\}^{1/2^k}} \leq \left\{\frac{1}{M_n^*(m)}\right\}^{1/2^k}$$

holds, assuming (30.2.1) (*if* $z_n \neq 0$).

30.3

The point of this corollary is that by fixing $m = m(n) \geq n$, k can be chosen, *depending on n only* so that the upper and lower bounds are as close to 1 as we wish. From the Newton–Girard formulas, the corollary immediately gives the following.

Rule I. Let k be an arbitrary fixed positive integer and let $m = m(n) \geq n$ be an integer. We form, by (30.1.3), the first k Graeffe transforms and let

$$f_k(x) = a_{0k} + \cdots + a_{nk}x^n \qquad (a_{nk} = 1). \tag{30.3.1}$$

Next we calculate the quantities $\sigma_1,\ldots,\sigma_m$ by the formulas

$$\begin{aligned} a_{nk}\sigma_1 + a_{n-1,k} &= 0 \\ a_{nk}\sigma_2 + a_{n-1,k}\sigma_1 + 2a_{n-1,k} &= 0 \\ a_{n,k}\sigma_n + \cdots + na_{0,k} &= 0 \\ a_{n,k}\sigma_{n+1} + \cdots + a_{0,k}\sigma_1 &= 0 \\ &\vdots \\ a_{n,k}\sigma_m + \cdots + a_{0,k}\sigma_{m-n} &= 0. \end{aligned} \tag{30.3.2}$$

Then the inequality

$$\left(\frac{1}{n}\right)^{1/2^k} \leq \frac{|z_n|}{\left\{\max\limits_{\nu=1,\ldots,m} |\sigma_\nu|^{1/\nu}\right\}^{1/2^k}} \leq \left(\frac{1}{M_n^*(m)}\right)^{1/2^k} \tag{30.3.3}$$

holds assuming (30.2.1).

The first such rule occurred in Turán [9] in 1951 with $m(n) = n$, using Theorem 3.3. Applying Theorem 3.4 instead of Theorem 3.3, with $m(n) = n$, we get the inequality

$$\left(\frac{1}{n}\right)^{1/2^k} \le \frac{|z_n|}{\left(\max\limits_{\nu=1,\dots,n} |\sigma_\nu|^{1/\nu}\right)^{1/2^k}} \le 6^{1/2^k}. \tag{30.3.4}$$

Using Theorem 3.2 with $m(n) = 2n - 1$, we get

$$\left(\frac{1}{n}\right)^{1/2^k} \le \frac{|z_n|}{\left\{\max\limits_{\nu=1,\dots,2n-1} |\sigma_\nu|^{1/\nu}\right\}^{1/2^k}} \le 1. \tag{30.3.5}$$

30.4

If we want 1% relative error at most in (30.3.5), then k must be chosen so that

$$k = 1 + \left[\frac{1}{\log 2} \log \frac{\log n}{\log (100/99)}\right],$$

which depends on n but is independent of the coefficients of f_0. Can the algorithm be changed so that k is independent of n? [The inequalities (30.3.4) and (30.3.5) already show that the right-hand side can be made arbitrarily close to 1 independently of n.] To achieve this let $m = m(n) \ge n$ be fixed and let $\mu_n(m)$ be defined by

$$\mu_n(m) = \min \max_{\nu=1,\dots,m} \left| \frac{\xi_1^\nu + \cdots + \xi_n^\nu}{n} \right|^{1/\nu} \tag{30.4.1}$$

where the minimum is taken over all $(\xi_1, \dots, \xi_n)$ with

$$\max_{\nu=1,\dots,n} |\xi_\nu| = 1, \tag{30.4.2}$$

and let $\mu_n^*(m)$ be any number with

$$\mu_n(m) \ge \mu_n^*(m). \tag{30.4.3}$$

Then, with the notations (30.2.1) and (30.2.5), we assert the following.

Theorem 30.2. *The inequality*

$$1 \le \frac{|z_n|}{\max\limits_{\nu=1,\dots,m} \left|\frac{s_\nu}{n}\right|^{1/\nu}} \le \frac{1}{\mu_n^*(m)}$$

holds (if $z_n \ne 0$).

Proof. Since

$$\frac{|s_\nu|}{n} \le |z_n|^\nu$$

for all ν,

$$|z_n| \ge \max_{1\le\nu\le m}\left(\frac{|s_\nu|}{n}\right)^{1/\nu}$$

which yields the lower bound. For the upper bound, apply (30.4.1)–(30.4.3) with

$$\xi_\nu = \frac{z_\nu}{z_n}, \qquad \nu = 1,\ldots,n.$$

If the maximum in (30.4.1) is attained for $\nu = \nu_1$, then we have

$$\left(\frac{|z_1^{\nu_1} + \cdots + z_n^{\nu_1}|}{n|z_n|^{\nu_1}}\right)^{1/\nu_1} \ge \mu_n^*(m),$$

that is,

$$|z_n| \le \frac{1}{\mu_n^*(m)} \max_{\nu=1,\ldots,m}\left|\frac{s_\nu}{n}\right|^{1/\nu}. \quad \blacksquare$$

30.5

Hence we have our second rule.

Rule II. Again let $k \ge 1$, $m = m(n) \ge n$ be fixed integers, form $f_k(x)$ as in (30.3.1) and the quantities $\sigma_1,\ldots,\sigma_m$ according to (30.3.2). Then with the notations (30.4.1)–(30.4.3), the inequality

$$(30.5.1) \qquad 1 \le \frac{|z_n|}{\left\{\max\limits_{\nu=1,\ldots,m}\left|\frac{\sigma_\nu}{n}\right|^{1/\nu}\right\}^{1/2^k}} \le \left(\frac{1}{\mu_n^*(m)}\right)^{1/2^k}$$

holds.

To get an effective form we can take $m(n) = n$ and use Theorem 3.6. This gives, with the notations of Rule II, the inequality

$$(30.5.2) \qquad 1 \le \frac{|z_n|}{\left\{\max\limits_{\nu=1,\ldots,n}\left|\frac{\sigma_\nu}{n}\right|^{1/\nu}\right\}^{1/2^k}} \le 5^{1/2^k}.$$

Thus if we prescribe the relative precision of $|z_n|$ to be 3% at most, this can be attained for $k = 6$ independently of n.

30.6

Very simple results from the theory of matrices permit elegant reformulation of all previous results of this section. We explicitly formulate only two such rules.

Rule III. Let A be an $n \times n$ matrix with complex entries and $(\lambda_1,\ldots,\lambda_n)$ its eigenvalues with

$$|\lambda_1| \leq \cdots \leq |\lambda_n|$$

and k an arbitrary positive integer. Then we form, by iterated squaring of the matrices, the sequence

$$A, A^2, A^{2^2},\ldots,A^{2^k} = B$$

and then the matrices

$$B, B^2,\ldots,B^n,\ldots,B^{2n-1}.$$

Then the inequality

$$\left(\frac{1}{n}\right)^{1/2^k} \leq \frac{|\lambda_n|}{\left\{\max_{\nu=1,\ldots,2n-1} |\operatorname{tr} B^\nu|^{1/\nu}\right\}^{1/2^k}} \leq 1$$

holds (if $\lambda_n \neq 0$).

Rule IV. With the notations of Rule III, forming only the matrices

$$B, B^2,\ldots,B^n$$

the inequality

$$1 \leq \frac{|\lambda_n|}{\left\{\max_{\nu=1,\ldots,n} \left|\frac{1}{n}\operatorname{tr} B^\nu\right|^{1/\nu}\right\}^{1/2^k}} \leq 5^{1/2^k}$$

holds (if $\lambda_n \neq 0$).

Again, prescribing the relative precision for the spectral radius $|\lambda_n|$ to be 3%, at most, this can be attained for $k = 6$, independent of n.

30.7

Next we turn to the analogous rule for the latent roots μ_j of the equation

$$\det K_0(\lambda) = 0 \tag{30.7.1}$$

where

$$K_0(\lambda) = \lambda^r E_n - \lambda^{r-1} A_1 - \cdots - A_r. \tag{30.7.2}$$

(Cf. Section 29.2.) Form recursively the sequence of λ-matrices.

$$K_{\nu+1}(\lambda) = (-1)^r K_\nu(\sqrt{\lambda}) K_\nu(-\sqrt{\lambda}). \tag{30.7.3}$$

Since

$$\det K_0(\lambda) = \prod_{j=1}^{rn} (\lambda - \mu_j),$$

supposing

$$\det K_\nu(\lambda) = \prod_{j=1}^{rn} \left(\lambda - \mu_j^{2^\nu}\right), \tag{30.7.4}$$

we have from (30.7.3)

$$\begin{aligned} \det K_{\nu+1}(\lambda) &= \det K_\nu(\sqrt{\lambda}) \cdot \det\left\{(-1)^r K_\nu(-\sqrt{\lambda})\right\} \\ &= \left\{\prod_{j=1}^{rn} \left(\sqrt{\lambda} - \mu_j^{2^\nu}\right)\right\}(-1)^{rn}\left\{\prod_{j=1}^{rn} \left(-\sqrt{\lambda} - \mu_j^{2^\nu}\right)\right\} \\ &= \prod_{j=1}^{rn} \left(\lambda - \mu_j^{2^{\nu+1}}\right), \end{aligned}$$

that is, (30.7.4) holds for all integers $\nu \geq 0$. So applying (30.5.2) and taking into account (29.2.4)–(29.2.6), we get our fifth rule for the roots μ_j of the equation of (30.7.1) and (30.7.2) with

$$|\mu_1| \leq |\mu_2| \leq \cdots \leq |\mu_{rn}|.$$

Rule V. Let $k \geq 1$ be a fixed integer and form recursively the matrices

$$S_1^{(k)}, S_2^{(k)}, \ldots, S_{rn}^{(k)},$$

applying (29.2.4)–(29.2.5) for $K(\lambda) = K_k(\lambda)$. Then the inequality

$$1 \leq \frac{|\mu_{rn}|}{\left\{ \max_{\nu=1,\ldots,rn} \left| \frac{1}{rn} \operatorname{tr} S_\nu^{(k)} \right|^{1/\nu} \right\}^{1/2^k}} \leq 5^{1/2^k}$$

holds (if $\mu_{rn} \neq 0$).

30.8

The corresponding rules for $|z_1|$ (resp. $|\lambda_1|$)—see (30.2.1) and (30.2.2)—can be derived easily from previous rules. Nevertheless, we formulate two such rules which are often referred to in the next section. Let the equation to be solved be

(30.8.1) $$f_0(z) \doteq a_0 + \cdots + a_n z^n = 0$$

where now

(30.8.2) $$a_0 a_n \neq 0.$$

We again form the $f_\nu(z)$ Graeffe transforms as in (30.1.3); if $k \geq 1$ is given and fixed, let

(30.8.3) $$f_k(z) = a_{0k} + a_{1k} z + \cdots + a_{nk} z^n$$

where

(30.8.4) $$a_{0k} a_{nk} \neq 0.$$

Let the zeros of $f_0(z) = 0$ be as in (30.2.1).

Rule VI. We form the sequence σ_ν^* recursively by

$$a_{0k}\sigma_1^* + a_{1k} = 0$$

$$a_{0k}\sigma_2^* + a_{1k}\sigma_1^* + 2a_{2k} = 0$$

$$\vdots$$

(30.8.5) $$a_{0k}\sigma_n^* + a_{1k}\sigma_{n-1}^* + \cdots + na_{nk} = 0.$$

Setting

(30.8.6) $$N = \frac{1}{\left\{ \max_{\nu=1,\ldots,n} \left| \frac{\sigma_\nu^*}{n} \right|^{1/\nu} \right\}^{1/2^k}}$$

we have the inequality

$$5^{-1/2^k} \le \frac{|z_1|}{N} \le 1. \tag{30.8.7}$$

30.9

For the $n \times n$ matrix A with complex elements we now require

$$\det A \neq 0. \tag{30.9.1}$$

Assume the eigenvalues are as in (30.2.2).

Rule VII. Let $k \geq 1$ be a fixed integer and form the matrices

$$A^{-1}, (A^{-1})^2, (A^{-1})^{2^2}, \ldots, (A^{-1})^{2^k} \doteq C,$$

$$C, C^2, \ldots, C^n.$$

Setting

$$N^* = \frac{1}{\left\{ \max\limits_{\nu=1,\ldots,n} \left| \frac{1}{n} \operatorname{tr} C^\nu \right|^{1/\nu} \right\}^{1/2^k}}$$

we have the inequality

$$5^{-1/2^k} \le \frac{|\lambda_1|}{N^*} \le 1.$$

31

APPLICATIONS TO NUMERICAL ALGEBRA III

Approximative Solution of Algebraic Equations

31.1

We define an elementary operation to be the following:

1. The addition, subtraction, multiplication, or division of complex numbers (when it is possible).
2. The determination of the positive kth root of positive numbers ($k = 2, 3, \ldots$).

Let the totality of these elementary operations be called the operation domain H. We define an algorithm U over H as a finite sequence of elementary operations from H, and we define $l(U)$, the length of the algorithm U, to be the number of elementary operations in U. Let

$$f(z) = a_0 + a_1 z + \cdots + a_n z^n \tag{31.1.1}$$

be an arbitrary polynomial with complex coefficients so that

$$a_0 a_n \neq 0. \tag{31.1.2}$$

Let the zeros of $f(z) = 0$ be $z_1, \ldots, z_n$.

Theorem 31.1. *Let* $0 < \varepsilon < 1$ *be given arbitrarily. Then we can give an algorithm* U^* *over* H *explicitly so that its length satisfies an inequality of the type*

$$l(U^*) \le c = c(n, \varepsilon) \tag{31.1.3}$$

When applied to an $f(z)$ *with* (31.1.1)–(31.1.2), *it yields a complex number* α *so that, for a suitable index* j *with* $1 \le j \le n$, *the inequality*

$$\left|\frac{z_j}{\alpha} - 1\right| \le \varepsilon$$

holds.

There is no need to emphasize that not only the length but also *the algorithm itself* is independent of the coefficients α_ν. The theoretical interest of this theorem is the following. Ruffini–Abel's classical theorem can be formulated crudely as the assertion that, for $n \ge 5$, there is no algorithm U^{**} over H with

$$l(U^{**}) \le c_1 = c_1(n)$$

so that applying it to each $f(z)$ with (31.1.1)–(31.1.2) it gives *exactly* one of the roots. So, briefly, Theorem 31.1 means that the analogue of the Ruffini–Abel theorem does not hold for *approximative* solutions of algebraic equations.

Before the proof is presented we note the identity

$$\max(a, b) = \frac{a+b}{2} + \sqrt{\left(\frac{a-b}{2}\right)^2} \qquad (a, b \text{ are real}) \tag{31.1.4}$$

which implies that the operation-domain H can include the operations

$$\max(a_1,\dots,a_n), \qquad \min(a_1,\dots,a_n) \tag{31.1.5}$$

(the a_ν's are real). In the following discussions we always include the operations (31.1.5) in the operation domain H.

31.2

An explicit description of one of our algorithms follows—many variants are possible. Since the beginning of the algorithm is somewhat different from the rest, we denote it as:

Zeroth step. Define

$$\xi^{(0)} = 0. \tag{31.2.1}$$

We apply Rule VI to the polynomial $f_0(z) = f(z)$ in (31.1.1)–(31.1.2) with $k = 4$ and denote the obtained N by $N^{(0)}$.

First step. Form the 12 numbers

$$\xi_j^{(1)} = \xi^{(0)} + \frac{19}{20} N^{(0)} e^{j(\pi i/6)}, \qquad j = 0, 1, \ldots, 11 \tag{31.2.2}$$

and consider for all these j the polynomials

$$f\left(\xi_j^{(1)} + w\right) = a_{01}^{(j)} + a_{11}^{(j)} w + \cdots + a_{n1}^{(j)} w^n. \tag{31.2.3}$$

If one of the numbers $f(\xi_j^{(1)})$ is 0, the procedure halts. If none of these is 0 we can apply to each polynomial of w in (31.2.3) Rule VI with $k = 4$. Denote the obtained N by $N_j^{(1)}$ and define the index μ_1 and the quantities

$$N^{(1)} \quad \text{and} \quad \xi^{(1)}$$

by

$$N^{(1)} \doteq N_{\mu_1}^{(1)} = \min_j N_j^{(1)} \tag{31.2.4}$$

and

$$\xi^{(1)} \doteq \xi_{\mu_1}^{(1)}. \tag{31.2.5}$$

Second step. Next we form the 12 numbers

$$\xi_j^{(2)} = \xi^{(1)} + \frac{19}{20} N^{(1)} e^{j(\pi i/6)}, \qquad j = 0, 1, \ldots, 11 \tag{31.2.6}$$

and consider, for all these j, the polynomials

$$f\left(\xi_j^{(2)} + w\right) = a_{02}^{(j)} + a_{02}^{(j)} w + \cdots + a_{n2}^{(j)} w^n. \tag{31.2.7}$$

If one of the numbers $f(\xi_j^{(2)})$ is 0, the procedure halts. If none of these is 0 we can apply Rule VI to each polynomial of w in (31.2.7) with $k = 4$. Denote the obtained N by $N_n^{(2)}$ ($j = 0, 1, \ldots, 11$); we define the index μ_2 and the quantities

$$N^{(2)} \quad \text{and} \quad \xi^{(2)}$$

by

$$\min_j N_j^{(2)} = N_{\mu_2}^{(2)} = N^{(2)} \tag{31.2.8}$$

and

(31.2.9) $$\xi^{(2)}_{\mu_2} = \xi^{(2)}.$$

Each succeeding step in our algorithm is like steps 1 and 2. If it stops we already have a zero of $f(z)$. If it does not stop then we get the numbers

$$N^{(0)}, N^{(1)}, \ldots$$

$$\xi^{(0)}, \xi^{(1)}, \ldots .$$

This is algorithm U^* of Theorem 31.1, and we give it a quantitative form.

Theorem 31.2. *If the algorithm U^* described previously does not stop in the first d steps ($d \geq 1$), then, for a suitable index j, the inequality*

$$\left| \frac{z_j}{\xi^{(d)}} - 1 \right| \leq 2 \cdot 3^{-d}$$

holds.

Hence the convergence is rather quick; for example, choosing $d = 4$, we get a relative error less than 2.5%. The assertion (31.1.3) is clearly satisfied and the length is, for a fixed ε, of order n^2.

As to the elimination of rounding-off errors, or "infinite precision integer arithmetic," see the interesting paper of G. E. Collins entitled "Computer algebra of polynomials and rationals functions."* The algorithm U^* has some formal analogies with one given by D. H. Lehmer [which does not satisfy the boundedness requirement (31.1.3)] in his paper "A machine method for solving polynomial equations."†

31.3

Before proving Theorem 31.2 we point out another nice feature of the algorithm U^*. Generally algorithms that work well with algebraic equations turn out to be unsatisfactory for eigenvalues of matrices. However, the algorithm U^*, formulated for eigenvalues of matrices, is even more elegant. For the sake of completeness we formulate it explicitly. A is our $n \times n$ matrix with det $A \neq 0$.

Zeroth step. Let

$$\eta^{(0)} = 0.$$

Apply Rule VII with $k = 4$ and denote the obtained quantity N^* by $N^{*(0)}$.

Amer. Math. Monthly*, **80, 725–754 (1973).

†*J. Assoc. Comp. Mach.*, **8**, 151–162 (1961).

First step. Form the 12 numbers

$$\eta_j^{(1)} = \eta^{(0)} + \tfrac{19}{20} N^{*(0)} e^{j(\pi i/6)}, \qquad j = 0, 1, \ldots, 11 \tag{31.3.1}$$

and consider, for all these j, the matrices

$$A - \eta_j^{(1)} E_n \tag{31.3.2}$$

(E_n is the $n \times n$ unit matrix). If one of these is singular, then the corresponding $\eta_j^{(1)}$ is an eigenvalue of A and the procedure halts. If not, then we can apply Rule VII to each matrix in (31.3.2) with $k = 4$. Denote the obtained N^* by $N_j^{*(1)}$ ($j = 0, 1, \ldots, 11$) and define the index ν_1 and the quantities

$$N^{*(1)} \quad \text{and} \quad \eta^{(1)}$$

by

$$\min_j N_j^{*(1)} = N_{\nu_1}^{*(1)} = N^{*(1)}, \tag{31.3.3}$$

and

$$\eta_{\nu_1}^{(1)} = \eta^{(1)} \tag{31.3.4}$$

There is no need to detail the later steps.

Theorem 31.3. *If the previously described algorithm does not stop in the first d steps then, for a suitable eigenvalue λ_j of A the inequality*

$$\left| \frac{\lambda_j}{\eta^{(d)}} - 1 \right| \leq 2 \cdot 3^{-d}$$

holds.

Again the length of the algorithm depends only upon n and the prescribed relative error. The difficulties in the regularity of a matrix disappear using infinite precision arithmetic. A similar algorithm can be given for a latent root of a λ-matrix; we do not go into details (see Turán [42]).

31.4

It is sufficient to prove Theorem 31.2. We observe that

$$\tfrac{9}{10} < 5^{-1/16} < \tfrac{19}{21},$$

and hence inequality (30.8.7) in Rule VI implies that

$$\tfrac{9}{10}N \leq |z_1| \leq N \tag{31.4.1}$$

for $k = 4$. This will be applied repeatedly.

The zeroth step in (31.2.1) implies that there is a zero [z_1 with the convention (30.2.1)] in the annulus

$$\tfrac{9}{10}N^{(0)} \leq |z| \leq N^{(0)}. \tag{31.4.2}$$

Divide this annulus into sectors $G_j^{(1)}$ by

$$\frac{2j-1}{12}\pi \leq \operatorname{arc} z < \frac{2j+1}{12}\pi; \tag{31.4.3}$$

(31.4.4) where $j = 0, 1, \ldots, 11$. Elementary geometry shows that the sector $G_j^{(1)}$ is covered by the disk

$$\left|z - \xi_j^{(1)}\right| \leq \vartheta N^{(0)}, \qquad j = 0, 1, \ldots, 11 \tag{31.4.5}$$

where ϑ is given by

$$\vartheta = \sqrt{\tfrac{1}{400} + \tfrac{19}{5}\sin^2\frac{\pi}{24}}\,. \tag{31.4.6}$$

Hence z_1 is contained in some of these disks; if $j = j_1$ is the corresponding index, this means that the inequality

$$\left|z_1 - \xi_{j_1}^{(1)}\right| \leq \vartheta N^{(0)} \tag{31.4.7}$$

holds. We may suppose $f(\xi_{j_1}^{(1)}) \neq 0$ and consider the annulus

$$\tfrac{9}{10}N_{j_1}^{(1)} \leq \left|z - \xi_{j_1}^{(1)}\right| \leq N_{j_1}^{(1)}. \tag{31.4.8}$$

If z' is of the zeros of $f(z)$ nearest to $\xi_{j_1}^{(1)}$, then by Rule VI and (31.4.7) we have

$$\tfrac{9}{10}N_{j_1}^{(1)} \leq \left|z' - \xi_{j_1}^{(1)}\right| \leq \left|z_1 - \xi_{j_1}^{(1)}\right| \leq \vartheta N^{(0)}.$$

So, by the minimum definition of $N^{(1)}$,

$$N^{(1)} \leq \frac{10\vartheta}{9}N^{(0)}. \tag{31.4.9}$$

Thus with the μ_1-index defined in (31.2.4) and denoting one of the zeros nearest to $\xi_{\mu_1}^{(1)} = \xi^{(1)}$ by $z^{(1)}$, (31.4.9) yields the inequality

$$\left|z^{(1)} - \xi^{(1)}\right| \leq N^{(1)} \leq \frac{10\vartheta}{9}N^{(0)}, \tag{31.4.10}$$

that is,

$$\left|\frac{z^{(1)}}{\xi^{(1)}} - 1\right| \leq \frac{10\vartheta}{9}\frac{N^{(0)}}{|\xi^{(1)}|}.$$

Since (31.2.2) shows that

(31.4.11) $$|\xi^{(1)}| = \tfrac{19}{20}N^{(0)},$$

we get

(31.4.12) $$\left|\frac{z^{(1)}}{\xi^{(1)}} - 1\right| \leq \frac{20}{19}\frac{(10/9)\vartheta}{1}.$$

31.5

Suppose with a $d \geq 2$, that our algorithm does not stop in the first $(d-1)$ steps; that is,

(31.5.1) $$N^{(0)}, N^{(1)}, \ldots, N^{(d-1)},$$

$$\xi^{(0)}, \xi^{(1)}, \ldots, \xi^{(d-1)}$$

are already defined and, for $\nu = 1, 2, \ldots, d-1$, the inequality

(31.5.2) $$N^{(\nu)} \leq \frac{10\vartheta}{9}N^{(\nu-1)}$$

holds. In addition to (31.4.11), we suppose in the case $d \geq 3$ that, for $\nu = 2, \ldots, d-1$, the inequality

(31.5.3) $$|\xi^{(\nu)}| \geq \tfrac{19}{20}N^{(0)}\left\{1 - \tfrac{10}{9}\vartheta - \left(\tfrac{10}{9}\vartheta\right)^2 - \cdots - \left(\tfrac{10}{9}\vartheta\right)^{\nu-1}\right\}$$

is already proved. Our construction shows that for a suitable zero $z^{(d-1)}$ of $f(z)$ the inequality

(31.5.4) $$\tfrac{9}{10}N^{(d-1)} \leq |z^{(d-1)} - \xi^{(d-1)}| \leq N^{(d-1)}$$

holds. According to our algorithm the points $\xi_j^{(d)}$ are defined by

(31.5.5) $$\xi_j^{(d)} = \xi^{(d-1)} + \tfrac{19}{20}N^{(d-1)}e^{j(\pi i/6)},$$

(31.5.6) where $j = 0, 1, \ldots, 11.$

Since (31.5.2) evidently gives

$$N^{(d-1)} \le \left(\tfrac{10}{9}\vartheta\right)^{d-1} N(0), \tag{31.5.7}$$

we get from (31.5.5) and (31.5.3), for $j = 0, 1, \ldots, 11$, that

$$\begin{aligned} \left|\xi_j^{(d)}\right| &\ge \left|\xi^{(d-1)}\right| - \tfrac{19}{20} N^{(d-1)} \\ &\ge \tfrac{19}{20} N^{(0)} \left\{1 - \tfrac{10}{9}\vartheta - \left(\tfrac{10}{9}\vartheta\right)^2 - \cdots - \left(\tfrac{10}{9}\vartheta\right)^{d-1}\right\}, \end{aligned} \tag{31.5.8}$$

that is, (31.5.3) holds for $\nu = d$ too. Next we form, from the annulus in (31.5.4), the 12 sectors $G_j^{(d)}$ with

$$\frac{2j-1}{12}\pi \le \operatorname{arc}\left(z - \xi^{(d-1)}\right) < \frac{2j+1}{12}\pi, \qquad j = 0, 1, \ldots, 11. \tag{31.5.9}$$

Again, as in (31.4.5)–(31.4.6) the disks

$$\left|z - \xi_j^{(d)}\right| \le \vartheta N^{(d-1)} \qquad (j = 0, 1, \ldots, 11)$$

cover the annulus in (31.5.4). Hence $z^{(d-1)}$ is contained in some of these disks; if $j = h_1$ is the corresponding index, this means that the inequality

$$\left|z^{(d-1)} - \xi_{h_1}^{(d)}\right| \le \vartheta N^{(d-1)} \tag{31.5.10}$$

holds. We may suppose $f(\xi_{h_1}^{(d)}) \ne 0$ and consider the annulus

$$\tfrac{9}{10} N_{h_1}^{(d)} \le \left|z - \xi_{h_1}^{(d)}\right| \le N_{h_1}^{(d)}.$$

If z'' is one of the zeros of $f(z)$ nearest to $\xi_{h_1}^{(d)}$, then by Rule VI and (31.5.10) we have

$$\tfrac{9}{10} N_{h_1}^{(d)} \le \left|z'' - \xi_{h_1}^{(d)}\right| \le \left|z^{(d-1)} - \xi_{h_1}^{(d)}\right| \le \vartheta N^{(d-1)}.$$

Then by the minimum definition of $N^{(d)}$ [as in (31.2.8)]

$$N^{(d)} \le \tfrac{10}{9}\vartheta N^{(d-1)}, \tag{31.5.11}$$

that is, (31.5.2) holds for $\nu = d$ too. Thus with the μ_d-index [defined in analogy with μ_2 in (31.2.8) and (31.2.9)] and denoting one of the zeros nearest to $\xi_{\mu_d}^{(d)} = \xi^{(d)}$ by $z^{(d)}$, (31.5.11) and (31.5.7) yield

$$\left|z^{(d)} - \xi^{(d)}\right| \le N^{(d)} \le \tfrac{10}{9}\vartheta N^{(d-1)} \le \left(\tfrac{10}{9}\vartheta\right)^d N^{(0)},$$

that is,

$$\left|\frac{z^{(d)}}{\xi^{(d)}} - 1\right| \le \left(\tfrac{10}{9}\vartheta\right)^d \cdot \frac{N^{(0)}}{|\xi^{(d)}|}.^* \tag{31.5.12}$$

Since (31.5.8) remains valid if $\xi_j^{(d)}$ is replaced by $\xi^{(d)}$, this together with (31.5.12) gives

$$\left|\frac{z^{(d)}}{\xi^{(d)}} - 1\right| \le \left(\tfrac{10}{9}\vartheta\right)^d \cdot \tfrac{20}{19}\left\{1 - \sum_{m=1}^{\infty}\left(\frac{10}{9}\vartheta\right)^m\right\}^{-1} \tag{31.5.13}$$

$$= \left(\frac{10}{9}\vartheta\right)^d \cdot \frac{20}{19} \cdot \frac{9 - 10\vartheta}{9 - 20\vartheta}.$$

Since we see from (31.4.6) that

$$\vartheta < \frac{81}{280}, \qquad \frac{20}{19} \cdot \frac{9 - 10\vartheta}{9 - 20\vartheta} < 2,$$

(31.5.13) yields

$$\left|\frac{z^{(d)}}{\xi^{(d)}} - 1\right| \le 2\left(\tfrac{9}{28}\right)^d < 2 \cdot 3^{-d}$$

as required.

*This estimates the *absolute* error, but it depends on the polynomial by way of $N^{(0)}$.

32

A REMARK ON FINITE HOMOGENEOUS MARKOV CHAINS

32.1

Let us consider a physical system which can be, for

$$t = 0, 1, 2, \ldots,$$

in one of the possible different states $S_1, \ldots, S_k$. Let $p_\mu^{(n)}$ be the probability that the system is, for $t = n$, in the state S_μ. If we write

$$(32.1.1) \qquad p^{(n)} \doteq \begin{pmatrix} p_1^{(n)} \\ \vdots \\ p_k^{(n)} \end{pmatrix},$$

then, for a homogeneous Markov chain, we have

$$(32.1.2) \qquad p^{(n)} = \mathbf{P}^n p^{(0)} \quad \text{for} \quad n = 1, 2, \ldots,$$

where $\mathbf{P} = (p_{\mu\nu})$ is a fixed transposed stochastic $k \times k$ matrix independent of n, such that

$$(32.1.3) \qquad \begin{aligned} \sum_{\mu=1}^{k} p_{\mu\nu} &= 1 \qquad \nu = 1, 2, \ldots, k \\ p_{\mu\nu} &\geq 0 \qquad 1 \leq \mu, \nu \leq k. \end{aligned}$$

(No rows consist exclusively of 0.) If the eigenvalues of **P** are

$$\lambda_1, \ldots, \lambda_k,$$

we have

$$\lambda_1 = 1, \qquad |\lambda_j| \le 1 \qquad (j = 2, \ldots, k). \tag{32.1.4}$$

32.2

We consider finite homogeneous Markov chains satisfying

$$|\lambda_j| \ge \vartheta, \qquad j = 2, \ldots, k \tag{32.2.1}$$

with $0 < \vartheta \le 1$.

Theorem 32.1. *For* $\mu = 0, 1, \ldots,$ *the inequality*

$$\max_{\substack{k^{\mu} \le \nu \le k^{\mu+1} \\ \nu \text{ integer}}} p_j^{(\nu)} > \left(\frac{\vartheta^{k^{\mu}}}{2} \right)^k \cdot p_j^{(0)}$$

holds for all $j = 1, \ldots, k$.

The interesting aspects of this inequality are the universality of the intervals $I_\mu = [k^\mu, k^{\mu+1}]$ and the weak dependence of the lower bound upon **P**. The content of the theorem is that the process behaves as if it "remembered" the initial state in each I_μ.

The case when $\vartheta = 1$ is particularly interesting since then the lower bound in the theorem is independent of μ. But in this case the classical theorem of Frobenius on eigenvalues of matrices with nonnegative elements implies that the eigenvalues have the structure

$$\exp \frac{2\pi i \nu_1}{h_1}, \exp \frac{2\pi i \nu_2}{h_2}, \ldots, \exp \frac{2\pi i \nu_l}{h_l}$$

with $\nu_1 = 1, \ldots, h_1$; $\nu_2 = 1, \ldots, h_2; \ldots;$ $\nu_l = 1, \ldots, h_l$, where

$$h_1 + \cdots + h_l = k.$$

So the process is not very general. It would be interesting to prove that for "almost all" stochastic $k \times k$ matrices, all eigenvalues are in the annulus

$$(1 \ge)|z| \ge 1 - \frac{c}{k}$$

for some numerical positive constant c.

32.3 PROOF OF THEOREM 32.1

Let

$$F(z) \doteq \frac{1}{\lambda_1\lambda_2\cdots\lambda_k}\det\{\mathbf{P} - z\mathbf{E}_k\} \tag{32.3.1}$$

where $\mathbf{E}_k$ is the $k \times k$ unit matrix. Then

$$F(z) = \prod_{j=1}^{k}\left(1 - \frac{z}{\lambda_j}\right) = 1 + A_1 z + \cdots + A_k z^k \tag{32.3.2}$$

with $A_\nu = A_\nu(\lambda_1,\ldots,\lambda_k)$. As in the proof of the first main theorem, form the polynomial

$$1 - F(z)s_m\left(\frac{1}{F}\right).$$

Then, for a nonnegative integer m,

$$1 - F(z)s_m\left(\frac{1}{F}\right) = \sum_{\nu=m+1}^{m+k} c_\nu z^\nu \tag{32.3.3}$$

with $c_\nu = c_\nu(\lambda_1,\ldots,\lambda_k)$, and also

$$\sum_{\nu=m+1}^{m+k} |c_\nu| < \binom{k+m}{m}\frac{2^k}{\vartheta^{m+k}}. \tag{32.3.4}$$

The identity (32.3.3) holds if z is replaced by $\mathbf{P}$; since $\mathbf{P}$ satisfies its characteristic equation, we get

$$\mathbf{E}_k = \sum_{\nu=m+1}^{m+k} c_\nu \mathbf{P}^\nu. \tag{32.3.5}$$

Multiplying by $p^{(0)}$ from right, we obtain the relation

$$\sum_{\nu=m+1}^{m+k} c_\nu p^{(\nu)} = p^{(0)}. \tag{32.3.6}$$

Taking jth components on both sides, we get

$$\sum_{\nu=m+1}^{m+k} c_\nu p_j^{(\nu)} = p_j^{(0)} \quad \text{for} \quad j = 1,\ldots,k. \tag{32.3.7}$$

32.4

Fixing j and using (32.3.4) this gives the inequality

$$\max_{\substack{m+1\le\nu\le m+k\\ \nu\text{ integer}}} p_j^{(\nu)} \ge \frac{p_j^{(0)}}{2^k}\vartheta^{m+k}\frac{1}{\binom{k+m}{k}}. \tag{32.4.1}$$

We apply it with $m = 0$. Since, for fixed $p^{(0)}$, $p_j^{(\nu)} = p_j^{(\nu)}(\mathbf{P})$, we have

$$\max_{\substack{1\le\nu\le k\\ \nu\text{ integer}}} p_j^{(\nu)}(\mathbf{P}) \ge p_j^{(0)}\left(\frac{\vartheta}{2}\right)^k. \tag{32.4.2}$$

Next, replace $\mathbf{P}$ by $\mathbf{P}^{k^\mu}$ and ϑ by ϑ^{k^μ}. Thus

$$\max_{\substack{1\le\nu\le k\\ \nu\text{ integer}}} p_j^{(\nu)}(\mathbf{P}^{k^\mu}) \ge p_j^{(0)}\left(\frac{\vartheta^{k^\mu}}{2}\right)^k. \tag{32.4.3}$$

From (32.4.3) and

$$p_j^{(\nu)}(\mathbf{P}^{k^\mu}) = p_j^{(k^\mu\nu)}(\mathbf{P})$$

we obtain

$$\max_{\substack{k^\mu\le\nu\le k^{\mu+1}\\ \nu\text{ integer}}} p_j^{(\nu)} \ge p_j^{(0)}\cdot\frac{\upsilon^{k^{\mu+1}}}{2^k}.$$

33

EXPONENTIAL SUMS AND RIEMANN'S CONJECTURE I

The Quasi-Riemann Conjecture

33.1

These investigations began with the following remark of Landau's in his "Handbuch" of 1909. "Die Tatsache, dass $\Sigma_\rho(x^\rho/\rho)$ gerade in der Nähe der Primzahlen und der höheren Primzahlpotenzen und sonst in der Nähe keiner Stelle > 1 ungleichmässig konvergiert, deutet auf einen arithmetischen Zusammenhang zwischen den komplexen Wurzeln ρ der Zetafunktion und den Primzahlen hin. Ich habe keine Ahnung, worin derselbe besteht." Less dramatically and much later (in 1930) Titchmarsh wrote in his Cambridge tract, after expounding on Landau's results: "It is clear that the numbers ρ are closely connected with the prime numbers. No more explicit relation between them that what is given by the above formulae has been discovered." These sentences reflect a rather puzzling situation. We know a lot about primes, and we know a lot about the ρ. We can even use the ρ's to deduce properties of primes, but nevertheless we do not understand why.

33.2

It is plausible to take the prime powers into consideration by way of the *finite* exponential sums (with real τ)

$$Z(\tau, N) = \sum_{N \le n \le 2N} \Lambda(n) e^{-i\tau \log n} \tag{33.2.1}$$

or more generally by way of

$$Z(\tau, N_1, N_2) = \sum_{N_1 \le n \le N_2} \Lambda(n) e^{-i\tau \log n} \tag{33.2.2}$$

with

$$N \le N_1 < N_2 \le 2N. \tag{33.2.3}$$

$\Lambda(n)$ will always stand for the von Mangoldt symbol

$$\Lambda(n) = \begin{cases} \log p & \text{for} \quad n = p^{\alpha} \\ 0 & \text{otherwise.} \end{cases} \tag{33.2.4}$$

From some knowledge about the

$$\rho = \beta + i\gamma \tag{33.2.5}$$

roots of $\zeta(s)$ it is easy to calculate an upper bound of $Z(\tau, N_1, N_2)$. Throughout this book c denotes a positive numerical constant (different constants on different occasions); when c depends on some parameters it is explicitly stated as $c(\varepsilon, \eta, \ldots)$.

Theorem 33.1. *Suppose that with a* $0 < B \le 1$, $\frac{1}{2} \le \Theta < 1$, *and* $T > 10$, $\zeta(s) \ne 0$ *in the parallelogram*

$$\sigma > \Theta, \qquad |t - T| \le \tfrac{1}{2} T^B. \tag{33.2.6}$$

Then the inequality

$$|Z(\tau_0, N_1, N_2)| < \frac{c(\delta)}{B} \frac{N \log^3 N}{t_0^B} \tag{33.2.7}$$

holds for all

$$|\tau_0 - T| \le \frac{1 - \delta}{2} T^B \tag{33.2.8}$$

and for all (N_1, N_2) *pairs with*

$$T^{B/(1-\Theta)} \le N \le N_1 < N_2 \le 2N \tag{33.2.9}$$

if only $T > c(\delta)$.

33.3

The proof is postponed to the end of this section. The result seems at first to be very weak compared to the strong hypothesis (33.2.6); for $N > \exp(\tau_0^{B/3})$ it is even trivial. So it seems hopeless to deduce from (33.2.7)–(33.2.9) the existence of any zero-free domain. All the same, we show in this and the next two sections that such deductions are possible, even in much stronger forms. In this chapter we prove the simplest such theorem.

Theorem 33.2. *Let a and* $0 < b \leq 1$ *be two positive constants so that for a real* $\tau_0 > 2$ *the inequality*

$$|Z(\tau_0, N_1, N_2)| < c\frac{N\log^{100}N}{\tau_0^b} \tag{33.3.1}$$

holds for all (N_1, N_2) *pairs with*

$$\tau_0^a \leq N \leq N_1 < N_2 \leq 2N. \tag{33.3.2}$$

Then $\zeta(s) \neq 0$ *on the segment*

$$\sigma \geq 1 - \frac{b^3}{a^2}, \qquad t = \tau_0 \tag{33.3.3}$$

if only $\tau_0 > c$.

Before the proof, we mention a number of comments and corollaries. Suppose now that, instead of the inequalities (33.3.1)–(33.3.2) "local" in τ_0, the inequality

$$|Z(\tau, N_1, N_2)| < c\frac{N\log^{100}N}{\tau^b} \tag{33.3.4}$$

(which is "semilocal" in τ_0) holds for *all* values τ with

$$|\tau - T| \leq \tfrac{1}{2}T^b \tag{33.3.5}$$

and all (N_1, N_2) with

$$\tfrac{1}{2}T^a \leq N \leq N_1 < N_2 \leq 2N. \tag{33.3.6}$$

Then the following is an immediate consequence of Theorem 33.2.

Corollary I. *The inequalities* (33.3.4)–(33.3.6) *imply that* $\zeta(s) \neq 0$ *in the parallelogram*

$$\sigma \geq 1 - \frac{b^3}{a^2}, \qquad |t - T| \leq \tfrac{1}{2}T^b$$

if only $T > c$.

This is already a sort of converse to Theorem 33.1.

Suppose, instead of the local inequalities (33.3.1)–(33.3.2), that these hold for all

(33.3.7) $$\tau_0 > c$$

(i.e., globally in τ_0).

Corollary II. *From* (33.3.1)–(33.3.2), *and* (33.3.7) *it follows that* $\zeta(s) \neq 0$ *in the domain*

(33.3.8) $$\sigma \geq 1 - \frac{b^3}{a^2}, \qquad |t| \geq c.$$

An assertion of type (33.3.8) can be called a quasi-Riemann conjecture.

33.4

Corollary II implies that $\zeta(s)$ has in the half-plane

$$1 - \frac{b^3}{a^2}$$

only finitely many zeros at most, and since $\zeta(s) \neq 0$ for $\sigma = 1$, there is a half-plane

$$\sigma > \vartheta, \qquad \vartheta < 1$$

which is zero free. But this implies the well-known formula

(33.4.1) $$\pi(x) = \int_2^x \frac{dr}{\log r} + O(x^\vartheta \log x).$$

Corollary III. *If with positive constants a and* $b < 1$ *the inequality*

(33.4.2) $$\left| \sum_{N_1 \leq p \leq N_2} e^{-i\tau \log p} \right| < c\frac{N \log^{100} N}{\tau^b}$$

holds for all (N_1, N_2) *with*

$$\left(\tfrac{1}{2}c^a \le\right)\tfrac{1}{2}\tau^a \le N \le N_1 < N_2 \le 2N, \tag{33.4.3}$$

then we have for some $\vartheta < 1$ *the strong prime number formula* (33.4.1).

The interest in this corollary lies in the fact that it deduces a "strong prime number theorem" from a premise *dealing exclusively with primes* (not involving the zeta function). A direct proof of this theorem (though ϑ is ineffective) would be extremely interesting. (A converse deduction, of (33.4.2)–(33.4.3) from (33.4.1), is easy.)

33.5

Let us investigate, instead of $Z(\tau, N_1, N_2)$, the sum

$$Z_d(\tau, N_1, N_2) \doteq \sum_{N_1 \le p \le N_2} \exp(-i\tau \log^d p) \tag{33.5.1}$$

with

$$\tfrac{1}{2} \le d \le 2, \qquad d \ne 1. \tag{33.5.2}$$

Then one can prove—using the trigonometrical sieve method of I. M. Vinogradov or by density theorems—the inequality

$$|Z_d(\tau, N_1, N_2)| < \frac{c}{\sqrt{d-1}} \frac{N \log^{10} N}{\sqrt{|\tau|}} \tag{33.5.3}$$

for

$$c < |\tau|^{10} \le N \le N_1 < N_2 \le 2N. \tag{33.5.4}$$

Define

$$Z_h^*(\tau, N_1, N_2) = \sum_{N_1 \le p \le N_2} \exp\left[-i\tau \log p (\log\log p)^h\right] \tag{33.5.5}$$

for

$$-\tfrac{1}{2} \le h \le \tfrac{1}{2}, \qquad h \ne 0, \tag{33.5.6}$$

then the inequality

$$|Z_h^*(\tau, N_1, N_2)| < \frac{c}{\sqrt{|h|}} \frac{N \log^{10} N}{\sqrt{|\tau|}} \tag{33.5.7}$$

holds for

$$c < |\tau|^{10} \le N \le N_1 < N_2 \le 2N. \tag{33.5.8}$$

(For these see Turán [5] and [36].)

33.6

We prove Theorem 33.2 in a more general framework. We define the class $\Delta = \Delta(A_1, A_2, A_3)$ of functions $f(s)$ by the following properties (the A_ν being positive numerical constants).

I.

$$f(s) = 1 + \sum_{n=2}^{\infty} \frac{a_n}{n^s}$$

is convergent for $\sigma > 1$ and satisfies the inequality

$$|f(s)| \ge \log^{-10}(3 + |t|).$$

II. $f(s)$ is meromorphic in the half-plane $\sigma \ge \frac{1}{2}$ and for $\sigma \ge \frac{1}{2}$, $|t| \ge 3$ the inequality

$$|f(s)| \le |t|^{A_1(1-\sigma)} \log^{10}|t|$$

holds.

Condition I implies that for sufficiently large σ_0 we have the representation

$$\frac{f'}{f}(s) = \sum_{n=2}^{\infty} \frac{b_n}{n^s}$$

valid for $\sigma > \sigma_0$. We further require the following condition.

III. The inequality

$$|b_n| \le c \log n$$

holds for $n \ge 2$.

Theorem 33.3. *Let $f(s) \in \Delta$ and moreover suppose that for*

$$c(A_1, A_2, A_3) < \tau_0^{A_2} \le N \le N_1 < N_2 \le 2N \tag{33.6.1}$$

the inequality

$$\left| \sum_{N_1 \le n \le N_2} b_n e^{-i\tau_0 \log n} \right| < c \frac{N \log^{100} N}{|\tau_0|^{A_3}} \tag{33.6.2}$$

holds. Then for an explicit

$$(0 <) A_4 = A_4(A_1, A_2, A_3) (< \tfrac{1}{2})$$

$f(s) \neq 0$ *on the segment*

$$\sigma > 1 - A_4, \qquad t = \tau_0.$$

We may suppose without loss of generality that

$$A_2 \geq A_3. \tag{33.6.3}$$

We can take A_4 to be the greatest number x satisfying (simultaneously) the inequalities

$$x \log \frac{A_3}{A_1 x} \leq \frac{1}{300 A_1} \cdot \frac{A_3^2}{A_2} \tag{33.6.4}$$

$$x \leq \frac{A_3}{4A_2} \exp\left(- \frac{4A_2 + A_3}{A_2} \right) \tag{33.6.5}$$

$$x \leq \frac{9}{1000} \frac{A_3}{A_1} \tag{33.6.6}$$

$$x \leq \frac{1}{1600} \frac{A_3}{A_2}. \tag{33.6.7}$$

33.7. PROOF

Suppose, contrary to the assertion of the theorem, that there exists a zero

$$s^* = \beta^* + i\tau_0$$

of $f(s)$ with

$$\beta^* > 1 - A_4 \tag{33.7.1}$$

and a τ_0 so large that

$$\tau_0 > \max\left\{e^{1600}, \exp\left(\exp\left(2\frac{4A_2 + A_3}{A_2}\right)\right)\right\} \tag{33.7.2}$$

Apply Lemma E.1 (see Appendix E) with $G(s) = f(s)$,

$$s_0 = 1 + \max\left[\log^{-1/2}\tau_0, 4\frac{A_2}{A_3}(1 - \beta^*)\right] + i\tau_0 \doteq (1 + \mu^* + i\tau_0) \tag{33.7.3}$$

and $R = \frac{1}{2}$. Conditions I and II in 33.6 imply

$$\left|\frac{f(s)}{f(s_0)}\right| < c\tau_0^{A_1/2}\log^{20}\tau_0 \quad \text{for} \quad |s - s_0| \le \tfrac{1}{2},$$

that is, for the M of this lemma we can choose

$$M = (1 + A_1)\log \tau_0. \tag{33.7.4}$$

Thus for all positive integers ν,

$$\left|\frac{f'}{f}(s)^{(\nu)} + (-1)^{\nu+1}\nu! \sum_{\substack{j \\ |s_j - s_0| \le 8\mu^*}} \frac{1}{(s_0 - s_j)^{\nu+1}}\right| \tag{33.7.5}$$

$$< c(A_1)(\nu + 1)!\left(\frac{1}{8\mu^*}\right)^{\nu+1} \log \tau_0.$$

33.8

Next we deduce an upper bound for

$$\left(\frac{f'}{f}(s)\right)^{(\nu)}_{s=s_0}$$

By Conditions II and III and (33.7.3) we have

$$\left|\left(\frac{f'}{f}(s)\right)^{(\nu)}_{s=s_0}\right| = (-1)^{\nu} \sum_{n=2}^{\infty} \frac{\log^{\nu} n}{n^{1+\mu^*}}\left(b_n e^{-i\tau_0 \log n}\right),$$

that is,

$$\left|\left(\frac{f'}{f}(s)\right)^{(\nu)}_{s=s_0}\right| \le c \sum_{n \le \tau_0^{A_2}} \frac{\log^{\nu+1} n}{n^{1+\mu^*}} \tag{33.8.1}$$

$$+ \sum_{n > \tau_0^{A_2}} \frac{\log^{\nu} n}{n^{1+\mu^*}} \left(b_n e^{-i\tau_0 \log n} \right) \doteq Z_1 + Z_2.$$

For ν we assume provisionally only the weaker restriction

$$A_2 \log \tau_0 \leq \nu \leq 2A_2 \log \tau_0. \tag{33.8.2}$$

Then since

$$\max_{x \geq 1} x^{-1-\mu^*} \log^{\nu+1} x = \left[\frac{\nu+1}{e(1+\mu^*)} \right]^{\nu+1}$$

we get

$$|Z_1| < \tau_0^{A_2} \left[\frac{\nu+1}{e(1+\mu^*)} \right]^{\nu+1} < c(A_2) \tau_0^{A_2} \nu! \log \tau_0. \tag{33.8.3}$$

33.9

To estimate $|Z_2|$ from above, using (33.6.1)–(33.6.2), split it into partial sums of the type

$$S_l = \sum_{\tau_0^{A_2} 2^l < n \leq \tau_0^{A_2} 2^{l+1}} \tag{33.9.1}$$

For convenience, set

$$x^{-1-\mu^*} \log^{\nu} x = \varphi(x) \tag{33.9.2}$$

$$\tau_0^{A_2} 2^{l+1} = K_l$$

and introduce

$$\Phi_l(x) = \sum_{K_l < n \leq x} b_n e^{-i\tau_0 \log n}.$$

From (33.9.1),

$$S_l = \int_{[(1/2) K_l]+1}^{K_l} \varphi(x)\, dS_l(x).$$

Partial integration and use of (33.6.1)–(33.6.2) yields

$$|S_l| < c \frac{K_l \log^{100} K_l}{\tau_0^{A_3}} \left(\max_{(1/2) K_l \leq x \leq K_l} \varphi(x) + \int_{(1/2) K_l}^{K_l} |\varphi'(x)| dx \right),$$

which in turn implies

$$|S_l| < c\frac{K_l \log^{100} K_l}{\tau_0^{A_3}} \max_{(1/2)K_l \le x \le K_l} \varphi(x). \tag{33.9.3}$$

For $x \geq 1$, $\varphi(x)$ attains its maximum when

$$x = \exp\left(\frac{\nu}{1+\mu^*}\right) \doteq x_0.$$

If l_0 is the (unique) index with

$$\tfrac{1}{2}K_{l_0} < x_0 \le K_{l_0},$$

then, for $l < l_0$,

$$|S_l| < c(A_2)\frac{K_l \log^{100} K_l}{\tau_0^{A_3}} \frac{\log^{\nu} K_l}{K_l^{1+\mu^*}} = c(A_2)\frac{\log^{\nu+100} K_l}{\tau_0^{A_3} K_l^{\mu^*}} \tag{33.9.4}$$

and for $l > l_0$,

$$|S_l| < c(A_2)\frac{K_l \log^{100} K_l}{\tau_0^{A_3}} \frac{\log^{\nu}\left(\frac{1}{2}K_l\right)}{\left(\frac{1}{2}K_l\right)^{1+\mu^*}} < c(A_2)\frac{\log^{\nu+100} K_l}{\tau_0^{A_3} K_l^{\mu^*}} 2^{\mu^*}.$$

But by (33.6.7) we have

$$4\frac{A_2}{A_3}A_4 \le \tfrac{1}{40}, \tag{33.9.5}$$

that is, (33.7.2) and (33.7.3) show that

$$\mu^* < \tfrac{1}{40}. \tag{33.9.6}$$

Hence for $l > l_0$ we have

$$|S_l| < c(A_2)\frac{\log^{\nu+100} K_l}{\tau_0^{A_3} K_l^{\mu^*}}. \tag{33.9.7}$$

Now use (33.8.2) to estimate S_{l_0}.

$$|S_{l_0}| < c\frac{K_{l_0} \log^{100} K_{l_0}}{\tau_0^{A_3}} \left[\frac{\nu+1}{e(1+\mu^*)}\right]^{\nu+1}$$

$$< c\frac{K_{l_0} \log^{100} K_{l_0}}{\tau_0^{A_3}} \nu! \sqrt{\nu} < c(A_2) \log \tau_0 \frac{K_{l_0} \log^{100} K_{l_0}}{\tau_0^{A_3}} \nu!.$$

Furthermore,

$$K_{l_0} < 2x_0 < 2\exp\frac{\nu}{1+\mu^*} < 2e^{\nu}$$

and

$$\log^{100} K_{l_0} < (\nu + \log 2)^{100} < c(A_2)\log^{100}\tau_0$$

so that

$$|S_{l_0}| < c(A_2)\frac{e^{\nu}\nu!\log^{101}\tau_0}{\tau_0^{A_3}}. \tag{33.9.8}$$

Hence

$$|Z_2| < \frac{c(A_2)}{\tau_0^{A_3}}\left\{e^{\nu}\nu!\log^{101}\tau_0 + \sum_{l=0}^{\infty}\frac{\log^{\nu+100}K_l}{K_l^{\mu^*}}\right\}.$$

The last sum on the right is smaller than

$$2\frac{(\nu+100)!}{(\mu^*)^{\nu+100}} + \int_0^{\infty}\frac{\log^{\nu+100}\{\tau_0^{A_2}2^{y}\}}{(\tau_0^{A_2}2^{y})^{\mu^*}}\,dy.$$

Since the last integral is smaller than

$$\frac{1}{\log 2}\int_1^{\infty}\frac{\log^{\nu+100}r}{r^{1+\mu^*}}\,dr = \frac{1}{\log 2}\frac{(\nu+100)!}{(\mu^*)^{\nu+101}} < c(A_2)\frac{\nu!\log^{100}\tau_0}{(\mu^*)^{\nu+101}}$$

we have

$$|Z_2| < c(A_2)\frac{\nu!\log^{101}\tau_0}{\tau_0^{A_3}}\left\{e^{\nu} + \frac{1}{(\mu^*)^{\nu+100}}\right\}.$$

From (33.6.7), (33.7.2), and (33.7.3),

$$e < \frac{1}{\mu^*}. \tag{33.9.9}$$

Thus by (33.7.3),

$$|Z_2| < c(A_2)\frac{\nu!\log^{101}\tau_0}{\tau_0^{A_3}(\mu^*)^{\nu+100}} \le c(A_2)\frac{\nu!\log^{151}\tau_0}{\tau_0^{A_3}(\mu^*)^{\nu}}. \tag{33.9.10}$$

33.10

Hence (33.8.1), (33.8.3), (33.9.10), and (33.7.5) give

$$(33.10.1)\qquad \left| \sum_{\substack{j \\ |s_j - s_0| \le 8\mu^*}} \frac{1}{(s_0 - s_j)^{\nu+1}} \right|$$

$$< c(A_1, A_2, A_3)\log^{151}\tau_0 \left\{ \left(\frac{1}{8\mu^*} \right)^{\nu+1} + \tau_0^{A_2} + \frac{1}{\tau_0^{A_3}(\mu^*)^{\nu}} \right\}.$$

On one hand, using (33.8.2) and (33.9.6), we have

$$(33.10.2)\qquad \frac{1}{\tau_0^{A_3}(\mu^*)^{\nu}} \ge \exp\left\{ \left(A_2 \log\frac{1}{\mu^*} - A_3 \right) \log\tau_0 \right\};$$

since (33.6.5) and (33.7.2) give, respectively,

$$4\frac{A_2}{A_3}(1 - \beta^*) \le 4\frac{A_2}{A_3}A_4 < \exp\left(-\frac{4A_2 + A_3}{A_2} \right)$$

and

$$\log^{-1/2}\tau_0 < \exp\left(-\frac{4A_2 + A_3}{A_2} \right).$$

So (33.7.3) implies

$$\mu^* < \exp\left(-\frac{4A_2 + A_3}{A_2} \right),$$

and we get

$$(33.10.3)\qquad A_2 \log\frac{1}{\mu^*} - A_3 > 4A_2.$$

Thus, from (33.10.2),

$$(33.10.4)\qquad \frac{1}{\tau_0^{A_3}(\mu^*)^{\nu}} > \tau_0^{4A_2} > \tau_0^{A_2}.$$

On the other hand, using (33.6.3) and (33.8.2) we get

$$(33.10.5)\qquad \frac{1}{(8\mu^*)^{\nu+1}} < \frac{1}{(\mu^*)^{\nu+1}} 8^{-A_2 \log\tau_0} < \frac{1}{\tau_0^{A_3}(\mu^*)^{\nu+1}}.$$

Hence (33.10.4), (33.10.5), and (33.10.1) yield the inequality

$$(33.10.6)\qquad \left| \sum_{\substack{j \\ |s_j - s_0| \le 8\mu^*}} \frac{1}{(s_j - s_0)^{\nu+1}} \right| < c(A_1, A_2, A_3) \frac{\log^{151}\tau_0}{\tau_0^{A_3}(\mu^*)^{\nu+1}}.$$

33.11

Multiplying both sides of (33.10.6) by

$$|s^* - s_0|^{\nu+1} = (1 + \mu^* - \beta^*)^{\nu+1}$$

[and applying (33.7.3) and (33.8.2)] this takes the form

$$(33.11.1)\qquad \left| \sum_{\substack{j \\ |s_j - s_0| \le 8\mu^*}} \left(\frac{s^* - s_0}{s_j - s_0} \right)^{\nu+1} \right|$$

$$< c(A_1, A_2, A_3) \frac{\log^{151}\tau_0}{\tau_0^{A_3}} \left(1 + \frac{1 - \beta^*}{\mu^*}\right)^{\nu+1}$$

$$\le c(A_1, A_2, A_3) \frac{\log^{151}\tau_0}{\tau_0^{A_3}} \left(1 + \frac{A_3}{4A_2}\right)^{2A_2 \log \tau_0}$$

$$< c(A_1, A_2, A_3) \frac{\log^{151}\tau_0}{\tau_0^{(1/2)A_3}},$$

which is the basic inequality.

33.12

On the left side of (33.11.1) we have a power sum of the fixed complex numbers

$$\frac{s_0 - s^*}{s_0 - s_j}$$

which are independent of ν. For the exponent ν we have the restriction (33.8.2). Now we narrow the interval for ν and give a lower bound for the remaining sum by an appropriate choice of ν. We will use the second main theorem

(Theorem 8.1) in the slightly weaker form: for

$$m \geq 0, \qquad n \leq N$$

there exists an integer $k = k(\xi_1, \ldots, \xi_n)$ satisfying

$$M \leq k \leq m + N$$

so that

$$|\xi_1^l + \cdots + \xi_n^l| \geq \left[\frac{N}{8e(m+N)}\right]^N \max_j |\xi_j|^l. \tag{33.12.1}$$

The role of the ξ_j is obviously played by the numbers

$$\frac{s_0 - s^*}{s_0 - s_j}.$$

We have to verify $\max_j |\xi_j| \geq 1$. This obviously holds if $|s_0 - s^*| \leq 8\mu^*$. But by (33.7.3) and (33.6.3) we have

$$7\mu^* \geq 7 \cdot 4\frac{A_2}{A_3}(1 - \beta^*) > 1 - \beta^*,$$

that is,

$$8\mu^* > 1 + \mu^* - \beta^* = |s_0 - s^*|.$$

We choose

$$m = A_2 \log \tau_0. \tag{33.12.2}$$

We will need an upper bound N for the number of zeros of $f(s)$ in the disk

$$|s - s_0| \leq 8\mu^*.$$

Using Jensen's upper bound for the number of zeros of $f(s)$—Condition II is applicable—we get

$$n \leq \frac{1}{\log 2} \max_{|s - s_0| \leq 16\mu^*} \log\left|\frac{f(s_0 + 16\mu^* e^{i\varphi})}{f(s_0)}\right|$$

$$< \frac{1}{\log 2} \log\left[\frac{(\tau_0 + 1)^{A_1 \cdot 15\mu^*} \log^{10}(1 + \tau_0)}{\log^{-10}(3 + \tau_0)}\right]$$

$$< 25(A_1 \mu^* \log \tau_0 + \log\log \tau_0)$$

with (33.7.3) in the form

$$\mu^* < 4\frac{A_2}{A_3}(1-\beta^*) + \log^{-1/2}\tau_0$$

we see that N can be chosen by

$$N = 100A_1\frac{A_2}{A_3}(1-\beta^*)\log\tau_0 + c(A_1)\sqrt{\log\tau_0}\,. \tag{33.12.3}$$

If we choose our $(\nu+1)$ from $(m, m+N)$ we must show that (33.8.2) is not violated by this choice. This will be evident if

$$N \le A_2\log\tau_0, \tag{33.12.4}$$

that is, if

$$100\frac{A_1}{A_3}A_4 < \frac{9}{10}$$

$$c(A_1)\sqrt{\log\tau_0} < \frac{1}{10}A_2\log\tau_0$$

which holds, by (33.6.6) for sufficiently large τ_0. Hence Theorem 8.1 assures the existence of a (ν_0+1) with

$$A_2\log\tau_0 \le \nu_0+1 \le A_2\log\tau_0 \tag{33.12.5}$$

$$+100\frac{A_2}{A_3}A_1(1-\beta^*)\log\tau_0 + c(A_1)\sqrt{\log\tau_0}\,,$$

so that the sum on the left of (33.11.1) is larger than

$$\left(\frac{1}{8e}\frac{N}{m+N}\right)^N > \left(\frac{1}{16e}\frac{N}{m}\right)^N > \left[\frac{1}{16e}100\frac{A_1}{A_3}(1-\beta^*)\right]^{100\frac{A_1A_2}{A_3}(1-\beta^*)\log\tau_0}$$

$$\times\left(\frac{1}{16e}\frac{1}{m}\right)^{c(A_1)\sqrt{\log\tau_0}}$$

$$> \tau_0^{-A_5}\exp\{-c(A_1, A_2, A_3)\log^{3/4}\tau_0\}$$

with

$$A_5 = 100\frac{A_1A_2}{A_3}(1-\beta^*)\log\left(\frac{A_3}{A_1}\frac{1}{1-\beta^*}\right).$$

From (33.6.4) we have

$$A_5 < \frac{A_3}{3},$$

and it follows from (33.11.1) that

$$\tau_0^{-(1/3)A_3}\exp\{-c(A_1, A_2, A_3)\log^{3/4}\tau_0\} < c(A_1, A_2, A_3)\frac{\log^{151}\tau_0}{\tau_0^{(1/2)A_3}}.$$

This is a contradiction, for all sufficiently large τ_0, and the proof is complete. ■

33.13

We now consider Theorems 33.1 and 33.2 from the point of view of Landau's problem in 33.1, concerning localization of the primes. As mentioned in 33.2 the supposition (33.2.6) on zero-free domains implies a *nontrivial* consequence, by way of the sums $Z(\tau, N_1, N_2)$, only for primes

$$T^{B/(1-\Theta)} < p < \exp(T^{B/3}).$$

In Theorem 33.2 the inequality (33.3.1) is [since $\Sigma_{n\le x}\Lambda(n) = O(x)$] trivial for

$$N > \exp(\tau_0^{b/100})$$

because in the proof of this theorem only the primes satisfying

$$\tau_0^a < p < \exp(\tau_0^{b/100}) \tag{33.13.1}$$

are used *nontrivially*. We did not mention this aspect of Theorem 33.2 for two reasons. First, it made the proof more transparent and second, in the next chapter we prove a much stronger localization concerning primes than in (33.13.1), by a less transparent argument.

33.14

We complete this section by proving Theorem 33.1. Although the proof does not use any facts from Part I, without this information one cannot grasp the real meaning of the results of the next sections. Consider, with an arbitrary integer N ≥ 3, the integral

$$K(N, \tau_0) = -\frac{1}{2\pi i}\int_{(l)} \frac{(N+\frac{1}{2})^w}{w}\frac{\zeta'}{\zeta}(w + i\tau_0)\,dw \tag{33.14.1}$$

where the path l, with $w = u + iv$, is the segment

$$u = 1 + \frac{1}{\log N}$$

$$(33.14.2) \qquad -\frac{\delta}{2}T^B \le v \le \frac{\delta}{2}T^B.$$

Replacing ζ'/ζ by its Dirichlet series in (33.14.2) we get

$$(33.14.3) \qquad K(N,\tau_0) = \sum_n \Lambda(n) e^{-i\tau_0 \log n} \frac{1}{2\pi i}\int_{(l)} \left(\frac{N+\frac{1}{2}}{n}\right)^w \frac{dw}{w}.$$

Completing the integral to

$$\frac{1}{2\pi i}\int_{1+(1/\log N)} \left(\frac{N+\frac{1}{2}}{n}\right)^w \frac{dw}{w},$$

estimating the "wings" by partial integration and using the fact that

$$(33.14.4) \qquad \sum_{n=1}^{\infty} \frac{1}{n^{1+(1/\log N)}\left|\log\left((N+\frac{1}{2})/n\right)\right|} < c\log^2 N$$

one obtains

$$(33.14.5) \qquad |Z(\tau_0, 1, N) - K(N,\tau_0)| < c(\delta)\frac{N\log^2 N}{T^B}.$$

Next we replace l in (33.14.2) by the directed parallelogram ABCD with the vertices

$$A = 1 + \frac{1}{\log N} - i\frac{\delta}{4}T^B$$

$$B = \Theta + \frac{1}{\log N} - i\frac{\delta}{4}T^B$$

$$C = \Theta + \frac{1}{\log N} + i\frac{\delta}{4}T^B$$

$$D = 1 + \frac{1}{\log N} + i\frac{\delta}{4}T^B$$

$\zeta'/\zeta(w + i\tau_0)$ is obviously regular in our parallelogram, and we apply simple classical inequalities from the theory of the zeta function to get

$$\left|\frac{1}{2\pi i}\int_{ABCD}\right| < c(\delta)\frac{N\log T \cdot \log N}{T^B} + N^{\Theta}\log^2 T\log N. \tag{33.14.6}$$

So far N and T are unrelated, we now use the restriction

$$N \geq T^{B/(1-\Theta)}, \tag{33.14.7}$$

so that $N^{1-\Theta} \geq T^B$,

$$N^{\Theta} \leq \frac{N}{T^B},$$

and hence

$$|Z(\tau_0, 1, N)| \leq c(\delta)\frac{N}{T^B}(\log^2 T \cdot \log N). \tag{33.14.8}$$

Since from (33.14.7)

$$\log T < \frac{1}{B}\log N$$

apply (33.14.8) with $N = N_1, N_2$ to complete the proof. ∎

Since 1909, progress has been substantial in all but one area. The notable exclusion is the last, namely, our understanding of the relationship between the ρ and the primes. In the next few sections our goal is to increase twofold this small amount of information.

34

SOME LEMMAS

34.1

To maintain continuity in the later applications, we collect (as in Part I) some lemmas we need in the sequel.

Lemma 34.1. *For an integer $k \geq 2$, real γ and positive λ let*

$$h_k(\gamma) = h_k(\gamma; \lambda) = \frac{1}{2\pi i}\int_{(10)}\left(\frac{e^{\lambda w} - e^{-\lambda w}}{2\lambda w}\right)^k e^{\gamma w}\, dw. \tag{34.1.1}$$

Then we have for $\gamma \geq k\lambda$ or for $\gamma \leq -k\lambda$

$$h_k(\gamma) = 0 \tag{34.1.2}$$

and for

$$-k\lambda \leq \gamma \leq k\lambda \tag{34.1.3}$$

the representation

$$h_k(\gamma) = \frac{1}{\lambda\pi}\int_0^\infty \left(\frac{\sin r}{r}\right)^k \cos\left(\frac{\gamma}{\lambda}r\right) dr. \tag{34.1.4}$$

The result (34.1.4) follows immediately after shifting the line of integration to $\mathrm{Re}\, w = 0$. The second half of (34.1.2) follows from the classical formula

$$\frac{1}{2\pi i}\int_{(10)} \frac{e^{xw}}{w^k}\, dw = 0 \quad \text{for} \quad x \leq 0.$$

The first half of (34.1.2) follows from

$$h_k(\gamma) = \frac{1}{2\pi i}\int_{(-1)}\left(\frac{e^{\lambda w} - e^{-\lambda w}}{2\lambda w}\right) e^{\gamma w}\, dw$$

and the formula

$$\frac{1}{2\pi i}\int_{(-1)}\frac{e^{xw}}{w^k}\, dw = 0 \quad \text{for} \quad x \geq 0.$$

Lemma 34.2. (Knapowski–Turán [9]). *For integers $k \geq 2$, $h_k(\gamma)$ is monotonically decreasing for $\gamma \geq 0$*

Proof. This proof is by induction with respect to k. For $k = 2$ the assertion is well known; suppose it is true for all $k < k_0$ (≥ 3). Then, from (34.1.4),

$$\frac{d}{d\gamma}h_{k_0}(\gamma) = \frac{1}{2\lambda^2\pi}\int_0^{\infty}\left(\frac{\sin r}{r}\right)^{k_0-1}\left(-2\sin r \sin\left(\frac{\gamma}{\lambda}r\right)\right) dr \tag{34.1.5}$$

$$= \frac{1}{2\lambda}\left\{h_{k_0-1}\left(1 + \frac{\gamma}{\lambda}\right) - h_{k_0-1}\left(1 - \frac{\gamma}{\lambda}\right)\right\}$$

$$= \frac{1}{2\lambda}\left\{h_{k_0-1}\left(\frac{\gamma}{\lambda} + 1\right) - h_{k_0-1}\left(\frac{\gamma}{\lambda} - 1\right)\right\}. \tag{34.1.6}$$

Now (34.1.5) completes the proof for $0 \leq \gamma \leq \lambda$ and (34.1.6) for $\gamma \geq \lambda$.

It is trivial that $h_k(\lambda)$ increases monotonically for $\gamma \leq 0$, and for real γ, $h_k(\gamma) \geq 0$ and

$$h_k(\gamma) \leq \frac{1}{\lambda\pi}\left(2 + \frac{1}{2^{k-1}}\right) < \frac{1}{\lambda}. \quad \blacksquare \tag{34.1.7}$$

34.2

Several of our applications of these methods in analytic number theory (in collaboration with S. Knapowski) used a certain construction based on Dirichlet's theorem on simultaneous approximation by rationals. J. Pintz noticed that this construction can be performed much more effectively by the following lemma.

Lemma 34.3. *Given the integers $q \geq 2$ and M, and the real numbers $\gamma_1, \ldots, \gamma_N$, there exist integers n_j with*

$$1 \leq n_1 < n_2 < \cdots < n_M \leq Mq^N \tag{34.2.1}$$

so that for every μ, ν *the distance of* $n_\mu \gamma_\nu$ *from the nearest integer is*

$$\leq \frac{1}{q}.$$

Proof. Consider the unit cube in N dimensional space and form the subcubes:

$$Q: \qquad \frac{\nu_j}{q} \leq x_j < \frac{\nu_j + 1}{q}, \qquad 0 \leq \nu_j < q; \qquad j = 1, 2, \ldots, N;$$

there are q^N of them. Consider the points

$$P_n = (n\gamma_1, n\gamma_2, \ldots, n\gamma_N), \qquad n = 1, 2, \ldots, Mq^N \tag{34.2.2}$$

reduced modulo the unit cube. If at least M of these reduced points fall in the cube

$$0 \leq x_j < \frac{1}{q} \qquad (j = 1, \ldots, N) \tag{34.2.3}$$

we are done. If they number less than M, then there is another such cube Q' containing at least $M + 1$ of the P_n, say with the indices:

$$1 \leq n_1' < n_2' < \cdots < n'_{M+1} \leq Mq^n.$$

But then the reduced points corresponding to

$$P_{n_2' - n_1'}, P_{n_3' - n_1'}, \ldots, P_{n'_{M+1} - n_1'}$$

belong to the cube

$$0 \leq |x_j| < \frac{1}{q} \tag{34.2.4}$$

which completes the proof. ■

This lemma is also contained in Titchmarsh [1] p. 153.

35

EXPONENTIAL SUMS AND RIEMANN'S CONJECTURE II

Strong Localization

35.1

In this chapter we show, roughly, that by restricting the (N_1, N_2) pairs only by

$$c < \tau_0^{\nu_1} \le N \le N_1 < N_2 \le 2N \le \tau_0^{\nu_2} \tag{35.1.1}$$

one can still deduce the nonvanishing of $\zeta(s)$ on a segment of the form

$$\sigma \ge \Theta = \Theta(b), \qquad t = \tau_0,$$

from (33.3.1).

Let

$$D \ge 4 \tag{35.1.2}$$

be fixed. Let $a_1, \ldots$ be positive, explicitly calculable numerical constants. (The numerical values of these constants would be of interest, but at present we have only large upper bounds.) Denote by c, positive, explicitly calculable numerical constants (different constants on different occasions) whose effective values are of no interest; if they depend on the D and b, parameters this will be represented by $c(D)$ or $c(b)$ or $c(D, b)$. At present we require only

$$0 < b \le \min\left(a_1, \frac{1}{64D^2}\right) \tag{35.1.3}$$

with an

$$a_1 \le \frac{1}{3^{20}}; \tag{35.1.4}$$

the final value of a_1 is given later.*

Theorem 35.1. *Suppose the inequality*

$$|Z(\tau_0, N_1, N_2)| < c\frac{N\log^{10}N}{\tau_0^b} \tag{35.1.5}$$

is valid for

$$c(D,b) \le \tau_0^D \le N \le N_1 < N_2 \le \min\left[2N, \tau_0^{D(1+4b^{1/20})}\right]. \tag{35.1.6}$$

Then $\zeta(s) \ne 0$ *on the segment*

$$\sigma \ge 1 - b^2, \qquad t = \tau_0. \tag{35.1.7}$$

35.2. PROOF

Assume the existence of a root

$$\rho^* = \beta^* + i\tau_0 \tag{35.2.1}$$

with

$$\beta^* > 1 - b^2, \tag{35.2.2}$$

and we want to obtain a contradiction if

$$0 < b < \min\left(a_1, \frac{1}{64D^2}\right), \qquad \tau_0 > c(b, D), \tag{35.2.3}$$

and a_1 is sufficiently small. For the sake of brevity, let

$$\vartheta = b^{1/20}, \tag{35.2.4}$$

$$s_0 = 1 + b^2 + i\tau_0 = 1 + \mu + i\tau_0, \tag{35.2.5}$$

*Beside (35.1.4) the final value of a_1 is affected only by (35.6.5), (35.6.8), and (35.6.12).

and

$$(35.2.6)\quad J(s_0) = -\frac{1}{2\pi i}\int_{(10)}\left(e^{w/b}\frac{e^{\vartheta w/b} - e^{-\vartheta w/b}}{2\vartheta w/b}\right)^k \frac{\zeta'}{\zeta}(s_0 + w)\,dw$$

where k is an integer restricted (for the time being) only by

$$(35.2.7)\qquad 2 \le \frac{bD}{1-\vartheta}\log\tau_0 \le k \le bD(1 + 2\vartheta)\log\tau_0.$$

Provided that

$$\tau_0 > c(D, b),$$

such an integer k obviously exists. Replace ζ'/ζ by its Dirichlet series in (35.2.6); we get—with the notation (34.1.1) and by (34.1.2) from Lemma 34.1 —the representation

$$(35.2.8)\quad J(s_0) = \sum_n \{\Lambda(n)e^{-i\tau_0\log n}\}\left\{n^{-1-\mu}h_k\left(\frac{k}{b} - \log n, \frac{\vartheta}{b}\right)\right\}.$$

The summation is extended over all n satisfying

$$(35.2.9)\qquad \exp\left(k\frac{1-\vartheta}{b}\right) \le n \le \exp\left(k\frac{1+\vartheta}{b}\right)$$

By (35.2.4)–(35.2.7), the range of n in (35.2.9) belongs to $(\tau_0^{\mathrm{D}}, \tau_0^{\mathrm{D}(1+4\vartheta)})$.

35.3

To use (35.1.5)–(35.1.6) we split the sum in (35.2.8) into, at most,

$$c\frac{k\vartheta}{b}$$

partial sums of the form

$$\sum_{N_1 \le n \le N_2}, \qquad N_2 \le 2N_1.$$

This gives

$$(35.3.1)\quad |J(s_0)| < c(D)\log\tau_0$$

$$\cdot \max \left| \sum_{N_1 \le n \le N_2} \{\Lambda(n) e^{-i\tau_0 \log n}\} \left\{ n^{-1-\mu} h_k\left(\frac{k}{b} - \log n, \frac{\vartheta}{b}\right)\right\} \right|$$

where the max refers to

$$\exp\left(k\frac{1-\vartheta}{b}\right) \le N_1 < N_2 \le \min\left\{2N_1, \exp\frac{k(1+\vartheta)}{b}\right\}. \tag{35.3.2}$$

Denote the inner sum in (35.3.1) by Z, and introduce

$$g(x) = \sum_{N_1 \le n \le x} \Lambda(n) e^{-i\tau_0 \log n} \tag{35.3.3}$$

to get the representation

$$Z = \int_{N_1}^{N_2} x^{-1-\mu} h_k\left(\frac{k}{b} - \log x, \frac{\vartheta}{b}\right) dg(x). \tag{35.3.4}$$

Integration by parts, and use of (34.1.7) and (35.1.5)–(35.1.6) give

$$|Z| < c\frac{N\log^{10}N}{\tau_0^b}\left\{\frac{1}{N_1^{1+\mu}} + \int_{N_1}^{N_2}\left|\frac{d}{dx}\left[x^{-1-\mu}h_k\left(\frac{k}{b} - \log x, \frac{\vartheta}{b}\right)\right]\right|dx\right\}$$

$$< c\frac{N\log^{10}N}{\tau_0^b}\left\{\frac{1}{N_1^{1+\mu}} + \frac{1}{N_1^{1+\mu}}\int_{N_1}^{N_2}\left|\frac{d}{dx}h_k\left(\frac{k}{b} - \log x, \frac{\vartheta}{b}\right)\right|dx\right\}.$$

By Lemma 34.2 the function $h_k(k/b - \log x)$ consists of at most two monotonic parts, and hence

$$|Z| < c\frac{\log^{10}N}{\tau_0^b}. \tag{35.3.5}$$

The remark at the end of 35.2 shows that

$$\log^{10}N < c(D)\log^{10}\tau_0,$$

and hence, from (35.3.1),

$$|J(s_0)| < c(D)\frac{\log^{11}\tau_0}{\tau_0^b}. \tag{35.3.6}$$

35.4

To obtain another analytic expression for $J(s_0)$ we shift the line of integration to

$$\text{Re}\, w = -2 - \mu. \tag{35.4.1}$$

Using a well-known inequality for ζ'/ζ the absolute value of new integral is

$$< c\int_{-\infty}^{\infty} \frac{\log(2 + |\tau_0 + \vartheta|)}{(1 + |\vartheta|)^k} b^{(19/20)k}\, d\vartheta < cb^{(19/20)k} \log \tau_0 \tag{35.4.2}$$

$$< \frac{c \log \tau_0}{4^k} < \frac{c \log \tau_0}{\tau_0^{2b}}$$

from (35.2.7) and (35.1.2). There is a residue at $w = 1 - s_0$. By (35.2.5), (35.2.4), (35.2.7), and (35.1.2) its absolute value contributes, at most,

$$\left(\frac{b^{19/20}}{\tau_0}\right)^k < \tau_0^{-bD \log \tau_0} < \tau_0^{-bD} < \tau_0^{-4b}.$$

This, (35.4.2), and (5.3.6), yield

(35.4.3)

$$\sum_\rho \left(\exp\left(\frac{\rho - s_0}{b}\right) \frac{\exp[\vartheta(s_0 - \rho)/b] - \exp[-\vartheta(s_0 - \rho)/b]}{2\vartheta(s_0 - \rho)} b\right)^k < c\frac{\log^{11}\tau_0}{\tau_0^b}$$

where $\rho = \beta + i\gamma$ runs over the nontrivial zeros of $\zeta(s)$.

35.5

The contribution of the ρ with

$$\tau_0 + \nu \le \gamma < \tau_0 + \nu + 1, \qquad \nu = 2, 3, \ldots$$

to the sum in (35.4.3) is, in absolute value,

$$< c \log(\tau_0 + \nu)\left(\frac{b^{19/20}}{\nu}\right)^k < \frac{c \log(\tau_0 + \nu)}{\nu^k},$$

that is, the total contribution of the ρ with $\gamma \ge \tau_0 + 2$ is

$$< c\frac{\log \tau_0}{2^k} < c\frac{\log \tau_0}{\tau_0^{2b}}, \tag{35.5.1}$$

as before. The same holds for the contribution of the ρ with

$$\gamma \leq \tau_0 - 2. \tag{35.5.2}$$

Consider the contribution of the ρ with

$$|\gamma - \tau_0| \leq 2, \qquad \frac{\rho - s_0}{b} \geq \frac{\pi}{2\vartheta} \tag{35.5.3}$$

to (35.4.3). This is evidently

$$< c \log \tau_0 M^k \tag{35.5.4}$$

where

$$M = \max\left| e^{-z/\vartheta} \frac{e^z - e^{-z}}{2^z} \right|, \tag{35.5.5}$$

and the max refers to

$$\operatorname{Re}(z) \geq 0, \qquad |z| \geq \frac{\pi}{2}. \tag{35.5.6}$$

Since $1/\vartheta > 1$,

$$|e^{-z[(1/\vartheta)-1]}| \leq 1, \qquad |e^{-z[(1/\vartheta)-1]}| \leq 1.$$

Thus

$$M \leq \frac{2}{\pi},$$

that is, the contribution in (35.5.4) is

$$< c \log \tau_0 \left(\frac{2}{\pi}\right)^k < c \frac{\log \tau_0}{\tau_0^b}. \tag{35.5.7}$$

For $\tau_0 > c(b)$ (35.5.1), (35.5.2), and (35.5.7) with (35.4.3) yield the inequality

(35.5.8)

$$\left| \sum{}' \left(\exp\left(\frac{\rho - s_0}{b}\right) \frac{\exp[\vartheta(s_0 - \rho)/b)] - \exp[-\vartheta(s_0 - \rho)/b)]}{2\vartheta(s_0 - \rho)} b \right)^k \right|$$

$$< c \frac{\log^{11} \tau_0}{\tau_0^b} < \tau_0^{-b/2}$$

where the range of summation is

$$|\rho - s_0| \leq \frac{\pi}{2\vartheta} b = \frac{\pi}{2} b^{19/20} \doteq r_0. \tag{35.5.9}$$

35.6

Inequality (35.5.8) holds for all integers k satisfying (35.2.7). Now we use the second main theorem in the form (33.12.1) to get a lower bound for the left-hand expression by appropriate choice of k. Choose

$$m = \frac{bD}{1-\vartheta}\log \tau_0. \tag{35.6.1}$$

We need an upper bound N for the number of terms in (35.5.8). This is done by using the inequality of Vinogradov–Korobov–Richert:

$$|\zeta(s)| < ct^{100(1-\sigma)^{3/2}}\log t \tag{35.6.2}$$

for

$$\tfrac{1}{2} \le \sigma \le 1, \qquad t \ge 2 \tag{35.6.3}$$

(see Richert [1]) and also by the classical inequality

$$\frac{1}{|\zeta(s)|} < c\log t \qquad \sigma \ge 1, \quad t \ge 2. \tag{35.6.4}$$

With the notation (35.5.9) we have, somewhat crudely,

$$\max_{|s-s_0|\le er_0} 100(1-\sigma)^{3/2} < 100\left(\frac{e\pi}{2}b^{19/20}\right)^{3/2} \doteq a_2 b^{57/40}. \tag{35.6.5}$$

For $\tau_0 > c(b)$ Jensen's inequality gives as an upper bound for the number of terms in (35.5.8) the quantity

$$a_2 b^{57/40}\log \tau_0 + 3\log\log \tau_0 < 2a_2 b^{57/40}\log \tau_0 \doteq N. \tag{35.6.6}$$

The second main theorem can be applied only if

$$[m, m+N] \subset \left[\frac{bD}{1-\vartheta}\log \tau_0,\ bD(1+2\vartheta)\log \tau_0\right],$$

which certainly holds if

$$N < \tfrac{1}{2}b^{21/20}D\log \tau_0 \tag{35.6.7}$$

since $\vartheta < a_1^{1/20} \le \frac{1}{3}$ by (35.1.4). But this in turn holds choosing a_1 in (35.1.3) so small that

$$a_1 \le \left(\frac{1}{a_2}\right)^{8/3}, \tag{35.6.8}$$

because in this case we have

$$b^{3/8} < a_1^{3/8} < \frac{1}{a_2} < \frac{D}{4a_2}$$

$$2a_2 b^{57/40} < \frac{1}{2} b^{21/20} D.$$

Thus (35.6.7) is fulfilled, and the theorem is applicable. Denote the resulting k exponent by k^*. We must also find a lower bound for

(35.6.9)

$$H \doteq \max_{|\rho - s_0| \le r_0} \left| \left(\exp\left[\frac{\rho - s_0}{b} \right] \frac{\exp[\vartheta(s_0 - \rho)/b)] - \exp[-\vartheta(s_0 - \rho)/b)]}{2\vartheta(s_0 - \rho)} b \right)^{k^*} \right|.$$

Since, by (35.2.5) and (35.2.2),

$$|\rho^* - s_0| = (1 + \mu - \beta^*) \le 2b^2 < r_0,$$

ρ^* is among our remaining ρ. Hence choosing $\rho = \rho^*$ in (35.6.9) we get for $x > 0$ the inequality

$$H \ge \left\{ \exp\left(\frac{\beta^* - 1 - \mu}{b} \right) \cdot \frac{\exp[\vartheta(1 + \mu - \beta^*)/b] - \exp[-\vartheta(1 + \mu - \beta^*)/b]}{[2\vartheta(1 + \mu - \beta^*)/b]} \right\}^{k^*}$$

$$> \exp\left(k^* \frac{\beta^* - 1 - \mu}{b} \right) > \exp(-2k^* b),$$

and using (35.2.7) and (35.1.3)

(35.6.10) $$H > \exp(-4b^2 D \log \tau_0) = \tau_0^{-4b^2 D} > \tau_0^{-(1/2)b^{3/2}}.$$

Hence, denoting the left-hand side of (35.5.8) by U, this choice of k^* gives the lower bound

(35.6.11) $$U > \tau_0^{-(1/2)b^{3/2}} \left(\frac{N}{8e \cdot 2m} \right)^N$$

$$= \exp\left\{ \log \tau_0 \left(-\tfrac{1}{2} b^{3/2} - 2a_2 b^{57/40} \log \frac{16eD}{a_2 b^{17/40}} \right) \right\}.$$

We require a_1 to be so small that for $0 < b < a_1$ the inequality

$$2a_2 \log \frac{16e}{8a_2 b^{37/40}} < \frac{1}{2}\left(\frac{1}{b}\right)^{1/40} \tag{35.6.12}$$

also holds. Then, since $b \le 1/64D^2$, the inequality

$$2a_2 \log \frac{16eD}{a_2 b^{17/40}} < \frac{1}{2}\left(\frac{1}{b}\right)^{1/40}$$

holds, and, from (35.6.11),

$$U > \exp\left\{\log \tau_0\left(\tfrac{1}{2}b^{3/2} - \tfrac{1}{2}b^{7/5}\right)\right\} > \tau_0^{-b^{7/5}}$$

But this and (35.5.8) show

$$-\frac{b}{2} \ge -b^{7/5}, \qquad a_1 \ge b \ge \left(\tfrac{1}{2}\right)^{5/2},$$

which contradicts (35.1.4). This proves Theorem 35.1. ■

35.7

Using the loose terminology in 33.3, Theorem 35.1 can be described as "local in τ_0" and "much stronger" locally in the primes than in (33.13.1). It will be of interest to state explicitly the "in τ_0 semilocal" corollary of this theorem.

Corollary I. *Let* $D \ge 4$, $0 < b \le \min(a_1, 1/36D^2)$. *Suppose that for* $T > c(D, b)$, *with*

$$\tfrac{1}{2}T^D \le N \le N_1 < N_2 \le \min\left[2N, T^{D(1+5b^{1/20})}\right] \tag{35.7.1}$$

and for all τ_0 *satisfying*

$$|\tau_0 - T| \le \tfrac{1}{2}T^b \tag{35.7.2}$$

the inequality

$$|Z(\tau_0, N_1, N_2)| < c\frac{N \log^{10} N}{\tau_0^b} \tag{35.7.3}$$

holds. Then $\zeta(s) \neq 0$ *in the parallelogram*

$$\sigma \geq 1 - b^2, \qquad |\tau - T| \leq \tfrac{1}{2}T^b. \tag{35.7.4}$$

From this corollary, Theorem 35.1, and Theorem 33.1 we could draw a number of qualitatively interesting but quantitatively weak conclusions for Landau's problem. But quantitatively sharp results can be obtained, if we assume the truth of the Lindelöf's conjecture, which asserts that for an arbitrarily small $\varepsilon > 0$, $\sigma \geq \frac{1}{2}$, and $|t| \geq 2$ the inequality

$$|\zeta(s)| \leq c(\varepsilon)|t|^{\varepsilon} \tag{35.7.5}$$

holds. With this assumption the whole situation is much more transparent, so in the next section we confine ourselves to this case, including some remarks on unconditional results.

36

EXPONENTIAL SUMS AND RIEMANN'S CONJECTURE III

Exact but Conditional Results

36.1

As we have explained, throughout this Section we assume the truth of (35.7.5). Let $0 < \delta < \frac{1}{4}$ be arbitrarily small and fixed. On a and b we impose a restriction that is a bit stronger than before, namely, that

$$6\delta^2 \log\frac{3}{\delta} \le b \le 1 \tag{36.1.1}$$

and that

$$\delta \le \frac{b}{a} \le \frac{1}{2}. \tag{36.1.2}$$

These imply

$$a \ge \delta^2. \tag{36.1.3}$$

and

$$a \le \frac{1}{\delta}. \tag{36.1.4}$$

Theorem 36.1. *Suppose, with a* $\tau_0 > 10$, *the inequality*

$$|Z(\tau_0, N_1, N_2)| < c\frac{N\log^{10}N}{\tau_0^b} \tag{36.1.5}$$

holds for

$$\tau_0^a \le N \le N_1 < N_2 \le \min\left[2N, \tau_0^{a(1+2\delta)}\right]. \tag{36.1.6}$$

Then $\zeta(s) \neq 0$ *on the segment*

$$\sigma \ge 1 - \frac{b}{a}(1 - 2\delta), \qquad t = \tau_0$$

if only $\tau_0 > c(\delta)$.

36.2. PROOF

Let

$$s_0 = 1 + \frac{\delta}{2}\frac{b}{a} + i\tau_0 \doteq 1 + \mu + i\tau_0, \tag{36.2.1}$$

and let k be an integer restricted for the time being by

$$\frac{b}{1 - \delta/3}\log\tau_0 \le k \le \left(1 + \frac{2\delta}{3}\right)b\log\tau_0. \tag{36.2.2}$$

The length of this interval is certainly ≥ 1 if $\tau_0 > c(\delta)$. Let

$$J(s_0) = -\frac{1}{2\pi i}\int_{(10)}\varphi(w)^k\frac{\zeta'}{\zeta}(s_0 + w)\,dw \tag{36.2.3}$$

where

$$\varphi(w) = e^{(a/b)w}\frac{3b}{\delta aw}\operatorname{sh}\left(\frac{\delta a}{3b}w\right). \tag{36.2.4}$$

Using the Dirichlet series for ζ'/ζ and Lemma 34.1, with the notation (34.1.1), we obtain

$$J(s_0) = \sum_{n\in I}\left[\Lambda(n)e^{-i\tau_0\log n}\right]\left[n^{-1-\mu}h_k\left(\frac{ka}{b} - \log n, \frac{\delta}{3}\frac{a}{b}\right)\right] \tag{36.2.5}$$

where I is the interval

$$\left\{\exp\left[k\frac{a}{b}\left(1-\frac{\delta}{3}\right)\right], \exp\left[k\frac{a}{b}\left(1+\frac{\delta}{3}\right)\right]\right\} \tag{36.2.6}$$

and

$$h_k(x,\lambda) = \frac{1}{\pi\lambda}\int_0^{\infty}\left(\frac{\sin r}{r}\right)^k \cos\frac{xr}{\lambda}\,dr. \tag{36.2.7}$$

By (36.2.2),

$$I \subset \left[\tau_0^a, \tau_0^{a(1+2\delta)}\right], \tag{36.2.8}$$

and so the reasoning in Section 35.3 can be repeated word for word using (36.1.5) and (36.1.6). This leads to

$$|J(s_0)| < c(\delta)\tau_0^{-b}\log^{11}\tau_0. \tag{36.2.9}$$

36.3

Next we shift the line of integration to

$$\operatorname{Re} w = -2-\mu. \tag{36.3.1}$$

Here we must be more careful than we were in Section 35.4. Again, using a well-known inequality for ζ'/ζ on the line $\operatorname{Re} z = -1$, this integral has the upper bound ($w = u + iv = -2 - \mu + iv$)

$$c\int_{-\infty}^{\infty} \log(2 + |\tau_0 + v|)\exp\left\{-k\frac{a}{b}(2+\mu)\right\}\cdot\left|\frac{3b}{\delta a w}\operatorname{sh}\left(\frac{\delta a}{3b}w\right)\right|^k dv. \tag{36.3.2}$$

We split this into three parts

$$J_1 = \int_{(100/\delta)(a/b)}^{\infty} \tag{36.3.3}$$

$$J_2 = \int_{-(100/\delta)(a/b)}^{(100/\delta)(a/b)}, \qquad J_3 = \int_{-\infty}^{-(100/\delta)(a/b)}; \tag{36.3.4}$$

it is enough to deal with J_1 and J_2. For J_2 we use the simple inequality

$$\left|\frac{e^z - e^{-z}}{2z}\right| = \left|\frac{1}{2}\int_{-1}^{1} e^{zr}\,dr\right| \le e^{|\operatorname{Re} z|}. \tag{36.3.5}$$

This gives, using 36.1, the estimate

$$|J_2| < c(\delta)\log\left(\tau_0 + \frac{100}{\delta}\frac{a}{b}\right)\exp\left\{\left(\frac{\delta}{3} - 1\right)k\frac{a}{b}\cdot 2\right\}$$

which is, by (36.2.2) and (36.1.2)

(36.3.6) $$< c(\delta)\log\tau_0 \cdot \tau_0^{-2a} \le c(\delta)\tau_0^{-4b}\log\tau_0$$

for $\tau_0 > c(\delta)$. To estimate $|J_1|$, write the last factor in the integrand of (36.3.2) as

$$|\ \ |^k = |\ \ |^{k-2} \cdot |\ \ |^2$$

and apply (36.3.5) to the first factor. Then

$$|J_1| < c \cdot \exp\left\{\left(\frac{\delta}{3} - 1\right)(k-2)\frac{a}{b}\cdot 2\right\}$$

$$\cdot \int_{(100/\delta)(a/b)}^{\infty} \log(2 + |\tau_0 + v|)\left|e^{(a/b)w}\frac{3b}{\delta a w}\operatorname{sh}\left(\frac{\delta a}{3b}w\right)\right|^2 dv$$

which by 36.1 and (36.2.2) is

$$< c(\delta)\tau_0^{-1.99a}\int_{-\infty}^{\infty}\frac{\log(2 + |\tau_0 + v|)}{4 + v^2}\,dv < c(\delta)\tau_0^{-3b}\log\tau_0.$$

Hence the integral along $u = -2 - \mu$ has absolute value

(36.3.7) $$< c(\delta)\tau_0^{-3b}\log\tau_0.$$

For the residue at $w = 1 - s_0$ we get $\varphi(1 - s_0)^k$ which by (36.2.4) is

$$< \left(\frac{\exp\{-[1-(\delta/3)](a/b)\mu\}}{(\delta/3)(a/b)\tau_0}\right)^k < \left(\frac{3b}{\delta a}\cdot\frac{1}{\tau_0}\right)^k < \left(\frac{3}{2\delta}\cdot\frac{1}{\tau_0}\right)^k.$$

For

$$\tau_0 > \frac{6}{\delta}$$

this is

(36.3.8) $$< 4^{-k} < \tau_0^{-b}$$

from (36.2.2). Hence setting

$$\varphi(\rho - s_0) = z_\rho, \tag{36.3.9}$$

we have obtained from (36.2.9), (36.3.7), and (36.3.8) the inequality

$$\left|\sum_\rho z_\rho^k\right| < c(\delta)\tau_0^{-b}\log^{11}\tau_0 \tag{36.3.10}$$

for $\tau_0 > c(\delta)$.

36.4

Next we consider the contribution of the z_ρ with

$$|\gamma - \tau_0| \geq \frac{18}{\delta}\frac{b}{a}.$$

We first investigate the quantity

$$\max\left|e^{-(3/\delta)z}\frac{e^z - e^{-z}}{2z}\right| = U(R) \tag{36.4.1}$$

where the max is taken over the domain

$$\operatorname{Re} z \geq 0, \qquad |z| \geq R. \tag{36.4.2}$$

But since $3/\delta > 1$ we have for $\operatorname{Re} z \geq 0$

$$|e^{-[(3/\delta)-1]z}| \leq 1$$

and

$$|e^{-[(3/\delta)+1]z}| \leq 1.$$

Hence

$$U(R) \leq \frac{1}{R}. \tag{36.4.3}$$

Then by (36.3.9) and (36.2.4), if z is the quantity

$$\frac{\delta}{3}\frac{a}{b}(s_0 - \rho)$$

and $R = \nu$ for any positive integer ν, that for all ρ with

$$|s_0 - \rho| \geq \frac{3}{\delta}\frac{b}{a}\nu \tag{36.4.4}$$

the inequality

$$|z_\rho| \leq \frac{1}{\nu} \tag{36.4.5}$$

holds. This implies that (36.4.5) holds for all ρ with

$$\gamma \geq \tau_0 + \frac{3b}{a\delta}\nu. \tag{36.4.6}$$

Hence

$$\sum_{\tau_0 + (3b/a\delta)\nu \leq \gamma \leq \tau_0 + (3b/a\delta)(\nu+1)} |z_\rho|^k < c\frac{3b}{a\delta}\log\left(\tau_0 + \frac{3b}{a\delta}(\nu + 1)\right)\frac{1}{\nu^k}$$
$$< c(\delta)\frac{\log(\tau_0 + \nu)}{\nu^k},$$

and thus from (36.2.2)

$$\sum_{\tau_0 + 18b/a\delta < \gamma} |z_\rho|^k < \frac{c(\delta)}{6^k}\log \tau_0 < c(\delta)\tau_0^{-b}\log \tau_0.$$

The argument is similar for the ρ with

$$\gamma < \tau_0 - \frac{18b}{a\delta};$$

hence from (36.3.10) we get

$$\left|\sum_{|\gamma - \tau_0| > (18/\delta)(b/a)} z_\rho^k\right| < c(\delta)\tau_0^{-b}\log^{11}\tau_0 \tag{36.4.7}$$

for $\tau_0 > c(\delta)$.

36.5

Next we eliminate the contribution of the ρ with

$$|\gamma - \tau_0| \leq \frac{18}{\delta}\frac{b}{a}, \qquad \beta \leq \frac{1}{2} + \frac{\delta}{2}\frac{b}{a}. \tag{36.5.1}$$

It is enough to observe that for $\operatorname{Re} z \geq V > 0$ (36.3.5) implies

$$\left| e^{-(3/\delta)z} \frac{e^{z} - e^{-z}}{2z} \right| \leq \exp\left\{ \left(1 - \frac{3}{\delta}\right) V \right\}. \tag{36.5.2}$$

Since in our case the role of z is played by the numbers

$$\frac{\delta}{3} \frac{a}{b} (s_0 - \rho)$$

we have from (36.2.1)

$$V = \frac{\delta}{3} \frac{a}{b} \left(1 + \mu - \frac{1}{2} - \frac{\delta}{2} \frac{b}{a}\right) = \frac{\delta}{3} \frac{a}{b} \cdot \frac{1}{2}.$$

Hence each z_ρ^k in (36.5.1) is, by (36.5.2), absolutely

$$\leq \exp\left\{ \left(\frac{\delta}{3} - 1\right) \frac{a}{b} \cdot \frac{1}{2} k \right\},$$

which in turn by (36.2.2) and (36.1.2) is

$$\leq \tau_0^{-a/2} \leq \tau_0^{-b}.$$

Since the number of terms in (36.5.1) is smaller than

$$c(\delta) \log \tau_0,$$

from (36.3.10) and (36.4.7) we get the basic inequality

$$Z \doteq \left| \sum{}' z_\rho^k \right| < c(\delta) \tau_0^{-b} \log^{11} \tau_0 \tag{36.5.3}$$

where the summation ranges over the ρ satisfying

$$\beta > \frac{1}{2} + \frac{\delta}{2} \frac{b}{a}, \qquad |\gamma - \tau_0| \leq \frac{18}{\delta} \frac{b}{a}. \tag{36.5.4}$$

36.6

To get a lower bound for the power sum on the left we use the second main theorem in the form of (33.12.1). We choose this time

$$m = \frac{b}{1 - \delta/3} \log \tau_0. \tag{36.6.1}$$

Suppose the theorem is false and there is a zero $\rho^* = \beta^* + i\tau_0$ such that

$$\beta^* \geq 1 - \frac{b}{a}(1 - 2\delta). \tag{36.6.2}$$

We again need N, the upper bound for n, the number of terms in the power sum. Here—and only here—we shall use the Lindelöf conjecture (or rather its well-known consequence) that if H_η is the number of ρ in the parallelogram

$$\sigma \geq \tfrac{1}{2} + \eta, \qquad T \leq t \leq T + 1 \tag{36.6.3}$$

(η positive, arbitrarily small) then

$$\lim_{T \to \infty} \frac{H_\eta(T)}{\log T} = 0. \tag{36.6.4}$$

This shows that for $\tau_0 > c(\delta)$ we may choose

$$N = \delta^3 \log \tau_0. \tag{36.6.5}$$

Again we must verify that this choice of k does not violate the restriction (36.2.2). This amounts to proving the inequality

$$\delta^3 < \left(1 + \frac{2\delta}{3} - \frac{1}{1 - \delta/3}\right) b$$

which is, by (36.1.1), true for all ρ with $0 < \delta < \frac{1}{4}$. Denote the so-obtained value of k by k^*, and we get

$$|Z| > \left\{ \frac{1}{8e\left[b/(1 - \delta/3)\delta^3 + 1\right]} \right\}^N \cdot H \tag{36.6.6}$$

where

$$H \doteq \max |z_\rho|^{k^*}, \tag{36.6.7}$$

with ρ in the domain (36.5.4). We obviously have

$$H \geq |z_{\rho^*}|^{k^*}. \tag{36.6.8}$$

Then

$$\frac{\delta}{3} \frac{a}{b}(s_0 - \rho^*) > 0$$

since for $x > 0$

$$\frac{e^x - e^{-x}}{2x} \geq 1,$$

(36.6.8) and (36.2.1) imply that

$$H \geq \exp\left\{k^* \frac{a}{b}\left(\beta^* - 1 - \frac{\delta}{2}\frac{b}{a}\right)\right\}.$$

Using (36.6.2) and (36.2.2) this gives

$$H > \tau_0^{-b+b(5/6)\delta}. \tag{36.6.9}$$

Since by (36.1.1)

$$8e\left(\frac{b}{(1-\delta/3)\delta^3} + 1\right) < 8e\left(\frac{1}{(11/12)\delta^3} + 1\right) < \left(\frac{3}{\delta}\right)^3,$$

we get, using (36.6.5),

$$\left\{\frac{1}{8e\left[b/(1-\delta/3)\delta^3 + 1\right]}\right\}^N > e^{-3\log(3/\delta)\cdot\delta^3\log\tau_0} = \tau_0^{-3\delta^3\log(3/\delta)}.$$

The exponent of τ_0 can be written as

$$-b\left(\frac{3\delta^3\log(3/\delta)}{b}\right)$$

which, by (36.1.1), is larger than or equal to

$$-b\frac{\delta}{2}.$$

Hence, from this, (36.6.9), and (36.6.6)

$$|Z| > \tau_0^{-b+b\delta/3}. \tag{36.6.10}$$

This together with (36.5.3) yields

$$\tau_0^{b\delta/3} < c(\delta)\log^{11}\tau_0,$$

and from (36.1.1)

$$\tau_0^{2\delta^3\log(3/\delta)} < c(\delta)\log^{11}\tau_0,$$

which is false for $\tau_0 > c(\delta)$. So Theorem 36.1 is proved. ■

36.7

Theorem 36.1 is, in our terminology, "local in τ_0" and for primes "very strongly localized". We formulate the (trivial) corollary of this theorem which is "semilocal" in τ_0.

Corollary I. *Suppose with the restrictions of* 36.1 *that the inequality* (36.1.5) *holds for all* τ_0 *with*

$$|\tau_0 - T| \le \tfrac{1}{2}T^b \tag{36.7.1}$$

and for all (N_1, N_2) *pairs with*

$$\tfrac{1}{2}T \le N \le N_1 < N_2 \le \min[2N, T^{a(1+3\delta)}]. \tag{36.7.2}$$

Then $\zeta(s) \ne 0$ *in the parallelogram*

$$\sigma \ge 1 - \frac{b}{a}(1 - 2\delta), \qquad |t - T| \le \tfrac{1}{2}T^b$$

if only $T > c(\delta)$.

The strength of Corollary I will be clearer if we formulate Theorem 33.1 with the choice

$$\Theta = 1 - \frac{b}{a}(1 - 2\delta), \qquad B = b.$$

Then for

$$0 < \frac{b}{a} \le \frac{1}{2(1 - 2\delta)}$$

it yields the following.

Corollary II. *Suppose for a* $T > 10$ *that* $\zeta(s) \ne 0$ *on the parallelogram*

$$\sigma \ge 1 - \frac{b}{a}(1 - 2\delta), \qquad |t - T| \le \tfrac{1}{2}T^b.$$

Then the inequality

$$|Z(\tau_0, N, N_2)| \le \frac{c(\delta)}{b}\,\frac{N\log^3 N}{\tau_0^b}$$

holds for all pairs N_1, N_2 *with*

$$T^{a/(1-2\delta)} \le N \le N_1 < N_2 \le 2N$$

and all τ_0 *with*

$$|\tau_0 - T| \leq \left(\frac{1-\delta}{2}\right) T^b$$

if only $T > c(\delta)$.

The comparison of these two corollaries shows an almost completely inverse character in every respect. Without supposing the truth of Lindelöf's conjecture, the comparison of Corollary I of Section 35 with Corollary II of this Section still gives *qualitatively* inverse theorems for small b, but no reliable quantitative correspondence can be drawn from it.

In applying Corollary I and II after each other an interesting phenomenon appears.

Corollary III. *Suppose that* δ *is an arbitrarily small fixed positive number such that for* a *and* b *the provisions of* 36.1 *hold. Assume, for a* $T > c(\delta)$, *the inequality*

$$|Z(\tau_0, N_1, N_2)| < c\frac{N \log^{10} N}{\tau_0^b} \tag{36.7.3}$$

holds for all

$$|\tau_0 - T| \leq \tfrac{1}{2} T^b \tag{36.7.4}$$

and all pairs (N_1, N_2) *with*

$$\tfrac{1}{2} T^a \leq N \leq N_1 < N_2 \leq \min[2N, T^{a(1+3\delta)}]. \tag{36.7.5}$$

Then the inequality

$$|Z(\tau_0, N_1, N_2)| < \frac{c(\delta)}{a} \frac{N \log^3 N}{\tau_0^b} \tag{36.7.6}$$

holds for all τ_0 *with*

$$|\tau_0 - T| \leq \frac{1-\delta}{2} T^b \tag{36.7.7}$$

and for all pairs N_1, N_2 *with*

$$T^{a/(1-2\delta)} \leq N \leq N_1 < N_2 \leq 2N. \tag{36.7.8}$$

This corollary expresses (supposing Lindelöf's conjecture) an interesting property of the important sums $Z(\tau_0, N_1, N_2)$. Briefly, if they satisfy the inequality (36.7.3) for the τ_0-range (36.7.4) and for the very short prime range (36.7.5), then they also satisfy the same inequality in the "unbounded" prime range (36.7.8), if we "shrink" the τ_0-range in accordance with (36.7.7). Corollary I from Section 35.1, and Corollary II of this section could produce *qualitatively* the same phenomenon, unconditionally for a small range of values. Remarkably enough, Corollary II is a statement concerning primes exclusively, and still we cannot see any possibility of proving it directly while remaining in the realm of integers.

36.8

So far δ has been arbitrarily small and fixed, and the *fixed* pair (a, b) was chosen so that the constraints of 36.1 would not be violated and so that (36.1.5) and (36.1.6) would hold. Now we also fix b so that (36.1.1) is satisfied, we let $S_1(\delta, b)$ be the set of a-values satisfying (36.1.2), and set

(36.8.1) $$\inf_{a \in S_1(\delta, b)} a \doteq a'.$$

Then if S_1 is nonempty, Theorem 36.1 implies that $\zeta(s) = 0$ on the segment

(36.8.2) $$\sigma > 1 - \frac{b}{a'}(1 - 2\delta), \qquad t = \tau_0.$$

We do not assert that this is the largest possible segment that can be derived from (36.1.5) and (36.1.6).

The situation changes somewhat if we consider instead of $S_1(\delta, b)$ the set $S_2(\delta, b)$, the set of such a-values that do not violate the constraints of 36.1 and for which (36.1.5), (36.7.1), and (36.7.2) holds. Setting

(36.8.3) $$\inf_{a \in S_2(\delta, b)} a \doteq a''$$

and supposing S_2 is nonempty, Corollary I gives at once that $\zeta(s) \neq 0$ in the parallelogram

(36.8.4) $$D_1 : \sigma > 1 - \frac{b}{a''}(1 - 2\delta), \qquad |t - T| \leq \tfrac{1}{2}T^b.$$

In this case, however, we can show that this conclusion is, in a sense, the sharpest if $a'' > (1 + \delta)a'$. Namely, suppose that from (36.1.5), (36.7.1), and (36.7.2) the stronger assertion

(36.8.5) $$\zeta(s) \neq 0 \quad \text{for} \quad s \in D_2$$

follows, where D_2 is the parallelogram

$$(36.8.6) \qquad \sigma > 1 - \frac{b}{a''}(1+\delta), \qquad |t - T| \le \frac{1}{2(1-\delta)}T^b.$$

Then we could apply Theorem 33.1 with

$$B = b + \frac{\log 1/(1-\delta)}{\log T}, \qquad \Theta = 1 - \frac{B}{a''}(1+\delta).$$

This could imply that for $T > c(\delta)$ the inequality

$$|Z(\tau_0, N_1, N_2)| < \frac{c(\delta)}{b}\frac{N\log^3 N}{\tau_0^b}$$

holds for all τ_0 with

$$|\tau_0 - T| \le \tfrac{1}{2}T^b$$

and for all pairs (N_1, N_2) with

$$T^{a''/(1+\delta)} \le N \le N_1 < N_2 \le 2N.$$

But by the definition of $S_2(\delta, b)$ and $a''/(1+\delta) \in S_1(\delta, b)$ this means that $a''/(1+\delta) \in S_2(\delta, b)$, and this contradicts (36.8.3). We have obtained another interesting corollary.

Corollary IV. *Assume Lindelöf's conjecture. Let δ be arbitrarily small and fixed and let a'' be defined as in* (36.8.3) *such that $a'' > (1+\delta)a'$, for a' in* (36.8.1). *Fix b, not violating* (36.1.1). *Then $\zeta(s)$ does not vanish in the parallelogram D_1 of* (36.8.4), *but does have a zero in $D_2 \setminus D_1$, where D_2 is defined in* (36.8.6).

Concerning Landau's problem, the following picture emerges from Sections 33, 35, and 36. An essential role is played by the finite sums $Z(\tau_0, N_1, N_2)$ involving only the primes in certain intervals of surprisingly small size depending on τ_0. From estimates for these sums, we conclude the existence of finite zero-free parallelograms, the horizontal length of which depends at fixed τ_0 on "the strength of the mod 1 interference behavior" of the numbers $(\tau_0/2\pi)\log p$. Moreover, for one of these zero-free parallelograms, any larger parallelogram is no longer zero-free. Thus the behavior of primes in a finite interval determines a ρ in some finite domain.

36.9

Let a and b be positive constants such that

$$0 \le b \le 1, \qquad \frac{b}{a} < \frac{1}{2}, \qquad a > 0 \tag{36.9.1}$$

and suppose the inequality

$$|Z(\tau_0, N_1, N_2)| < c\frac{N\log^{100}N}{|\tau_0|^b}, \quad \text{all} \quad |\tau_0| > 10 \tag{36.9.2}$$

holds for all pairs (N_1, N_2) with

$$|\tau_0|^a \le N \le N_1 < N_2 \le 2N. \tag{36.9.3}$$

Let S_3 be the set of such pairs (a, b). Then it can be shown—supposing the Lindelöf conjecture—that if

$$1 - \sup_{a,\, b \in S_3} \frac{b}{a} \doteq T, \tag{36.9.3}$$

the half-plane $\sigma > T + \varepsilon$ contains, at most, finitely many ρ for arbitrarily small $\varepsilon > 0$, whereas the half-plane $\sigma > T - \varepsilon$ contains infinitely many. The same conclusion follows for the class of functions defined in Section 33.6, replacing requirement II by the corresponding Lindelöf hypothesis. It would be of great interest to prove the results of this section without this conjecture; in Turán [23] we succeeded in proving (36.9.3), assuming a much weaker form of the Lindelöf conjecture. We describe this result in the next section.

37

DENSITY THEOREMS I

A Weaker Form of Lindelöf's Hypothesis

37.1

To prove the necessary estimate for the exponential sums occurring in the preceding theorems, thereby obtaining nontrivial zero-free half-planes, seems to be hopeless at present. The situation is different, however, concerning "almost everywhere" estimates that lead to estimates for the number of zeros in certain domains. The next few paragraphs are devoted to the study of

(37.1.1)

$$N(\alpha, T) = \text{the number of zeros of } \zeta(s) \text{ in the rectangle } \alpha \leq \operatorname{Re} s < 1, \; |\operatorname{Im} s| \leq T,$$

and the "density hypothesis"

$$N(\alpha, T) < T^{(2+\varepsilon)(1-\alpha)} \log^c T \qquad \left(T > c(\varepsilon), \tfrac{1}{2} \leq \alpha \leq 1\right) \tag{37.1.2}$$

which sometimes serves as a substitute for the Riemann hypothesis in number theory. Ingham deduced this from the Lindelöf hypothesis (to be abbreviated L.H. hereafter). More precisely, he proved that

$$|\zeta(\sigma + it)| < |t|^a \qquad \left(\sigma \geq \tfrac{1}{2}, |t| > c(a)\right) \tag{37.1.3}$$

implies

$$N(\alpha, T) < T^{2(1+2a)(1-\alpha)} \log^5 T. \tag{37.1.4}$$

The early success of the method of this book in improving Ingham's result in the neighborhood of $\alpha = 1$—where it is the sharpest (see Turán [13])—suggested, on one hand, that the L.H. should be replaced by a considerably weaker assumption and, on the other hand, that bounds for the zeta-function have deeper consequences than was thought after Ingham's result. The latter expectation will be shown to be correct; here we deduce the density hypothesis from a weak assumption.

37.2

Backlund and Littlewood showed that the L.H. is equivalent to

$$N(\alpha, T+1) - N(\alpha, T) = o(\log T) \tag{37.2.1}$$

for each $\alpha > \frac{1}{2}$ as $T \to \infty$ (part of which has already been made use of in Section 36); that is, that for large enough T the concentration of zeros in the rectangle $\alpha \leq \operatorname{Re} s < 1$, $T \leq \operatorname{Im} s \leq T+1$ is small, compared to the total number in $0 \leq \operatorname{Re} s \leq 1$, $T \leq \operatorname{Im} s \leq T+1$ (which is $\sim c \log T$ in average). In the following statement the sparseness of zeros is assumed to be due only to the smallness of the considered domain and not due to the largeness of T.

Weak L.H. *For every $A > 0$ there is a $0 < \delta = \delta(A) < A$ such that the number of zeros of $\zeta(s)$ in the circle $|s - s_0| \leq \delta$ is less than*

$$A\delta \log \tau_0 \qquad (s_0 = \sigma_0 + i\tau_0)$$

whenever

$$\zeta(s) \neq 0 \quad \text{in} \quad \operatorname{Re} s > \sigma_0, \qquad |\operatorname{Im} s - \tau_0| < \log \tau_0 \tag{37.2.2}$$

and $\sigma_0 \geq \frac{1}{2} + 10\delta$, $\tau_0 \geq c(A, \delta)$.

This follows trivially from the equivalent form (37.2.1) of the classical L.H. To prove this even for $A\delta$ replaced, for example, by $\delta^{0,01}$ would be just as difficult as proving the L.H. itself, if we omitted the restriction (37.2.2). In this form, however, it follows from general function theoretic considerations with $A = 0.71$ for all $\delta < \frac{1}{20}$ (see Turán [22]), and we have actually proved it with arbitrary $A > 0$ for $\sigma_0 = 1$ in (35.6.6) using the tangential vanishing of the Lindelöf μ-function for $\zeta(s)$ at $\sigma = 1$. So this "weak L.H." calls for a proof that the "local μ-function" vanishes tangentially not only at the boundary $\sigma = 1$ but also at any s_0 inside the critical strip whenever there is a large zero-free rectangle to the right of s_0. This is much weaker than the L.H. and does not seem to be hopeless.

Theorem 37.1. *Suppose that the weak L.H. holds with an $A < 0.01$. Then*

$$N(\alpha, T) < T^{2(1-\alpha)+5\sqrt{A}} \qquad \left(\tfrac{1}{2} \leq \alpha \leq 1,\ T > c(A)\right).$$

This is essentially the theorem of Turán [22] in a quantitative form except that here we have not bothered with values α near 1 where our statement is comparatively weak, because we return to them in the next paragraph without any hypothesis.*

37.3. PROOF

We again consider

$$J(s) = -\frac{1}{2\pi i}\int_{(10)} \varphi^k(w)\frac{\zeta'}{\zeta}(s+w)\,dw \tag{37.3.1}$$

where

$$\varphi(w) = \frac{e^{\lambda w} - e^{-\lambda w}}{2\lambda w}e^{\mu w}. \tag{37.3.2}$$

With the notation (34.1.1) Lemma 34.1 gives

$$J(s) = \sum_n{}' \frac{\Lambda(n)}{n^s}h_k(\mu k - \log n, \lambda)$$

where the summation Σ_n ranges over $|\log n - \mu k| < k\lambda$. Our formula has the additional advantage here that we can work with s in the critical strip: let $s = \sigma + it$ with

$$\frac{1}{2} + \frac{1}{\log T} < \sigma < 1 - \frac{1}{\log T}, \qquad T \le t \le 2T,\ T > c(A), \tag{37.3.3}$$

and, with the $\delta = \delta(A)$ of the weak L.H., let

$$\lambda = \frac{3}{\delta}, \qquad \mu = \frac{2}{\delta\sqrt{A}}, \qquad \frac{\log T}{\mu} \le k \le \frac{\log T}{\mu}(1 + 2\sqrt{A}). \tag{37.3.4}$$

Using $|\sin r| \le r$ $(r \le 1)$ and $|\sin r| \le 1$ $(r \ge 1)$ in (34.1.4) we see immediately that

$$|h_k(\mu k - \log n, \lambda)| < \frac{1}{\lambda} < 1. \tag{37.3.5}$$

*With more careful calculation the order of magnitude of A could be improved to $cA\log 1/A$. In Turán [43] this is stated with cA, but there seems to be an oversight in the calculation.

37.4

Now,

$$J_k(\sigma, T) \doteq \int_T^{2T} |J(\sigma + it)|^2 \, dt$$

$$< T + \sum_n{}' \frac{\Lambda(n)}{n^{2\sigma}} + 4\sum_n{}' \frac{\Lambda(n)}{n^{\sigma}} \sum_{m<n}{}' \frac{\Lambda(m)}{m^{\sigma}|\log n/m|},$$

carrying out the squaring and integration as usual and using (37.3.5). The first sum, by (37.3.3) and (37.3.4), is

$$< 4\log^2 T \sum_{n > e^{(\mu-\lambda)k}} \frac{1}{n^{2\sigma}} < 5\log^3 T e^{(\mu-\lambda)k(1-2\sigma)} < T^{1-2\sigma+2\sqrt{A}}.$$

In the second sum $|\log n/m| \geq (n-m)/n$ and the contribution of terms $n/2 \leq m < n$ to the inner sum is

$$< 8\log^2 T n^{1-\sigma}$$

and that of $m \leq n/2$ is

$$< 2\log^2 T n^{1-\sigma}.$$

The complete second sum is

$$< 20\log^3 T \sum_n{}' n^{1-2\sigma} < 21\log^4 T e^{(\mu+\lambda)k(2-2\sigma)} < T^{2-2\sigma+3.8\sqrt{A}}.$$

We have thus proved

$$J_k(\sigma, T) < T^{2-2\sigma+3.9\sqrt{A}}.$$

This implies, for $s = \sigma + it$, $T \leq t \leq 2T$, σ fixed,

$$|J(s)| < T^{\beta-\alpha+2\sqrt{A}} \tag{37.4.1}$$

with a β to be defined later, unless t belongs to a set of measure smaller than

$$T^{2(1-\beta)}\log^{-8} T. \tag{37.4.2}$$

37.5

Consider now

$$\tfrac{1}{2} + \frac{1}{\log T} \leq \sigma = \sigma_i = 1 - i/\log T (i = 1, \ldots)$$

and the strips

$$l_j: \frac{j}{\log T} \leq t \leq \frac{j+1}{\log T}$$

belonging to $T \leq t \leq 2T$. For each i we can find an $s = s_{ij} = \sigma_i + t_{ij} \in l_j$ satisfying (37.4.1) with the exception of, at most,

$$T^{2(1-\beta)} \log^{-7} T$$

indices j, otherwise the exceptional measure would be greater than (37.4.2). There are, at most, $\log T$ indices i and $\log T$ values of the parameter k, and we see that dropping at most

(37.5.1) $$T^{2(1-\beta)} \log^{-5} T$$

"bad" strips l_j we can find an $s = s_{ij} = \sigma_i + it_{ij} \in l_j$ satisfying

(37.5.2) $$|J(s_{ij})| < T^{\beta - \sigma_i + 2\sqrt{A}}$$

simultaneously for each i, k, and for each remaining "good" strip l_j belonging to (37.3.4) and (37.3.3) respectively.

37.6

Returning to the original definition of $J(s)$ in (37.3.1), the residue theorem yields

(37.6.1)

$$J(s) = \varphi^k(1-s) - \sum_{\rho} \varphi^k(\rho - s) - \frac{1}{2\pi i} \int_{(-1-\sigma)} \varphi^k(w) \frac{\zeta'}{\zeta}(s+w)\, dw.$$

On the line $(-1-\sigma)$, $\varphi(w)$ of (37.3.2) satisfies

$$|\varphi^k(w)| \leq \left| \frac{e^{-(\mu+\lambda)(1+\sigma)} + e^{-(\mu-\lambda)(1+\sigma)}}{2w} \right|^k \leq \frac{e^{-3/2(\mu-\lambda)k}}{|w|^2},$$

and the integral satisfies

(37.6.2) $$\int_{(-1-\sigma)} < ce^{-3/2(\mu-\lambda)k} \int_{-\infty}^{\infty} \frac{\log(T+|u|)}{u^2+1}\, du < T^{-1/4}.$$

(We have also used (37.3.3), (37.3.4), and trivial estimates for ζ'/ζ.)

37.7

We call a zero ρ of $\zeta(s)$ belonging to the rectangle $\alpha \le \operatorname{Re}\rho < 1$, $T+1 \le \operatorname{Im}\rho \le 2T-1$ extremal if $\zeta(s) \ne 0$ in the rectangle $\operatorname{Re}\rho + 1/\log T \le \sigma \le 1$, $|t - \operatorname{Im}\rho| \le \log^3 T$. If ρ is not extremal, we can choose a zero in the latter rectangle; if this is not extremal either, we can repeat this step, and so on. Since the real parts of zeros thus chosen increase by at least $1/\log T$ and there are no zeros in $\sigma > 1$, the procedure must break down after at most the $[\frac{1}{2}\log T]$-th step yielding an extremal zero, provided that the zero we have started with belongs to $\alpha \le \operatorname{Re}\rho < 1$, $T + \log T \log 3T < \operatorname{Im}\rho < 2T - \log T \log 3T$. We conclude that the zeros in $\alpha \le \sigma < 1, T \le t \le 2T$ are fully covered by horizontal strips of width $2\log T \log 3T$ built around extremal zeros, plus two such strips at the lower and upper end. Each such strip contains at most $2\log T\log 3Tc\log T < c\log^3 T$ zeros, and we have as many strips as there are extremal zeros, plus two. We show that with a β suitably chosen, an extremal zero cannot belong to a "good" strip l_j as defined in 37.5. The number of bad strips has been estimated in (37.5.1), each containing at most $c\log T$ zeros. All in all, this will show

$$N(\alpha, 2T) - N(\alpha, T) < (T^{2(1-\beta)}\log^{-5}T + 2)c\log^4 T < c\frac{T^{2(1-\beta)}}{\log T} \qquad (T > c(A))$$

if, for example,

$$1 - \beta > c(A) \tag{37.7.1}$$

and using this for $T/2, T/4, \ldots$ and adding

$$N(\alpha, T) < T^{2(1-\beta)} \qquad (T > c(A)). \tag{37.7.2}$$

37.8

So, let ρ_0 be an extremal zero defined in 37.7 belonging to a "good" strip as defined in 37.5. For the crucial step of the proof we now deduce a contradiction.

It is known that the rectangle $\sigma \ge 1 - 2/\log T, T \le t \le 2T$ is zero-free, thus $\operatorname{Re}\rho_0 < 1 - 2/\log T$, and according to 37.5 we can find an s_{ij} with

$$\sigma_i \ge Re\,\rho_0 + \frac{1}{\log T}, \qquad |\rho_0 - s_{ij}| < \frac{3}{\log T}, \tag{37.8.1}$$

with ρ_0 in a good strip, satisfying (37.5.2). Since ρ_0 is also extremal,

$$\zeta(s) \ne 0 \quad \text{in} \quad |\operatorname{Im} s - t_{ij}| \le \log t_{ij}. \tag{37.8.2}$$

We use (37.6.1) with $s = s_{ij}$. The integral has been dealt with in (37.6.2). The contribution of a ρ with $|\mathrm{Im}\,\rho - t_{ij}| > \log t_{ij}$ is, by (37.3.4),

$$< \frac{e^{(\mu+\lambda)k}}{|\rho - s|^k} < \frac{T^2}{(\log T)^{k-2}|\rho - s|^2} < \frac{1}{T^2|\rho - s|^2},$$

also valid for the residue $\varphi^k(1 - s)$. Dividing into strips of unit width and summing, the total contribution of these terms is

$$(37.8.3) \qquad < \frac{1}{T^2} \sum_{l=1}^{\infty} \frac{\log(T + l)}{l^2} < T^{-1}.$$

Since

$$|\varphi(w)| < \frac{|e^{(\mu+\lambda)w}| + |e^{(\mu-\lambda)w}|}{2\lambda|w|} < \frac{1}{\lambda|w|}$$

for $\mathrm{Re}\,w \le 0$, the contribution of the ρ with $\mathrm{Re}\,\rho \le \sigma_i$, $|\mathrm{Im}\,\rho - t_{ij}| \le \log t_{ij}$, $|\rho - s_{ij}| > \delta$ [where δ is $\delta(A)$ of the weak L.H.] is

$$(37.8.4) \qquad < c \log U 2T \frac{1}{(\lambda\delta)^k} < e^{-k} < T^{-\frac{1}{2}\delta\sqrt{A}}$$

by (37.3.4). From (37.8.2), there remain zeros only in $\mathrm{Re}\,s \le \sigma_i$, $|s - s_{ij}| \le \delta$. Collecting our estimates (37.6.2), (37.8.3), (37.8.4), and (37.6.1), and using also (37.5.2), we get

$$(37.8.5) \qquad \left| \sum_{\substack{|\rho - s_{ij}| < \delta \\ \mathrm{Re}\,\rho \le \sigma_i}} \varphi^k(\rho - s_{ij}) \right| < T^{\beta - \sigma_i + 2\sqrt{A}} + 2T^{-\frac{1}{2}\delta\sqrt{A}}$$

for all k satisfying (37.3.4).

37.9

Now we apply the second main theorem in the form (33.12.1) to get a lower bound for the left-hand side for a specific $k = k_0$. We need an upper bound for n, the number of terms, and it is here that we make use of the weak L.H. By (37.8.2), its conditions are also fulfilled if $\sigma_i \ge \frac{1}{2} + 10\delta$, that is, since $\alpha \le \sigma_i$, if

$$(37.9.1) \qquad \alpha \ge \tfrac{1}{2} + 10\delta$$

and we infer $n < A\delta \log T \doteq N$. Let $m = \log T/\mu$. The second main theorem gives us a $k = k_0$ in

$$m = \frac{\log T}{\mu} \le k_0 \le m + N = \frac{\log T}{\mu} + A\delta \log T = \frac{\log T}{\mu}(1 + 2\sqrt{A})$$

such that

$$\left[\frac{N}{8e(m+N)}\right]^N \left|\varphi^{k_0}(\rho_0 - s_{ij})\right| < T^{\beta - \sigma_i + 2\sqrt{A}} + 2T^{-\frac{1}{2}\delta\sqrt{A}}$$

since, for our k_0, (37.3.4)—and hence also (37.8.5)—hold. Here

$$\left[\frac{N}{8e(m+N)}\right]^N = T^{-A\delta \log[8e(1+1/2\sqrt{A})]} > e^7 T^{-\frac{1}{2}\delta\sqrt{A}}$$

for $A < 0.01$ and since $|\varphi'(w)| \le \lambda + \mu$ for $\operatorname{Re} w \le 0$, $\varphi(0) = 1$ (37.8.1) implies

$$\left|\varphi^{k_0}(\rho_0 - s_{ij})\right| \ge \left(1 - 3\frac{\lambda + \mu}{\log T}\right)^{k_0} > e^{-5}.$$

Hence

$$T^{-\frac{1}{2}\delta\sqrt{A}} < T^{\beta - \sigma_i + 2\sqrt{A}}, \qquad \beta > \sigma_i - 2,5\sqrt{A} \ge \alpha - 2,5\sqrt{A},$$

which is a contradiction if we choose $\beta = \alpha - 2,5\sqrt{A}$. This choice also guarantees that (37.7.1), and thus (37.7.2), is valid,

$$N(\alpha, T) < T^{2(1-\alpha) + 5\sqrt{A}}$$

at least for (37.9.1), but otherwise the exponent here is larger than 1 because $\delta < A$ and the inequality is trivial. This proves Theorem 37.1. ■

38

DENSITY THEOREMS II

Strong Conditional and Unconditional Estimates

38.1

We now turn to the second problem mentioned in 37.1. The Lindelöf conjecture, being equivalent to the weak (37.2.1), had not seemed to have much of a connection with the vertical distribution of zeros, and Ingham's strong deduction (37.1.3) and (37.1.4) from it came as a surprise in 1937. Much more is true, however. The following theorem was proved in collaboration with G. Halász [1].

Theorem 38.1. *The Lindelöf hypothesis, that is,* (37.1.3) *for any* $a > 0$, *implies, with the notation* (37.1.1),

$$N(\alpha, T) < T^{\varepsilon}$$

for each $\varepsilon > 0$ *and* $\alpha > \frac{3}{4}$.

This has been conjectured in Turán [13] for each $\alpha > \frac{1}{2}$, and the problem is still open. At any rate, the theorem shows that as far as the half-plane $\sigma > \frac{3}{4}$ is concerned, the Lindelöf hypothesis is almost as strong as the Riemann hypothesis.

Replacing the Lindelöf hypothesis by the rather strong known estimates of the zeta function in the vicinity of $\sigma = 1$ led—in the same joint paper with G. Halász [1]—to the following.

Theorem 38.2. *With the notation* (37.1.1)

$$N(\alpha, T) < T^{1.2 \cdot 10^5 (1-\alpha)^{3/2}} \log^c T \qquad (T > c).$$

In other words, the exponent of T for $N(\alpha, T)$ vanishes tangentially at $\alpha = 1$ just as the known exponent in the estimate of $\zeta(\alpha + iT)$ does. In the neighborhood of $\alpha = 1$ this is even stronger than the density hypothesis (37.1.2). This was the first result showing the truth of the latter, at least for $c < \alpha < 1$, with a numerical $c < 1$.

38.2

These theorems will appear again as special cases of more general results. Let the class $\Delta = \Delta(A, \gamma, T)$ $[A < 10(1 - \gamma), \gamma \geq \frac{1}{2}]$ of functions $f(s)$ be defined this time by the following properties.

I.

$$f(s) = 1 + \sum_{n=2}^{\infty} \frac{a_n}{n^s}, \qquad (\sigma > 1)$$

$$\frac{f'}{f}(s) = \sum_{n=2}^{\infty} \frac{b_n}{n^s}, \qquad (\sigma > 1)$$

with $|b_n| < c \log n$,

$$|f(s)| > c(\sigma - 1)^{10} \qquad (1 < \sigma \leq 2).$$

II. $f(s)$ can be continued to a meromorphic function in $\sigma \geq \gamma$, where it satisfies

$$|f(s)| \leq T^A \qquad (a < |t| \leq 3T).$$

Theorem 38.3 (see also Turán–Halász [1]). *Assume the Lindelöf hypothesis (37.1.3) for $\zeta(s)$ with any $a > 0$ and let $f(s) \in \Delta(A, \frac{3}{4}, T)$ for any $A > 0$ and $T > c(A)$. Then, if $N(\alpha, T, f)$ is the number of zeros of $f(s)$ in the rectangle $\alpha \leq \sigma < 1$, $|t| \leq T$,*

$$N(\alpha, T, f) < T^{\varepsilon}$$

for each $\varepsilon > 0$ and $\alpha > \frac{3}{4}$.

Note that the special function $\zeta(s)$ plays a role for general $f(s)$. The assumption on the former seems to be even more important than on the latter: more generally, (37.1.3) for $\sigma \geq \gamma (\geq \frac{1}{2})$ and $f(s) \in \Delta(A, (1 + \gamma)/2, T)$ would imply the same conclusion for $\alpha > (1 + \gamma)/2$. The following theorem holds for arbitrary γ but having a specific application to Theorem 38.2 in mind, we do not state it in such a precise form.

Theorem 38.4. *Suppose that both* $\zeta(s)$ *and* $f(s)$ *belong to* $\Delta(A, \gamma, T)$. *Then, with the notation of Theorem* 38.3,

$$N(\alpha, T, f) \leq T^{31A} \log^c T \qquad (T > c)$$

for $\alpha = 1 - (1 - \gamma)/11$.

38.3. PROOF OF THEOREM 38.3

Let

$$(38.3.1) \quad s_0 = \sigma_0 + it = 2 + it, \qquad (c \leq |t| \leq T), \qquad \tfrac{3}{4} < \alpha_1 < \alpha,$$

and let integer ν be confined for the time being to

$$(38.3.2) \qquad b(\alpha)\varepsilon \log T < \nu + 1 < 2b(\alpha)\varepsilon \log T$$

where $b(\alpha)$ is defined by the restrictions from below to be imposed in the course of the proof.

There will be no need for the intricate kernel function of the preceding proofs, and we again consider

$$(38.3.3) \qquad \left[\frac{f'}{f}(s)\right]^{\nu}_{s=s_0} = (-1)^{\nu} \sum_{n=2}^{\infty} \frac{b_n \log^{\nu} n}{n^{s_0}}.$$

We use Lemma V.1 with $R = \sigma_0 - \frac{3}{4}$, $r = \sigma_0 - \alpha_1$. As follows from properties I and II, 38.2, we can take $M = c \log T$, and using (38.3.2) we obtain

$$(38.3.4) \qquad \left| \sum_{|\rho - s_0| \leq \sigma_0 - \alpha_1} \frac{1}{(\rho - s_0)^{\nu+1}} \right| \leq \frac{1}{\nu!} \left| \left[\frac{f'}{f}(s)\right]^{(\nu)}_{s=s_0} \right|$$

$$+ c(\alpha) \frac{\nu \log T}{(\sigma_0 - \alpha_1)^{\nu}} < \frac{\log^3 T}{(\sigma_0 - \alpha_1)^{\nu}}$$

provided

$$(38.3.5) \qquad \left| \left[\frac{f'}{f}(s)\right]^{(\nu)}_{s=s_0} \right| \leq \frac{\nu!}{(\sigma_0 - \alpha_1)^{\nu}};$$

ρ runs through the zeros of $f(s)$.

38.4

Suppose that $-T \le t_1 < \cdots < t_K \le T$, $|t_i - t_j| \ge c$, and

$$\left|\left[\frac{f'}{f}(s)\right]^{(\nu)}_{s=\sigma_0+it_j}\right| > \frac{\nu!}{(\sigma_0-\alpha_1)^{\nu}} \qquad (j=1,\ldots,K). \tag{38.4.1}$$

We need an estimate for K. In Section 37 we calculated the square integral of the function for this purpose. Now we use an approach found by G. Halász in his investigations of the mean values of multiplicative functions. Recalling (38.3.3), (38.4.1) can be written as

$$\eta_j \sum_{n=2}^{\infty} \frac{b_n \log^{\nu} n}{n^{2+it_j}} > \frac{\nu!}{(\sigma_0-\alpha_1)^{\nu}}$$

with an $|\eta_j| = 1$. Summing over j and using property I, 38.2,

$$\frac{K\nu!}{(\sigma_0-\alpha_1)^{\nu}} < c\sum_{n=2}^{\infty}\frac{\log^{\nu+1} n}{n^2}\left|\sum_{j=1}^{K}\eta_j n^{-it_j}\right|.$$

Now, Cauchy's inequality yields

$$\frac{K^2\nu!^2}{(\sigma_0-\alpha_1)^{2\nu}} < c\sum_{n=2}^{\infty}\frac{1}{n\log^2 n}\sum_{n=2}^{\infty}\frac{\log^{2\nu+4} n}{n^{2\sigma_0-1}}\left|\sum_{j=1}^{K}\eta_j n^{-it_j}\right|^2 \tag{38.4.2}$$

$$< c\sum_{n=2}^{\infty}\frac{\log^{2\nu+4} n}{n^{2\sigma_0-1}}\sum_{l,j=1}^{K}\bar{\eta}_l\eta_j n^{-i(t_j-t_l)}$$

$$< cK\left|\zeta^{(2\nu+4)}(2\sigma_0-1)\right|$$

$$+cK^2 \max_{c(\varepsilon,\alpha)\le|t|\le 2T}\left|\zeta^{(2\nu+4)}(2\sigma_0-1+it)\right|.$$

Using Cauchy's estimate for the derivative in the circle $|s-(2\sigma_0-1)| \le (1-1/\nu)(2\sigma_0-2)$, we obtain

(38.4.3)

$$\left|\zeta^{(2\nu+4)}(2\sigma_0-1)\right| < \frac{c\nu}{\sigma-1}\frac{(2\nu+4)!}{[(1-1/\nu)(2\sigma_0-2)]^{2\nu+4}} < \frac{c(2\nu+5)!}{(2\sigma_0-2)^{2\nu+5}}.$$

In the circle $|s-(2\sigma_0-1+it)|<2\sigma_0-1-\frac{1}{2}$ we can make use of our assumption (37.1.3) with $a=\varepsilon$ and obtain

$$(38.4.4)\qquad \left|\zeta^{(2\nu+4)}(2\sigma_0-1+it)\right|<cT^{\varepsilon}\frac{(2\nu+4)!}{\left(2\sigma_0-\frac{3}{2}\right)^{2\nu+4}}.$$

This last inequality shows that the multiplier of K^2 on the extreme right of (38.4.2) is smaller than the multiplier on the extreme left if

$$\frac{\nu!^2}{(\sigma_0-\alpha_1)^{2\nu}}>cT^{\varepsilon}\frac{(2\nu+4)!}{\left(2\sigma_0-\frac{3}{2}\right)^{2\nu+4}}$$

with a large enough c, or, using Stirling's formula and (38.3.2), if

$$\left(\frac{\sigma_0-\frac{3}{4}}{\sigma_0-\alpha_1}\right)^{2\nu}>cT^{\varepsilon}\log^4T.$$

By (38.3.1) the quotient on the left is a constant >1, and the last inequality in fact follows from (38.3.2), giving the first lower restriction on $b(\alpha)$. Thus (38.4.3) and (38.4.2) imply

$$(38.4.5)\quad K<\frac{c(2\nu+5)!(\sigma_0-\alpha_1)^{2\nu}}{(2\sigma_0-2)^{2\nu+5}\nu!^2}<c\log^5T\left(\frac{\sigma_0-\alpha_1}{\sigma_0-1}\right)^{2\nu}<T^{4b(\alpha)\varepsilon},$$

again using (38.3.2).

38.5

We call a zero ρ in the rectangle $\alpha\le\sigma\le1$, $|t|\le T$ "good" if (38.5.5) holds for $s_0=\sigma_0+i\operatorname{Im}\rho$ and for all ν in (38.3.2), otherwise "bad." Each bad zero ρ belongs to at least one ν in the sense that (38.3.5) is not satisfied for this ν and $s_0=\sigma_0+i\operatorname{Im}\rho$. For each ν let us choose a sequence ρ_j of bad zeros belonging to it with $|\operatorname{Im}(\rho_i-\rho_j)|>c$, as required at the beginning of 38.4, and with maximal number of terms $K=K_\nu$. For this (38.4.5) holds. All the bad zeros belonging to ν will then be covered by horizontal strips of width c built around these K_ν zeros, each strip containing at most $c(\alpha)\log T$ bad zeros, as we learn from Jensen's inequality. We show that there exist no good zeros, and since there are, at most, $\log T$ values ν, this will imply

$$N(\alpha,T,f)<c(\alpha)\log T\sum_{\nu}K_\nu<T^{5b(\alpha)\varepsilon},$$

and the proof will be complete.

38.6

Suppose, on the contrary, that $\rho_0 = \beta_0 + it_0$ is a good zero. By definition, $\beta_0 \geq \alpha$; also condition (38.3.5) is fulfilled with $s_0 = \sigma_0 + it_0$ so that (38.3.4) also holds. To obtain a lower bound for the left-hand side of the latter we need an upper bound N for the number of its terms. Using the fact that $f(s) \in \Delta(\varepsilon, \frac{3}{4}, T)$ (see properties I and II, 38.2), Jensen's inequality gives

$$N = c(\alpha)\varepsilon \log T.$$

The second main theorem in the form (33.12.1) ensures the existence of a ν_0 with

$$(38.6.1)\quad \left| \sum_{|\rho - s_0| < \sigma_0 - \alpha_1} \frac{1}{(\rho - s_0)^{\nu_0 + 1}} \right| > \left[\frac{N}{8e(m+N)} \right]^N \frac{1}{|\rho_0 - s_0|^{\nu_0+1}}$$

$$> \left[\frac{N}{8e(m+N)} \right]^N \frac{1}{(\sigma_0 - \alpha)^{\nu_0+1}}$$

in the interval

$$m \overset{\text{def}}{=} b(\alpha)\varepsilon \log T \leq \nu_0 + 1 \leq m + c(\alpha)\varepsilon \log T.$$

This contradicts (38.3.4) if

$$\left[\frac{N}{8e(m+N)} \right]^N \frac{1}{(\sigma_0 - \alpha)^{\nu_0+1}} > \frac{\log^3 T}{(\sigma_0 - \alpha_1)^{\nu_0}},$$

that is,

$$\left(\frac{\sigma_0 - \alpha_1}{\sigma_0 - \alpha} \right)^{b(\alpha)} > c \left\{ \frac{8e[b(\alpha) + c(\alpha)]}{c(\alpha)} \right\}^{c(\alpha)},$$

yielding a second lower restriction on $b(\alpha)$ [the quotient on the left is > 1, and $c(\alpha)$ is fixed]. This proves Theorem 38.3. ■

38.7

The proof of Theorem 38.4 runs along the same lines with the new definitions and restrictions

$$(38.7.1)\quad \delta = 1 - \alpha, \qquad s_0 = \sigma_0 + it = 1 + \delta + it, \qquad \alpha_1 = 1 - 5\delta,$$
$$\gamma = 1 - 11\delta$$

$$(38.7.2)\quad m = 7A \log T \leq \nu + 1 \leq 8.5A \log T, \qquad A \geq 300 \frac{\log\log T}{\log T}.$$

Since we have assumed $A < 10(1-\gamma) = 110\delta$, the last inequality implies

$$\log\frac{1}{\delta} < 0.004A\log T \qquad (T > c). \tag{38.7.3}$$

Using Lemma V.1 with $R = \sigma_0 - \gamma = 12\delta$, $r = \sigma_0 - \alpha_1 = 6\delta$, we again get (38.3.4). In 38.4 we must replace the right-hand side of (38.4.4) by

$$T^A\frac{(2\nu+4)!}{(2\sigma_0 - 1 - \gamma)^{2\nu+4}},$$

and the condition for including this in the coefficient of K^2 on the left of (38.4.2) becomes

$$\frac{\nu!^2}{(\sigma_0-\alpha_1)^{2\nu}} > cT^A\frac{(2\nu+4)!}{(2\sigma_0-1-\gamma)^{2\nu+4}}.$$

In view of (38.7.1) and (38.7.2), this is equivalent to

$$\left[\frac{\sigma_0-(1+\gamma)/2}{\sigma_0-\alpha_1}\right]^{2\nu} = \left(\frac{6.5}{6}\right)^{2\nu} > cT^A\nu^8 \qquad (T > c)$$

which is in fact a consequence of (38.7.2). This means that in (38.4.5)

$$K < \frac{c(2\nu+5)!(\sigma_0-\alpha_1)^{2\nu}}{(2\sigma_0-2)^{2\nu+5}\nu!^2} < c\nu^{10}\left(\frac{\sigma_0-\alpha_1}{\sigma_0-1}\right)^{2\nu} = c\nu^{10}6^{2\nu} < T^{31A}$$

by (38.7.2) and (38.7.1). Using Jensen's inequality for circles $|s-(\sigma_0+it)| < 12\delta$ and noting that $f(s) \in \Delta(A,\gamma,T)$ (see properties I and II, 38.2), the number of zeros in any circle $|s - \sigma_0 + it| \le \sigma_0 - \alpha_1 = 6\delta$ does not exceed

$$\frac{A\log T + 10\log c/\delta}{\log 2} < 1.5A\log T \doteq N \qquad (T > c) \tag{38.7.4}$$

by (38.7.2) and (38.7.3). A rough consequence is that the number of zeros in any rectangle $\alpha \le \sigma \le 1$, $|t - t_0| \le c$ is less than $c/\delta \log T < \log^3 T$. As in 38.5, we obtain

$$N(\alpha, T, f) < T^{31A}\log^4 T, \tag{38.7.5}$$

provided that we can show the nonexistence of good zeros.

(38.7.4) has been shown to be a bound for the number of terms in (38.6.1), and (38.6.1) contradicts (38.3.4) if

$$\left[\frac{N}{8e(m+N)}\right]^N \frac{1}{(\sigma_0-\alpha)^{\nu_0+1}} > \frac{\log^3 T}{(\sigma_0-\alpha_1)^{\nu_0}},$$

or in other words, recalling (38.7.2), if

$$\left(\frac{\sigma_0 - \alpha_1}{\sigma_0 - \alpha}\right)^{7A \log T} = 3^{7A \log T} > \log^3 T \left(\frac{8e8.5}{1.5}\right)^{1.5A \log T}$$

This, in fact, is true by (38.7.2). If the condition that $A \geq 300 \log \log T/\log T$ in the latter is not fulfilled, we can use $A = 300 \log \log T/\log T$, giving a power of $\log T$ in (38.7.5) so that Theorem 38.4 holds in any case.

Theorem 38.2 is a direct consequence of this and (35.6.2). We remark that these proofs also work for $\sigma \leq \frac{3}{4}$. We show how in a similar situation in the next section. However, in 38.4 we are then unable to use any deep (even hypothetical) properties of the zeta function, and in Theorem 38.1 we only get the density hypothesis from Lindelöf's hypothesis.

39

DENSITY THEOREMS III

A "q-Analogue" of the Lindelöf Hypothesis for L Functions

39.1

The most important functions for which Theorems 38.3 and 38.4 are applicable are the functions $L(s, q, \chi)$ corresponding to a Dirichlet character $\chi(n)$ belonging to the modulus q, and these theorems imply sharp estimate for fixed q. Quite often the dependence on q is also important and even more so than the dependence on T. In this section we investigate a situation similar to that in Section 38 for fixed T and varying q.

A "q-analogue" of the Lindelöf hypothesis is

(39.1.1) $$|L(s, q, \chi)| < c(\varepsilon, T)q^{\varepsilon} \qquad (\sigma \geq \tfrac{1}{2}, |t| \leq T).$$

A q-analogue of the density hypothesis in our situation is the following special case of Bombieri's conjecture:

(39.1.2)

$$S(\alpha, Q, T) \doteq \sum_{q \leq Q} \sum_{\chi}{}^{*} N(\alpha, T, q, \chi) < Q^{4(1-\alpha)+\varepsilon} \qquad \tfrac{1}{2} \leq \alpha < 1, Q > c(\varepsilon, T)$$

where $N(\alpha, T, q, \chi)$ denotes the number of zeros of $L(s, q, \chi)$ in the rectangle $\alpha \leq \sigma < 1$, $|t| \leq T$ and the star in the inner sum indicates that χ runs through primitive characters mod q only. The relationship between these two hypotheses is similar to that for the t-analogue.

Theorem 39.1 (Turán–Halász [2]). *Inequality* (39.1.1) *implies*

$$S(\alpha, Q, T) < Q^{\varepsilon}$$

for $\alpha > \frac{3}{4}$ *and* (39.1.2) *for* $\frac{1}{2} < \alpha \le \frac{3}{4}$.

We cannot state a result corresponding to Theorem 38.2 because nothing like a q-analogue of 35.6.2 is known for general q.

39.2. PROOF

We may assume $\alpha > \frac{1}{2}$ since the case $\alpha = \frac{1}{2}$ is trivial. Let α_1,

(39.2.1) $\quad \frac{1}{2} < \alpha_1 < \alpha \quad \text{if} \quad \frac{1}{2} < \alpha < 1 \quad \text{and} \quad \frac{3}{4} < \alpha_1 < \alpha \quad \text{if} \quad \alpha > \frac{3}{4},$

be arbitrarily near α. Temporarily we make the following restriction on the parameters:

(39.2.2)
$$\sigma_0 = \sigma_0(\alpha, \alpha_1, T) \doteq 1 + \frac{T^2}{\alpha - \alpha_1}, \qquad T > 10$$

implying

(39.2.3)
$$\left| \frac{\sigma_0 - \alpha_1}{\sigma_0 - (\alpha + iT)} \right| > 1$$

(39.2.4)
$$m \le \nu + 1 \le m + N$$

where

$$b_1(\alpha, \alpha_1, T)\varepsilon \log Q < m < \log^2 Q, \qquad N = b_2(\alpha, \alpha_1, T)\varepsilon \log Q,$$

and the constants b_1 and b_2 will be specified by further restrictions in the course of the proof.

39.3

We use Lemma V.1 with $G(s) = L(s, q, \chi)$ where $1 < q \le Q$ [$Q > c(\varepsilon, \alpha, \alpha_1, T)$, χ primitive], $s_0 = \sigma_0$, $R = \sigma_0$, and $r = \sigma_0 - \alpha_1$. We can take $M = \log Q$ giving

(39.3.1)
$$\left| \sum_{|\rho - \sigma_0| \le \sigma_0 - \alpha_1} \frac{1}{(\rho_0 - \sigma_0)^{\nu+1}} \right| < \frac{\log^4 Q}{(\sigma_0 - \alpha_1)^{\nu}}$$

provided that

$$\left|\left[\frac{L'}{L}(s, q, \chi)\right]^{(\nu)}_{s=\sigma_0}\right| < \frac{\nu!}{(\sigma_0 - \alpha_1)^\nu} \tag{39.3.2}$$

where ρ runs through the zeros of $L(s, q, \chi)$, by (39.2.4).

We call a χ "good" if (39.3.2), and hence also (39.3.1), holds for every ν in (39.2.4) and show first that the corresponding L-function does not vanish for $\alpha \le \sigma < 1$, $|t| < T$.

39.4

Suppose, to the contrary, that ρ_0 is a zero with $\operatorname{Re}\rho_0 \ge \alpha$, $|\operatorname{Im}\rho_0| \le T$. We use Jensen's inequality for the circle $|s - \sigma_0| \le \sigma_0 - \frac{1}{2}$ and our assumption (39.1.1) to show that the number of terms in (39.3.1) is at most

$$c(\alpha, \alpha_1, T)\varepsilon \log Q \overset{\text{def}}{=} b_2 \varepsilon \log Q = N.$$

[This defines b_2 in (39.2.4).] Now the second main theorem in the form (33.12.1) gives us a $\nu + 1 = \nu_0 + 1$ in (39.2.4) such that

$$\left|\sum_{|\rho - \sigma_0| \le \sigma_0 - \alpha_1} \frac{1}{(\rho - \sigma_0)^{\nu_0 + 1}}\right| \ge \left[\frac{N}{8e(m + N)}\right]^N \frac{1}{|\rho_0 - \sigma_0|^{\nu_0 + 1}}. \tag{39.4.1}$$

By (39.2.3), ρ_0 occurs in the sum since $|\rho_0 - \sigma_0| \le |\sigma_0 - (\alpha + iT)|$. For the same reason (39.4.1) contradicts (39.3.1) if

$$\left|\frac{\sigma_0 - \alpha_1}{\sigma_0 - (\alpha + iT)}\right|^m > Q^\varepsilon \left[8e\left(\frac{m}{N} + 1\right)\right]^N,$$

$$\left|\frac{\sigma_0 - \alpha_1}{\sigma_0 - (\alpha + iT)}\right|^{b_1/b_2} > e^{1/b_2} 8e\left(\frac{b_1}{b_2} + 1\right),$$

recalling (39.2.4). As the quotient on the left-hand side is > 1 this will be satisfied for $b_1 \ge c(\alpha, \alpha_1, T)$ because b_2 has already been defined.

39.5

An L-function corresponding to a "good" χ has no zero in the rectangle in question, and we thus count the "bad" characters.

Fix a ν and let $\chi_1,\ldots,\chi_K$ be the characters that do not satisfy (39.3.2), that is,

$$\eta_j \sum_{n=2}^{\infty} \frac{\Lambda(n)\chi_j(n)\log^\nu n}{n^{\sigma_0}} \geq \frac{\nu!}{(\sigma_0 - \alpha_1)^\nu}$$

with an η_j of modulus 1. The same argument that led to (38.4.2) now gives

$$\text{(39.5.1)} \quad \frac{K^2\nu!^2}{(\sigma_0 - \alpha_1)^{2\nu}} \leq cK\left|\zeta^{(2\nu+4)}(2\sigma_0 - 1)\right|$$

$$+ cK^2 \max_{\substack{i,\, j=1,\ldots,K \\ i\neq j}} \sum_{n=2}^{\infty} \frac{\log^{(2\nu+4)} n \cdot \chi_i(n)\bar{\chi}_j(n)}{n^{2\sigma_0-1}},$$

χ_i and χ_j in the sum being different primitive characters. $\chi = \chi_i\bar{\chi}_i$ is a nonprincipal character belonging to a modulus $q \leq Q^2$, and the sum of $\left|L^{(2\nu+4)}(2\sigma_0\text{-}1, q, \chi)\right|$. To estimate it for $\alpha > \frac{3}{4}$ we are able to make another use of (39.1.1) by applying Cauchy's estimate for the derivative in the circle $|s - (2\sigma_0 - 1)| \leq 2\sigma_0 - 1 - \frac{1}{2}$:

$$\left|L^{(2\nu+4)}(2\sigma_0 - 1, q, \chi)\right| < Q^{2\varepsilon}\frac{(2\nu + 4)!}{(2\sigma_0 - \frac{3}{2})^{2\nu+4}}.$$

If

$$\frac{\nu!^2}{(\sigma_0 - \alpha_1)^{2\nu}} > cQ^{2\varepsilon}\frac{(2\nu + 4)!}{(2\sigma_0 - \frac{3}{2})^{2\nu+4}}$$

with c sufficiently large, that is, using (39.2.4) if

$$\left(\frac{\sigma_0 - \frac{3}{4}}{\sigma_0 - \alpha_1}\right)^{2\nu} > Q^{3\varepsilon}$$

[which is satisfied, owing to (39.2.1), if we assume that in (39.2.4) $b_1 \geq c(\alpha, \alpha_1, T)$; we can now define b_1 as the maximum of this threshold and the one at the end of 39.4], then we get from (39.5.1), and (38.4.3), that

$$\text{(39.5.2)} \quad K < \frac{c(2\nu + 5)!}{\nu!^2}\left(\frac{\sigma_0 - \alpha_1}{2\sigma_0 - 2}\right)^{2\nu} < Q^{c(\alpha, \alpha_1, T)\varepsilon} \qquad \left(\alpha > \tfrac{3}{4}\right).$$

Here we have made use of (39.2.4) with $m = b_1(\alpha, \alpha_1, T)\varepsilon \log Q$.

For $\alpha \le \frac{3}{4}$ we use the circle $|s-(2\sigma_0-1)| \le 2\sigma_0 + 2\sqrt{\sigma_0}$ and the fact that $|L(s,q,\chi)| < c(\alpha,\alpha_1,T)q^{2\sqrt{\sigma_0}+2} < Q^{4\sqrt{\sigma_0}+4.5}$, a direct consequence of the functional equation. As above,

$$\left|L^{(2\nu+4)}(2\sigma_0-1,q,\chi)\right| < Q^{4\sqrt{\sigma_0}+4.5}\frac{(2\nu+4)!}{\left(2\sigma_0+2\sqrt{\sigma_0}\right)^{2\nu+4}}$$

and the condition

$$\frac{\nu!^2}{\left(\sigma_0-\alpha_1\right)^{2\nu}} > cQ^{4\sqrt{\sigma_0}+4.5}\frac{(2\nu+4)!}{\left(2\sigma_0+2\sqrt{\sigma_0}\right)^{2\nu+4}}$$

will be satisfied if

$$\left(\frac{\sigma_0+\sqrt{\sigma_0}}{\sigma_0-\alpha_1}\right)^{2\nu} > e^{2\nu}\frac{\sqrt{\sigma_0}}{\sigma_0+\sqrt{\sigma_0}} > Q^{4\sqrt{\sigma_0}+5},$$

that is, if we set in (39.2.4)

$$m = \left(2\sigma_0+4\sqrt{\sigma_0}\right)\log Q. \tag{39.5.3}$$

We can then infer from (39.5.1), using (38.4.3), that

$$\begin{aligned} K &< c\frac{(2\nu+5)!}{\nu!^2}\left(\frac{\sigma_0-\alpha_1}{2\sigma_0-2}\right)^{2\nu} < \log^{11}Q\left(\frac{\sigma_0-\alpha_1}{\sigma_0-1}\right)^{2\nu} \\ &< \log^{11}Q\exp\left(2\nu\frac{1-\alpha_1}{\sigma_0-1}\right) \\ &< \log^{11}Q\cdot Q^{2\frac{2\sigma_0+5\sqrt{\sigma_0}}{\sigma_0-1}(1-\alpha_1)} \qquad \left(\alpha\le\tfrac{3}{4}\right). \end{aligned} \tag{39.5.4}$$

We have used (39.5.3) and (39.2.4), choosing $\varepsilon = \varepsilon(\alpha,\alpha_1,T)$ sufficiently small here.

39.6

In either of the cases $\alpha > \frac{3}{4}$ and $\alpha \le \frac{3}{4}$ we have at most $N = b_2\varepsilon\log Q$ values ν, and to get a bound for the number of "bad" χ defined in 39.3, we only have to multiply this by (39.5.2) and (39.5.4), respectively.

Each character χ yields at most $c(T)\log Q$ zeros, and only the "bad" characters yield zeros in $\sigma \ge \alpha$, $|t| \le T$. Thus Theorem 39.1 is obtained after the following observation. We may choose α_1 sufficiently close to α, so that σ_0 in (39.2.2) is large enough to make the exponent in (39.5.4) arbitrarily close to $4(1 - \alpha)$. We then include the zeros of $\zeta(s)$, less than $c(T)$ in number, which we have disregarded up to this point.

39.7

Here we also used the device of 38.4 for $\alpha \leq \frac{3}{4}$ in a situation where we previously (in Section 37) used square integration. In Turán–Halász [2] it is shown how the q-analogue of the latter, the large-sieve inequality of Bombieri, leads to the same result. Of course we could have combined any two, or indeed all, of the principles discussed in the last three sections, but for clarity we preferred to present them separately.

We end this topic with some historical remarks. The investigation of the vertical distribution of zeros goes back to Bohr, Landau, and Carlson, but it was Hoheisel who discovered its arithmetical significance in 1930, proving for the first time the existence of primes in $(x, x + x^{\vartheta})$ with a $\vartheta < 1$. Ingham, giving a better ϑ, also observed that the density hypothesis implies this for any $\vartheta > \frac{1}{2}$, and nothing stronger follows even from the Riemann hypothesis. Later, many more applications were discovered, especially of "hybrid t and q" results. We only mention the names Linnik and Rényi in connection with the Goldbach problem and Bombieri in connection with his important "prime number theorem." In Turán–Halász [2] a (hypothetical) application is given of Theorem 39.1 to the least prime in arithmetic progressions. We return to these primes later in the book, giving a complete proof of Linnik's celebrated theorem.

Beginning with the proofs of Sections 38 and 39, Montgomery replaced their power sum methods by a more classical approach, improving, in particular, the best-known value of Hoheisel's constant ϑ. This was further improved by Huxley, Jutila, and others. In the same way, Bombieri combined the principles of Sections 37 and 38 with sharp estimates.

40

THE REMAINDER TERM OF THE PRIME NUMBER FORMULA I

Connection with the Zero-Free Region of $\zeta(s)$

40.1

In this section we investigate the connection between zeros of $\zeta(s)$ and the remainder term of the prime number formula. Define

(40.1.1) $$\Delta_1(x) \doteq \pi(x) - \operatorname{li} x \doteq \sum_{p \le x} 1 - \int_0^x \frac{dr}{\log r}$$

(40.1.2) $$\Delta_2(x) \doteq \Pi(x) - \operatorname{li} x \doteq \sum_{p^m \le x} \frac{1}{m} - \operatorname{li} x$$

(40.1.3) $$\Delta_3(x) \doteq \Theta(x) - x \doteq \sum_{p \le x} \log p - x$$

(40.1.4) $$\Delta_4(x) \doteq \psi(x) - x \doteq \sum_{n \le x} \Lambda(n) - x.$$

In 1899 de la Vallée Poussin showed that $\zeta(s) \neq 0$ in the domain

(40.1.5) $$\sigma > 1 - \frac{c_1}{\log |t|}, \qquad |t| \ge c_2.$$

For the remainder term, this implies the estimate

$$\Delta_i(x) = O\left[x\exp\left(-c_3\sqrt{\log x}\right)\right] \qquad (1 \le i \le 4). \tag{40.1.6}$$

It turned out that if a larger domain of this type with $\zeta(s) \neq 0$ can be assured, then a smaller remainder term in (40.1.6) can be proven. A general theorem of this type was formulated by Ingham ([2], Theorem 22, p. 63). We confine ourselves to the special case when $\zeta(s)$ is assumed to be nonvanishing in the domain

$$\sigma > 1 - \frac{c_4}{\log^{\alpha}|t|}, \qquad |t| \ge c_5 \tag{40.1.7}$$

with

$$0 < \alpha \le 1. \tag{40.1.8}$$

Ingham's theorem gives

$$\Delta_i(x) = O\left[x\exp\left(-c_6\log^{1/(1+\alpha)}x\right)\right] \qquad (1 \le i \le 4). \tag{40.1.9}$$

The best-known value for α in (40.1.7), according to the results of Korobov–Vinogradov, is $\frac{2}{3} + \varepsilon$.

We ask, of course, whether or not better estimates for the remainder terms could be inferred from the premisses (40.1.7)–(40.1.8). The question is obviously equivalent to whether or not one can assure from (40.1.9) a zero-free domain of $\zeta(s)$ of the type (40.1.7)–(40.1.8). The question is interesting in itself, since in the affirmative case it would mean a new contribution to the mysterious connection between the primes and the zeros of $\zeta(s)$. Its importance would be crucial if the upper limit of the real parts of the zeros of $\zeta(s)$ were 1.

The power sum method allows us to answer our question affirmatively, that is, to prove the complete converse of Ingham's theorem in the aforementioned case (40.1.7)–(40.1.9) (see Turán [8]).

Theorem 40.1. *If for an α with $0 < \alpha \le 1$ and for an i $(1 \le i \le 4)$ we have*

$$\Delta_i(x) = O\left[x\exp\left(-c_7\log^{1/(1+\alpha)}x\right)\right], \tag{40.1.10}$$

then $\zeta(s)$ does not vanish in the domain

$$\sigma > 1 - \frac{c_8}{\log^{\alpha}|t|}, \qquad |t| \ge c_9(\alpha). \tag{40.1.11}$$

40.2

As the premisses (40.1.10) are equivalent for any i $(1 \le i \le 4)$ (perhaps with a different value for c_7), we can suppose it for $i = 4$. That is, we assume

$$\psi(x) = \sum_{n \le x} \Lambda(n) = x + O\left[x \exp\left(-c_7 \log^{1/(1+\alpha)} x\right)\right]. \tag{40.2.1}$$

From (40.2.1) we first draw some conclusions for the expression

$$f(t) = \sum_{N' \le n \le N''} \Lambda(n) e^{-it \log n} \tag{40.2.2}$$

where the integers N', N'', and $t > 2$ are restricted by

$$1 + t^2 \le \exp\left\{\left(\frac{2}{c_7} \log t\right)^{1+\alpha}\right\} \le N \le N' < N'' \le 2N. \tag{40.2.3}$$

Since

$$\Lambda(n) = 1 + \Delta_4(n) - \Delta_4(n-1), \tag{40.2.4}$$

we have

$$\begin{aligned} |f(t)| \le &\left| \sum_{N' \le n \le N''} e^{-it \log n} \right| \\ &+ \left| \sum_{N' \le n \le N''} [\Delta_4(n) - \Delta_4(n-1)] e^{-it \log n} \right| \doteq J_1 + J_2. \end{aligned} \tag{40.2.5}$$

With partial summation we get

$$\begin{aligned} J_2 = &\,| - \Delta_4(N'-1) e^{-it \log N'} + \Delta_4(N'') e^{-it \log N''} \\ &+ \sum_{n=N'}^{N''-1} \Delta_4(n) \left[e^{-it \log n} - e^{-it \log(n+1)}\right]| \\ \le &\,|\Delta_4(N'-1)| + |\Delta_4(N'')| \\ &+ \sum_{n=N'}^{N''-1} |\Delta_4(n)| |1 - e^{-it \log(1+1/n)}| \\ < &\, c_{10} \exp\left(-c_7 \log^{1/(1+\alpha)} N\right) \left(N + \sum_{n=N'}^{N''-1} n |1 - e^{-it \log(1+1/n)}|\right) \\ < &\, c_{11} N t \exp\left(-c_7 \log^{1/(1+\alpha)} N\right). \end{aligned} \tag{40.2.6}$$

From (40.2.3)

$$\exp\left(-c_7 \log^{1/(1+\alpha)} N\right) \le \frac{1}{t^2}, \tag{40.2.7}$$

and (40.2.6) implies

$$J_2 < c_{11}\frac{N}{t}. \tag{40.2.8}$$

For the estimate of J_1 we note that

$$\begin{aligned} &\left|(n+1)^{1-it} - n^{1-it} - (1-it)n^{-it}\right| \\ &= n\left|\left(1+\frac{1}{n}\right)^{1-it} - 1 - (1-it)\frac{1}{n}\right| \\ &\le n\left|\left(1+\frac{1}{n}\right)^{1-it} - 1 - (1-it)\log\left(1+\frac{1}{n}\right)\right| \\ &\quad + n|1-it|\left|\log\left(1+\frac{1}{n}\right) - \frac{1}{n}\right|. \end{aligned} \tag{40.2.9}$$

Since for $|z| \le 1$

$$|e^z - 1 - z| \le |z|^2, \tag{40.2.10}$$

we get, from (40.2.9) and (40.2.10) [with $z = (1-it)\log(1+1/n)$],

(40.2.11)

$$\left|(n+1)^{1-it} - n^{1-it} - (1-it)n^{-it}\right| \le \frac{1+t^2}{n} + \frac{\sqrt{1+t^2}}{n} < \frac{2(1+t^2)}{n}.$$

Summing this for $N' \le n \le N''$ we get

$$\begin{aligned} &\left|\sum_{N' \le n \le N''} \left\{(n+1)^{1-it} - n^{1-it} - (1-it)n^{-it}\right\}\right| \\ &\qquad \le 2(1+t^2) \sum_{N' \le n \le N''} \frac{1}{n} < 2(1+t^2). \end{aligned} \tag{40.2.12}$$

On the other hand,

$$\text{(40.2.13)}\quad \left|\sum_{N'\le n\le N''}\left\{(n+1)^{1-it}-n^{1-it}-(1-it)n^{-it}\right\}\right|$$

$$=\left|(N''+1)^{1-it}-(N')^{1-it}-(1-it)\sum_{N'\le n\le N''}n^{-it}\right|$$

$$\ge|1-it|\left|\sum_{N'\le n\le N''}n^{-it}\right|-\left|(N''+1)^{1-it}-(N')^{1-it}\right|$$

$$\ge\sqrt{1+t^2}\left|\sum_{N'\le n\le N''}n^{-it}\right|-4N.$$

From (40.2.12) and (40.2.13) it follows that

$$\text{(40.2.14)}\qquad \left|\sum_{N'\le n\le N''}n^{-it}\right|\le\frac{4N+2(1+t^2)}{\sqrt{1+t^2}}.$$

From this, again using (40.2.3), we get

$$\text{(40.2.15)}\qquad J_1\le\frac{6N}{\sqrt{1+t^2}}<\frac{6N}{t}.$$

Together with (40.2.8) this yields

$$\text{(40.2.16)}\qquad |f(t)|<c_{12}\frac{N}{t}.$$

Using (40.2.16) and partial summation,

$$\text{(40.2.17)}\qquad \left|\sum_{N'\le n\le N''}\frac{\Lambda(n)}{n^s}\right|<c_{13}\frac{N^{1-\sigma}}{t},\quad\text{for}\quad\sigma>1$$

if (40.2.3) is satisfied. Furthermore (40.2.17) remains valid when N' and N'' are not integers.

40.3

Let

$$\text{(40.3.1)}\qquad \eta\ge\exp\left\{\left(\frac{2}{c_7}\log t\right)^{1+\alpha}\right\}$$

and apply (40.2.17) with the choice

$$N' = \eta \cdot 2^j, \qquad N'' = \eta \cdot 2^{j+1}, \qquad j = 0, 1, 2, \ldots. \tag{40.3.2}$$

Thus

$$\left| \sum_{n \geq \eta} \frac{\Lambda(n)}{n^s} \right| < \frac{c_{13}}{t} \eta^{1-\sigma} \sum_{j=0}^{\infty} 2^{(1-\sigma)j} < \frac{c_{14}\eta^{1-\sigma}}{(\sigma - 1)t}. \tag{40.3.3}$$

Finally, if

$$\xi \geq \exp\left\{ \left(\frac{2}{c_7} \log t \right)^{1+\alpha} \right\}, \tag{40.3.4}$$

then for the integral

$$I(s, \xi) = \int_{\xi}^{\infty} \left(\sum_{n \geq \eta} \frac{\Lambda(n)}{n^s} \right) \frac{\log^k (n/\xi)}{\eta} \, d\eta \tag{40.3.5}$$

we have on one hand

$$|I(s, \xi)| \leq \frac{c_{14}}{(\sigma - 1)t} \int_{\xi}^{\infty} \eta^{-\sigma} \log^k \frac{\eta}{\xi} \, d\eta \tag{40.3.6}$$

$$= \frac{c_{14}\xi^{1-\sigma}}{t(\sigma - 1)} \int_0^{\infty} e^{-(\sigma-1)v} \cdot v^k \, dv = \frac{c_{14}\xi^{1-\sigma} k!}{t(\sigma - 1)^{k+2}}$$

and on the other

$$I(s, \xi) = \sum_{n \geq \xi} \frac{\Lambda(n)}{n^s} \int_{\xi}^{n} \frac{1}{\eta} \log^k \frac{\eta}{\xi} \, d\eta \tag{40.3.7}$$

$$= \sum_{n \geq \xi} \frac{\Lambda(n)}{n^s} \int_1^{n/\xi} \frac{\log^k v}{v} \, dv = \frac{1}{k+1} \sum_{n \geq \xi} \frac{\Lambda(n)}{n^s} \log^{k+1} \frac{n}{\xi}.$$

From this and (40.3.6) we have

$$\left| \sum_{n \geq \xi} \frac{\Lambda(n)}{n^s} \log^{k+1} \frac{n}{\xi} \right| \leq \frac{c_{14}\xi^{1-\sigma}(k+1)!}{t(\sigma - 1)^{k+2}}. \tag{40.3.8}$$

Apply the relation

$$\text{(40.3.9)}\qquad \frac{1}{(k+1)!}\sum_{n\geq\xi}\frac{\Lambda(n)}{n^s}\log^{k+1}\frac{n}{\xi}$$

$$=\frac{\xi^{1-s}}{(s-1)^{k+2}}-\sum_{\rho}\frac{\xi^{\rho-s}}{(s-\rho)^{k+2}}+O\left[\frac{\log(|t|+2)}{\xi^2}\right]$$

(see Lemma F.1 of the Appendix), for $1 \leq \sigma \leq 2$. Then from (40.2.3), (40.3.4), (40.3.8), and (40.3.9),

$$\text{(40.3.10)}\qquad \left|\sum_{\rho}\frac{\xi^{\rho-s}}{(s-\rho)^{k+2}}\right| \leq \frac{c_{15}\xi^{1-\sigma}}{t(\sigma-1)^{k+2}}.$$

40.4

Assume, contrary to (40.1.11), that $\zeta(s)$ has a zero $\rho_0 = \beta_0 + i\gamma_0 = 1 - \delta_0 + i\gamma_0$ with

$$\text{(40.4.1)}\qquad \delta_0 < \frac{c_0}{\log^{\alpha}\gamma_0}, \qquad \gamma_0 > C_0$$

where c_0 is a positive constant [to be chosen later in (40.5.3)] and C_0 is a sufficiently large positive constant both depending on $c_7, c_8, \ldots, c_{15}$ and α. (They do not depend on the c_ν with $\nu > 15$.)

Let k be any integer restricted at present only by

$$\text{(40.4.2)}\qquad \log\gamma_0 \leq k+2 \leq \tfrac{11}{10}\log\gamma_0.$$

Let

$$\text{(40.4.3)}\qquad s_0 = \sigma_0 + i\gamma_0 = 1 + 5\delta_0 + i\gamma_0.$$

Thus (40.3.10) can be written in the form

$$\text{(40.4.4)}\qquad \left|\sum_{\rho}\xi^{\rho-s_0}\left(\frac{5\delta_0}{s_0-\rho}\right)^{k+2}\right| \leq c_{15}\frac{\xi^{1-\sigma_0}}{\gamma_0}.$$

There are, at most,

$$\text{(40.4.5)}\qquad c_{16}\log T$$

zeros with the property

$$T \le |\gamma| \le T + 1, \tag{40.4.6}$$

so the contribution of zeros with

$$|\rho - s_0| \ge 20\delta_0 \tag{40.4.7}$$

to the infinite sum in (40.4.4) is, at most,

$$\sum_{n=4}^{\infty} \xi^{1-\sigma_0} \cdot \left(\frac{1}{n}\right)^{k+2} \cdot c_{16} \log(\gamma_0 + 5n\delta) \le \frac{c_{17} \xi^{1-\sigma_0} \log \gamma_0}{4^k} \le c_{18} \frac{\xi^{1-\sigma_0}}{\gamma_0}. \tag{40.4.8}$$

Hence

$$W = \left| \sum_{|s_0 - \rho| < 20\delta_0} \xi^{\rho - s_0} \left(\frac{5\delta_0}{s_0 - \rho}\right)^{k+2} \right| < c_{19} \frac{\xi^{1-\sigma_0}}{\gamma_0}. \tag{40.4.9}$$

By Jensen's inequality the number of terms in (40.4.9) is at most

$$c_{20}\delta_0 \log \gamma_0 + c_{21} \log\log \gamma_0 \tag{40.4.10}$$

which is by (40.4.1)

$$< \tfrac{1}{10} \log \gamma_0. \tag{40.4.11}$$

Let

$$\xi = e^{\omega(k+2)} \tag{40.4.12}$$

where

$$\omega = \left(\frac{2}{c_7}\right)^{1+\alpha} (\log \gamma_0)^{\alpha}. \tag{40.4.13}$$

Then (40.3.4) is obviously satisfied due to (40.4.2), and our task reduces to the lower estimate of the finite power sum W in (40.4.9).

40.5

Now we can apply the second main theorem in the form given in Section 33.12. By (40.4.11) there exists a k with

$$\log \gamma_0 \le k + 2 \le \tfrac{11}{10} \log \gamma_0 \tag{40.5.1}$$

for which [using the relations (40.4.1)–(40.4.3) and (40.4.11)–(40.4.13)]

$$(40.5.2)\quad W = \left| \sum_{|s_0-\rho|<20\delta_0} \left[e^{\omega(\rho-s_0)} \cdot \frac{5\delta_0}{s_0-\rho} \right]^{k+2} \right|$$

$$\geq e^{-6\delta_0\omega(k+2)} \cdot \left(\frac{5}{6}\right)^{k+2} \cdot \left(\frac{1}{8e\cdot 11}\right)^{(1/10)\log\gamma_0}$$

$$\geq \xi^{-6\delta_0} \cdot \exp\left(-0.3 \cdot \tfrac{11}{10}\log\gamma_0\right) \cdot \exp\left(-5.7 \cdot \tfrac{1}{10}\log\gamma_0\right)$$

$$= \xi^{1-\sigma_0} \cdot \xi^{-\delta_0} \cdot e^{-0.9\log\gamma_0}$$

$$= \frac{\xi^{1-\sigma_0}}{\gamma_0} \exp\left[\frac{1}{10}\log\gamma_0 - \delta_0\omega(k+2)\right]$$

$$\geq \frac{\xi^{1-\sigma_0}}{\gamma_0} \exp\left\{\frac{1}{10}\log\gamma_0 - \frac{c_0}{\log^\alpha\gamma_0} \cdot \log^\alpha\gamma_0 \left(\frac{2}{c_7}\right)^{1+\alpha} \cdot \frac{11}{10}\log\gamma_0\right\}$$

$$= \frac{\xi^{1-\sigma_0}}{\gamma_0} \exp\left\{\frac{1}{10}\log\gamma_0 - \frac{11}{10}\left(\frac{2}{c_7}\right)^{1+\alpha} c_0\log\gamma_0\right\}$$

$$= \frac{\xi^{1-\sigma_0}}{\gamma_0} \cdot \gamma_0^{\frac{1}{10}-\frac{11}{10}\left(\frac{2}{c_7}\right)^{1+\alpha}c_0}.$$

But this contradicts (40.4.9) if we choose

$$(40.5.3)\qquad c_0 = \frac{1}{12}\left(\frac{c_7}{2}\right)^{1+\alpha}$$

and if C_0 is chosen sufficiently large; thus the proof of Theorem 40.1 is complete. ■

40.6

A generalization of Theorem 40.1 was proved by W. Stás [5]. Another, more complete, generalization of Theorem 40.1 is due to J. Pintz [1]. Both proofs use the method of this book. To formulate this result we introduce the reduced remainder terms of the prime number formula

$$(40.6.1)\qquad \bar{\Delta}_i(x) = \begin{cases} \dfrac{\Delta_i(x)}{x} & \text{for } i = 3,4 \\[2ex] \dfrac{\Delta_i(x)}{(x/\log x)} & \text{for } i = 1,2 \end{cases},$$

and for an arbitrary continuous real function $g(u) \geq 0$ defined for $u \geq 0$, its corresponding $\omega_g(x)$ function is

$$\omega_g(x) \doteq \min_{u \geq 0} [g(u) \log x + u]. \tag{40.6.2}$$

Let ε be an arbitrary real number with $0 < \varepsilon < 1$.

Theorem (Pintz). *Suppose $\zeta(s)$ has no zero in the domain*

$$\sigma > 1 - g(\log|t|), \qquad |t| > 1 \tag{40.6.3}$$

where $g(u)$, for $u \geq 0$, is a continuous decreasing function with $0 < g(u) \leq \frac{1}{2}$; then

$$\bar{\Delta}_i(x) = O\left[\frac{1}{e^{(1-\varepsilon)\omega_g(x)}}\right] \qquad 1 \leq i \leq 4. \tag{40.6.4}$$

On the other hand suppose $\zeta(s)$ has an infinity of zeros in the domain (40.6.3), *and in addition to the preceding properties, $g(u)$ also satisfies*

$$g'(u) \nearrow 0 \quad \textit{for} \quad u \to \infty. \tag{40.6.5}$$

[*Condition* (40.6.5) *means that $g'(u)$ tends to zero monotonically increasing for $u \geq c_g$, and if $\lim_{u \to \infty} g(u) = 0$ then $g'(u)$ tends to zero strictly monotonically increasing for $u \geq c_g$.*] *Then*

$$\bar{\Delta}_i(x) = \Omega_{\pm}\left[\frac{1}{e^{(1+\varepsilon)\omega_g(x)}}\right]. \tag{40.6.6}$$

We note that the relatively natural conditions for $g(u)$ are satisfied in the most interesting special case, when

$$g(u) = \begin{cases} \dfrac{\log^B u (\log \log u)^C}{u^A} & \text{if} \quad u \geq u_0 = u_0(A, B, C) \\ g(u_0) & \text{if} \quad 0 \leq u \leq u_0 \end{cases} \tag{40.6.7}$$

where $A > 0$, B, C are arbitrary, or $A = 0$, $B < 0$, C is arbitrary, or $A = B = 0$ and $C < 0$.

41

THE REMAINDER TERM OF THE PRIME NUMBER FORMULA II

Effective Estimation of the Sign Changes of $\pi(x) - \operatorname{li} x$

41.1

Riemann stated in his famous memoir from 1859 [1] the inequality

$$\pi(x) < \operatorname{li} x \quad \text{that is} \quad \Delta_1(x) < 0 \tag{41.1.1}$$

for $x > 2$. There were opposing opinions regarding its truth for more than 50 years. Erhard Schmidt [1] exhibited in 1903 infinitely many sign changes of the closely related functions

$$\Delta_2(x) = \Pi(x) - \operatorname{li} x \tag{41.1.2}$$

and

$$\Delta_4(x) = \psi(x) - x \tag{41.1.3}$$

as $x \to \infty$ [for the notations see (40.1.1)–(40.1.4)].

However, the corresponding problem for $\Delta_1(x)$ was much deeper. As Schmidt remarked, the truth of Riemann's assertion (41.1.1) would imply the truth of the famous Riemann conjecture, that all the nontrivial zeros of $\zeta(s)$ are on the line $\operatorname{Re} s = \frac{1}{2}$. This remark naturally made the problem even more interesting, as it seemed to open a new way to prove the Riemann conjecture.

The truth of (41.1.1) seemed to be supported by the calculations of D. N. Lehmer [1], who verified the truth of (41.1.1) in 1914 for all $x < 10^7$, a bound considered at that time so incredibly large as to permit no doubt as to the general validity of (41.1.1).

But Littlewood [2] disproved (41.1.1) in the same year, showing that $\Delta_1(x)$ changes sign infinitely often as $x \to \infty$. He even proved

$$\Delta_1(x) = \Omega_{\pm}\left(\frac{\sqrt{x}}{\log x}\log_3 x\right) \tag{41.1.4}$$

which trivially implies the truth of the same relation for the case $i = 2$ too, and analogously showed

$$\Delta_i(x) = \Omega_{\pm}\left(\sqrt{x}\log_3 x\right) \quad \text{for} \quad i = 3, 4. \tag{41.1.5}$$

41.2

Nevertheless, Lehmer's work made it clear that the first sign change occurs "very far," and it would be of interest to find an explicit numerical upper bound for it. Curiously enough, Littlewood's method in its original form turned out to be incapable of furnishing such an upper bound. His proof was divided into two parts: the first one dealt with the case when Riemann's conjecture was assumed false and the second one when it was assumed true, and both cases were treated ineffectively. The first case was settled quickly using a theorem of Landau; the second was considered to be the deep part, and here the ineffectiveness was caused by the use of the Phragmén–Lindelöf theorem.

Still it was the second case that was made effective earlier than the first. After 20 years, in 1933 Skewes [1] found

$$10^{10^{10^{34}}} \tag{41.2.1}$$

as an upper bound for the first sign change of $\Delta_1(x)$, assuming the truth of Riemann's conjecture.

So the emphasis was laid on the first case; there is some interesting commentary in a paper written by Littlewood in 1937 [1]. Though the possibility of effectivization was shown by Kreisel, the first effective value of this constant

$$e_4(7.705)^* \tag{41.2.2}$$

was given only in 1955 by Skewes [2]. For an insight into the difficulties of the proof, we mention that it required the numerical values of all ζ-zeros in $|t| < 500$ $(s = \sigma + it)$.

*We recall for the reader the notation $e_{\nu+1}(x) = \exp[e_\nu(x)]$, $e_1(x) = \exp x = e^x$ and $\log_{\nu+1}(x) = \log\log_\nu(x)$, $\log_1(x) = \log x$.

In 1966 Sherman Lehman [1] found the much smaller upper bound

$$1.65 \cdot 10^{1165} \tag{41.2.3}$$

for the first sign change; he needed the numerical values of the ζ-roots for $|t| \leq 12{,}000$ to seven decimals (six-decimal accuracy was not enough!).

To complete the picture of this remarkable phenomenon, we remark that Rosser–Schoenfeld [1] verified the assertion (41.1.1) for $x < 10^8$, and Lehman commented at the end of his paper [1] that it is probably even valid for $x < 10^{20}$.

41.3

Littlewood's refutation of Riemann's assertion (41.1.1) naturally led to the general study of the oscillatory nature of remainder terms in the various forms of the prime number formula. This work started with a paper of Pólya [2].

Let

$$V_i(Y) \quad \text{for} \quad Y > 2 \qquad (1 \leq i \leq 4) \tag{41.3.1}$$

denote the number of sign changes of $\Delta_i(x)$ in the interval $[2, Y]$. [For the definition of $\Delta_i(x)$ see (40.1.1)–(40.1.4).] Then Pólya's result is

$$\overline{\lim_{Y \to \infty}} \frac{V_4(Y)}{\log Y} > 0. \tag{41.3.2}$$

The corresponding problem for $V_1(Y)$ is, of course, much deeper. Ingham [1] attacked the problem in 1936. He proved the strong theorem that $\Delta_1(x)$ changes sign in every interval of type

$$[Y, c_0 Y]. \tag{41.3.3}$$

However, his proof depended on the unproved assertion of the existence of a Θ such that

$$\zeta(s) \neq 0 \quad \text{for} \quad \sigma > \Theta, \tag{41.3.4}$$

but for a suitable real t_0

$$\zeta(\Theta + it_0) = 0. \tag{41.3.5}$$

This is satisfied, for example, if Riemann's conjecture is true. However the depth of the condition is obvious if we recall that because $\zeta(s)$ has no zeros on the line $\sigma = 1$, the Θ in (41.3.5) must be less than 1, and Ingham's condition implies the truth of the so called quasi-Riemann conjecture.

Ingham's result also contains the assertion that if the condition is satisfied,

$$\varlimsup_{Y\to\infty} \frac{V_1(Y)}{\log Y} > 0. \tag{41.3.6}$$

Apart from the conditional character another disadvantage of Ingham's beautiful theorem is that the constant c_0 in (41.3.3) is ineffective, and therefore also the least lower bound of the left-hand side of (41.3.6) is ineffective—even in the case when the Riemann conjecture is assumed true.

41.4

The first unconditional lower bound for $V_1(Y)$ was given by S. Knapowski in 1961 and 1962 [4, 12] using one-sided power-sum theorems. He gave stronger ineffective and weaker effective lower bounds. His unconditional inequality is

$$\varlimsup_{Y\to\infty} \frac{V_1(Y)}{\log_2 Y} > 0. \tag{41.4.1}$$

This is not effective, but he also found the effective inequality

$$V_1(Y) > e^{-35}\log_4 Y \tag{41.4.2}$$

valid for $Y > e_5(35)$. For (41.4.2) he needed, instead of the numerical values of the zeros "low" in the critical strip, merely that for $|t| < 10^4$ they are all on the line $\sigma = \frac{1}{2}$.

These results were improved by the use of two-sided power-sum theorems (namely, the second main theorem) in 1974 and 1976 (see Knapowski–Turán [16, 17]).

The first work (Knapowski–Turán [16]) contained the proof of the following unconditional and ineffective result.

Theorem. *For $Y > Y_0$ (an ineffective constant) the interval*

$$\left[Y, Y\exp\left\{\log^{3/4} Y(\log_2 Y)^4\right\}\right] \tag{41.4.3}$$

contains a sign change of $\Delta_1(x)$, and thus

$$\varlimsup_{Y\to\infty} \frac{V_1(Y)}{(\log Y)^{1/4}(\log_2 Y)^{-4}} > 0. \tag{41.4.4}$$

In the second work (Knapowski–Turán [17]) the unconditional effective improvement of (41.4.2) was proved.

Theorem. *There are effectively computable constants c_1 and c_2 such that for $Y > c_1$ the inequality*

$$V_1(Y) > c_2 \log_3 Y \tag{41.4.5}$$

holds.

41.5

In this section we prove the following.

Theorem 41.1 (Pintz [2]). *For $Y > Y_i$ $(1 \le i \le 4)$ (ineffective constants) the interval*

$$\left[Y, Y \exp\left\{63\sqrt{\log Y} \log_2 Y\right\}\right] \tag{41.5.1}$$

contains a sign change of $\Delta_i(x)$ $(1 \le i \le 4)$.

This theorem already implies the ineffective inequality (see 41.10)

$$\varliminf_{Y \to \infty} \frac{V_i(Y)}{\sqrt{\log Y} (\log_2 Y)^{-1}} > 0. \tag{41.5.2}$$

We note that the constants Y_2 and Y_4 corresponding to the cases $i = 2$ and $i = 4$ could be made effective with slight modification of the proof, and thus we get (41.5.2) also in an effective form in these cases. Thus Theorem 41.2 (to be formulated shortly) becomes necessary only in the more difficult cases $i = 1$ and 3. For the slight changes required in the proof of Theorem 41.1 we refer to Section 42.11, where we indicate the necessary changes for the proof of the similar Theorem 42.1.

However, using a more explicit form of Theorem 41.1, it will be possible to deduce from it the inequality (41.5.2) in an effective form as follows:

Theorem 41.2 (Pintz [2]). *There exist effectively computable constants c_3 and c_4 such that for $Y > c_3$ the inequality*

$$V_i(Y) > c_4 \frac{\sqrt{\log Y}}{\log_2 Y} \qquad (1 \le i \le 4) \tag{41.5.3}$$

holds.

The reason we stated the seemingly weaker (because for $i = 1$ and $i = 3$ ineffective) Theorem 41.1 separately is that it contains a localization for the sign changes of $\Delta_i(x)$, whereas Theorem 41.2 gives only a lower bound for the total number of sign changes without any localization. (More precisely, there is a very weak but effective localization contained in the proof, namely, the interval $[\exp(c\sqrt{\log Y}(\log_2 Y)^{-1}), Y]$.)

The following section is devoted to the proof of a partially ineffective improvement of (41.5.2): namely, that

$$V_i(Y) > 10^{-11}\frac{\log Y}{(\log_2 Y)^3} \quad \text{for} \quad Y > Y_0. \tag{41.5.4}$$

We give the proof only for the deepest and most interesting case $i = 1$. Our proof implicitly contains the case $i = 2$. The cases $i = 3$ and 4 could be proved similarly and are even easier.

41.6

Theorem 41.1 will be an immediate consequence of the following lemma.

Lemma 41.1. *If for a* $Z > c_5$ *(an effective constant) the function* $\zeta(s)$ *has a zero* $\rho^* = \beta^* + i\gamma^*$ *with*

$$\beta^* \geq \frac{1}{2} + \frac{10\log\gamma^*}{4\sqrt{\log Z}(\log_2 Z)^{-1}} \tag{41.6.1}$$

$$0 < \gamma^* \leq \exp\left[\tfrac{1}{10}\sqrt{\log Z}(\log_2 Z)^{-1}\right]$$

then the interval

$$I(Z) = \left[Z\exp\left(-31\sqrt{\log Z}\log_2 Z\right), Z\exp\left(31\sqrt{\log Z}\log_2 Z\right)\right] \tag{41.6.2}$$

contains a sign change of $\Delta_1(x)$.

It is easy to see that if the Riemann conjecture is not true, then any zero $\rho^* = \beta^* + i\gamma^*$ with $\beta^* > \frac{1}{2}$ satisfies (41.6.1) if $Z > Z_0(\rho^*)$, and thus in this case Theorem 41.1 follows from Lemma 41.1.

On the other hand, if the Riemann conjecture is true, then Ingham's theorem (see 41.3.3) shows the validity of Theorem 41.1 in an even stronger form. (For the sake of completeness we note that Ingham's theorem is also true for the cases $i = 2, 3, 4$. The proof is essentially the same.)

In the proof of our lemma we use the following abbreviations:

$$L \doteq \log Z \tag{41.6.3}$$

$$M \doteq 100(\log_2 Z)^2 = 100\log^2 L \tag{41.6.4}$$

$$\lambda \doteq \frac{\sqrt{\log Z}}{10\log_2 Z} = \sqrt{\frac{L}{M}}. \tag{41.6.5}$$

Let k be a real number to be determined later, which, for the time being, will be restricted only by

$$M \le k \le M\left(1 + \frac{1}{L}\right). \tag{41.6.6}$$

Let

$$\mu \doteq k\lambda^2 \tag{41.6.7}$$

$$A \doteq \exp(\mu - 3k\lambda) \tag{41.6.8}$$

$$B \doteq \exp(\mu + 3k\lambda) \tag{41.6.9}$$

$$f(x) \doteq \Pi(x) - \lg x \pm \sqrt{x} \doteq \Pi(x) - \sum_{2 \le n \le x} \frac{1}{\log n} \pm \sqrt{x} \tag{41.6.10}$$

$$H(s) \doteq \frac{\zeta'}{\zeta}(s) + \zeta(s) - 1 \pm \frac{1}{2(s - \frac{1}{2})^2}. \tag{41.6.11}$$

Here we must either take the upper signs in both (41.6.10) and (41.6.11), or the lower signs in both.

Assume that $f(x)$ has no sign change in the interval

$$I' \doteq [A, B] \subset I(Z). \tag{41.6.12}$$

We show that this assumption leads to contradiction, and thus proves the existence of

$$x', x'' \in I' \subset I(Z) \tag{41.6.13}$$

for which the inequalities

$$\Pi(x') - \lg x' > \sqrt{x'} \tag{41.6.14}$$

and

$$\Pi(x'') - \lg x'' < -\sqrt{x''} \tag{41.6.15}$$

hold. The inequalities

$$\pi(x') - \operatorname{li} x' > \tfrac{1}{2}\sqrt{x'} \tag{41.6.16}$$

and

$$\pi(x'') - \operatorname{li} x'' < -\tfrac{1}{2}\sqrt{x''} \tag{41.6.17}$$

then hold, in view of the trivial estimates

$$\Pi(x) - \pi(x) = O\left(\frac{\sqrt{x}}{\log x}\right) \tag{41.6.18}$$

and

$$\lg x = \operatorname{li} x + O(1). \tag{41.6.19}$$

41.7

We distinguish the following two cases.

CASE A. There exists a zero $\rho_0 = \beta_0 + i\gamma_0$ such that

$$\beta_0 \geq \frac{1}{2} + \frac{\log \gamma_0}{4\lambda}, \qquad 0 < \gamma_0 \leq \lambda^5. \tag{41.7.1}$$

Then let $\rho_1' = \beta_1' + i\gamma_1'$ be the zero with the maximal real part β_1' among those satisfying (41.7.1). Define recursively ρ_{n+1}' to be the zero with the maximal real part β_{n+1}' satisfying

$$\gamma_n' \leq \gamma_{n+1}' \leq \gamma_n' + 2\lambda; \qquad \beta_{n+1}' \geq \beta_n' + \frac{1}{\mu} \tag{41.7.2}$$

if such a zero exists. Thus after, at most, $[\mu/2]$ steps we can find a zero $\rho_N' = \beta_N' + i\gamma_N' \doteq \rho_1 = \beta_1 + i\gamma_1$

$$\beta_1 \geq \frac{1}{2} + \frac{1}{\lambda}, \qquad 0 < \gamma_1 \leq 2\lambda^5 \tag{41.7.3}$$

for which the domains

$$|t| \leq \lambda^5, \qquad \sigma > \beta_1 \tag{41.7.4}$$

and

$$|t - \gamma_1| \leq 2\lambda, \qquad \sigma > \beta_1 + \frac{1}{\mu} \tag{41.7.5}$$

are zero free.

CASE B. There is no zero satisfying (41.7.1). Then let $\rho_1 = \beta_1 + i\gamma_1$ be any zero satisfying (41.6.1). With our new notations, we have in this case

$$\beta_1 \geq \frac{1}{2} + \frac{\log \gamma_1}{4\lambda}, \qquad \lambda^5 < \gamma_1 \leq e^{\lambda}. \tag{41.7.6}$$

41.8

In both cases our starting formula is

$$\int_1^{\infty} f(x)\frac{d}{dx}(x^{-s}\log x)\,dx = H(s), \tag{41.8.1}$$

which can easily be proved by partial summation for $\sigma > 1$.

We will use the formula ($A > 0$, B arbitrary complex)

(41.8.2)

$$\frac{1}{2\pi i}\int_{(2)} e^{As^2+Bs}\,ds = \exp\left(-\frac{B^2}{4A}\right)\cdot\frac{1}{2\pi i}\int_{(2)}\exp\left(\sqrt{A}\,s+\frac{B}{2\sqrt{A}}\right)^2 ds$$

$$= \exp\left(-\frac{B^2}{4A}\right)\frac{1}{\sqrt{A}}\cdot\frac{1}{2\pi i}\int_{(0)} e^{z^2}\,dz$$

$$= \frac{1}{2\sqrt{\pi A}}\exp\left(-\frac{B^2}{4A}\right).$$

Replace s by $s + i\gamma_1$ in (41.8.1), multiply by $e^{ks^2+\mu s}$, and integrate with respect to s along the line $\sigma = 2$ (the change of order of integrations is obviously justified). Then

(41.8.3)

$$U = \frac{1}{2\pi i}\int_{(2)} H(s+i\gamma_1)e^{ks^2+\mu s}\,ds$$

$$= \frac{1}{2\pi i}\int_{(2)}\int_1^{\infty} f(x)\frac{d}{dx}\left(x^{-s-i\gamma_1}\log x\cdot e^{ks^2+\mu s}\right)dx\,ds$$

$$= \int_1^{\infty} f(x)\frac{d}{dx}\left\{x^{-i\gamma_1}\log x\cdot\frac{1}{2\pi i}\int_{(2)}\exp[ks^2+(\mu-\log x)s]\,ds\right\}dx$$

$$= \frac{1}{2\sqrt{\pi k}}\int_1^{\infty} f(x)\frac{d}{dx}\left\{x^{-i\gamma_1}\log x\exp\left[-\frac{(\log x-\mu)^2}{4k}\right]\right\}dx$$

$$= \frac{1}{2\sqrt{\pi k}}\int_1^{\infty}\frac{f(x)}{x}x^{-i\gamma_1}\exp\left[-\frac{(\log x-\mu)^2}{4k}\right]$$

$$\cdot\left\{-i\gamma_1\log x+1+\log x\frac{\mu-\log x}{2k}\right\}dx.$$

This will be our starting point.

41.9

The basic idea of the proof is the following. If $f(x)$ does not change its sign in I', one can deduce an upper bound for the absolute value of the right-hand side of (41.8.3); on the other hand, one can give a lower bound for the absolute value of the left-hand side of (41.8.3) by an appropriate choice of k satisfying (41.6.6), and these two estimations will contradict each other.

In the estimate of the right-hand side of (41.8.3) owing to the presence of the kernel function $\exp\{-(\log x - \mu)^2/4k\}$ only the integral over the interval I' is essential. Therefore we can use the supposition that $f(x)$ has no sign change in I'.

41.10

To perform the first part of this program, we split the integral U on the right-hand side of (41.8.3) into

$$U = U_1 + U_2 + U_3 \tag{41.10.1}$$

where

$$U_1 = \int_1^A, \qquad U_2 = \int_A^B, \qquad U_3 = \int_B^\infty. \tag{41.10.2}$$

Considering our notations (41.6.7)–(41.6.12) and (41.8.3), we have

$$\begin{aligned}
|U_2| &\le \frac{1}{2\sqrt{\pi k}} \int_A^B \frac{|f(x)|\log x}{x} \exp\left[-\frac{(\log x - \mu)^2}{4k}\right] \\
&\quad \cdot \left(\gamma_1 + \frac{1}{\log x} + \frac{|\mu - \log x|}{2k}\right) dx \\
&\le \frac{1}{2\sqrt{\pi k}} \int_A^B \frac{|f(x)|\mu(1 + 3/\lambda)}{x} \exp\left[-\frac{(\log x - \mu)^2}{4k}\right] \\
&\quad \cdot \left(\gamma_1 + 1 + \frac{3k\lambda}{2k}\right) dx \\
&\le \frac{2\mu(\gamma_1 + \lambda)}{2\sqrt{\pi k}} \int_A^B \frac{|f(x)|}{x} \exp\left[-\frac{(\log x - \mu)^2}{4k}\right] dx \\
&= \frac{2\mu(\gamma_1 + \lambda)}{2\sqrt{\pi k}} \left| \int_A^B \frac{f(x)}{x} \exp\left[-\frac{(\log x - \mu)^2}{4k}\right] dx \right|,
\end{aligned} \tag{41.10.3}$$

since $f(x)$ does not change its sign in $[A, B]$.

On the other hand, we can estimate in a trivial way

$$(41.10.4)\qquad |U_4| \doteq \left| \int_B^\infty \frac{f(x)}{x} \exp\left[-\frac{(\log x - \mu)^2}{4k} \right] dx \right|$$

$$\leq \int_B^\infty \exp\left[-\frac{(\log x - \mu)^2}{4k} \right] dx$$

$$= \int_{3k\lambda}^\infty \exp\left(\mu + y - \frac{y^2}{4k} \right) dy$$

$$\leq \int_{3k\lambda}^\infty \exp(\mu + y - 2y - 2k\lambda^2)\, dy \leq \int_0^\infty e^{-\mu - y} = e^{-\mu}$$

and

$$(41.10.5)\qquad |U_5| \doteq \left| \int_1^A \frac{f(x)}{x} \exp\left[-\frac{(\log x - \mu)^2}{4k} \right] dx \right|$$

$$\leq \int_1^{e^{\mu - 3k\lambda}} \exp\left(-\frac{9k^2\lambda^2}{4k} \right) = e^{\mu - 3k\lambda - (9/4)\mu} \leq e^{-\mu}.$$

Similarly,

$$(41.10.6)\qquad |U_1| \leq e^{-\mu} \quad \text{and} \quad |U_3| \leq e^{-\mu}.$$

Thus using (41.10.1)–(41.10.6) we can change the intervals on the left- and right-hand side of (41.10.3) from $[A, B]$ to $[1, \infty]$ and so with

$$(41.10.7)\qquad K \doteq \frac{1}{2\sqrt{\pi k}} \int_1^\infty \frac{f(x)}{x} \exp\left[-\frac{(\log x - \mu)^2}{4k} \right] dx$$

we get

$$(41.10.8)\qquad |U| = |U_2| + o(1) = 2\mu(\gamma_1 + \lambda)|K| + o(1).$$

41.11

Now we estimate $|K|$ from above. For $\sigma > 1$ we get, with a constant h, the formula

(41.11.1)

$$\int_1^\infty \frac{f(x)}{x^{s+1}} dx = \frac{1}{s} \left\{ \int_2^s \left[\frac{\zeta'}{\zeta}(z) + \zeta(z) - 1 \right] dz + h \right\} \pm \frac{1}{s - \frac{1}{2}} \doteq \varphi(s) \pm \frac{1}{s - \frac{1}{2}}.$$

Multiplying on both sides by $e^{ks^2+\mu s}$ and integrating along $\sigma = 2$ we get

$$(41.11.2) \qquad K = \pm\frac{1}{2\pi i}\int_{(2)}\frac{e^{ks^2+\mu s}}{s-\frac{1}{2}}\,ds + \frac{1}{2\pi i}\int_{(2)}\varphi(s)e^{ks^2+\mu s}\,ds$$

$$\doteq \pm K_1 + K_2.$$

Shift the line of integration to $\sigma = 0$ so that

$$(41.11.3) \quad K_1 = e^{k/4+\mu/2} + \frac{1}{2\pi}\int_{-\infty}^{+\infty}\frac{e^{-kt^2+\mu it}}{it-\frac{1}{2}}\,dt = e^{k/4+\mu/2} + O(1).$$

41.12

In Case A we transform the path of integration in K_2 to the broken line l, defined for $t \geq 0$ by

$$(41.12.1) \qquad l_1\colon \sigma = \tfrac{5}{4} \qquad \text{for} \quad t \geq \lambda$$

$$l_2\colon \beta_1 + \frac{1}{\mu} \leq \sigma \leq \frac{5}{4} \qquad \text{for} \quad t = \lambda$$

$$l_3\colon \sigma = \beta_1 + \frac{1}{\mu} \qquad \text{for} \quad 10 \leq t \leq \lambda$$

$$l_4\colon \frac{1}{4} \leq \sigma \leq \beta_1 + \frac{1}{\mu} \qquad \text{for} \quad t = 10$$

$$l_5\colon \sigma = \tfrac{1}{4} \qquad \text{for} \quad 0 \leq t \leq 10,$$

and for $t \leq 0$ by reflection with respect to the real axis. By the choice of ρ_1 [see (41.7.3)–(41.7.5)] $\varphi(s)$ is regular to the right of l and on l. Thus

$$(41.12.2) \qquad K_2 = \frac{1}{2\pi i}\int_{(l)}\varphi(s)e^{ks^2+\mu s}\,ds.$$

We also use the fact that if $\zeta(s)$ has no zero in the domain

$$(41.12.3) \qquad \sigma \geq \beta \quad (>\tfrac{1}{2}), \qquad |t| \leq T+1$$

then for

$$(41.12.4) \qquad \sigma \geq \beta + \eta, \qquad 2 < |t| \leq T$$

$$(41.12.5) \qquad \left|\frac{\zeta'}{\zeta}(z)\right| = O\left(\frac{\log|t|}{\eta}\right).$$

[This follows easily from (G.1.2) of the Appendix in case of $q = 1$.] We use the classical estimate

$$\zeta(z) = O\left(\sqrt{|t|}\right) \quad \text{for} \quad \sigma \geq \tfrac{1}{2},\ |t| \geq 10. \tag{41.12.6}$$

From (41.12.5) and (41.12.6) an easy computation for the integral J_i on the interval I_i $(1 \leq i \leq 5)$ yields the estimates

$$\begin{aligned}
&|J_1| = O\left[\exp\left(k \cdot \tfrac{25}{16} - k\lambda^2 + \tfrac{5}{4}\mu\right)\right] \leq e^{\mu/3} \\
&|J_2| = O\left[\log\lambda \cdot \mu \exp\left(k \cdot \tfrac{25}{16} - k\lambda^2 + \tfrac{5}{4}\mu\right)\right] \leq e^{\mu/3} \\
&|J_3| = O\left\{\log\lambda \cdot \mu \exp\left[-98k + \mu\left(\beta_1 + \frac{1}{\mu}\right)\right]\right\} \leq e^{-97k+\mu\beta_1} \\
&|J_4| = O\left\{\exp\left[-98k + \mu\left(\beta_1 + \frac{1}{\mu}\right)\right]\right\} \leq e^{-97k+\mu\beta_1} \\
&|J_5| = O\left(\exp\left(\frac{k}{16} + \frac{\mu}{4}\right)\right) \leq e^{\mu/3}.
\end{aligned} \tag{41.12.7}$$

Thus, considering (41.11.2)–(41.11.3) and (41.12.7), we have the upper bound

$$|K| = O\left(e^{-97M+\mu\beta_1} + e^{k/4+\mu/2}\right). \tag{41.12.8}$$

By (41.7.3), (41.10.8), and (41.12.8) we get

$$|U| \leq e^{-96M+\mu\beta_1} + \exp\left(\frac{k}{4} + \mu\beta_1 - \frac{\mu}{\lambda} + \lambda\right) \leq \frac{e^{k\beta_1^2+\mu\beta_1}}{e^{96M}}. \tag{41.12.9}$$

41.13

To perform the second part of the program, sketched in 41.9, we estimate the absolute value of the integral U on the left of (41.8.3) by appropriate choice of k. Shifting the line of integration to $\sigma = -\frac{1}{2}$ we get

(41.13.1)

$$U = \sum_{\rho} \exp\left\{k\left[(\rho - \gamma_1)^2 + \lambda^2(\rho - i\gamma_1)\right]\right\}$$

$$\pm \frac{1}{2}\frac{d}{ds}\left(e^{ks^2+\mu s}\right)_{s=1/2-i\gamma_1} + \frac{1}{2\pi i}\int_{(-1/2)} H(s + i\gamma_1)e^{ks^2+\mu s}\,ds.$$

It is easy to see the last integral is $O(1)$ and the second residue is, in absolute value, at most

$$\tfrac{1}{2}\left(2k\left|\tfrac{1}{2} - i\gamma_1\right| + \mu\right)\exp\left(\frac{k}{4} + \frac{\mu}{2} - k\gamma_1^2\right) \leq \frac{e^{k\beta_1^2+\mu\beta_1}}{e^{99M}}. \tag{41.13.2}$$

We need the fact that the number of zeros of $\zeta(s)$ in $T \leq t \leq T+1$ is

$$(41.13.3) \qquad < c \log T \quad \text{where} \quad c = 15 \quad \text{for} \quad T > T_0$$

(see W. J. Ellison–M. Mendès France [1] p. 165). The number of zeros with

$$(41.13.4) \qquad 10 \leq |\gamma - \gamma_1| \leq 2\lambda$$

is, by (41.13.3) and (41.7.3),

$$(41.13.5) \qquad \leq 2 \cdot 2\lambda \cdot c \cdot 6 \log \lambda \leq \mu .$$

Thus for the contribution of such zeros to the infinite power sum, (41.7.5) gives the upper bound

$$(41.13.6) \quad \mu \exp\left\{ k\left(\beta_1 + \frac{1}{\mu}\right)^2 - 100k + \mu\left(\beta_1 + \frac{1}{\mu}\right)\right\} \leq \frac{e^{k\beta_1^2 + \mu\beta_1}}{e^{99M}} .$$

Again by (41.13.3), for the contribution of zeros with $|\gamma - \gamma_1| > 2\lambda$, we have the upper estimate

$$(41.13.7) \qquad 2 \sum_{n=[2\lambda]}^{\infty} c \log(\gamma_1 + n) \exp(k - kn^2 + \mu) = O(1).$$

So the number of remaining zeros with

$$(41.13.8) \qquad |\gamma - \gamma_1| < 10$$

is, by (41.13.3) and (41.6.5),

$$(41.13.9) \quad n < 20 \cdot c \log(\gamma_1 + 10) < 300 \log(2\lambda^5 + 10) < 900 \log L .$$

Now apply the continuous form of the second main theorem for the numbers

$$(41.13.10) \qquad \alpha_j = (\rho_j - i\gamma_1)^2 + \lambda^2(\rho_j - i\gamma_1)$$

with

$$(41.13.11) \qquad |\gamma_j - \gamma_1| < 10$$

with the choice

$$(41.13.12) \qquad a = M, \qquad d = \frac{M}{L} .$$

This yields a t_0 with $a \leq t_0 \leq a + d$ such that for arbitrary complex numbers

b_j, α_j ($\operatorname{Re}\alpha_1 \geq \operatorname{Re}\alpha_2 \geq \cdots \geq \operatorname{Re}\alpha_n$) the inequality

$$\left|\sum_{j=1}^{n} b_j e^{\alpha_j t_0}\right| \geq 2\left[\frac{1}{8e(a/d+1)}\right]^n |e^{\alpha_1 t_0}| \min_{1\leq j\leq n} |b_1 + \cdots + b_j| \tag{41.13.13}$$

holds. This can be derived from the second main theorem (Theorem 8.1) analogously to (6.3.1)–(6.3.7), which is the continuous form of the first main theorem.

So there exists a k satisfying (41.6.6) for which

(41.13.14)

$$W_1 = \left|\sum_{|\gamma_j-\gamma_1|<10} \exp\left\{k\left[(\rho_j - i\gamma_1)^2 + \lambda^2(\rho_j - i\gamma_1)\right]\right\}\right|$$

$$\geq \frac{\exp(k\beta_1^2 + k\lambda^2\beta_1)}{(30L)^{900\log L}} \geq \frac{\exp(k\beta_1^2 + \mu\beta_1)}{e^{1000\log^2 L}} = \frac{\exp(k\beta_1^2 + \mu\beta_1)}{e^{10M}}.$$

Thus (41.13.2), (41.13.6), (41.13.7), and (41.13.14) together imply

$$|U| \geq \frac{\exp(k\beta_1^2 + \mu\beta_1)}{2e^{10M}} \tag{41.13.15}$$

which contradicts (41.12.9) and proves the lemma in Case A.

41.14

In Case B transform the path of integration in K_2 to the broken line l, defined for $t \geq 0$ by

$$l_1:\ \sigma = \tfrac{5}{4} \qquad \text{for } t \geq \lambda \tag{41.14.1}$$

$$l_2:\ \alpha + \frac{1}{\mu} \leq \sigma \leq \frac{5}{4} \qquad \text{for } t = \lambda$$

$$l_3:\ \sigma = \alpha + \frac{1}{\mu} \qquad \text{for } 0 \leq t \leq \lambda$$

$$l_4:\ \frac{1}{4} \leq \sigma \leq \alpha + \frac{1}{\mu} \qquad \text{for } t = 10$$

$$l_5:\ \sigma = \tfrac{1}{4} \qquad \text{for } 0 \leq t \leq 10$$

where

$$\alpha = \frac{1}{2} + \frac{\log\lambda}{2\lambda}, \tag{41.14.2}$$

and for $t \le 0$ by reflection with respect to the real axis. Since there is no zero with (41.7.1), $\varphi(s)$ is regular to the right of l and on l.

Observe that the only change in (41.14.1) compared to (41.12.1), is that β_1 is replaced by α. As in (41.12.7) we get the estimate

$$(41.14.3) \qquad |K_2| = O\left(e^{-97M+\mu\alpha} + e^{\mu/3}\right) \le e^{k\beta_1^2+\mu\alpha}.$$

Taking into account (41.7.6) and (41.14.2),

$$(41.14.4) \quad \beta_1 - \alpha \ge \frac{\log\gamma_1}{4\lambda} - \frac{\log\lambda}{2\lambda} \ge \frac{\log\gamma_1}{4\lambda} - \frac{\frac{1}{5}\log\gamma_1}{2\lambda} > \frac{\log\gamma_1}{8\lambda}.$$

Estimating $|U|$, we get from (41.10.8), (41.14.3), and (41.14.4) the inequality

$$(41.14.5) \quad |U| \le e^{2\lambda}\cdot\frac{e^{k\beta_1^2+\mu\beta_1}}{\exp[\mu(\log\gamma_1/8\lambda)]} \le \frac{e^{k\beta_1^2+\mu\beta_1}}{\exp[(1/9)M\lambda\log\gamma_1]}.$$

To get a lower bound for $|U|$, observe that in (41.13.1) the integral is again $O(1)$; and by (41.7.6) the residue is at most

(41.14.6)

$$\tfrac{1}{2}\left(2k\left|\tfrac{1}{2} - i\gamma_1\right| + \mu\right)\exp\left(\frac{k}{4} + \frac{\mu}{2} - k\gamma_1^2\right) \le e^{2\lambda}\cdot\exp\left(\frac{k}{4} + \frac{\mu}{2} - k\lambda^{10}\right)$$
$$\le e^{\mu-\lambda^8\mu} < 1.$$

Furthermore, the infinite power sum corresponding to the zeros with $|\gamma - \gamma_1| \ge 2\lambda$ is $O(1)$, as given by (41.13.7).

Again, only the behavior of the finite power sum belonging to the zeros with

$$(41.14.7) \qquad |\gamma - \gamma_1| < 2\lambda$$

is interesting. Here the number of terms is,

$$(41.14.8) \qquad 1 \le n < 4\lambda\cdot 15\cdot\log(\gamma_1 + 2\lambda) < 61\lambda\log\gamma_1,$$

by (41.13.3), and proceeding as in (41.13.10)–(41.13.14),

$$(41.14.9) \quad W_2 = \left|\sum_{|\gamma_j-\gamma_1|<2\lambda}\exp\left\{k\left[(\rho_j - i\gamma_1)^2 + \lambda^2(\rho_j - i\gamma_1)\right]\right\}\right|$$
$$\ge \frac{\exp\left(k\beta_1^2 + k\lambda^2\beta_1\right)}{(30L)^{61\lambda\log\gamma_1}} \ge \frac{e^{k\beta_1^2+\mu\beta_1}}{\exp(62\lambda\log\gamma_1\cdot\log L)}$$
$$\ge \frac{e^{k\beta_1^2+\mu\beta_1}}{\exp\left(\lambda\log\gamma_1\log^2 L\right)} = \frac{e^{k\beta_1^2+\mu\beta_1}}{\exp[(1/100)\lambda M\log\gamma_1]}.$$

This implies essentially the same estimate for $|U|$, namely, in view of the upper estimate for the integral, the residue, and the contribution of the zeros with $|\gamma_j - \gamma_1| > 2\lambda$ we have

$$|U| \geq \frac{e^{k\beta_1^2 + \mu\beta_1}}{2\exp[(1/100)\lambda M \log \gamma_1]} \tag{41.14.10}$$

contradicting (41.14.5). This completes the proof of Case B.

41.15. PROOF OF THEOREM 41.2

Let

$$\lambda_0 \doteq \frac{1}{10}\frac{\sqrt{\log\sqrt{Y}}}{\log_2\sqrt{Y}}\left(> \frac{1}{20}\frac{\log Y}{\log_2 Y}\right). \tag{41.15.1}$$

In the course of the proof we distinguish the following two cases.

CASE I. There is a zero $\rho^* = \beta^* + i\gamma^*$ with

$$\beta^* \geq \frac{1}{2} + \frac{\log \gamma^*}{4\lambda_0}, \qquad 0 < \gamma^* \leq e^{\lambda_0}. \tag{41.15.2}$$

This case is already settled by Lemma 41.1. Namely, the zero ρ^*, for which (41.15.2) holds, satisfies the condition of Lemma 41.1 for any $Z \geq \sqrt{Y}$ (if $Y > c_6$). Thus for any Z with

$$\sqrt{Y} \leq Z \leq Y \tag{41.15.3}$$

there is at least one sign change of $\Delta_1(x)$ in the interval

(41.15.4)

$$I^*(Z) = \left[\exp\left(\log Z - 31\sqrt{\log Y}\cdot \log_2 Y\right), \exp\left(\log Z + 31\sqrt{\log Y}\cdot \log_2 Y\right)\right]$$

because, by (41.15.3), $I^*(Z)$ contains the interval $I(Z)$ given by (41.6.2).

Apply this for

$$Z_\nu = \sqrt{Y}\cdot \exp\left(62\nu\sqrt{\log Y}\cdot \log_2 Y\right), \qquad 1 \leq \nu \leq \left[\frac{\sqrt{\log Y}}{124\log_2 Y}\right] - 1 \tag{41.15.5}$$

to obtain at least

$$\left[\frac{\sqrt{\log Y}}{124\log_2 Y}\right] - 1 > \frac{1}{125}\frac{\sqrt{\log Y}}{\log_2 Y} \tag{41.15.6}$$

disjoint intervals contained in

(41.15.7) $$[\sqrt{Y}, Y]$$

such that every interval contains at least one sign change of $\Delta_1(x)$. Thus the total number of sign changes in [2, Y] is

(41.15.8) $$V_1(Y) > \frac{1}{125} \frac{\sqrt{\log Y}}{\log_2 Y},$$

which proves Theorem 41.2 in Case I.

41.16

The second case can be formulated in the following way.

Case II. All zeros $\rho = \frac{1}{2} + \delta + i\gamma$ of $\zeta(s)$ with

(41.16.1) $$|\gamma| \le e^{\lambda_0}$$

satisfy

(41.16.2) $$|\delta| < \frac{\log|\gamma|}{4\lambda_0}.$$

In this case we prove that there are at least

(41.16.3) $$c_7\lambda_0$$

sign changes of $\Delta_1(x)$ in $[2, e^{\lambda_0/2}]$, where c_7 is a positive effectively computable constant. Thus

(41.16.4) $$V_1(Y) \ge V_1(e^{\lambda_0/2}) \ge c_7\lambda_0 > \frac{c_7}{20} \frac{\sqrt{\log Y}}{\log_2 Y},$$

which will prove Theorem 41.2 in Case II.

The treatment of this case uses ideas of Littlewood, Ingham, and Skewes.

41.17

We show that, under the conditions (41.16.1)–(41.16.2), the investigation of $\Delta_1(x)$ can be reduced to the investigation of the more easily manageable $\Delta_4(x)$ [for notation see (40.1.1)–(40.1.4)].

We introduce the abbreviations

$$\Delta(x) = \int_1^x \Delta_4(\vartheta)\, d\vartheta \tag{41.17.1}$$

$$\Delta_1^*(r) = \frac{\Delta_1(r)}{\sqrt{r}/\log r} \tag{41.17.2}$$

$$\Delta_4^*(r) = \frac{\Delta_4(r)}{\sqrt{r}} \tag{41.17.3}$$

and use two well-known lemmas. (All the constants as well as those implied by the O and o symbols will be absolute, effective constants.)

Lemma 41.2.

$$\Delta(u) = -\sum_{|\gamma| \le u^2} \frac{u^{\rho+1}}{\rho(\rho+1)} + O(u). \tag{41.17.4}$$

The proof follows easily from Theorem 28 (p. 73) of Ingham [2].

Lemma 41.3. *For $r \to \infty$,*

(41.17.5) $\Delta_1^*(r) - \Delta_4^*(r) + 1 + o(1)$

$$= \frac{\Delta(r)}{r^{3/2}\log r} + \frac{\log r}{\sqrt{r}} \int_2^r \Delta(u) \frac{\log u + 2}{u^2 \log^3 u}\, du.$$

For the proof see Ingham [2], formula (33) in Theorem 35 (p. 104).

41.18

So far (41.16.1)–(41.16.2) are not used here. These together with Lemma 41.2, yield the following.

Lemma 41.4. *Under the conditions* (41.16.1)–(41.16.2) *for*

$$u \le e^{\lambda_0/2} \tag{41.18.1}$$

the inequality

$$|\Delta(u)| \le c_8 u^{3/2} \tag{41.18.2}$$

holds.

Proof. By (41.17.4), (41.18.1), and (41.16.1)–(41.16.2),

(41.18.3)

$$|\Delta(u) + O(u)| \le \sum_{|\gamma| \le u^2} \frac{u^{3/2+\delta}}{\gamma^2} \le \sum_{|\gamma| \le u^2} \frac{u^{3/2} \cdot e^{(\lambda_0/2)\delta}}{\gamma^2}$$

$$\le u^{3/2} \sum_{|\gamma| \le u^2} \frac{\exp[(\lambda_0/2) \cdot (\log|\gamma|/4\lambda_0)]}{\gamma^2} \le u^{3/2} \sum_{\gamma} \frac{|\gamma|^{1/8}}{\gamma^2} = O(u^{3/2})$$

which proves the lemma. ■

Lemma 41.5. *Under the condition* (41.16.1)–(41.16.2), *for*

(41.18.4) $$r \le e^{\lambda_0/2}$$

we have the relation

(41.18.5) $$\Delta_1^*(r) = \Delta_4^*(r) - 1 + o(1).$$

[The $o(1)$ symbol means that the corresponding quantity is absolutely less than ε if $r > r_0(\varepsilon)$ and r satisfies (41.18.4).]

By Lemma 41.3 it is enough to prove that the right-hand side of (41.17.5) is $o(1)$. This is trivially true for the first term by (41.18.2). Again, using (41.18.2), we obtain for the integral on the right-hand side of (41.17.5) the upper bound

(41.18.6) $$\int_2^r c_8 u^{3/2} \frac{\log u + 2}{u^2 \log^3 u} du \le c_9 \frac{\sqrt{r}}{\log^2 r} = o\left(\frac{\sqrt{r}}{\log r}\right),$$

and the lemma is proved.

41.19

By Lemma 41.5 $\Delta_1(r)$ certainly has a sign change in an interval

(41.19.1) $$J \subset \left[c_{10}, e^{\lambda_0/2}\right]$$

if we can show that

(41.19.2) $$\max_{r \in J} \Delta_4^*(r) > \tfrac{3}{2}$$

and

(41.19.3) $$\min_{r \in J} \Delta_4^*(r) \le -\tfrac{3}{2}.$$

The advantage of this approach lies in the fact that by the finite form of the Riemann–von Mangoldt exact prime number formula we have

$$(41.19.4)\qquad \Delta_4^*(r) = -\sum_{|\gamma| \le e^{\lambda_0}} \frac{r^{\delta+i\gamma}}{\rho} + o(1), \quad \text{for} \quad r \le e^{\lambda_0/2}$$

[see, e.g., Ingham [2] Theorem 29 (p. 77)]. Thus $\Delta_4^*(r)$ can be handled more easily than $\Delta_1^*(r)$. If

$$(41.19.5)\qquad G(v) \doteq \sum_{|\gamma| \le e^{\lambda_0}} \frac{e^{(\delta+i\gamma)v}}{\rho},$$

$\Delta_1(r)$ certainly has a sign change in an interval

$$(41.19.6)\qquad \left[e^{a_1}, e^{a_2}\right] \subset \left[c_{11}, e^{\lambda_0/2}\right]$$

if we can show that

$$(41.19.7)\qquad \max_{a_1 \le v \le a_2} G(v) > 2$$

and

$$(41.19.8)\qquad \min_{a_1 \le v \le a_2} G(v) < -2.$$

[Since the zeros of $\zeta(s)$ are symmetric to the real axis, $G(v)$ is real.]

41.20

The required inequalities (41.19.7)–(41.19.8) will be obtained—following an idea of Bohr and Littlewood—by appropriately using Dirichlet's theorem in the modified form given by Lemma 34.3. Without any modifications in this lemma the number of terms in the sum (41.19.5) would be too large. To avoid this problem we use a "term-shortening" idea of Ingham [1] which boils down to an appropriate use of the Fourier transform of the Fejér kernel

$$(41.20.1)\qquad \int_{-\infty}^{+\infty} \left(\frac{\sin(y/2)}{y/2}\right)^2 e^{iuy} = \begin{cases} 2\pi(1-|u|) & \text{for} \quad |u| \le 1 \\ 0 & \text{for} \quad |u| \ge 1. \end{cases}$$

Let $A > 20$ and B be sufficiently large effective numerical constants, B an integer ≥ 8, to be determined later, and let ω be any real number satisfying

$$\log c_{11} + 1 \le \omega \le \frac{\lambda_0}{2} - 1.$$

We remark that the constants c_ν with $12 \le \nu \le 22$ will be absolute effective positive constants whose values do not depend on A, B either.

Using the notation (41.19.5) we define the integral

$$(41.20.2)\quad I_1(\omega) \doteq \int_{-A/4}^{A/4} \left[\frac{\sin(y/2)}{y/2}\right]^2 G\left(\omega + \frac{y}{A}\right) dy$$

$$= \sum_{|\gamma| \le e^{\lambda_0}} \frac{e^{(\delta + i\gamma)\omega}}{\rho} \int_{-A/4}^{A/4} \left(\frac{\sin(y/2)}{y/2}\right)^2 e^{iy(\gamma/A)} \cdot e^{\delta y/A}\, dy.$$

Also, define

$$(41.20.3)\quad I_2(\omega) \doteq \sum_{|\gamma| \le e^{\lambda_0}} \frac{e^{(\delta + i\gamma)\omega}}{\rho} \int_{-A/4}^{A/4} \left(\frac{\sin(y/2)}{y/2}\right)^2 e^{iy(\gamma/A)}\, dy$$

and

$$(41.20.4)\quad I_3(\omega) \doteq \sum_{|\gamma| \le e^{\lambda_0}} \frac{e^{(\delta + i\gamma)\omega}}{\rho} \int_{-\infty}^{+\infty} \left(\frac{\sin(y/2)}{y/2}\right)^2 e^{iy(\gamma/A)}\, dy.$$

We will prove that

$$(41.20.5)\quad |I_1(\omega) - I_3(\omega)| \le c_{12}.$$

Using the Fejér kernel we can show that for the investigation of the average of $G(v)$ in the interval $[\omega - \frac{1}{4}, \omega + \frac{1}{4}]$ in (41.20.2) only the contribution of the low zeros, those with $|\gamma| < A$, is essential.

By partial integration

$$(41.20.6)\quad \left|\int_{-A/4}^{A/4} \left(\frac{\sin(y/2)}{y/2}\right)^2 (e^{\delta y/A} - 1) e^{iy(\gamma/A)}\, dy\right|$$

$$\le \left|\left[\frac{A}{i\gamma} e^{iy(\gamma/A)} \left(\frac{\sin(y/2)}{y/2}\right)^2 (e^{\delta y/A} - 1)\right]_{-A/4}^{A/4}\right|$$

$$+ \left|\int_{-A/4}^{A/4} \frac{A}{i\gamma} e^{iy(\gamma/A)} \cdot \frac{d}{dy}\left\{\left(\frac{\sin(y/2)}{y/2}\right)^2 (e^{\delta y/A} - 1)\right\} dy\right|$$

$$\le 2 \cdot \frac{A}{|\gamma|} \cdot \frac{1}{(A/8)^2} (e^{|\delta|/4} - 1)$$

$$+ \frac{A}{|\gamma|}\left[\frac{|\delta|}{A} e^{|\delta|/4} \cdot 2\pi + c_{13}(e^{|\delta|/4} - 1)\right] \le c_{14} \frac{A|\delta|}{|\gamma|}.$$

Thus

(41.20.7)

$$
\begin{aligned}
&|I_1(\omega) - I_2(\omega)| \\
&\qquad \le \sum_{|\gamma| \le e^{\lambda_0}} \left| \frac{e^{(\delta + i\gamma)\omega}}{\rho} \right| \left| \int_{-A/4}^{A/4} \left(\frac{\sin(y/2)}{y/2} \right)^2 e^{iy(\gamma/A)} (e^{\delta y/A} - 1)\, dy \right| \\
&\qquad \le \sum_{|\gamma| \le e^{\lambda_0}} \frac{e^{\delta\omega}}{|\gamma|} \cdot c_{14} \frac{A|\delta|}{|\gamma|} \\
&\qquad \le c_{14} A \sum_{|\gamma| \le e^{\lambda_0}} \frac{\exp[(\log|\gamma|/4\lambda_0) \cdot (\lambda_0/2)]}{|\gamma|} \cdot \frac{\log|\gamma|}{4\lambda_0} \cdot \frac{1}{|\gamma|} \\
&\qquad \le \frac{c_{14} A}{4\lambda_0} \sum_{\gamma} \frac{\log|\gamma| \cdot |\gamma|^{1/8}}{|\gamma|^2} = \frac{c_{15} A}{4\lambda_0} \le c_{15}.
\end{aligned}
$$

Again by partial integration

(41.20.8)

$$
\begin{aligned}
\left| \int_{A/4}^{\infty} \left(\frac{\sin(y/2)}{y/2} \right)^2 e^{iy(\gamma/A)}\, dy \right| &\le \left| \left[\frac{A}{i\gamma} e^{iy(\gamma/A)} \left(\frac{\sin(y/2)}{y/2} \right)^2 \right]_{A/4}^{\infty} \right| \\
&\quad + \left| \frac{A}{i\gamma} \int_{A/4}^{\infty} e^{iy(\gamma/A)} \cdot \frac{d}{dy} \left[\left(\frac{\sin(y/2)}{y/2} \right)^2 \right] dy \right| \\
&\le \frac{A}{|\gamma|} \cdot \frac{1}{(A/8)^2} + \frac{A}{|\gamma|} \cdot \frac{c_{16}}{A} < \frac{c_{17}}{|\gamma|},
\end{aligned}
$$

and the same holds for the corresponding integral in $(-\infty, A/4]$.

From this we have, as in (41.20.7),

$$
|I_3(\omega) - I_2(\omega)| \le \sum_{|\gamma| \le e^{\lambda_0}} \frac{|\gamma|^{1/8}}{|\gamma|} \cdot \frac{2c_{17}}{|\gamma|} < c_{18}. \tag{41.20.9}
$$

So by (41.20.7) and (41.20.9) find that (41.20.5) is valid with the choice $c_{12} = c_{15} + c_{18}$.

41.21

We now investigate the integral given by (41.20.4) which—according to (41.20.1)—can be written in the form

$$I_3(\omega) = \sum_{|\gamma|<A} \frac{e^{(\delta+i\gamma)\omega}}{\frac{1}{2}+\delta+i\gamma} \cdot 2\pi\left(1-\frac{|\gamma|}{A}\right). \tag{41.21.1}$$

Since the zeros of $\zeta(s)$ are symmetrical to the real axis, an easy computation shows that

(41.21.2)

$$I_3(\omega) = 2\pi \sum_{0<\gamma<A} \frac{e^{\delta\omega}\{(1+2\delta)\cos(\gamma\omega)+2\gamma\sin(\gamma\omega)\}}{\left(\frac{1}{2}+\delta\right)^2+\gamma^2}\left(1-\frac{\gamma}{A}\right).$$

If we restrict ω, in addition to $\log c_{11}+1 \le \omega \le \lambda_0/2-1$, by

$$\log c_{11}+1 \le \omega \le \frac{\lambda_0}{2\log A}, \tag{41.21.3}$$

then for the zeros with $0<\gamma<A$

$$|\delta|\omega \le \frac{\log A}{4\lambda_0}\cdot\frac{\lambda_0}{2\log A} = \frac{1}{8}, \tag{41.21.4}$$

and so

$$0.8 < e^{\delta\omega} < 1.2. \tag{41.21.5}$$

We abbreviate:

(41.21.6)

$$J_\omega(\eta) = 2\pi \sum_{0<\gamma<A} \frac{e^{\delta\omega}\{(1+2\delta)\cos(\gamma\eta)+2\gamma\sin(\gamma\eta)\}}{\left(\frac{1}{2}+\delta\right)^2+\gamma^2}\left(1-\frac{\gamma}{A}\right).$$

Then

$$I_3(\omega) = J_\omega(\omega). \tag{41.21.7}$$

Choose, in $J_\omega(\eta)$,

$$\eta = \frac{1}{A} \quad \text{and} \quad \eta = -\frac{1}{A} \text{ resp.} \tag{41.21.8}$$

Then $J_\omega(\eta)$ can be made "large positive" and "large negative" (resp.) for any ω in (41.21.3), if we choose A sufficiently large.

Using (41.21.5), the well-known fact

$$(41.21.9)\quad N(T) = \frac{T}{2\pi}\log\frac{T}{2\pi e} + O(\log T) > \frac{T}{7}\log T \quad \text{for} \quad T > T_1$$

(see, e.g., Ingham [2], Theorem 25), and the inequality

$$(41.21.10)\qquad \sin t \geq \frac{2}{\pi}t \quad \text{for} \quad 0 \leq t \leq \frac{\pi}{2}$$

we get, since

$$(41.21.11)\qquad \gamma > 14,$$

for any $A > 2T_1$ the inequality

$$(41.21.12)\quad J_\omega\left(\frac{1}{A}\right) > 2\pi \sum_{0<\gamma<A} \frac{0.8 \cdot 2\gamma \cdot \frac{2}{\pi} \cdot \frac{\gamma}{A}}{1.01\gamma^2}\left(1 - \frac{\gamma}{A}\right)$$

$$> \frac{6}{A}\sum_{0<\gamma<A}\left(1 - \frac{\gamma}{A}\right) \geq \frac{6}{A}N\left(\frac{A}{2}\right)\cdot\frac{1}{2}$$

$$> \frac{1}{2}\cdot\frac{6}{A}\cdot\frac{A}{2\cdot 7}\log\frac{A}{2} > \frac{1}{5}\log\frac{A}{2} > \frac{1}{10}\log A.$$

Similarly, for $A > 2T_1$

(41.21.13)

$$J_\omega\left(-\frac{1}{A}\right) < 2\pi\left\{\sum_{0<\gamma<A}\frac{1.2\cdot 2}{\gamma^2} - \sum_{0<\gamma<A}\frac{0.8\cdot 2\gamma\cdot(2/\pi)\cdot(\gamma/A)}{1.01\gamma^2}\left(1 - \frac{\gamma}{A}\right)\right\}$$

$$< 4.8\pi\sum_{\gamma>0}\frac{1}{\gamma^2} - \frac{6}{A}\sum_{0<\gamma<A}\left(1 - \frac{\gamma}{A}\right) < c_{19} - \tfrac{1}{10}\log A.$$

41.22

Now we apply Lemma 34.3 for the numbers

$$(41.22.1)\qquad \frac{\gamma}{2\pi} \quad \text{such that} \quad 0 < \gamma < A,$$

their total number being

$$N = N(A) < A \log A \quad \text{for} \quad A > c_{20}. \tag{41.22.2}$$

Using (41.21.9) and $A > 20$, we also have

$$N(A) \geq 1. \tag{41.22.3}$$

Choose in Lemma 34.3

$$q = B \doteq [\log^2 A] \tag{41.22.4}$$

and

$$M = \left[\frac{(\lambda_0/2\log A) - 1}{B^{N(A)}}\right] \geq c(A)\lambda_0 \tag{41.22.5}$$

where $c(A)$ is an effectively computable constant depending only on A.

Thus, denoting the distance of a real number x from the nearest integer by $\|x\|$, we get the existence of positive integers n_ν with

$$1 \leq n_1 < n_2 < \cdots < n_M \leq \frac{\lambda_0}{2\log A} - 1 \tag{41.22.6}$$

for which all the relations

$$\left\| \frac{\gamma_j}{2\pi} n_\nu \right\| \leq \frac{1}{B} \qquad (1 \leq j \leq N, 1 \leq \nu \leq M) \tag{41.22.7}$$

hold.

This implies

$$\left|\sin\left\{\gamma_j\left(n_\nu \pm \frac{1}{A}\right)\right\} - \sin\left(\pm\frac{\gamma_j}{A}\right)\right| \leq 2\left|\sin\frac{\gamma_j n_\nu}{2}\right| < \frac{2\pi}{B} \tag{41.22.8}$$

and

$$\left|\cos\left\{\gamma_j\left(n_\nu \pm \frac{1}{A}\right)\right\} - \cos\left(\pm\frac{\gamma_j}{A}\right)\right| \leq 2\left|\sin\frac{\gamma_j n_\nu}{2}\right| < \frac{2\pi}{B} \tag{41.22.9}$$

for $1 \leq j \leq N$, $1 \leq \nu \leq M$. In the preceding formulas we must always take either the upper signs in both terms or the lower signs in both terms.

Choose ω_ν' and ω_ν'' by

$$\omega_\nu^{(i)} = n_\nu + \frac{(-1)^i}{A}. \tag{41.22.10}$$

The inequality (41.21.3) is satisfied due to (41.22.6) so by (41.22.8)–(41.22.9), for $\nu > \log c_{11} + 2$

(41.22.11)

$$\left| J_{\omega_\nu^{(i)}}\left[\omega_\nu^{(i)}\right] - J_{\omega_\nu^{(i)}}\left[\frac{(-1)^i}{A}\right]\right| \le 2\pi \sum_{0<\gamma<A} \frac{1.2(2+2\gamma)\cdot 2\pi/B}{\gamma^2}$$

$$< \frac{110}{B} \sum_{0<\gamma<A} \frac{1}{\gamma} < \frac{110}{B}\cdot c_{21}\log^2 A < c_{22}.$$

Here we used (41.22.4) and the inequality

$$\sum_{0<\gamma<A} \frac{1}{\gamma} < c_{21}\log^2 A, \tag{41.22.12}$$

which is an easy consequence of (41.21.9).

Using (41.21.12)–(41.21.13), we have

$$J_{\omega'_\nu}(\omega'_\nu) = I_3(\omega'_\nu) < -\tfrac{1}{10}\log A + c_{19} + c_{22} \tag{41.22.13}$$

and

$$J_{\omega''_\nu}(\omega''_\nu) = I_3(\omega''_\nu) > \frac{1}{10}\log A - c_{22}. \tag{41.22.14}$$

Combining this with (41.20.5) we already get the needed results for the average of $G(v)$ in $[\omega - \frac{1}{4}, \omega + \frac{1}{4}]$, namely,

$$I_1(\omega'_\nu) \le -\frac{1}{10}\log A + c_{19} + c_{22} + c_{12} \tag{41.22.15}$$

and analogously

$$I_1(\omega''_\nu) > \tfrac{1}{10}\log A - c_{22} - c_{12}. \tag{41.22.16}$$

41.23

Now fix A as

$$A = \max\{\exp[10(c_{22} + c_{19} + c_{12} + 4\pi)], 2T_1, c_{20}, 20\}. \tag{41.23.1}$$

Recalling the definition of $I_1(\omega)$ in (41.20.2), formulas (41.22.15) and (41.22.16) yield

$$\int_{-A/2}^{A/2} \left(\frac{\sin(y/2)}{y/2}\right)^2 G\left(\omega'_\nu + \frac{y}{A}\right) dy < -4\pi \tag{41.23.2}$$

and

$$\int_{-A/4}^{A/4}\left(\frac{\sin(y/2)}{y/2}\right)^2 G\left(\omega_\nu'' + \frac{y}{A}\right)dy > 4\pi. \tag{41.23.3}$$

Since, by (41.20.1),

$$\int_{-\infty}^{+\infty}\left(\frac{\sin(y/2)}{y/2}\right)^2 = 2\pi, \tag{41.23.4}$$

(41.23.2) and (41.23.3) immediately give

$$\min_{\omega_\nu' - 1/4 \le v \le \omega_\nu' + 1/4} G(v) < -2 \tag{41.23.5}$$

and

$$\max_{\omega_\nu'' - 1/4 \le v \le \omega_\nu'' + 1/4} G(v) > 2. \tag{41.23.6}$$

Since $1/A < 1/20$, (41.22.10) implies

$$\left[\omega_\nu^{(i)} - \tfrac{1}{4}, \omega_\nu^{(i)} + \tfrac{1}{4}\right] \subset \left[n_\nu - \tfrac{1}{3}, n_\nu + \tfrac{1}{3}\right]. \tag{41.23.7}$$

Thus, by (41.19.6)–(41.19.8), $\Delta_1(r)$ has at least one sign change in every interval

$$I_\nu \doteq \left[e^{n_\nu - 1/3}, e^{n_\nu + 1/3}\right] \subset \left[c_{11}, e^{\lambda_0/2}\right] \tag{41.23.8}$$

(if $\nu > \log c_{11} + 2$).

As the n_ν are positive integers, these intervals are all disjoint, and their total number is, by (41.22.5), at least

$$\begin{aligned} M - (\log c_{11} + 2) &\ge c(A)\lambda_0 - \log c_{11} - 2 \\ &\ge c_{23}\lambda_0 - \log c_{11} - 2 \ge c_7\lambda_0. \end{aligned} \tag{41.23.9}$$

(Since A is already fixed, $c(A) = c_{23}$ is an effectively computable absolute positive constant.)

Thus we have in Case II in the interval $[2, e^{\lambda_0/2}]$ at least $c_7\lambda_0$ sign changes of $\Delta_1(x)$, and by (41.16.4) Case II is settled too. This completes the proof of Theorem 41.2. ■

42

THE REMAINDER TERM OF THE PRIME NUMBER FORMULA III

Ineffective Estimation for the Sign Changes of $\pi(x) - \operatorname{li} x$

42.1

We continue our investigations concerning the oscillatory properties of the remainder term of the prime number formula.

Theorem 42.1 (Pintz [3]). *There are absolute constants* Y_i $(1 \le i \le 4)$ *such that for* $Y > Y_i$ *the inequality*

$$V_i(Y) > \frac{1}{10^{11}} \cdot \frac{\log Y}{(\log_2 Y)^3} \qquad (1 \le i \le 4) \tag{42.1.1}$$

holds, where the constants Y_i *are effectively computable for* $i = 2, 4$ *but not for* $i = 1, 3$.

[As mentioned, $V_i(Y)$ denotes the number of sign changes of $\Delta_i(x)$ defined by (40.1.1)–(40.1.4) in [2, Y].] This result is already very close to Ingham's [1] conditional (and also ineffective) result

$$V_i(Y) > c \log Y \quad \text{for} \quad Y > Y_0 \tag{42.1.2}$$

[see (41.3.3)–(41.3.6)]. However, here we can not give a corresponding localization of a sign change.

42.2

In the proof we use the following "kernel function":

$$J_{k,\mu}(u)=\frac{1}{2\pi i}\int_{(2)}\left(\frac{e^{s}-e^{-s}}{2s}\right)^{k}\exp\left(us+\frac{s^{2}}{\mu}\right)ds \tag{42.2.1}$$

where $k\geq 1$ is an integer, $\mu\geq 1$ is real, and u is real.

First we state some properties of the function $J_{k,\mu}(u)$.

Lemma 42.1. *The function* $J_{k,\mu}(u)$, *defined by* (42.2.1), *satisfies*

$$J_{k,\mu}(u)=J_{k,\mu}(-u) \tag{42.2.2}$$

$$J_{k,\mu}(u)\geq 0 \tag{42.2.3}$$

If $|u|\geq k+2$ *then*

$$\left|J_{k,\mu}(u)\right|\leq\frac{1}{e^{(|u|-k-1)\mu}}. \tag{42.2.4}$$

To prove (42.2.2) shift the line of integration to $\sigma=0$, to obtain

$$J_{k,\mu}(u)=\frac{1}{\pi}\int_{0}^{\infty}\left(\frac{\sin t}{t}\right)^{k}\exp\left(-\frac{t^{2}}{\mu}\right)\cos ut\,dt, \tag{42.2.5}$$

from which (42.2.2) follows.

It is sufficient to prove (42.2.4) for $u\geq k+2$. Shifting the line of integration to $\sigma=-\mu$, we get

(42.2.6)

$$\begin{aligned}
\left|J_{k,\mu}(u)\right| &=\left|\frac{1}{2\pi i}\int_{(-\mu)}\left(\frac{e^{s}-e^{-s}}{2s}\right)^{k}\exp\left(us+\frac{s^{2}}{\mu}\right)ds\right|\\
&=\left|\frac{1}{2^{k}}\sum_{d=0}^{k}(-1)^{d}\binom{k}{d}\cdot\frac{1}{2\pi i}\int_{(-\mu)}\frac{\exp\{(k-2d+u)s+s^{2}/\mu\}}{s^{k}}ds\right|\\
&\leq\max_{0\leq d\leq k}\left|\frac{1}{2\pi i}\int_{(-\mu)}\frac{\exp\{(k-2d+u)s+s^{2}/\mu\}}{s^{k}}ds\right|\\
&\leq\frac{1}{2\pi}\int_{-\infty}^{+\infty}\frac{\exp\{(k-2k+u)(-\mu)+(\mu^{2}-t^{2})/\mu\}}{\mu^{k}}dt\\
&\leq\frac{\exp\{-\mu(u-k-1)\}}{\mu^{k}}\cdot\frac{1}{2\pi}\int_{-\infty}^{+\infty}\exp\left(-\frac{t^{2}}{\mu}\right)dt<\frac{1}{e^{\mu(u-k-1)}}.
\end{aligned}$$

Thus

$$\lim_{u\to\infty} J_{k,\mu}(u) = 0. \tag{42.2.7}$$

In view of (42.2.7) and (42.2.2), (42.2.3) will be proved if we show that $J_{k,\mu}(u)$ is monotonically decreasing for $u \geq 0$.

This will be proved by induction with respect to k. For $k = 1$ (41.8.2) implies

(42.2.8)

$$\frac{dJ_{1,\mu}(u)}{du} = \frac{1}{2\pi i}\int_{(2)} \frac{e^s - e^{-s}}{2s} \cdot \exp\left(\frac{s^2}{\mu}\right) \cdot s \cdot e^{us}\, ds$$

$$= \frac{1}{2}\left\{\frac{1}{2\pi i}\int_{(2)} \exp\left[\frac{s^2}{\mu} + (u+1)s\right] ds - \frac{1}{2\pi i}\int_{(2)} \exp\left[\frac{s^2}{\mu} + (u-1)s\right] ds\right\}$$

$$= \frac{\sqrt{\mu}}{4\sqrt{\pi}}\left[\exp\left(-\frac{\mu}{4}(u+1)^2\right) - \exp\left(-\frac{\mu}{4}(u-1)^2\right)\right] \leq 0.$$

If we assume that the assertion is true for $k = k_0 - 1$, then, as in Lemma 34.2,

(42.2.9)

$$\frac{dJ_{k_0,\mu}(u)}{du} = \frac{1}{\pi}\int_0^\infty \left(\frac{\sin t}{t}\right)^{k_0} (-t\sin(ut))\exp\left(-\frac{t^2}{\mu}\right) dt$$

$$= \frac{1}{\pi}\int_0^\infty \left(\frac{\sin t}{t}\right)^{k_0-1} \cdot \frac{1}{2}\{\cos[(u+1)t] - \cos[(u-1)t]\}\exp\left(-\frac{t^2}{\mu}\right) dt$$

$$= \tfrac{1}{2}\left(J_{k_0-1,\mu}(u+1) - J_{k_0-1,\mu}(u-1)\right) \leq 0$$

or alternatively

$$= \tfrac{1}{2}\left[J_{k_0-1,\mu}(1+u) - J_{k_0-1,\mu}(1-u)\right] \leq 0. \tag{42.2.10}$$

Thus (42.2.9) proves the assertion in the case $u \geq 1$, and (42.2.10) proves the case $0 \leq u \leq 1$.

42.3

We first treat the (ineffective) case $i = 1$. If the Riemann hypothesis is true, then the quoted theorem of Ingham [1] [see (42.1.2) and (41.3.6)] already settles this case (for $1 \le i \le 4$). Thus we assume that the Riemann hypothesis is false.

Let $\rho_0 = \beta_0 + i\gamma_0$ be the zero with $\beta_0 > \frac{1}{2}$, and with minimal $\gamma_0 > 0$. If there are several such zeros, then let $\rho_1' = \beta_1' + i\gamma_1'$ be that zero with maximal real part. If there is only one such zero then let $\rho_1' = \rho_0$.

Define recursively $\rho_{n+1}' = \beta_{n+1}' + i\gamma_{n+1}'$ to be the zero with maximal real part among those satisfying

(42.3.1) $$\gamma_n' < \gamma \le \gamma_n' + 2\log Y, \qquad \beta \ge \beta_n' + \frac{1}{\log Y}$$

if such a zero exists.

Thus after, at most, $[\log Y/2]$ steps we get a zero $\rho_N' = \beta_N' + i\gamma_N' \doteq \rho_1 = \beta_1 + i\gamma_1$ with

(42.3.2) $$\beta_1 > \tfrac{1}{2}, \qquad 0 < \gamma_1 < 2\log^2 Y$$

(because $\log^2 Y + \gamma_1' < 2\log^2 Y$ if $Y > Y_0$, an ineffective constant), such that the domains

(42.3.3) $$0 \le t \le \gamma_1, \qquad \sigma > \beta_1$$

and

(42.3.4) $$|t - \gamma_1| \le 2\log Y, \qquad \sigma \ge \beta_1 + \frac{1}{\log Y}$$

are zero free.

42.4

We introduce the following notation. Let

(42.4.1) $$\mu = \log Y, \qquad L = \log_2 Y.$$

Let k be any positive integer to be chosen later, for which

(42.4.2) $$4000L \le k \le 4400L.$$

Let λ be any real number satisfying

(42.4.3) $$\frac{\mu}{10^4 L} \le \lambda \le \frac{2\mu}{10^4 L}.$$

Let

$$A \doteq \exp\{k(\lambda - 2)\} \tag{42.4.4}$$

$$B \doteq \exp\{k(\lambda + 2)\} \tag{42.4.5}$$

$$g_{k,D}(u, s) \doteq \left(\frac{e^s - e^{-s}}{2s}\right)^k \exp\left(us + \frac{s^2}{D}\right) \tag{42.4.6}$$

$$f(x) \doteq \Pi(x) - \lg x \pm \sqrt{x} \doteq \Pi(x) - \sum_{2 \le n \le x} \frac{1}{\log n} \pm \sqrt{x} \tag{42.4.7}$$

$$H(s) \doteq \frac{\zeta'}{\zeta}(s) + \zeta(s) - 1 \mp \frac{1}{2(s - \frac{1}{2})^2} \tag{42.4.8}$$

where the choice of signs in (42.4.7) and in (42.4.8) must be consistent.

We choose

$$\mu' = \frac{\mu \cdot 4400L}{k}. \tag{42.4.9}$$

Thus by (42.4.2)

$$\mu \le \mu' \le \tfrac{11}{10}\mu. \tag{42.4.10}$$

We prove that for every real λ satisfying (42.4.3), there exists an integer k satisfying (42.4.2) such that $f(x)$ has a sign change in $[A, B]$.

Let λ be fixed in (42.4.3) and assume, contrary to our assertion, that with any k in (42.4.2) $f(x)$ does not change sign in $[A, B]$.

42.5

As in 41.8, we now start with the formula (valid for $\sigma > 1$)

$$\int_1^\infty f(x) \frac{d}{dx}(x^{-s} \log x)\, dx = H(s). \tag{42.5.1}$$

Replace s by $s + i\gamma_1$ in (42.5.1), multiply by $g_{k,\mu'}(k\lambda, s)$, and integrate with respect to s along the line $\sigma = 2$. (The change of order of integrations is obviously justified.) By (42.2.9),

(42.5.2)

$$U = \frac{1}{2\pi i}\int_{(2)} H(s+i\gamma_1) g_{k,\mu'}(k\lambda, s)\, ds$$

$$= \frac{1}{2\pi i}\int_{(2)}\int_1^\infty f(x)\frac{d}{dx}\left(x^{-s-i\gamma_1}\log x \cdot g_{k,\mu'}(k\lambda, s)\right) dx\, ds$$

$$= \int_1^\infty f(x)\frac{d}{dx}\left\{x^{-i\gamma_1}\log x \cdot \frac{1}{2\pi i}\int_{(2)} g_{k,\mu'}(k\lambda - \log x, s)\, ds\right\} dx$$

$$= \int_1^\infty f(x)\frac{d}{dx}\left\{x^{-i\gamma_1}\log x \cdot J_{k,\mu'}(k\lambda - \log x)\right\} dx$$

$$= \int_1^\infty f(x)\left\{-i\gamma_1\frac{x^{-i\gamma_1}}{x}\log x \cdot J_{k,\mu'}(\log x - k\lambda) + \frac{x^{-i\gamma_1}}{x}J_{k,\mu'}(\log x - k\lambda)\right.$$

$$\left. + x^{-i\gamma_1}\log x \cdot J'_{k,\mu'}(\log x - k\lambda)\frac{1}{x}\right\} dx$$

$$= \int_1^\infty \frac{f(x)\log x \cdot x^{-i\gamma_1}}{x}\left\{J_{k,\mu'}(\log x - k\lambda)\left(-i\gamma_1 + \frac{1}{\log x}\right)\right.$$

$$\left. + \frac{1}{2}J_{k-1,\mu'}(\log x - k\lambda + 1) - \frac{1}{2}J_{k-1,\mu'}(\log x - k\lambda - 1)\right\} dx.$$

Now we give an upper bound for the right-hand side of (42.5.2) using the proved properties of the kernel function $J_{k,\mu}(u)$ and the fact that $f(x)$ does not change sign in $[A, B]$ [defined by (42.4.4)–(42.4.5)]. On the other hand, we show that the left-hand side can essentially be reduced to a finite power sum, for which using the second main theorem we can give a non trivial lower estimate by suitable choice of k in (42.4.2). This contradicts the upper estimate above.

42.6

To estimate U from above we again split the integral U into three parts

(42.6.1) $$U = U_1 + U_2 + U_3$$

where

$$U_1 = \int_1^A, \qquad U_2 = \int_A^B, \qquad U_3 = \int_B^\infty. \tag{42.6.2}$$

Considering (42.3.2), (42.5.2), and (42.2.3), (42.4.2)–(42.4.5), and the fact that $f(x)$ does not change sign in $[A, B]$, we get

$$\begin{aligned}|U_2| &\le \int_A^B \frac{|f(x)|k(\lambda+2)}{x}\left\{\frac{11}{10}\gamma_1 J_{k,\mu'}(\log x - k\lambda)\right. \\ &\quad \left. + \tfrac{1}{2}J_{k-1,\mu'}(\log x - k\lambda + 1) + \tfrac{1}{2}J_{k-1,\mu'}(\log x - k\lambda - 1)\right\} dx \\ &\le \mu^3 \int_A^B \frac{|f(x)|}{x}\{J_{k,\mu'}(\log x - k\lambda) + J_{k-1,\mu'}(\log x - k\lambda + 1) \\ &\quad + J_{k-1,\mu'}(\log x - k\lambda - 1)\}\, dx \\ &= \mu^3 \left| \int_A^B \frac{f(x)}{x}\{J_{k,\mu'}(\log x - k\lambda) + J_{k-1,\mu'}(\log x - k\lambda + 1)\right. \\ &\quad \left. + J_{k-1,\mu'}(\log x - k\lambda - 1)\}\, dx \right|.\end{aligned} \tag{42.6.3}$$

Similarly, introducing the new variable $u = \log x - k\lambda$, and using (42.4.2)–(42.4.3) and (42.2.4), $|f(x)| \le |x|$ shows that

(42.6.4)

$$\begin{aligned}|U_3| &\le \mu^3 \int_B^\infty \{J_{k,\mu'}(\log x - k\lambda) + J_{k-1,\mu'}(\log x - k\lambda + 1) \\ &\quad + J_{k-1,\mu'}(\log x - k\lambda - 1)\}\, dx \\ &= \mu^3 \int_{2k}^\infty \{J_{k,\mu'}(u) + J_{k-1,\mu'}(u+1) + J_{k-1,\mu'}(u-1)\} e^{u+k\lambda}\, du \\ &\le e^\mu \int_{2k}^\infty \{e^{-(u-k-1)\mu'} + e^{-(u+1-k-1)\mu'} + e^{-(u-1-k-1)\mu'}\} e^u\, du \\ &\le 3\int_{2k}^\infty e^{-(u-k-2)\mu' + u}\, du = o(1),\end{aligned}$$

and similarly

$$|U_1| = o(1). \tag{42.6.5}$$

Also

$$(42.6.6)\quad U_4 \doteq \mu^3 \int_B^\infty \frac{f(x)}{x}\{J_{k,\mu'}(\log x - k\lambda) + J_{k-1,\mu'}(\log x - k\lambda + 1)$$
$$+J_{k-1,\mu'}(\log x - k\lambda - 1)\}\, dx$$
$$= o(1)$$

and

$$(42.6.7)\quad U_5 \doteq \mu^3 \int_1^A \frac{f(x)}{x}\{J_{k,\mu'}(\log x - k\lambda) + J_{k-1,\mu'}(\log x - k\lambda + 1)$$
$$+J_{k-1,\mu'}(\log x - k\lambda - 1)\}\, dx$$
$$= o(1).$$

With the notation

(42.6.8)

$$K \doteq \int_1^\infty \frac{f(x)}{x}\{J_{k,\mu'}(\log x - k\lambda) + J_{k-1,\mu'}(\log x - k\lambda + 1)$$
$$+J_{k-1,\mu'}(\log x - k\lambda - 1)\}\, dx$$

(42.6.3)–(42.6.8) immediately give the relation

$$(42.6.9)\qquad |U| = |U_2| + o(1) \le \mu^3|K| + o(1).$$

42.7

Now, as in 41.12, we estimate K from above using the following formula, valid for $\sigma > 1$ with a constant h.

$$(42.7.1)\quad \int_1^\infty \frac{f(x)}{x^{s+1}}\,dx = \frac{1}{s}\left\{\int_2^s\left[\frac{\zeta'}{\zeta}(z) + \zeta(z) - 1\right]dz + h\right\} \pm \frac{1}{s-\frac{1}{2}}$$
$$\doteq \varphi(s) \pm \frac{1}{s-\frac{1}{2}}.$$

Multiply on both sides by $(1/2\pi i)G(s)$, where

$$(42.7.2)\quad G(s) \doteq g_{k,\mu'}(k\lambda, s) + g_{k-1,\mu'}(k\lambda - 1, s) + g_{k-1,\mu'}(k\lambda + 1, s)$$

and integrate along the line $\sigma = 2$. Then, by (42.4.6) and (42.6.8),

$$K = \pm \frac{1}{2\pi i}\int_{(2)} \frac{G(s)}{s-\frac{1}{2}}\,ds + \frac{1}{2\pi i}\int_{(2)} G(s)\varphi(s)\,ds \tag{42.7.3}$$
$$\doteq \pm K_1 + K_2.$$

Shift the line of integration in K_1 to $\sigma = -\mu'$, and an easy computation shows that

(42.7.4)

$$K_1 = G\left(\frac{1}{2}\right) + \frac{1}{2\pi i}\int_{(-\mu')} \frac{G(s)}{s-\frac{1}{2}}\,ds$$
$$= O(e^{k+k\lambda/2}) + O\left(\int_{-\infty}^{+\infty} \frac{\exp\left\{\mu' k - (k\lambda - 1)\mu' + \mu' - \frac{t^2}{\mu'}\right\}}{(\mu')^k}\,dt\right)$$
$$= O(e^{k+k\lambda/2}) + O(e^{-\mu'(k\lambda-k-2)})$$
$$= O(e^{k+k\lambda/2}).$$

To estimate the integral K_2, we transform it on the broken line l defined for $t \geq 0$ by

(42.7.5)

$$l_1: \sigma = \tfrac{5}{4} \qquad \text{for} \quad t \geq 2\mu$$
$$l_2: \beta_1 + \frac{2}{\mu} \leq \sigma \leq \frac{5}{4} \qquad \text{for} \quad t = 2\mu$$
$$l_3: \sigma = \beta_1 + \frac{2}{\mu} \qquad \text{for} \quad 10 \leq t \leq 2\mu$$
$$l_4: \frac{1}{4} \leq \sigma \leq \beta_1 + \frac{2}{\mu} \qquad \text{for} \quad t = 10$$
$$l_5: \sigma = \tfrac{1}{4} \qquad \text{for } 0 \leq t \leq 10$$

and for $t \leq 0$ by reflection with respect to the real axis. Hence

$$K_2 = \frac{1}{2\pi i}\int_{(l)} G(s)\varphi(s)\,ds, \tag{42.7.6}$$

because (42.3.3)–(42.3.4) imply that $\varphi(s)$ is regular to the right of l and on l.

Now using the well-known properties of $\varphi(s)$ mentioned in (41.12.3)–(41.12.6), (42.3.2)–(42.3.4), and definitions (42.4.6) and (44.7.2), we have the following estimates for the integrals over I_ν $(1 \le \nu \le 5)$:

$$(42.7.7)\quad |J_1| = O\left[\left(\frac{2}{\mu}\right)^{k-2} \exp\left(\frac{5}{4}k\lambda - \frac{4\mu^2}{\mu'}\right)\right] = O(1)$$

$$|J_2| = O\left[\mu \log \mu \left(\frac{2}{\mu}\right)^{k-1} \exp\left(\frac{5}{4}k\lambda - \frac{4\mu^2}{\mu'}\right)\right] = O(1)$$

$$|J_3| = O\left(\mu \log \mu \left(\tfrac{1}{5}\right)^{k-2} \exp\left[\left(\beta_1 + \frac{2}{\mu}\right)k\lambda\right]\right) = O(e^{\beta_1 k\lambda - k})$$

$$|J_4| = O\left(\left(\tfrac{1}{5}\right)^{k-1} \exp\left[\left(\beta_1 + \frac{2}{\mu}\right)k\lambda\right]\right) = O(e^{\beta_1 k\lambda - k})$$

$$|J_5| = O(e^{(1/4)k\lambda + 2k}).$$

Hence

$$(42.7.8)\qquad |K_2| = O(e^{(\beta_1\lambda - 1)k}) + O(e^{(\lambda/4+2)k}).$$

By (42.7.3)–(42.7.4) we get

$$(42.7.9)\qquad |K| = O(e^{(\beta_1\lambda - 1)k}) + O(e^{(\lambda/2+1)k}).$$

Now, by the definition of $\rho_0 = \beta_0 + i\gamma_0$ and $\rho_1 = \beta_1 + i\gamma_1$ (given in 42.3), and using (42.4.1) and (42.4.3), we find that

$$(42.7.10)\quad \beta_1 - \frac{1}{2} \ge \beta_0 - \frac{1}{2} \ge \frac{2 \cdot 10^4 \log_2 Y}{\log Y} = \frac{2 \cdot 10^4 L}{\mu} \ge \frac{2}{\lambda}$$

for $Y > Y_1$, an ineffective constant.

Thus

$$(42.7.11)\qquad e^{(\beta_1\lambda - 1)k} \ge e^{(\lambda/2+1)k}$$

and so

$$(42.7.12)\qquad |K| = O(e^{\beta_1\lambda k - k}).$$

Combining this with (42.6.9) and (42.4.1) we get the required upper estimate for the integral U in (42.5.2), namely,

$$(42.7.13)\qquad |U| \le \mu^3 |K| + o(1) = O(e^{\beta_1\lambda k - k + 3L}).$$

42.8

Now we turn to the lower estimate of the left-hand side of U in (42.5.2) (with the choice of a suitable k), and we follow the scheme of Sections 41.13–41.14. Shift the line of integration to $\sigma = -\frac{1}{2}$ to get

$$U = \sum_{\rho} g_{k,\mu'}(k\lambda, \rho - i\gamma_1) \tag{42.8.1}$$

$$\mp \frac{1}{2}\frac{d}{ds}\left[g_{k,\mu'}(k\lambda, s)\right]_{s=1/2-i\gamma_1}$$

$$+ \frac{1}{2\pi i}\int_{(-1/2)} H(s + i\gamma_1) g_{k,\mu'}(k\lambda, s)\, ds.$$

An easy computation shows that this last integral is $O(1)$, and that the second term is

$$\pm \frac{1}{2}\left(\frac{e^{1/2-i\gamma_1} - e^{-1/2+i\gamma_1}}{1 - 2i\gamma_1}\right)^k \exp\left[k\lambda\left(\frac{1}{2} - i\gamma_1\right) + \frac{(1/2 - i\gamma_1)^2}{\mu}\right] \tag{42.8.2}$$

$$\cdot\left\{k\frac{(d/ds)[(e^s - e^s)/2s]_{s=1/2-i\gamma_1}}{(e^{1/2-i\gamma_1} - e^{-1/2+i\gamma_1})/(1 - 2i\gamma_1)} + k\lambda + \frac{1 - 2i\gamma_1}{\mu}\right\}$$

$$= O(k\lambda e^{k\lambda/2}) = O(e^{k\lambda/2+k}) = O(e^{\beta_1\lambda k - k}),$$

using the inequality (42.7.11).

In the first term [in (42.8.1)] we can estimate in a trivial way the sum containing the infinitely many zeros with

$$|\gamma - \gamma_1| \ge 2\mu. \tag{42.8.3}$$

Namely, for the contribution of these zeros, the upper estimate

$$2\sum_{n\ge[2\mu]} c\log(\gamma_1 + n)\left(\frac{2}{n}\right)^k \exp\left(k\lambda + \frac{1 - n^2}{\mu'}\right) = O(1) \tag{42.8.4}$$

holds. We estimate similarly the sum corresponding to the zeros with

$$6 \le |\gamma - \gamma_1| < 2\mu. \tag{42.8.5}$$

The number of these zeros is, by (41.13.3) and (42.3.2), at most,

$$4\mu \cdot c\log(\gamma_1 + 2\mu) \le 4c\mu\log(\mu^2 + 2\mu) = O(\mu^2). \tag{42.8.6}$$

Furthermore, by (42.3.4), for the zeros satisfying (42.8.5), we have

$$\beta \le \beta_1 + \frac{1}{\mu}, \tag{42.8.7}$$

and so for these zeros

(42.8.8)

$$|g_{k,\mu'}(k\lambda, \rho - i\gamma_1)| \le \left(\frac{e+1}{2\cdot 6}\right)^k \exp\left[k\lambda\left(\beta_1 + \frac{1}{\mu}\right) + \frac{1}{\mu'}\right] < e^{k\lambda\beta_1 - k}.$$

Thus for the contributions of zeros satisfying (42.8.5) to the sum (42.8.1) we have the upper bound

$$O(\mu^2 e^{k\lambda\beta_1 - k}) = O(e^{k\lambda\beta_1 - k + 2L}). \tag{42.8.9}$$

42.9

These estimates were independent of the choice of k in (42.4.2). So the essential part of U is the finite power sum, containing the zeros with

$$|\gamma - \gamma_1| < 6. \tag{42.9.1}$$

The number n of such zeros satisfies, in view of (41.13.3) and (42.3.2),

(42.9.2)
$$1 \le n \le 2\cdot 6\cdot c\log(\gamma_1 + 6) \le 180\log(2\log^2 Y + 6)$$
$$\le 400\log_2 Y = 400L \doteq N.$$

So for the zeros of (42.9.1) we use the second main theorem in the form given by Section 33.12.

Choosing

$$m = 4000L, \tag{42.9.3}$$

we get a positive integer k satisfying (42.4.2) for which

(42.9.4)

$$|W| = \left|\sum_{|\gamma-\gamma_1|<6} g_{k,\mu'}(k\lambda, \rho - i\gamma_1)\right|$$

$$= \left|\sum_{|\gamma-\gamma_1|<6}\left\{\frac{e^{\rho-i\gamma_1} - e^{-(\rho-i\gamma_1)}}{2(\rho - i\gamma_1)}\cdot\exp\left[\lambda(\rho - i\gamma_1) + \frac{(\rho-i\gamma_1)^2}{4400L\mu}\right]\right\}^k\right|$$

$$\ge \left(\frac{1}{8e\cdot 12}\right)^{400L}\cdot\left(\frac{e^{\beta_1} - e^{-\beta_1}}{2\beta_1}\right)^k \exp\left(k\lambda\beta_1 + \frac{\beta_1^2 k}{4400L\mu}\right)$$

$$> \left(\frac{1}{e^6}\right)^{400L}\cdot e^{k\lambda\beta_1} = e^{k\lambda\beta_1 - 2400L}$$

(because for real $x > 0$, $e^x - e^{-x} > 2x$).

Now combining the $O(1)$ estimate for the integral, the upper bound (42.8.2) for the residue in (42.8.1), the inequalities (42.8.4) and (42.8.9) concerning the zeros with $|\gamma - \gamma_1| > 6$, and the lower bound (42.9.4) and (42.4.2), we obtain

$$|U| \geq \tfrac{1}{2}\exp(k\lambda\beta_1 - 2400L) \tag{42.9.5}$$

which contradicts (42.7.13). Thus we have shown that to every λ with (42.4.3) there exists an integer k with (42.4.2), such that $f(x)$ in (42.4.7) and thus also $\Delta_1(x)$ and $\Delta_2(x)$, has a sign change in the interval

$$[A, B] = [e^{k\lambda - 2k}, e^{k\lambda + 2k}]. \tag{42.9.6}$$

This in itself does not yet give the required inequality (42.1.1).

42.10

However, because the total Lebesgue measure of the interval (42.4.3) is

$$\frac{\mu}{10^4 L} \tag{42.10.1}$$

and k can take at most $[400L] + 1$ values, there must exist a fixed k_0 with (42.4.2) for which there are λ with Lebesgue measure at least

$$\frac{\mu}{10^4 L} \cdot \frac{1}{401L} = \frac{\mu}{4.01 \cdot 10^6 L^2} \tag{42.10.2}$$

such that $\Delta_1(x)$ has a sign change in

$$[A_\lambda, B_\lambda] = [e^{k_0\lambda - 2k_0}, e^{k_0\lambda + 2k_0}] \subset \left(e^{k_0\lambda - 10^4 L}, e^{k_0\lambda + 10^4 L}\right). \tag{42.10.3}$$

Since the Lebesgue measure of the set of λ belonging to this fixed k_0 is not less than the quantity given by (42.10.2), we can choose at least

$$N = \left[\frac{\mu}{4.01 \cdot 10^6 L^2} \cdot \frac{1}{2 \cdot 10^4 L}\right] > \frac{\mu}{10^{11} L^3} \tag{42.10.4}$$

λ_j $(1 \leq j \leq N)$, such that the difference of any two of them is

$$|\lambda_j - \lambda_\nu| \geq 2 \cdot 10^4 L \qquad (1 \leq \nu < j \leq N). \tag{42.10.5}$$

So for the corresponding $[A_{\lambda_j}, B_{\lambda_j}]$ we have, by (42.10.3),

$$\left[A_{\lambda_j}, B_{\lambda_j}\right] \cap \left[A_{\lambda_\nu}, B_{\lambda_\nu}\right] = \varnothing \qquad (1 \leq \nu < j \leq N). \tag{42.10.6}$$

By (42.4.2)–(42.4.5), to every k and λ satisfying (42.4.2) and (42.4.3) (resp.) the corresponding interval

$$(42.10.7) \qquad [A, B] = [e^{k(\lambda-2)}, e^{k(\lambda+2)}] \subset [e^{0.3\mu}, e^{0.9\mu}] = [Y^{0.3}, Y^{0.9}] \subset [2, Y].$$

So considering (42.10.4), (42.10.6), and (42.10.7) we conclude that there exist at least

$$(42.10.8) \qquad \frac{\mu}{10^{11}L^3} = \frac{\log Y}{10^{11}(\log_2 Y)^3}$$

disjoint intervals contained in $[2, Y]$ such that $\Delta_1(x)$ changes its sign in every one of these intervals, and the proof of Theorem 42.1 is complete. ■

42.11

In the case $i = 2$ the following modifications are necessary in the proof to attain the inequality (42.1.1) for $Y > Y_2$ (an effective constant).

Here we do not distinguish between the case when the Riemann hypothesis is assumed true and when it is not, and choose $\rho_0 = \frac{1}{2} + i\gamma_0$ as the zero with the minimal imaginary part ($\gamma_0 \approx 14.13$). Then there exists a zero $\rho_1 = \beta_1 + i\gamma_1$ with the properties described in (42.3.2)–(42.3.4), the only change being that instead of $\beta_1 > \frac{1}{2}$ we only have $\beta_1 \geq \frac{1}{2}$. In (42.4.7), change the exponent $\frac{1}{2}$ of x to $\frac{1}{4}$ and correspondingly define $H(s)$ in (42.4.8) with $\frac{1}{4}(s - \frac{1}{4})^{-2}$ in the last term instead of $\frac{1}{2}(s - \frac{1}{2})^{-2}$. Thus for the K_1 in (42.7.4)

$$(42.11.1) \qquad |K_1| = O(e^{k+k\lambda/4}),$$

and so without (42.7.10) and (42.7.11) we immediately get

$$(42.11.2) \qquad K = O(e^{\beta_1\lambda k-k}).$$

For the residue in (42.8.1) at the point $s = \frac{1}{4} - i\gamma_1$, the upper estimate $O(e^{k+k\lambda/4}) = O(e^{\beta_1\lambda k-k})$ holds. The other parts of the proof remain valid without any change.

The cases $i = 3$ and $i = 4$ can be treated similarly. They are even easier, so we do not discuss them.

43

THE LEAST PRIMES IN ARITHMETIC PROGRESSIONS I

A Special Type of Density Estimations

43.1

In the three paragraphs that follow we are concerned with a quantitative version of Dirichlet's classical theorem on the existence of primes in arithmetic progressions. For $(l, q) = 1$ let $P(q, l)$ denote the smallest prime p such that $p \equiv l(\bmod q)$. In 1944 Linnik proved the basic estimate

$$P(q, l) < q^c. \tag{43.1.1}$$

Chowla had remarked earlier that this follows from the generalized Riemann hypothesis with any $c > 2$ for q large enough, and we deduced (43.1.1) from the weaker assumption that the L-functions have no zeros in an universally fixed neighborhood of 1.

Linnik obtained his theorem from two main lemmas concerning these possibly existing "small" zeros, a density estimation and a zero-free region. The proofs of both, even in Rodosski's simplified form, were extremely complicated. In Turán [28] our method was used to give a much simpler proof for the first of these lemmas, and the conviction was expressed that the same method might also be applicable to the second. This has been shown to be the case by Knapowski [13] so that we are now able to give a transparent, relatively short proof for Linnik's theorem. Beyond the obvious advantage of such a proof, it can also yield better values of c in (43.1.1). We do not pursue this problem here and refer to Jutila [1, 2] who worked out these simpler proofs improving earlier values of c to $c = 550$ for large enough q.

43.2

In this section we describe Linnik's density result.

Let $L(s, \chi)$ denote an L-function where χ is a Dirichlet character belonging to the modulus q. We do not specify q in this notation because it will be thought of as fixed (and, without mentioning it, sufficiently large).

It is known that there is a universal constant A such that for the number of zeros $N(r, s_0, \chi)$ of $L(s, \chi)$ in the circle $|s - s_0| < r$

$$(43.2.1) \qquad N(r, s_0, \chi) < Ar\log(q + r - 30)$$

where

$$(43.2.2) \qquad s_0 = 1 + it_0, \qquad |t_0| < q^3, \qquad r\log q > c$$

with

$$(43.2.3) \qquad A > \tfrac{1}{4}.$$

Theorem 43.1. *With the preceding notation*

$$(43.2.4) \qquad \sum_{\chi} N(\delta, s_0, \chi) < q^{387A\delta}$$

for

$$(43.2.5) \qquad \frac{c}{\log q} < \delta \le 2$$

where the summation is over the $\varphi(q)$ *characters* mod q.

The following is the form we use in applications.

Corollary (Linnik). *Let* $N(\alpha)$ *be the total number of zeros of all the* $\varphi(q)$ *L-functions* mod q *in the rectangle*

$$\alpha \le \sigma < 1, \qquad |t| \le \min\left(\frac{q^3}{\log q}, \frac{q^{b(1-\alpha)}}{b\log q}\right).$$

Then

$$N(\alpha) < q^{c(1-\alpha)+b(1-\alpha)} \qquad (0 \le \alpha < 1)$$

uniformly in the parameter $b \ge 4$.

Without the absence of a power of log q as a factor in the last inequality, the result would be easily shown. It is also easy to prove for the range $1 - \alpha > \log\log q/\log q$ as we showed by classical methods prior to Linnik's work.

Using our method, Fogels [1] obtained corresponding results for taller rectangles, and Gallagher [1] was able to sum for q and, with the notation (39.1.2), prove that $S(\alpha, Q, T) \leq [Q(T+1)]^{c(1-\alpha)}$, again without a log factor.

43.3. PROOF OF THEOREM 43.1

We require at first only

$$\lambda = \frac{1}{28}, \qquad 6\delta \log q < k \leq 200A\delta \log q \tag{43.3.1}$$

and consider

$$J(s_o, \chi) = \sum_n{}' \frac{\Lambda(n)\chi(n)}{n^{s_o}} h_k(2\lambda k - \log n, \lambda) \tag{43.3.2}$$

with the notation (34.1.1) where $\sum'$ means summation for $e^{\lambda k} < n < e^{3\lambda k}$.

43.4

Let e_n be the coefficient of $\chi(x)$ in $\sum'$.

$$\sum_\chi |J(s_o\chi)|^2 = \sum_\chi \sum_n{}' \sum_m{}' e_n \bar{e}_m \chi(n)\bar{\chi}(m) \tag{43.4.1}$$

$$= \varphi(q) \sum_{\substack{\ell=1 \\ (\ell, q)=1}}^{q-1} \left| \sum_{\substack{n \\ n \equiv \ell \bmod q}}{}' e_n \right|^2$$

$$< \frac{\varphi(q)}{\lambda^2} \sum_{\substack{l=1 \\ (l, q)=1}}^{q-1} \left(\sum_{\substack{n \\ n \equiv l (\bmod q)}}{}' \frac{\Lambda(n)}{n} \right)^2$$

where, in the last step, we used the first inequality of (37.3.5). As a very weak consequence of the so-called Brun–Titchmarsh theorem we have

$$\sum_{\substack{x < n < x^3 \\ n \equiv l (\bmod q)}} \frac{\Lambda(n)}{n} < c \frac{\log x}{\varphi(q)}$$

for $q^3 < x$. This condition is satisfied for $x = e^{\lambda k}$ according to (43.3.1), and thus (43.4.1) implies

$$\sum_{\chi} |J(s_0, \chi)|^2 < ck^2.$$

Hence, by (43.3.1),

$$|J(s_0, \chi)| < q^{-192A\delta} \tag{43.4.2}$$

with the exception of, at most,

$$ck^2 q^{384A\delta} < q^{385A\delta} \qquad (\delta \log q > c)$$

characters. Since, according to (43.3.1), there are, at most, $200A\delta \log q$ values of the parameter k, (43.4.2) will hold for all k excluding not more than

$$200A\delta \log q \cdot q^{385A\delta} + 1 < q^{386A\delta}$$

characters. (We have added 1 to exclude also the principal character.) By (43.2.1) each such character contributes at most, $A\delta \log q$ to (43.2.4). Theorem 43.1 will follow if we can show that for the remaining characters, that is, the nonprincipal characters satisfying (43.4.2) for all our k, $L(s, \chi) \neq 0$ for $|s - s_0| < \delta$.

43.5

We repeatedly apply the argument that follows so it is presented in a somewhat general form, with the properties needed of our quantities stated explicitly.

Suppose that

$$g(s) = \sum_{n=1}^{\infty} \frac{b_n}{n^s} \qquad (\sigma > 1)$$

where $g(s)$ can be continued to a meromorphic function having simple poles with positive integers as residues, the sum of residues $N(r, s_0, g)$ in $|s - s_0| < r$ satisfying

$$N(r, s_0, g) < Ar \log(q + r - 30) \tag{43.5.1}$$

for a parameter $q > c$ with

$$s_0 = 1 + it_0, \qquad |t_0| < q^3, \qquad r \log q > c, \qquad A > \tfrac{1}{2}. \tag{43.5.2}$$

Let

$$\varphi(w) = e^{2\lambda w}\frac{e^{\lambda w} - e^{-\lambda w}}{2\lambda w} \tag{43.5.3}$$

and

$$J(s_0, g) = \frac{1}{2\pi i}\int_{(10)} \varphi^k(w) g(s_0 + w)\, dw = \sum_n{}' \frac{b_n}{n^{s_0}} h_k(2\lambda k - \log n, \lambda) \tag{43.5.4}$$

with the notation of (34.1.1), where, according to Lemma 34.1, Σ' means summation over $e^{\lambda k} < n < e^{3\lambda k}$. Let

$$m = 108A\delta \log q \le k \le 120A\delta \log q, \qquad \frac{c}{\log q} < \delta < 2, \qquad \lambda = \frac{1}{28}, \tag{43.5.5}$$

and

$$|J(s_0, g)| < q^{-192A\delta} \tag{43.5.6}$$

for all k in (43.5.5). Suppose that we may push the line of integration in (43.5.4) to (-8.5) and

$$|g(-7.5 + it)| < c \log\{(q|t| + 1)\}. \tag{43.5.7}$$

This yields

$$J(s_0, g) = \sum_\rho \varphi^k(\rho - s_0) + \int_{(-8.5)} \varphi^k(w) g(s_0 + w)\, dw. \tag{43.5.8}$$

By (43.5.5)

$$|\varphi^k(w)| \le \left(\frac{e^{-3\lambda 8.5} + e^{-\lambda 8.5}}{2|w|\lambda}\right)^k < 16\frac{e^{-8.5\lambda k}}{|w|^2} < \frac{q^{-400A}}{|w|^2} \qquad (\operatorname{Re} w = -8.5).$$

So for the last integral,

$$\left|\int_{-8.5}\right| < cq^{-400A}\int_{-\infty}^{\infty} \frac{\log q(|t| + |t_0| + 1)}{t^2 + 1}\, dt < q^{-199A\delta}, \tag{43.5.9}$$

also taking (43.5.7) and (43.5.2) into account. The variable ρ runs through the poles of $g(s)$ in $-8.5 < \sigma \leq 1$, each occurring with the corresponding residue as multiplicity. To complete the proof we show that under these conditions these ρ satisfy $|\rho - s_0| > \delta$.

43.6

For $\varphi(w)$ in (43.5.3) we have the upper bound $1/\lambda w$ for $\operatorname{Re} w \leq 0$, and so by (43.5.1) and (43.5.5) we can estimate the contribution of $|\rho - s_0| \geq 12\delta$ in the sum of (43.5.8) as

$$\left| \sum_{|\rho - s_0| \geq 12\delta} \varphi^k(\rho - s_0) \right| \leq \int_{12\delta}^{\infty} \frac{1}{(\lambda r)^k} dN(r, s_0, q)$$

$$\leq \frac{k}{\lambda k} \int_{12\delta}^{\infty} \frac{N(r, s_0, q)}{r^{k+1}} dr < \frac{kA}{\lambda k} \int_{12\delta}^{\infty} \frac{\log(q + r)}{r^k}$$

$$< c \frac{A \log q}{\lambda^k (12\delta)^{k-1}} < c \frac{A\delta \log q}{6^k} < q^{-192A\delta}$$

$$(\delta \log q > c).$$

Putting this and (43.5.9) into (43.5.8) and at the same time using (43.5.6), we obtain

$$\left| \sum_{|\rho - s_0| < 12\delta} \varphi^k(\rho - s_0) \right| < q^{-191A\delta} \tag{43.6.1}$$

for all k in (43.5.5).

43.7

Suppose now, contrary to what has been stated, there is a $\rho = \rho_0$ with $|\rho_0 - s_0| < \delta$. We use the second main theorem in the form (33.12.1). For the number of terms N in (43.6.1) we have, by (43.5.1), the upper bound

$$N = A12\delta \log q$$

since $\delta \leq 2$. For the term corresponding to ρ_0 we have the lower bound

$$|\varphi^k(\rho_0 - s_0)| \geq \min_{|w| = 1/2} \left| e^{2w} \frac{e^w - e^{-w}}{2w} \right| > e^{-1.05k} \geq q^{-126A\delta}$$

by (43.5.5). The second main theorem shows that there exists a k in (43.5.5) such that

$$\left| \sum_{|\rho - s_0| < 12\delta} \varphi^k(\rho - s_0) \right| \geq \left(\frac{A \cdot 12\delta \log q}{8e \cdot 120 A\delta \log q} \right)^{12A\delta \log q} q^{-126A\delta}$$

$$> q^{-191A\delta},$$

which contradicts (43.6.1). ■

43.8

The corollary follows directly from Theorem 43.1 by covering the rectangle in question by circles of radius $\delta = 2(1 - \alpha)$. If the condition $\delta > c/\log q$ in (43.2.5) is not satisfied, we cover by circles of radius $\delta = c/\log q$.

44

THE LEAST PRIMES IN ARITHMETIC PROGRESSIONS II

The Effect of Exceptional Zeros

44.1

It is known that among all the $\varphi(q)$ L-functions mod q there is, at most, one that can vanish in the region $\delta > 1 - c/\log\{q(|t| + 1)\}$ for a suitable $c > 0$. The corresponding χ is necessarily real, and the corresponding $L(s, \chi)$ has only a simple real zero in this region, so-called the "exceptional" or "Siegel" zero.

There are remarkable connections between the exceptional zeros and the other zeros called Deuring–Heilbronn phenomenon after Deuring and Heilbronn who connected the exceptional zeros to the zeros of the ζ-function and the zeros of the L-functions, respectively. The following theorem shows that the closer the exceptional zero is to 1, the larger the zero-free region.

Theorem 44.1 (Linnik). *Suppose that for a real character* χ_1 mod q, $L(\beta, \chi_1) = 0$ *with* $1 - \beta < c/\log q (c < \frac{1}{2})$. *Then none of the L-functions* mod q *vanishes in the rectangle*

$$1 - \frac{\log[1/(1 - \beta)\log q]}{385A \log q} < \sigma \qquad |t| < q^3$$

with the exception of $L(s, \chi_1)$ *which has only the simple zero* β *there.* [*A is defined in* (43.2.1)].

The proof, as we mentioned, essentially follows Knapowski [13].

44.2

Suppose first that $\chi \neq \chi_1$. Then $\chi\chi_1$ is not the principal character, and we can take for $g(s)$ in 43.5

$$g(s) = \frac{L'}{L}(s, \chi) + \frac{L'}{L}(s, \chi \cdot \chi_1). \tag{44.2.1}$$

With the notation (43.5.4) and $x = e^{\lambda k}$ we have

$$|J(s_0, g)| = \left| \sum_{x<n<x^3} \frac{\Lambda(n)}{n^{s_0}} \chi(n)[1 + \chi_1(n)]\, h_k(2\lambda k - \log n, \lambda) \right| \tag{44.2.2}$$

$$< \frac{\log x^3}{\lambda} \sum_{x<p^l<x^3} \frac{1 + \chi_1(p^l)}{p^l} = 3k \sum_{x<p^l<x^3} \frac{1 + \chi_1(p^l)}{p^l},$$

using (43.5.2) and the first inequality in (37.3.5).

We make only the following restrictions on the parameters:

$$4 \log q < \log x = \lambda k \leq 240A \log q, \qquad q > c. \tag{44.2.3}$$

44.3

For the contribution of $l \geq 2$ in the sum of (44.2.2) we have

$$\sum_{\substack{p^l > x \\ l \geq 2}} \frac{1 + \chi_1(p^l)}{p^l} < \sum_{p \leq \sqrt{x}} \frac{4}{x} + \sum_{p > \sqrt{x}} \frac{4}{p^2} < \frac{9}{\sqrt{x}}. \tag{44.3.1}$$

To estimate the contribution of terms with $l = 1$ we introduce

$$a_n = \sum_{d|n} \chi_1(d).$$

This is a nonnegative multiplicative function with the corresponding Dirichlet series

$$G(s) = \sum_{n=1}^{\infty} \frac{a_n}{n^s} = \zeta(s) L(s, \chi_1). \tag{44.3.2}$$

Let

$$S = \sum_{n \leq x} \frac{a_n}{n}. \tag{44.3.3}$$

Then

$$\text{(44.3.4)} \qquad S \sum_{x<p<x^3} \frac{a_p}{p} \le \sum_{x<n<x^4} \frac{a_n}{n}$$

since an n in the latter range has, at most, one representation $n = mp$, $m \le x$, $x < p < x^3$.

44.4

Let

$$\text{(44.4.1)} \qquad \gamma = 1 + \frac{1}{\log x}$$

and consider, with (44.3.2),

$$\text{(44.4.2)} \qquad U = \frac{1}{2\pi i}\int_{(10)} \frac{x^s}{s}\left(\frac{\gamma-1}{s-\gamma} + \frac{1-\beta}{s-\beta}\right)G(s)\,ds.$$

Insert the series representation of $G(s)$ in the integral. Calculating

$$\frac{1}{2\pi i}\int_{(10)} \frac{a^s}{s(s-K)}\,ds = \begin{cases} \dfrac{a^K-1}{K} < \dfrac{a^K}{K} & \text{for } a > 1 \\ 0 & \text{for } 0 < a \le 1 \end{cases}$$

we obtain (observe $\beta < 1$)

$$\text{(44.4.3)} \qquad U < \frac{\gamma-1}{\gamma} x^\gamma \sum_{n<x} \frac{a_n}{n^\gamma} + \frac{1-\beta}{\beta} x \sum_{n<x} \frac{a_n}{n}.$$

On the other hand, push the line of integration in (44.4.2) to the line $(\frac{1}{2})$ and observe that $G(\beta) = 0$ [see (44.3.2)] while the pole of $G(s)$ at $s = 1$ is cancelled by the zero at $s = 1$ of the function in the bracket in (44.4.2). We then obtain

$$U = \frac{\gamma-1}{\gamma} x^\gamma G(\gamma) + \frac{1}{2\pi i}\int_{(1/2)}$$

where, by use of the well-known bounds for the ζ- and L-function, the last integral is majorized by

$$c(qx)^{1/2} < x^{3/4}$$

by (44.2.3). With (44.3.2) and (44.4.3), this implies

$$\frac{\gamma - 1}{\gamma} x^{\gamma} \sum_{n \geq x} \frac{a_n}{n^{\gamma}} < \frac{1 - \beta}{\beta} x \sum_{n < x} \frac{a_n}{n} + x^{3/4},$$

or, by use of (44.4.1), (44.3.3), and (44.2.3),

$$\sum_{x < n < x^4} \frac{a_n}{n} < e^4 \sum_{n > x} \frac{a_n}{n^{\gamma}} < 2e^3(1 - \beta) S \log x + x^{-1/4} \log x$$

$$< cS(1 - \beta) \log q.$$

In the last step we also made use of $S \geq a_1 = 1$ and the classical $1 - \beta > 1/q^{2/3}$. In (44.3.4) this yields

$$\sum_{x < p < x^3} \frac{a_p}{p} < c(1 - \beta) \log q,$$

and in (44.2.2), also by use of (44.3.1) and $a_p = 1 + \chi_1(p)$,

(44.4.4) $$|J(s_0, g)| < 3kc(1 - \beta) \log q + \frac{27k}{\sqrt{x}} < ck(1 - \beta) \log q.$$

44.5

Now define

$$\delta = \frac{\log[1/(1 - \beta) \log q]}{385A \log q}.$$

Restrict the parameters as in 43.5 and (43.5.5) and we have

$$|J(s_0, q)| < c\delta(1 - \beta) \log^2 q = c\delta \log q \cdot q^{-385A\delta} < q^{-384A\delta}.$$

The conditions of Section 43.5 will be satisfied for $(1 - \beta)\log q < c$ with (44.2.1) if, in accordance with the latter, we replace A by $2A$. Thus $g(s)$, hence also $L'/L(s, \chi)$, has no poles in $|s - s_0| \leq \delta$, and since s_0 was arbitrary, satisfying only (43.5.2), Theorem 44.1 follows for $\chi \neq \chi_1$.

44.6

To cancel the pole coming from the pole of the principal L-function when $\chi = \chi_1$, define

$$g(s) = \frac{L'}{L}(s, \chi_1) + \frac{\zeta'}{\zeta}(s + 1 - \beta).$$

Here, in place of (44.2.2), we get

$$J(s_0, g) \leq \frac{\log x^3}{\lambda} \sum_{x<p^l<x^3} \frac{1+\chi_1(p^l)}{p^l} + \frac{1}{\lambda} \sum_{x<n<x^3} \frac{\Lambda(n)(1-n^{1-\beta})}{n},$$

and by (44.2.3) this will give only the additional term

$$\frac{1}{\lambda} \sum_{x<n<x^3} \frac{\Lambda(n)(1-\beta)\log n}{n} < c\frac{(1-\beta)\log^2 x}{\lambda} < c(1-\beta)k\log q$$

in (44.4.4), otherwise the proof remains unchanged. $(\zeta'/\zeta)(s+1-\beta)$ decreases the residue of $(L'/L)(s, \chi_1)$ by 1 at β, and that is why we can get information only about the remaining zero.

45

THE LEAST PRIMES IN ARITHMETIC PROGRESSIONS III

Completion of the Proof

45.1

In possession of the two main lemmas proved in the preceding two sections we now complete the proof of Linnik's theorem (43.1.1). This part of the proof does not use the power sum method. We follow a simplified version given by Jutila [2].

45.2

Let

$$\lambda = b \log q \tag{45.2.1}$$

where the constant b will be defined by subsequent lower restrictions. With the notation (34.1.1) we have

$$-\frac{1}{2\pi i}\int_{(10)} \frac{1}{\varphi(q)} \sum_{\chi} \bar{\chi}(l) \frac{L'}{L}(w+1, \chi)\varphi^2(w)\, dw \tag{45.2.2}$$

$$= \sum_{n \equiv l (\mathrm{mod}\ q)} \frac{\Lambda(n) h_\lambda(4\lambda - \log n, \lambda)}{n}$$

(by Lemma 34.1); Σ' means summation over $e^{2\lambda} < n < e^{6\lambda}$ and

(45.2.3) $$\varphi(w) = e^{2\lambda w}\frac{e^{\lambda w} - e^{-\lambda w}}{2\lambda w}.$$

The contribution of $n = p^s$, $s \geq 2$ in (45.2.2) is at most (for $b > 3$)

(45.2.4) $$\frac{6}{\lambda}\sum_{\substack{p^s > e^{2\lambda}\\ s\geq 2}} \frac{1}{p^s} < 30e^{-\lambda} < q^{-3},$$

proved in the same way as in (44.2.2) and (44.3.1). We must show that there will be remaining terms.

45.3

Shift the path of integration to $(-\frac{3}{2})$. For $b > 1$ the integral becomes

(45.3.1) $$\left|\int_{(-3/2)}\right| < cq^{-3b}\int_{-\infty}^{\infty}\frac{\log q(|t| + 1)}{|t|^2 + 1}\,dt < q^{-3},$$

using

$$|\varphi(w)|^2 \leq \left|\frac{e^{-9/2\lambda} + e^{-3/2\lambda}}{2\lambda w}\right|^2 < \frac{e^{-3\lambda}}{|w|^2} = \frac{q^{-3b}}{|w|^2}$$

and well-known estimations for $L'/L(s, \chi)$. For the residues,

(45.3.2) $$\frac{1}{\varphi(q)}\left[1 - \sum_{\chi}\bar{\chi}(l)\sum_{\rho\chi}\varphi^2(\rho_\chi - 1)\right] \geq \frac{1}{\varphi(q)}\left(1 - \sum_{\rho}|\varphi^2(\rho - 1)|\right),$$

where ρ_χ and ρ run through the zeros in $\sigma \geq 0$ of $L(s, \chi)$ and of all the L-functions mod q, respectively.

45.4

We first treat the ρ in $|t| \leq q^3/\log q$, $0 \leq \sigma < 1$. For a suitable $c < \frac{1}{2}$ there is at most one such ρ in $1 - (c/\log q) < \sigma < 1$, the exceptional zero β in the sense of Section 44. Theorem 44.1 and classical properties of zeros imply that for

$$\alpha_0 = 1 - c\frac{\log[1/(1 - \beta_0)\log q]}{\log q} < \sigma < 1, \qquad |t| \leq \frac{q^3}{\log q}$$

there is at most one zero, namely, β_0 where $\beta_0 = \beta$ if β exists and $\beta_0 = 1 - 0.5/\log q$ otherwise. From (45.2.3)

$$|\varphi(w)| = \left| e^{\lambda w} \frac{e^{2\lambda w} - 1}{2\lambda w} \right| \leq q^{b(\alpha - 1)}$$

(Re $w = \alpha - 1$ or Im $w = \pm q^{b(1-\alpha)}/b \log q$, Re $w \leq 0$).

Using the corollary to Theorem 43.1 and (45.2.1) with b sufficiently large, we infer

(45.4.1)

$$\begin{aligned}
\sum_{|\mathrm{Im}\,\rho| \leq q^3/\log q} |\varphi^2(\rho - 1)| &\leq q^{2b(\beta_0 - 1)} + \int_{\alpha_0}^{0} q^{2b(\alpha-1)} dN(\alpha) \\
&< 1 - b \log q(1 - \beta_0) + N(0) q^{2b} \\
&\quad + 2b \log q \int_0^{\alpha_0} q^{2b(\alpha-1)} N(\alpha)\, d\alpha \\
&< 1 - b \log q(1 - \beta_0) + q^{c-b} \\
&\quad + 2b \log q \int_0^{\alpha_0} q^{b(\alpha-1)} q^{c(1-\alpha)}\, d\alpha \\
&< 1 - b \log q(1 - \beta_0) + q^{-1} + 3q^{(c-b)(1-\alpha_0)} \\
&< 1 - b \log q(1 - \beta_0) + q^{-1} + (1 - \beta_0) \log q \\
&< 1 - (1 - \beta_0) \log q + q^{-1}.
\end{aligned}$$

45.5

Since

$$|\varphi(w)| \leq \frac{1}{|w|} \qquad (\mathrm{Re}\, w \leq 0),$$

for the remaining part we have

(45.5.1)
$$\begin{aligned}
\sum_{\rho} |\varphi^2(\rho - 1)| &< \sum_{s=[q^3/\log q]}^{\infty} \frac{1}{s^2} \sum_{s < |\mathrm{Im}\,\rho| \leq s+1} 1 \\
&< \sum_{s=[q^3/\log q]}^{\infty} \frac{cq \log s}{s^2} < q^{-1}
\end{aligned}$$

45.6

Combining (45.5.1) and (45.4.1) and using $1 - \beta_0 > q^{-1}$, the residue sum (45.3.2) has the lower bound

$$\frac{1}{\varphi(q)}\left[(1 - \beta_0)\log q - 2q^{-1}\right] > \frac{1}{q^2}.$$

This, together with (45.3.1) and (45.2.4), implies

$$\sum_{\substack{q^{2b}<p<q^{6b} \\ p\equiv l(\mathrm{mod}\ q)}} \frac{\log p\, h_\lambda(4\lambda - \log p, \lambda)}{p} > 0,$$

and this proves (43.1.1) with $c = 6b$.

46

SIEGEL'S THEOREM

After the results of Sections 43 and 44 we are in a position to quickly obtain a theorem of great importance, referred to in the title. It states that $L(s, q, \chi) \neq 0$ for $s > 1 - c(\varepsilon)q^{-\varepsilon}$ for any $\varepsilon > 0$ and $\chi \bmod q$. This is the best that is known today.

The following proof is again due to Knapowski. Siegel actually proved that $L(\beta, q, \chi) = 0$ for $\beta > 1 - q^{\varepsilon}$ implies $L(s, q', \chi') \neq 0$ for $s > 1 - c\varepsilon$ for any real character χ' and $q' < c(q)$ where $c(q) \to \infty$ as $q \to \infty$, from which the above theorem obviously follows. Knapowski [11] essentially proved the following theoretically stronger result.

Theorem 46.1. *Suppose that* $L(\beta, q, \chi) = 0$ *with* $1 - \beta < c/\log q$ ($c < \frac{1}{2}$ *is sufficiently small*). *Let* χ' *be any other character belonging to a modulus* $q' \leq q$ *but not equivalent to* χ. *Then* $L(s, q', \chi')$ *does not vanish in the rectangle*

$$1 - c\frac{\log \dfrac{1}{(1-\beta)\log q}}{\log q} < \sigma \leq 1, \qquad |t| < q^3.$$

The restriction to nonequivalent characters is not important: in the alternative case, Theorem 44.1 gives the information. The present theorem is in fact an extension of the latter to the case of different moduli, and the same proof works.

46.2

Since χ and χ' are nonequivalent, $\chi\chi'$ is not a principal character, and Sections 44.2–44.5 are valid if we replace χ by χ' and χ_1 by χ; the fact that χ (in the notation of Section 44.2) would then belong to a different modulus is inessential for Sections 44.2–44.4 because only the fact that $|\chi| \leq 1$ was used there. The only difference is that A in Section 44.4 must be defined differently. What is needed is the assumption (43.5.1) with $g(s) = L'/L(s, \chi') + L'/L(s, \chi\chi')$; this is satisfied in the present case with a constant. The proof of Theorem 46.1 is now complete. ■

47

THE REMAINDER TERM OF THE PRIME NUMBER FORMULA FOR ARITHMETIC PROGRESSION

47.1

In this section we compare the order of magnitude of

$$(47.1.1)\quad \Delta(x, q, l) \doteq \psi(x, q, l) - \frac{x}{\varphi(q)} \doteq \sum_{\substack{n \equiv l(q) \\ n \leq x}} \Lambda(n) - \frac{x}{\varphi(q)}$$

and that of $\Delta(x, q, 1)$ for fixed $q \geq 3$.

It can be shown by classical methods that if q is fixed and with an $\frac{1}{2} \leq \vartheta < 1$

$$(47.1.2)\qquad \Delta(x, q, 1) = O(x^{\vartheta+\varepsilon})$$

for every $\varepsilon > 0$, then

$$(47.1.3)\qquad \Delta(x, q, l) = O(x^{\vartheta+\varepsilon})$$

for every $\varepsilon > 0$ and all l.

However, these methods can not reduce the relation (47.1.2)–(47.1.3) to an explicit inequality. A result of this type was proved by Knapowski [1], using our power sum method.

Theorem 47.1 (Knapowski). *Let*

$$\alpha(t) = \max_{|\rho| \le t} \operatorname{Re} \rho \tag{47.1.4}$$

where ρ runs through all zeros of Dirichlet $\mathscr{L}$*-functions* mod q, *and let*

$$\delta(T) = \alpha(\sqrt{T}) - \alpha\left[\exp\left(\sqrt{\log_2 T}\right)\right]. \tag{47.1.5}$$

Then for

$$T \ge \max\left[c_0, \exp(q^{13})\right] \tag{47.1.6}$$

(*where* c_0 *is an effective constant*) *we have*

$$\max_{x \le T} |\Delta(x, q, l)| \le T^{\delta(T) + 9/\sqrt{\log_2 T}} \max_{x \le T} |\Delta(x, q, 1)|. \tag{47.1.7}$$

We note that the exponent of T in (47.1.7) is $o(1)$ for $T \to \infty$. Furthermore, under the assumption of the Riemann–Piltz conjecture, $\delta(T) \equiv 0$ (for $T > T_0$).

By partial summation our theorem is valid for

$$\Delta_1(x, q, l) \doteq \pi(x, q, l) - \frac{\operatorname{li} x}{\varphi(q)}, \tag{47.1.8}$$

and for the corresponding $\Delta_1(x, q, l)$ with the slight modification

$$\max_{x \le T} |\Delta_1(x, q, l)| \le T^{\delta(T) + 10/\sqrt{\log_2 T}} \left(\max_{x \le T} |\Delta_1(x, q, 1)| + \sqrt{T} \right). \tag{47.1.9}$$

It would certainly be of interest to show that $\Delta(x, q, 1)$ could be replaced by $\Delta(x, q, l')$ with any $(l', q) = 1$.

47.2

The crucial point of the proof will naturally be the lower estimate of $\Delta(x, q, 1)$ in terms of the zeros. The upper estimation of $\Delta(x, q, l)$ is carried out in the usual way.

Define

$$\begin{aligned} F(s) &\doteq \sum_{n \equiv 1(q)} \frac{\Lambda(n)}{n^s} - \frac{1}{\varphi(q)} \sum_n \frac{1}{n^s} \\ &= -\frac{1}{\varphi(q)} \left[\sum_\chi \frac{\mathscr{L}'}{\mathscr{L}}(s, \chi) + \zeta(s) \right], \end{aligned} \tag{47.2.1}$$

$$J(\xi) \doteq \frac{1}{2\pi i} \int_{(2)} \frac{\xi^s}{s^k} F(s)\, ds, \tag{47.2.2}$$

where k is a positive integer to be determined later, for which

$$\frac{5\log T}{\log_2 T} - (\log T)^{1/3} \le k \le \frac{5\log T}{\log_2 T} \tag{47.2.3}$$

holds, and

$$\xi \doteq \exp\left(\frac{\log_2 T \cdot k}{5}\right) \qquad (\le T). \tag{47.2.4}$$

By the integral formula

$$\frac{1}{2\pi i}\int_{(2)} \frac{x^s}{s^k}\,ds = \begin{cases} \dfrac{\log^{k-1} x}{(k-1)!} & \text{for} \quad x \ge 1 \\ 0 & \text{for} \quad 0 < x \le 1 \end{cases} \tag{47.2.5}$$

we have, from (47.2.1) and (47.2.2),

$$J(\xi) = \sum_{\substack{n\equiv 1(q)\\ n\le\xi}} \frac{\Lambda(n)\log^{k-1}(\xi/n)}{(k-1)!} - \frac{1}{\varphi(q)}\sum_{n\le\xi}\frac{\log^{k-1}(\xi/n)}{(k-1)!}. \tag{47.2.6}$$

Hence by Abel's inequality,

$$|J(\xi)| \le \frac{\log^{k-1}\xi}{(k-1)!}\max_{x\le T}|\Delta(x,q,1)|. \tag{47.2.7}$$

47.3

Shift the line of integration for the infinite broken line V defined by Lemma G.1 of Appendix G to obtain

$$J(\xi) = \frac{1}{2\pi i}\int_{(V)} \frac{\xi^s}{s^k}F(s)\,ds - \frac{1}{\varphi(q)}\sum{}' \frac{\xi^\rho}{\rho^k}, \tag{47.3.1}$$

where Σ' means that the summation is extended over all zeros of all $\mathscr{L}(s,\chi)$ functions mod q, lying to the right of V. By Lemma G.1 of Appendix G, and by (47.1.6), (47.2.3), and (47.2.4), the integral is

$$O(\xi^{1/100} \cdot 200^k \cdot q\log^2 q) = O(T^{1/50}). \tag{47.3.2}$$

The contribution of a zero $\rho = \beta + i\gamma$ with

$$|\gamma| > (\log T)^{1/4} \tag{47.3.3}$$

to the right of the broken line V is, by (47.2.3) and (47.2.4),

$$O\left(\frac{T}{\exp\left[\frac{1}{4}(\log_2 T)(k-2)\right]}\frac{1}{\gamma^2}\right)=0\left(\frac{1}{\gamma^2\log q}\right). \tag{47.3.4}$$

Thus the total contribution of all zeros with imaginary part satisfying (47.3.3) is $O(1)$, because the number of zeros of an $\mathscr{L}$-function mod q in the range $t \le |\gamma| \le t+1$ is

$$O[\log q(t+2)]. \tag{47.3.5}$$

From (47.3.1)–(47.3.4),

$$J(\xi)=-\frac{1}{\varphi(q)}\sum_{|\gamma|\le(\log T)^{1/4}}\frac{\xi^\rho}{\rho^k}+O(T^{1/50}). \tag{47.3.6}$$

47.4

The number of remaining zeros in (47.3.6) is

$$O\left\{\varphi(q)(\log T)^{1/4}\log\left[q(\log T)^{1/4}\right]\right\}<(\log T)^{1/3} \tag{47.4.1}$$

by (47.3.5) and (47.1.6).

Now we can use the second main theorem in the form given by (33.12.1) for

$$z_j=\frac{\exp(\rho_j\log_2 T/5)}{\rho_j} \tag{47.4.2}$$

their total number being at most $(\log \mathrm{T})^{1/3}$ by (47.4.1). By the definition of $\alpha(t)$ in (47.1.4), there exists a zero ρ_0 with

$$\beta_0=\alpha\left[\exp(\log_2 T)^{1/2}\right],\qquad |\rho_0|\le\exp(\log_2 T)^{1/2}\left[<(\log T)^{1/4}\right]. \tag{47.4.3}$$

So we have a k satisfying (47.2.3) for which

(47.4.4)

$$\frac{1}{\varphi(q)}\left|\sum_{|\gamma|\le(\log T)^{1/4}}{}'\frac{\xi^\rho}{\rho^k}\right|\ge\frac{1}{\varphi(q)}\left(\frac{1}{\log T}\right)^{(\log T)^{1/3}}\left(\frac{\exp(\beta_0\log_2 T/5)}{|\rho_0|}\right)^k$$

$$>\exp\left\{-\log q-(\log T)^{1/3}\log_2 T\right.$$

$$+\left(\frac{\beta_0\log_2 T}{5}-(\log_2 T)^{1/2}\right)$$

$$\left.\cdot\left(\frac{5\log T}{\log_2 T}-(\log T)^{1/3}\right)\right\}>\exp\left\{\beta_0\log T-6\frac{\log T}{(\log_2 T)^{1/2}}\right\}.$$

Thus we have for this ξ

$$|J(\xi)| > T^{\beta_0}\exp\left(-\frac{7\log T}{(\log_2 T)^{1/2}}\right) \tag{47.4.5}$$

by (47.3.6).

47.5

By (47.1.6), (47.2.3), and (47.2.4) we have

$$\frac{\log^{k-1}\xi}{(k-1)!} < \left[\frac{\log T}{(k-1)/4}\right]^{k-1} < (\log_2 T)^{k-1} \tag{47.5.1}$$

$$< \exp\left(\log_3 T \frac{5\log T}{\log_2 T}\right) < \exp\left[\frac{\log T}{(\log_2 T)^{1/2}}\right].$$

Expressions (47.2.7), (47.4.3), and (47.4.5), with (47.5.1) together imply

$$\max_{x\le T}|\Delta(x,q,1)| > T^{\alpha\exp[\sqrt{(\log_2 T)}]}\exp\left[-8\frac{\log T}{\sqrt{(\log_2 T)}}\right]. \tag{47.5.2}$$

On the other hand, the finite explicit prime number formula for the arithmetic progression

$$\Delta(x,q,l) = -\frac{1}{\varphi(q)}\sum_{\chi}\bar{\chi}(l)\sum_{|\rho|\le\sqrt{x}}\frac{x^{\rho}}{\rho} + O\left[\sqrt{x}\log^2(qx)\right] \tag{47.5.3}$$

(see Prachar [1], Satz 4.4, p. 228) and (47.3.5) imply the estimate

$$\max_{x\le T}|\Delta(x,q,l)| = O\left[T^{\alpha(\sqrt{T})}\log^2(qT)\right]. \tag{47.5.4}$$

Recalling the notation (47.1.5), the theorem is an obvious consequence of (47.5.2) and (47.5.4).

48

COMPARATIVE PRIME NUMBER THEORY I

Sign Changes of $\psi(x, q, l_1) - \psi(x, q, l_2)$

48.1

The investigations in the theory of distribution of primes in various residue classes mod q lead in two directions. The first—and prevailing—direction attempts to exhibit uniformity of the distribution. As its first main result the relation

$$\pi(x, q, l) \sim \frac{\operatorname{li} x}{\varphi(q)} \tag{48.1.1}$$

can be considered, where $(l, q) = 1$ and $\pi(x, q, l)$ denotes the number of primes $p \leq x$ with $p \equiv l \pmod{q}$.

The second direction intends to exhibit discrepancies and is due to Chebyshev [1] as early as in 1853. He asserted*—imprecisely expressed—that "there are more primes congruent to 3 (mod 4) than ≡ 1 (mod 4)." The exact statement of Chebyshev's assertion is that

$$\lim_{r \to +0} \sum_{p>2} (-1)^{(p-1)/2} e^{-pr} = -\infty \tag{48.1.2}$$

which, if true, would allow one to assert only that the preponderance of the primes ≡ 3 (mod 4) holds only in the sense of "Abel summation." During the century that has passed since Chebyshev's announcement only three papers

*But never published his proof.

were written—Landau [1, 2] and Hardy–Littlewood [1]—that dealt with conjecture (48.1.2). None of these papers decided the falsity or truth of it (the problem is open to this day). What they did prove was its equivalence with the (very deep) conjecture that the function

$$\mathscr{L}(s) = \sum_{n=1}^{\infty} \frac{(-1)^n}{(2n+1)^s} \tag{48.1.3}$$

does not vanish in the half-plane $\sigma > \frac{1}{2}$, a special case of the Riemann–Piltz conjecture.

Littlewood disproved [2] the direct interpretation of Chebyshev's assertion that the function

$$\pi(x,4,1) - \pi(x,4,3) \tag{48.1.4}$$

is negative for all $x \geq x_0$. This was the only unconditional result until 1960 concerning the discrepancies of the distribution of primes in various arithmetical progressions.

A systematic study of these questions and subsequent development of comparative prime number theory was made in two series of papers written by S. Knapowski and the author of this book (Knapowski–Turán [1–15]). The main tool of this theory was our power sum method, releasing the theory from its apparent deadlock. In Sections 48–56 we discuss the main results of the comparative prime number theory. These are all contained (with the exception of results in Section 52) in Knapowski–Turán [1–15].

48.2

For the exact formulation of the problems we use the notations

$$\pi(x,q,l) = \sum_{\substack{p \leq x \\ p \equiv l(q)}} 1 \tag{48.2.1}$$

$$\Pi(x,q,l) = \sum_{\substack{n \leq x \\ n \equiv l(q)}} \frac{\Lambda(n)}{\log n}$$

$$\Theta(x,q,l) = \sum_{\substack{p \leq x \\ p \equiv l(q)}} \log p$$

$$\psi(x,q,l) = \sum_{\substack{n \leq x \\ n \equiv l(q)}} \Lambda(n),$$

and, as in (40.1.1)–(40.1.4),

(48.2.2)
$$\Delta_1(x,q,l_1,l_2) = \pi(x,q,l_1) - \pi(x,q,l_2)$$
$$\Delta_2(x,q,l_1,l_2) = \Pi(x,q,l_1) - \Pi(x,q,l_2)$$
$$\Delta_3(x,q,l_1,l_2) = \Theta(x,q,l_1) - \Theta(x,q,l_2)$$
$$\Delta_4(x,q,l_1,l_2) = \psi(x,q,l_1) - \psi(x,q,l_2).$$

The main problems are the following (in Sections 48–56 we always assume that $l_1 \not\equiv l_2(q)$, $(l_1, q) = (l_2, q) = 1$):

PROBLEM 1

Do all the functions $\Delta_i(x,q,l_1,l_2)$ have sign changes? If so, give an upper bound for the first sign change.

PROBLEM 2

Do all the functions $\Delta_i(x,q,l_1,l_2)$ have an infinity of sign changes?

If so, denote the number of sign changes of $\Delta_i(x,q,l_1,l_2)$ in $[1, Y]$ by $V_i(Y,q,l_1,l_2)$. The next natural question is the following:

PROBLEM 3

What is the asymptotic behavior of $V_i(Y,q,l_1,l_2)$?

To give an exact characterisation of the asymptotic behavior of $V_i(Y,q,l_1,l_2)$ would be completely hopeless; we would be content with lower estimates on the number of sign changes.

A solution to Problem 4 would also furnish an answer to the previous three problems.

PROBLEM 4

Find a function $h_i(T,q)$ such that every function $\Delta_i(x,q,l_1,l_2)$ has a sign change in every interval of the form $[h_i(T,q), T]$ for $T > c_i(q)$.

All these problems can be considered under some imposed hypothesis concerning the zeros of Dirichlet's $\mathscr{L}$-functions (e.g., assuming the Riemann–Piltz conjecture). As with the behavior of $\pi(x) - \operatorname{li} x$, discussed in Sections 41–42, the problem of finding effective estimates plays an essential role in these questions. It would be important to find effective, unconditional answers for the preceding problems.

48.3

A brief consideration of the formula

$$(48.3.1)\quad \Delta_4(x,q,l_1,l_2)=\frac{1}{\varphi(q)}\sum_{\chi}[\bar{\chi}(l_2)-\bar{\chi}(l_1)]\sum_{|\gamma_\rho|\le x^3}\frac{x^\rho}{\rho}$$

$$+O\left[\log^2(qx)\right]$$

shows us that no affirmative answer to Problem 2 can be given without some unproved hypothesis for $\mathscr{L}$-functions. (At the same time one naturally conjectures that the answer for Problem 2 is positive; this will be supported by the results of Sections 48–56 too.) Namely, assume there exists a real zero $\beta>\frac{1}{2}$ of an $\mathscr{L}(s,\chi')$ function mod q, such that all other zeros of all $\mathscr{L}$-functions mod q are in the half-plane $\sigma\le\beta-\delta$ for $\delta>0$ (a possibility which is not excluded so far). This χ' must obviously be real. In this case (48.3.1) implies the relation

$$(48.3.2)\quad \Delta_4(x,q,l_1,l_2)=\frac{1}{\varphi(q)}[\chi'(l_2)-\chi'(l_1)]\frac{x^\beta}{\beta}+0\left[x^{\beta-\delta}\log^2(qx)\right],$$

which means that if l_1 and l_2 are chosen so that $\chi'(l_1)\ne\chi'(l_2)$, then $\Delta_4(x,q,l_1,l_2)$ does not change sign if x is sufficiently large. (And the same is true for $1\le i\le 3$.)

Thus we always assume that for all $\mathscr{L}(s,\chi)$ mod q

$$(48.3.3)\quad \mathscr{L}(s,\chi)\ne 0\quad\text{for}\quad 0<\sigma<1,\qquad |t|\le A(q)\ (<1).$$

In other words, we suppose that no $\mathscr{L}$-function mod q vanishes in the real segment $0<\sigma<1$. This assumption we call the "Haselgrove condition," because Haselgrove was the first to calculate the number $A(q)$ for some values of q (in particular for $q\le 10$). We know the values of $A(q)$ for all $q<25$ by the calculations of R. Spira [1], and for $q<25$ the Haselgrove condition is certainly fulfilled; thus some of our theorems will be unconditional for $q<25$. On the other hand, we do not even know whether there are infinitely many q satisfying this condition [of course for any given q there is, in principle, no difficulty checking the condition, or determining $A(q)$].

48.4

Assuming only the Haselgrove condition, Problem 4 can be settled for the easiest case $i = 4$. As always in the following sections, we use the notations (48.2.2) in the formulation of our results, and we assume the natural conditions $(l_1, q) = (l_2, q) = 1$, $l_1 \not\equiv l_2(q)$. In the following sections a sum of the form $\Sigma_\chi[\bar{\chi}(l_2) - \bar{\chi}(l_1)]\Sigma_\rho f(\rho)$ occurs often. Here the sum is extended over all $\chi \bmod q$ with $\chi(l_1) \neq \chi(l_2)$, and $\rho = \beta + i\gamma$ denotes the nontrivial zeros of the corresponding $\mathscr{L}(s, \chi)$ function. $A(q)$ denotes the number in (48.3.3) appearing in the Haselgrove condition. The symbol c_ν always denotes positive, absolute, explicitly calculable constants.

Theorem 48.1. *If q satisfies the Haselgrove condition* (48.3.3) *and*

$$\omega \geq \max\left[\exp(q^{c_1}), \exp\left(\frac{40}{A(q)}\right)\right] \tag{48.4.1}$$

then $\Delta_4(x, q, l_1, l_2)$ has a sign change in the interval

$$\omega \leq x \leq e^{2\sqrt{\omega}}. \tag{48.4.2}$$

This theorem naturally gives a positive answer to Problems 1 and 2, namely, we have as a direct consequence the following:

Corollary I. *If q satisfies the Haselgrove condition* (48.3.3), *then $\Delta_4(x, q, l_1, l_2)$ has a sign change in the interval*

$$1 \leq x \leq \max\left[e_2(q^{c_1}), e_2\left(\frac{30}{A(q)}\right)\right]. \tag{48.4.3}$$

From Theorem 48.1 we immediately get the following.

Corollary II. *If q satisfies the Haselgrove condition* (48.3.3), *then $\Delta_4(x, q, l_1, l_2)$ changes sign infinitely often for $1 \leq x < \infty$.*

Naturally Theorem 48.1 also implies a (very weak) lower bound for $V_4(Y, q, l_1, l_2)$ and so answers Problem 3.

48.5

In the proof an important role is played by the following result of S. Knapowski [6]:

Lemma 48.1 (Knapowski). *For* $r > \exp(q^{c_2})$

$$\int_{r\exp(-\log^{0.8} r)}^{r} \frac{|\Delta_4(x, q, l_1, l_2)|}{x} > r^{1/4}. \tag{48.5.1}$$

We use this in the weakened form

$$\int_{\omega}^{\omega^5/2} \frac{|\Delta_4(x, q, l_1, l_2)|}{x} > c_3 \omega^{5/4}, \tag{48.5.2}$$

if $\omega > \exp(q^{c_2})$.

We sketch the proof. For a detailed proof see the original paper (Knapowski [6]). The proof uses our power sum method and requires the next result.

Lemma 48.2. *There exists a number* D *with*

$$\max(c_4, q^5) \le D \le \max(c_5, q^{c_6}) \tag{48.5.3}$$

such that

$$\left| \frac{1}{\varphi(q)} \sum_{\chi} [\bar{\chi}(l_2) - \bar{\chi}(l_1)] \sum_{\rho} D^{\rho} \left(\frac{e^{\eta\rho} - e^{-\eta\rho}}{2\eta\rho} \right)^2 \right| \ge D \log D \tag{48.5.4}$$

where

$$\eta = \frac{1}{3D}. \tag{48.5.5}$$

By Linnik's theorem there exists a prime number $D \equiv l_1(q)$ satisfying (48.5.3). [Although Linnik's theorem asserts only the upper bound for D, from the proof (see, e.g., Section 45) one can infer that the lower restriction in (48.5.3) is satisfied too.]

We start with the integral

$$\begin{aligned} I(\chi) &\doteq \frac{1}{2\pi i} \int_{(10)} D^s \left(\frac{e^{\eta s} - e^{-\eta s}}{2\eta s} \right)^2 \frac{\mathscr{L}'}{\mathscr{L}}(s, \chi)\, ds \\ &= -\sum_{n=1}^{\infty} \chi(n)\Lambda(n) \cdot \frac{1}{2\pi i} \int_{(10)} \left(\frac{D}{n} \right)^s \left(\frac{e^{\eta s} - e^{-\eta s}}{2\eta s} \right)^2 ds. \end{aligned} \tag{48.5.6}$$

But by Lemma 34.1 and (48.5.5) all the terms with

$$n \notin (De^{-2\eta}, De^{2\eta}) \subset (D - 1, D + 1) \tag{48.5.7}$$

have coefficient zero, whereas the term with $n = D$ has the coefficient

$$(48.5.8)\quad \frac{1}{2\pi i}\int_{(10)}\left(\frac{e^{\eta s}-e^{-\eta s}}{2\eta s}\right)^2 ds = \frac{1}{2\pi i}\int_{(0)}\left(\frac{e^{\eta s}-e^{-\eta s}}{2\eta s}\right)^2 ds$$

$$= \frac{1}{2\pi}\int_{-\infty}^{\infty}\left(\frac{\sin(\eta t)}{\eta t}\right)^2 dt = \frac{1}{2\eta}.$$

Shift the line of integration to $\sigma = -\frac{1}{2}$. Using Cauchy's theorem and the fact that $|(\mathcal{L}'/\mathcal{L})(s,\chi)| = O\{\log[q(|t|+2)]\}$ for $\sigma = -\frac{1}{2}$ [which follows easily from (G.1.2) of Appendix G] an easy computation shows

$$(48.5.9)\quad I(\chi) = \sum_{\rho} D^{\rho}\left(\frac{e^{\eta\rho}-e^{-\eta\rho}}{2\eta\rho}\right)^2 + O(\sqrt{D}\log D)$$

for $\chi \neq \chi_0$. Together with (48.5.6)–(48.5.8), (48.5.9) gives

$$(48.5.10)\quad \frac{-\chi(l_1)\log D\cdot 3D}{2} = \sum_{\rho} D^{\rho}\left(\frac{e^{\eta\rho}-e^{-\eta\rho}}{2\eta\rho}\right)^2 + O(\sqrt{D}\log D).$$

Now multiply by

$$\frac{1}{\varphi(q)}[\bar{\chi}(l_2)-\bar{\chi}(l_1)]$$

and sum over the χ to get (48.5.4), which proves Lemma 48.2.

Proof of Lemma 48.1. Let

$$(48.5.11)\quad A = \tfrac{1}{5}\log_2 r$$

$$(48.5.12)\quad B = (\log r)^{-1/4}$$

$$(48.5.13)\quad m = \frac{5\log r}{\log_2 r} - \log^{3/4} r$$

and let ν be an integer to be chosen later which satisfies

$$(48.5.14)\quad m \le \nu \le m + \log^{3/5} r.$$

With the abbreviation

$$(48.5.15)\quad F(s) \doteq \frac{1}{\varphi(q)}\sum_{\chi}[\bar{\chi}(l_2)-\bar{\chi}(l_1)]\frac{\mathcal{L}'}{\mathcal{L}}(s,\chi)$$

we consider

(48.5.16)

$$G(\nu) \doteq \frac{1}{2\pi i}\int_{(10)} D^s\left(\frac{e^{\eta s}-e^{-\eta s}}{2\eta s}\right)^2\left(e^{As}\frac{e^{Bs}-e^{-Bs}}{2Bs}\right)^\nu F(s)\,ds$$

$$= \sum_{n\equiv l_1(q)}\Lambda(n)\int_{(10)}\left(\frac{D}{n}\right)^s\left(\frac{e^{\eta s}-e^{-\eta s}}{2\eta s}\right)^2\left(e^{As}\frac{e^{Bs}-e^{-Bs}}{2Bs}\right)^\nu ds$$

$$- \sum_{n\equiv l_2(q)}\Lambda(n)\int_{(10)}\left(\frac{D}{n}\right)^s\left(\frac{e^{\eta s}-e^{-\eta s}}{2\eta s}\right)^2\left(e^{As}\frac{e^{Bs}-e^{-Bs}}{2Bs}\right)^\nu ds.$$

Noting that the terms of the series in (48.5.16) are nonzero only in

(48.5.17) $$X_1 \doteq De^{A\nu-B\nu-2\eta} < n < De^{A\nu+B\nu+2\eta} \doteq X_2,$$

we obtain, after integration by parts,

(48.5.18)

$$G(\nu) = -\int_{X_1}^{X_2}\Delta_4(x,q,l_1,l_2)$$

$$\cdot\frac{d}{dx}\left\{\frac{1}{2\pi i}\int_{(10)}\left(\frac{D}{x}\right)^s\left(\frac{e^{\eta s}-e^{-\eta s}}{2\eta s}\right)^2\left(e^{As}\frac{e^{Bs}-e^{-Bs}}{2Bs}\right)^\nu ds\right\}dx$$

$$= -\int_{X_1}^{X_2}\Delta_4(x,q,l_1,l_2)$$

$$\cdot\frac{d}{dx}\left\{\frac{1}{2\pi}\int_{-\infty}^{+\infty}\cos[t(\log D + A\nu - \log x)]\left[\frac{\sin(\eta t)}{\eta t}\right]^2\left(\frac{\sin(Bt)}{Bt}\right)^\nu dt\right\}dx$$

$$= -\int_{X_1}^{X_2}\frac{\Delta_4(x,q,l_1,l_2)}{x}\cdot\frac{1}{2\pi}\int_{-\infty}^{+\infty} t\cdot\sin[t(\log D + A\nu - \log x)]$$

$$\cdot\left[\frac{\sin(\eta t)}{\eta t}\right]^2\left[\frac{\sin(Bt)}{Bt}\right]^\nu dt.$$

From this,

(48.5.19) $$|G(\nu)| \le \int_{X_1}^{X_2}\frac{|\Delta_4(x,q,l_1,l_2)|}{x}\cdot\frac{1}{2\pi}\int_{-\infty}^{+\infty}\left|t\left(\frac{\sin(Bt)}{Bt}\right)^\nu\right|dt$$

$$< \frac{1}{B^2}\int_{X_1}^{X_2}\frac{|\Delta_4(x,q,l_1,l_2)|}{x}\,dx.$$

Considering $r \geq \exp(q^{c_2})$, (48.5.11)–(48.5.14), and (48.5.17), we have $[X_1, X_2] \subset [r\exp(-\log^{0.8} r), r]$, and so (48.5.19) implies

(48.5.20) $$\int_{r\exp(-\log^{0.8} r)}^{r} \frac{|\Delta_4(x, q, l_1, l_2)|}{x}\,dx > \frac{|G(\nu)|}{\sqrt{\log r}}.$$

We must estimate $|G(\nu)|$ from below by an appropriate choice of ν.

Applying Cauchy's theorem (48.5.15)–(48.5.16) yield

(48.5.21)

$$G(\nu) = \frac{1}{\varphi(q)} \sum_{\chi} [\bar{\chi}(l_2) - \bar{\chi}(l_1)] \sum_{\rho}{}' D^{\rho} \left(\frac{e^{\eta\rho} - e^{-\eta\rho}}{2\eta\rho} \right)^2 \left(e^{A\rho} \frac{e^{B\rho} - e^{-B\rho}}{2B\rho} \right)^{\nu}$$

$$+ \frac{1}{2\pi i} \int_{(V)} D^{s} \left(\frac{e^{\eta s} - e^{-\eta s}}{2\eta s} \right)^2 \left(e^{As} \frac{e^{Bs} - e^{-Bs}}{2Bs} \right)^{\nu} F(s)\,ds$$

where Σ' indicates that the sum is extended only over zeros to the right of V, defined by Lemma G.1 of Appendix G.

By Lemma G.1 and (48.5.11)–(48.5.14) (after some computation) the integral has the upper bound

(48.5.22) $$O(r^{1/50}).$$

For the contribution of zeros with

(48.5.23) $$|\gamma| > \sqrt{\log r}$$

we get, again after some computation, $O(1)$. Thus we only have to consider the finite power sum

(48.5.24) $$K(\nu) = \sum_{j=1}^{n} b_j z_j^{\nu}$$

where

(48.5.25) $$b_j = \frac{1}{\varphi(q)} [\bar{\chi}(l_2) - \bar{\chi}(l_1)]\, D^{\rho} \left(\frac{e^{\eta\rho} - e^{-\eta\rho}}{2\eta\rho} \right)^2$$

(48.5.26) $$z_j = e^{A\rho} \frac{e^{B\rho} - e^{-B\rho}}{2B\rho} \qquad \left(|\gamma| \leq \sqrt{\log r}\right),$$

and by (47.3.5)

(48.5.27) $$n < c_7 \varphi(q) \log\left[q\left(\sqrt{\log r} + 2\right)\right] \sqrt{\log r} < \log^{3/5} r.$$

Let ρ_1 be the zero for which the corresponding z_1 has the maximal absolute value* (among those with $|\gamma| \le \sqrt{\log r}$) and let ρ_2 be the zero for which the corresponding z_λ has the minimal absolute value among those with

$$\beta \ge \tfrac{7}{27}, \qquad |\gamma| \le D^{5/2} \qquad (\le \log^{1/6} r) \tag{48.5.28}$$

(if $c_2 \ge 15c_6$). Then by (48.5.11)–(48.5.13) and (48.5.27)–(48.5.28),

$$\frac{|z_\lambda|}{|z_1|} > \frac{e^{(7/27)A}\cdot\frac{1}{2}}{e^A\cdot 2} > \frac{1}{(\log r)^{1/5}} > \frac{4n}{m+n}. \tag{48.5.29}$$

Now we apply, with $\lambda_1 = n$, the power sum theorem of S. Knapowski mentioned without proof in Section 8.6 with the slight modification that we do not normalize the power sum, that is, $\max|z_j| \ne 1$. First we estimate the expression

$$B_1 = \min_{\lambda\le j\le n} |b_1 + b_2 + \cdots + b_j| \tag{48.5.30}$$

that occurs in the inequality.

Since all the b_j belonging to zeros with (48.5.28) are contained in B_1, Lemma 48.2 implies

$$\begin{aligned} B_1 > D\log D &- \frac{2}{\varphi(q)}\sum_\chi \sum_{|\gamma|>D^{5/2}} D\left|\frac{e^{\eta\rho}-e^{-\eta\rho}}{2\eta\rho}\right|^2 \\ &- \frac{2}{\varphi(q)}\sum_\chi \sum_{\beta<7/27} D^{7/27}\left|\frac{e^{\eta\rho}-e^{-\eta\rho}}{2\eta\rho}\right|^2. \end{aligned} \tag{48.5.31}$$

By (47.3.5) it is easy to see that the first sum has the upper bound

$$\frac{c_8\varphi(q)}{\varphi(q)} D^3 \sum_{n\ge[D^{5/2}]} \frac{\log[q(n+3)]}{n^2} < c_9\sqrt{D}\log D. \tag{48.5.32}$$

For an estimate of the second sum we use the density theorem according to which the total number of zeros of all $\mathscr{L}$-functions mod q with $\beta \le \alpha < \frac{1}{2}$, $|\gamma| \le T$ $(T \ge 1)$ is at most

$$c_{10}(qT)^{3\alpha/(1+\alpha)}\log^9(qT) \tag{48.5.33}$$

if $\frac{1}{5} \le \alpha \le \frac{1}{2}$ (see, e.g., Montgomery [1], Theorem 12.1). In particular, for $\alpha = \frac{7}{27}$ the number of these zeros is at most

$$c_{11}\varphi(q)T^{2/3}. \tag{48.5.34}$$

*As always, in the following if there are several such zeros we choose any one of them.

This implies the upper estimate

$$(48.5.35)\qquad \frac{c_{12}\varphi(q)D^{2/3}\cdot D^{7/27}}{\varphi(q)}+\frac{c_{13}\varphi(q)D^{7/27}\cdot D^{2}}{\varphi(q)}\int_D^\infty\frac{(x^{2/3})'}{x^2}dx<D$$

for the second sum in (48.5.31) (if $D>c_4$), and so from (48.5.31), (48.5.32), and (48.5.35) we get

$$(48.5.36)\qquad B_1>D\log D-2D>1\quad\text{for}\quad D>c_4.$$

So using the two-sided theorem of Knapowski [see Section 8.6] we get an integer ν_0 satisfying (48.5.14) [by (48.5.27)] such that by (48.5.11), (48.5.13), (48.5.27), and (48.5.36),

$$(48.5.37)\qquad |K(\nu_0)|>\left(\frac{1}{\log r}\right)^{\log^{3/5}r}\left(\frac{|z_\lambda|}{2}\right)^{\nu_0}$$

$$>\exp(-\log^{3/4}r)\left(\frac{e^{(7/27)A}}{4}\right)^{\nu_0}$$

$$>r^{-1/1000}\exp\left(\frac{7}{27}\cdot\frac{\log_2 r}{5}\cdot 4.99\frac{\log r}{\log_2 r}\right)$$

$$=r^{(7/27)(1-1/500)-1/1000}>r^{0.257}.$$

This, together with (48.5.20)–(48.5.24), proves Lemma 48.1. ■

48.6

From Lemma 48.1 we can deduce the following:

Lemma 48.3. *Suppose $\Delta(x,q,l_1,l_2)$ does not change sign in the interval $\omega\le x\le\omega^5$, where $\omega\ge\exp(q^{c_{14}})$. Then we have the inequality*

$$(48.6.1)\qquad S\doteq\left|\frac{1}{\varphi(q)}\sum_\chi[\bar\chi(l_1)-\bar\chi(l_2)]\sum_\chi{}'\frac{(\omega^5)^\rho}{\rho^{1000}}\right|>c_{15}\omega^{5/4}$$

where the dash indicates that the sum is extended over the nontrivial zeros to the right of the broken line V defined in Lemma G.1 of Appendix G.

Proof. Consider the integral

$$(48.6.2)\qquad H(\omega)\doteq\frac{1}{2\pi i}\int_{(2)}\frac{(\omega^5)^s}{s^{1000}}\cdot\frac{1}{\varphi(q)}\sum_\chi[\bar\chi(l_1)-\bar\chi(l_2)]\frac{\mathscr{L}'}{\mathscr{L}}(s,\chi)\,ds.$$

Inserting the Dirichlet-series representation we get

(48.6.3)

$$H(\omega) = \frac{1}{999!}\left(\sum_{\substack{n\le\omega^5\\ n\equiv l_2(q)}} \Lambda(n)\log^{999}\frac{\omega^5}{n} - \sum_{\substack{n\le\omega^5\\ n\equiv l_1(q)}} \Lambda(n)\log^{999}\frac{\omega^5}{n} \right)$$

$$= -\frac{1}{999!}\int_1^{\omega^5}\log^{999}\frac{\omega^5}{x}\,d_x[\Delta_4(x,q,l_1,l_2)]$$

$$= \frac{1}{999!}\int_1^{\omega^5}\Delta_4(x,q,l_1,l_2)\frac{d}{dx}\left(\log^{999}\frac{\omega^5}{x}\right)dx$$

$$= -\frac{1}{998!}\int_1^{\omega^5}\frac{\Delta_4(x,q,l_1,l_2)}{x}\log^{998}\frac{\omega^5}{x}\,dx$$

$$= -\frac{1}{998!}\int_\omega^{\omega^5}\frac{\Delta_4(x,q,l_1,l_2)}{x}\log^{998}\frac{\omega^5}{x}\,dx$$

$$+O(\omega\log^{998}\omega).$$

Since $\Delta_4(x,q,l_1,l_2)$ does not change sign in the interval $\omega\le x\le\omega^5$, we get

(48.6.4) $$|H(\omega)| = \frac{1}{998!}\int_\omega^{\omega^5}\frac{|\Delta_4(x,q,l_1,l_2)|}{x}\log^{998}\frac{\omega^5}{x}\,dx + O(\omega\log^{998}\omega)$$

$$\ge c_{16}\int_\omega^{\omega^5/2}\frac{|\Delta_4(x,q,l_1,l_2)|}{x}\,dx - c_{17}\omega\log^{998}\omega,$$

and so by (48.5.2)

(48.6.5) $$|H(\omega)| > c_{18}\omega^{5/4}.$$

On the other hand, by Cauchy's integral theorem

(48.6.6)

$$H(\omega) = \frac{1}{\varphi(q)}\sum_\chi[\bar\chi(l_1)-\bar\chi(l_2)]\sum_\rho{}'\frac{(\omega^5)^\rho}{\rho^{1000}}$$

$$+\frac{1}{2\pi i}\int_{(V)}\frac{(\omega^5)^s}{s^{1000}}\cdot\frac{1}{\varphi(q)}\sum_\chi[\bar\chi(l_1)-\bar\chi(l_2)]\frac{\mathcal{L}'}{\mathcal{L}}(s,\chi)\,ds.$$

By Lemma G.1 of Appendix G and (48.4.1) the integral is absolutely

$$(48.6.7) \qquad < c_{19}\omega^{1/20} q \log^2 q < c_{20}\omega^{1/10},$$

and this together with (48.6.5) completes the proof. ■

48.7. PROOF OF THEOREM 48.1

Suppose that $\Delta(x, q, l_1, l_2)$ has no sign change in the interval

$$(48.7.1) \qquad [\omega, \omega^5 e^{\sqrt{\omega}}] \subset [\omega, e^{2\sqrt{\omega}}]$$

with an ω satisfying (48.4.1).

Let y be a number, to be chosen later, such that

$$(48.7.2) \qquad \tfrac{1}{40}\log\omega \le y \le \tfrac{1}{20}\log\omega$$

and let ν be an integer such that

$$(48.7.3) \qquad 0.99\frac{\sqrt{\omega}}{y} \le \nu \le \frac{\sqrt{\omega}}{y}.$$

Consider the integral

$$(48.7.4) \quad J_\nu \stackrel{\text{def}}{=} \frac{1}{2\pi i}\int_{(2)} \frac{(\omega^5 e^{\nu y})^s}{s^{\nu+1000}} \cdot \frac{1}{\varphi(q)} \sum_{\chi} [\bar{\chi}(l_1) - \bar{\chi}(l_2)] \frac{\mathscr{L}'}{\mathscr{L}}(s, \chi)\, ds.$$

As in (48.6.3), inequalities (48.7.2), (48.7.3), and an easy computation show that

$$(48.7.5) \quad J_\nu = -\frac{1}{(\nu+998)!}\int_{\omega}^{\omega^5 e^{\nu y}} \frac{\Delta_4(x, q, l_1, l_2)}{x} \log^{\nu+998}\frac{e^{\nu y}\omega^5}{x}\, dx$$
$$+ O\left[\exp\left(41\frac{\sqrt{\omega}\,\log_2\omega}{\log\omega}\right)\right].$$

Thus we arrive at a contradiction if we can prove the existence of a ν_1 and ν_2 satisfying (48.7.3) with the property

$$(48.7.6) \qquad J_{\nu_1} > e^{\sqrt{\omega}/100}$$

and

$$(48.7.7) \qquad J_{\nu_2} < -e^{\sqrt{\omega}/100}.$$

48.8

Apply Cauchy's integral theorem to J_ν to obtain

$$(48.8.1)\quad J_\nu = \frac{1}{\varphi(q)} \sum_\chi [\bar{\chi}(l_1) - \bar{\chi}(l_2)] \sum_\rho{}' \frac{(\omega^5)^\rho}{\rho^{1000}} \left(\frac{e^{y\rho}}{\rho} \right)^\nu$$

$$+ \frac{1}{2\pi i} \int_{(V)} \frac{(\omega^5 e^{\nu y})^s}{s^{\nu+1000}} \cdot \frac{1}{\varphi(q)} \sum_\chi [\bar{\chi}(l_1) - \bar{\chi}(l_2)] \frac{\mathscr{L}'}{\mathscr{L}}(s, \chi)\, ds.$$

Using Lemma G.1 of Appendix G, and (48.4.1), (48.7.2), and (48.7.3), the absolute value of the integral has the upper bound

$$(48.8.2)\qquad c_{21}(e^{2\sqrt{\omega}})^{1/100} \cdot 200^\nu q \log^2 q < c_{22} e^{\sqrt{\omega}/40}.$$

The total number of all nontrivial zeros of an $\mathscr{L}$-function with imaginary parts between r and $r + 1$, counting multiplicity, is

$$(48.8.3)\qquad < c_{23} \log[q(|r| + 2)].$$

By (48.4.1), (48.7.2), and (48.7.3) the contribution of zeros with

$$(48.8.4)\qquad |\gamma| > \omega^{1/19} + 1$$

to the infinite sum in (48.8.1) has the upper estimate

$$(48.8.5)\qquad c_{24} \sum_{n \geq \omega^{1/19}} \frac{\log(qn)}{n^{\nu+1000}} \omega^5 e^{\sqrt{\omega}} < c_{25} \frac{\log(q\omega)}{\omega^{(1/19)(\nu+999)}} \omega^5 e^{\sqrt{\omega}} < 1.$$

48.9

Thus we must investigate the finite power sum

$$(48.9.1)\qquad D(\nu) \doteq \frac{1}{\varphi(q)} \sum_\chi [\bar{\chi}(l_1) - \bar{\chi}(l_2)] \sum_\rho{}'' \frac{(\omega^5)^\rho}{\rho^{1000}} \left(\frac{e^{y\rho}}{\rho} \right)^\nu$$

where Σ'' means that the summation is extended over the zeros to the right of V such that

$$(48.9.2)\qquad |\gamma| \leq \omega^{1/19} + 1.$$

Let $\rho_1 = \beta_1 + i\gamma_1$ be one of the zeros satisfying (48.9.2) for which

$$(48.9.3)\qquad \left| \frac{e^{y\rho}}{\rho} \right|$$

is maximal, and let

$$Z(\nu) \doteq \frac{1}{\varphi(q)} \sum_{\chi} [\bar{\chi}(l_1) - \bar{\chi}(l_2)] \sum_{\rho}{}'' \frac{(\omega^5)^\rho}{\rho^{1000}} \left(\frac{e^{y(\rho-\beta_1)}}{\rho} |\rho_1| \right)^\nu . \tag{48.9.4}$$

Now we apply (the one-sided) Corollary II of Section 13 with the choices

$$b_j = \frac{1}{\varphi(q)} [\bar{\chi}(l_1) - \bar{\chi}(l_2)] \frac{(\omega^5)^\rho}{\rho^{1000}} \tag{48.9.5}$$

$$z_j = \frac{e^{y(\rho-\beta_1)}}{\rho} |\rho_1| \tag{48.9.6}$$

$$m = \left[0.99 \frac{\sqrt{\omega}}{y} \right]. \tag{48.9.7}$$

By (48.8.3) and (48.4.1) the number n of terms satisfies

$$n < c_{26} \varphi(q) \log[q(\omega^{1/19} + 3)](\omega^{1/19} + 1) < \omega^{1/18}. \tag{48.9.8}$$

Considering (48.4.1) and by Corollary I of Appendix H, we have

$$\pi \geq |\text{arc } z_j| > \frac{1}{8n} > \frac{1}{8\omega^{1/18}}, \tag{48.9.9}$$

with a suitable y satisfying (48.7.2). So we may choose

$$\kappa \doteq \frac{1}{8\omega^{1/18}}. \tag{48.9.10}$$

Thus by (48.7.2), (48.9.7), (48.9.8), and (48.9.10),

$$\left[m + 1, m + n\left(3 + \frac{\pi}{\kappa}\right) \right] \subset \left[0.99 \frac{\sqrt{\omega}}{y}, \frac{\sqrt{\omega}}{y} \right]. \tag{48.9.11}$$

Let z_{h_1} in Corollary II of Section 13 be defined by

$$\frac{e^{y(\rho_2-\beta_1)}}{\rho_2} |\rho_1|,$$

where $\rho_2 = \beta_2 + i\gamma_2$ minimizes

$$\left| \frac{e^{y\rho}}{\rho} \right| \tag{48.9.12}$$

among all ρ in the domain

$$\sigma \geq \tfrac{1}{5}, \qquad |t| \leq \omega^{1/250}, \tag{48.9.13}$$

and let $h_2 = n$. Then from (48.7.2)

$$|z_{h_1}| \geq \frac{e^{-(4/5)y}}{2\omega^{1/250}} \cdot \frac{1}{200} \geq \frac{\omega^{-1/25-1/250}}{400} \geq \omega^{-1/20} \tag{48.9.14}$$

Hence by the minimum property in the definition of z_{h_1}, the B in (13.8.4) satisfies

$$B = \left| \frac{1}{\varphi(q)} \cdot \operatorname{Re} \sum_{\chi} [\bar{\chi}(l_1) - \bar{\chi}(l_2)] \sum_{\rho}{}''' \frac{(\omega^5)^\rho}{\rho^{1000}} \right| \tag{48.9.15}$$

where the summation is extended over all ρ in the domain (48.9.13). To obtain a positive lower bound for B we use (48.6.1). The symmetry of V shows that the sum S is real and we have

$$\begin{aligned} B &\geq S - \frac{2}{\varphi(q)} \sum_{\chi} \sum_{1/200 \leq \beta \leq 1/5} \frac{(\omega^5)^{1/5}}{|\rho|^{1000}} \\ &\quad - \frac{2}{\varphi(q)} \sum_{\chi} \sum_{|\gamma| > \omega^{1/250}} \frac{\omega^5}{|\rho|^{1000}} \\ &> c_{15}\omega^{5/4} - c_{27}\left(\omega \log q + \frac{\omega^5 \log(q\omega)}{\omega^{999/250}} \right) > 1. \end{aligned} \tag{48.9.16}$$

48.10

Apply Corollary II of Section 13 to obtain a ν_1 satisfying (48.7.3) such that

$$Z(\nu_1) > |z_{h_1}|^{\nu_1} \cdot |z_{h_1}|^{n(12+\pi/\kappa)} \cdot 2^{-m-n(12+\pi/\kappa)} \tag{48.10.1}$$

$[Z(\nu)$ is real$]$ and from this, by (48.7.2), (48.9.7)–(48.9.10), and (48.9.14),

$$Z(\nu_1) > \exp\left(-40 \frac{\sqrt{\omega}}{\log \omega} \right) \left(\frac{e^{y(\beta_2 - \beta_1)}}{|\rho_2|} |\rho_1| \right)^{\nu_1}. \tag{48.10.2}$$

That is, for the real sum $D(\nu)$ in (48.9.1):

$$D(\nu_1) > \exp\left(-40 \frac{\sqrt{\omega}}{\log \omega} \right) \left(\frac{e^{y\beta_2}}{|\rho_2|} \right)^{\nu_1}. \tag{48.10.3}$$

But since ρ_2 is in the domain (48.9.13), formulas (48.7.2) and (48.7.3) imply

$$\left(\frac{e^{y\beta_2}}{|\rho_2|}\right)^{\nu_1} > \left(\frac{e^{(1/5)y}}{2\omega^{1/250}}\right)^{\nu_1} > \left(\frac{e^{(1/5)y}}{e^{(1/6)y}}\right)^{0.99\sqrt{\omega}/y} = e^{0.033\sqrt{\omega}}. \tag{48.10.4}$$

This, with (48.10.3), yields

$$D(\nu_1) > e^{\sqrt{\omega}/31} \tag{48.10.5}$$

which together with (48.8.1), (48.8.2), and (48.8.5) proves (48.7.6). Similarly, we obtain (48.7.7), and the proof is complete. ■

49

COMPARATIVE PRIME NUMBER THEORY II

Comparison of the Progressions $\equiv 1 \pmod q$ and $\equiv l \pmod q$

49.1

In Section 48 we answered Problems 1–4 of 48.2 for the case $i = 4$ under Haselgrove's condition (48.3.3). In the present section we answer these problems (under the same condition) for all i $(1 \le i \le 4)$ in the case $l_1 = 1$. The proofs show that the condition $\chi(l_1) = \chi(1) = 1$ for all χ makes this case easier to handle.

Theorem 49.1. *If Haselgrove's condition* (48.3.3) *holds for some* q *and there is a zero* $\rho_0 = \beta_0 + i\gamma_0$ *of an* $\mathscr{L}(s, \chi')$ mod q *with* $\chi'(l) \neq 1$ *for which*

(49.1.1) $$\beta_0 \ge \tfrac{1}{2},$$

then for

(49.1.2) $$T > \max\left\{c_1, e_2\left(\frac{|\gamma_0|}{2}\right), \exp(q^3), e_2\left[\frac{40}{A(q)}\right]\right\}$$

the inequalities

(49.1.3) $$\max_{T^{1/3} \le x \le T} \Delta_i(x, q, 1, l) > T^{\beta_0} \exp\left(-41 \frac{\log T \log_3 T}{\log_2 T}\right)^*$$

*The exponent $\frac{1}{3}$ can be replaced by $\frac{1}{2} - \varepsilon$ [see (49.5.10)–(49.5.11)].

and

$$(49.1.4) \qquad \min_{T^{1/3} \le x \le T} \Delta_i(x, q, 1, l) < -T^{\beta_0} \exp\left(-41 \frac{\log T \log_3 T}{\log_2 T}\right)^*$$

hold for $i = 2$ and $i = 4$.

Since the proofs for $i = 2$ and $i = 4$ are similar, but the first case is a bit more difficult, we will only prove it for $i = 2$.

The existence of a ρ_0 with (49.1.1) is not an essential condition, since Siegel [3] proved that every $\mathscr{L}(s, \chi)$ function belonging to a primitive character χ mod q has at least one zero in the domain

$$(49.1.5) \qquad \sigma \ge \tfrac{1}{2}, \qquad |t| \le \frac{c_2}{\log_3(q + e^{100})}.$$

So the following provides an answer to Problem 4 in Section 48.2.

Corollary I. *Under Haselgrove's condition for the modulus q we have for*

$$(49.1.6) \qquad T > \max\left\{c_1, \exp(q^3), e_2\left[\frac{40}{A(q)}\right]\right\}$$

the relations

$$(49.1.7) \qquad \max_{T^{1/3} \le x \le T} \Delta_i(x, q, 1, l) > \sqrt{T} \exp\left(-41 \frac{\log T \log_3 T}{\log_2 T}\right)$$

and

$$(49.1.8) \qquad \min_{T^{1/3} \le x \le T} \Delta_i(x, q, 1, l) < -\sqrt{T} \exp\left(-41 \frac{\log T \log_3 T}{\log_2 T}\right)$$

when $i = 2$ and $i = 4$.

Corollary II. *If q satisfies the Haselgrove condition, then for $i = 2$ and $i = 4$ $\Delta_i(x, q, 1, l)$ has a sign change in the interval*

$$(49.1.9) \qquad 1 \le x \le \max\left\{c_1, \exp(q^3), e_2\left[\frac{40}{A(q)}\right]\right\}.$$

Corollary III. *Under Haselgrove's condition for q and in the case $i = 2$ and $i = 4$, $\Delta_i(x, q, 1, l)$ has an infinity of sign changes for $1 \le x < \infty$.*

From Corollary I, an easy computation shows that the number $V_i(Y, q, 1, l)$ of sign changes of $\Delta_i(x, q, 1, l)$ in $[1, Y]$ is given by the following

Corollary IV. *If for q the Haselgrove condition holds and*

$$Y > \max\left\{ c_3, \exp(q^6), e_2\left[\frac{80}{A(q)}\right]\right\}, \tag{49.1.10}$$

then

$$V_i(Y, q, 1, l) \geq \frac{1}{2\log 3}\log_2 Y \qquad (i = 2, 4). \tag{49.1.11}$$

For $i = 4$ Corollaries II and III are already contained in the more general results of Section 48, but the quantitative results of Corollaries I and IV are (for $i = 4$) much better than those given in the previous section.

49.2

The question becomes more difficult in the cases $i = 1$ and $i = 3$.

Theorem 49.2. *If Haselgrove's condition* (48.3.3) *is valid for q, then for*

$$Y > \max\left\{ e_4(q^{c_4}), e_2\left[\frac{50}{A(q)}\right]\right\} \tag{49.2.1}$$

there exist x_i' and x_i'' in the interval

$$\left[(\log_2 Y)^{\exp(-q^{c_5})}, Y\right] \tag{49.2.2}$$

for which the inequalities

$$\Delta_1(x_1', q, 1, l) > 2\frac{\sqrt{x_1'}}{\log x_1'} \tag{49.2.3}$$

$$\Delta_1(x_1'', q, 1, l) < -2\frac{\sqrt{x_1''}}{\log x_1''}$$

and

$$\Delta_3(x_3', q, 1, l) > 2\sqrt{x_3'} \tag{49.2.4}$$

$$\Delta_3(x_3'', q, 1, l) < -2\sqrt{x_3''}$$

hold.

This already gives an upper bound for the first sign change and implies a very weak lower bound for the number of sign changes. But it is possible to prove a sharper estimate for $V_i(Y, q, 1, l)$.

Theorem 49.3.* *Under the Haselgrove condition* (48.3.3) *the inequality*

$$V_i(Y,q,1,l) > \exp(-q^{c_7})\log_3 Y \tag{49.2.5}$$

holds for $i = 1$ *and* 3, *and for*

$$Y > \max\left\{e_4(q^{c_6}), e_2\left[\frac{50}{A(q)}\right]\right\}. \tag{49.2.6}$$

Theorem 49.2 contains the solution of Problems 1 and 2 for the cases $i = 1, 3$.

Corollary V. *If* q *satisfies the Haselgrove condition, then for* $i = 1, 3$, $\Delta_i(x, q, 1, l)$ *has a sign change in the interval*

$$1 \le x \le \max\left\{e_4(q^{c_4}), e_2\left[\frac{50}{A(q)}\right]\right\}. \tag{49.2.7}$$

Corollary VI. *If the Haselgrove condition for* q *holds, then for* $i = 1$ *and* 3, $\Delta_i(x, q, 1, l)$ *changes sign infinitely often in* $1 \le x < \infty$.

Thus Problems 1–4 of Section 48.2 are settled in all cases $1 \le i \le 4$ if $l_1 = 1$, even with relatively satisfactory quantitative results [except for Problem 4 in the case $i = 1$ and 3, because the localization (49.2.2) in Theorem 49.2 is very weak].

All the results of this section can be proved for

$$\pi(x,q,1) - \frac{\pi(x)}{\varphi(q)} \tag{49.2.8}$$

and for

$$\pi(x,q,1) - \frac{\operatorname{li} x}{\varphi(q)} \tag{49.2.9}$$

instead of

$$\Delta_1(x,q,1,l) = \pi(x,q,1) - \pi(x,q,l). \tag{49.2.10}$$

The same remark applies to the cases $i = 2, 3$, and 4.

*This theorem is contained in a weaker form in Knapowski–Turán [3] with $\log_4 Y$ instead of $\log_3 Y$. The improvement, proved here, is due to the application of Lemma 34.3.

49.3

For the proof of Theorem 49.1 (in the more complicated case of $i = 2$) let y be a number, to be chosen later, for which

$$\tfrac{1}{20}\log_2 T \le y \le \tfrac{1}{10}\log_2 T. \tag{49.3.1}$$

Let ν be an integer restricted by

$$\frac{\log T}{y} - \log^{0.9} T \le \nu \le \frac{\log T}{y}. \tag{49.3.2}$$

Let

$$\begin{aligned} F(s) &\doteq \frac{1}{\varphi(q)} \sum_{\chi} [1 - \bar{\chi}(l)] \frac{\mathcal{L}'}{\mathcal{L}}(s, \chi) \\ &= \sum_{n \equiv l(q)} \frac{\Lambda(n)}{n^s} - \sum_{n \equiv 1(q)} \frac{\Lambda(n)}{n^s} \end{aligned} \tag{49.3.3}$$

and

$$J(\nu) \doteq \frac{1}{2\pi i} \int_{(2)} \left(\frac{e^{ys}}{s} \right)^{\nu} F(s)\, ds. \tag{49.3.4}$$

Then using the integral formula (47.2.5) we obtain

$$\begin{aligned} J(\nu) &= \frac{1}{(\nu - 1)!} \left\{ \sum_{\substack{n \le e^{\nu y} \\ n \equiv l(q)}} \Lambda(n) \log^{\nu - 1} \frac{e^{\nu y}}{n} - \sum_{\substack{n \le e^{\nu y} \\ n \equiv 1(q)}} \Lambda(n) \log^{\nu - 1} \frac{e^{\nu y}}{n} \right\} \\ &= \frac{-1}{(\nu - 1)!} \int_1^{e^{\nu y}} \log x \cdot \log^{\nu - 1} \frac{e^{\nu y}}{x}\, d_x \Delta_2(x, q, 1, l) \qquad (49.3.5) \\ &= \frac{1}{(\nu - 1)!} \int_1^{e^{\nu y}} \Delta_2(x, q, 1, l) \frac{d}{dx} \left\{ \log x \cdot \log^{\nu - 1} \frac{e^{\nu y}}{x} \right\} dx. \end{aligned}$$

Since an easy computation shows that

$$\frac{d}{dx} \left(\log x \cdot \log^{\nu - 1} \frac{e^{\nu y}}{x} \right) \gtreqless 0 \quad \text{for} \quad x \lesseqgtr e^{y}, \tag{49.3.6}$$

we split $J(\nu)$ in two parts

$$J_1(\nu) = \frac{1}{(\nu - 1)!} \int_1^{e^y}, \qquad J_2(\nu) = \frac{1}{(\nu - 1)!} \int_{e^y}^{e^{\nu y}}. \tag{49.3.7}$$

Here we can estimate $J_1(\nu)$ trivially:

(49.3.8) $|J_1(\nu)|$

$$\leq \max_{1 \leq x \leq e^y} |\Delta_2(x, q, 1, l)| \cdot \frac{1}{(\nu-1)!} \int_1^{e^y} \frac{d}{dx}\left(\log x \cdot \log^{\nu-1} \frac{e^{\nu y}}{x}\right) dx$$

$$\leq e^y \cdot \frac{1}{(\nu-1)!} \cdot y[(\nu-1)y]^{\nu-1}.$$

On the other hand, we have

(49.3.9) $|J_2(\nu)|$

$$\leq \min_{e^y \leq x \leq e^{\nu y}} \Delta_2(x, q, 1, l) \cdot \frac{1}{(\nu-1)!} \int_{e^y}^{e^{\nu y}} \frac{d}{dx}\left(\log x \cdot \log^{\nu-1} \frac{e^{\nu y}}{x}\right) dx$$

$$= - \min_{e^y \leq x \leq e^{\nu y}} \Delta_2(x, q, 1, l) \cdot \frac{1}{(\nu-1)!} y[(\nu-1)y]^{\nu-1}.$$

Thus with (49.3.1), (49.3.7), (49.3.8), and (49.3.9) we get

$$\min_{1 \leq x \leq T} \Delta_2(x, q, 1, l) \leq -J(\nu) \frac{(\nu-1)!}{y[(\nu-1)y]^{\nu-1}} + \log^{1/10} T. \tag{49.3.10}$$

49.4

Shift the line of integration in (49.3.4) to the broken line V defined in Lemma G.1 of Appendix G:

$$J(\nu) = \frac{1}{\varphi(q)} \sum_{\chi} [1 - \bar{\chi}(l)] \sum_{\rho}{}' \left(\frac{e^{\rho y}}{\rho}\right)^{\nu} + \frac{1}{2\pi i} \int_{(V)} \left(\frac{e^{ys}}{s}\right)^{\nu} F(s)\, ds \tag{49.4.1}$$

where the dash indicates that the summation runs over the zeros to the right of V.

By Lemma G.1 of Appendix G, (49.1.2), (49.3.1), and (49.3.2) the integral is here absolutely

$$< c_8 T^{1/100} \cdot 200^{\nu} \cdot q \log^2 q < T^{1/50}. \tag{49.4.2}$$

And by (49.1.2), (49.3.1), and (49.3.2) the contribution of zeros with

$$(49.4.3)\qquad |\gamma| \geq \log^{1/9} T$$

to $J(\nu)$ is

$$(49.4.4)\qquad < c_9 T \frac{\log\left[q\left(\log^{1/9} T + 2\right)\right]}{\left(\log^{1/9} T - 1\right)^{\nu - 1}} < 1.$$

49.5

Thus the essential part of $J(\nu)$ is the finite real power-sum

$$(49.5.1)\qquad Z(\nu) = \frac{1}{\varphi(q)} \sum_{\chi} [1 - \bar{\chi}(l)] \sum_{|\gamma| < \log^{1/9} T}{}' \left(\frac{e^{\rho y}}{\rho}\right)^{\nu}.$$

Now we apply Corollary I of Section 13. The number of terms n, from (48.3.3) and (49.1.2), satisfies

$$(49.5.2)\qquad n < c_{10} \varphi(q) \log\left[q\left(\log^{1/9} T + 2\right)\right] \log^{1/9} T < \frac{\log^{9/20} T}{8}$$

and is certainly ≥ 1, since by (49.1.2)

$$(49.5.3)\qquad |\gamma_0| \leq 2 \log_2 T < \log^{1/9} T.$$

For the numbers

$$b_j = \frac{1}{\varphi(q)} [1 - \bar{\chi}(l)]$$

and for $\mu = 1, 2, \ldots, n$, we have

$$(49.5.4)\qquad \operatorname{Re}\left(b_1 + b_2 + \cdots + b_\mu\right) \geq \frac{1}{\varphi(q)} \left[1 - \cos \frac{2\pi}{\varphi(q)}\right] > \frac{1}{q^3}.$$

Moreover, Corollary I of Appendix H gives the existence of a y contained in the interval of (49.3.1) such that by (49.5.2)

$$(49.5.5)\qquad \pi \geq |\operatorname{arc} z_j| = \left|\operatorname{arc} \frac{e^{i\gamma_\rho y}}{\rho}\right| > \frac{1}{8n} > \log^{-9/20} T \stackrel{\text{def}}{=} \kappa.$$

Now we choose

$$(49.5.6)\qquad m = \left[\frac{\log T}{y} - \log^{0.9}T\right].$$

Thus, by (49.5.2), (49.5.5), and (49.5.6), the interval $[m+1, m+n(3+\pi/\kappa)]$ is contained in the interval of (49.3.2).

Thus, applying Corollary I of Section 13 for $Z(\nu)$ in (49.5.1), we get a ν_1 satisfying (49.3.2) such that

$$(49.5.7)\qquad Z(\nu_1) = \operatorname{Re} Z(\nu_1) > \frac{1}{q^3}\cdot\frac{1}{\frac{3}{8}\log^{9/20}T}\left(\frac{1}{\log T}\right)^{(1/4)\log^{9/20}T} \cdot\left(\frac{e^{\beta_0 y}}{|\rho_0|}\right)^{\nu_1}$$

$$> \exp(-\log^{1/2}T)\frac{\exp\left(\beta_0\log T - \beta_0\log^{0.9}T\cdot\frac{1}{10}\log_2 T\right)}{(3\log_2 T)^{20\log T/\log_2 T}}$$

$$> 2T^{\beta_0}\exp\left(-21\frac{\log T}{\log_2 T}\log_3 T\right),$$

and so by (49.4.1), (49.4.2), and (49.4.4),

$$(49.5.8)\qquad J(\nu_1) > \tfrac{3}{2}T^{\beta_0}\exp\left(-21\frac{\log T}{\log_2 T}\log_3 T\right).$$

Using (49.3.1) and (49.3.2), we have

(49.5.9)

$$\frac{(\nu-1)!}{y[(\nu-1)y]^{\nu-1}} > \left(\frac{\nu/4}{\nu y}\right)^{\nu} > \frac{1}{(\log_2 T)^{\nu}} \geq \exp\left(-20\frac{\log T}{\log_2 T}\log_3 T\right).$$

This, together with (49.3.10) and (49.5.8), implies

$$(49.5.10)\qquad \min \Delta_2(x,q,1,l) < -T^{\beta_0}\exp\left(-41\frac{\log T}{\log_2 T}\log_3 T\right).$$

But because $\beta_0 \geq \frac{1}{2}$ and we have the trivial estimate

$$(49.5.11)\qquad |\Delta_2(x,q,1,l)| \leq x,$$

(49.1.4) follows at once from (49.5.11) even for the interval $[T^{1/2-\varepsilon}, T]$. Similarly, we have (49.1.3), and this completes the proof of Theorem 49.1. ■

49.6

We now prove Theorems 49.2 and 49.3 in the case $i = 1$. (The case $i = 3$ is similar.) We distinguish two cases as follows:

CASE I. There is an $\mathscr{L}(s, \chi')$ mod q with $\chi'(l) \neq 1$ which has a zero $\rho_0 = \beta_0 + i\gamma_0$ with

$$\beta_0 > \tfrac{1}{2} + 42\frac{\log_3 Y}{\log_2 Y}, \qquad |\gamma_0| \leq \log_2 Y. \tag{49.6.1}$$

In this case (49.1.2) is satisfied with $T = Y^{3^{-\nu}}$ where

$$\nu = 0, 1, \ldots, [\log_3 Y]. \tag{49.6.2}$$

Applying Theorem 49.1 in the case $i = 2$, we get by easy computation from (49.1.3) and (49.1.4) that

$$\max_{Y^{3^{-\nu-1}} \leq x \leq Y^{3^{-\nu}}} \Delta_2(x, q, 1, l) > \sqrt{Y^{3^{-\nu}}} \tag{49.6.3}$$

and

$$\min_{Y^{3^{-\nu-1}} \leq x \leq Y^{3^{-\nu}}} \Delta_2(x, q, 1, l) < -\sqrt{Y^{3^{-\nu}}} \tag{49.6.4}$$

Now (49.6.2)–(49.6.4) immediately give (49.2.3), which proves Theorem 49.2 in Case I. Furthermore,

$$V_1(Y) > \log_3 Y, \tag{49.6.5}$$

which proves Theorem 49.3 in Case I (for $i = 1$).

Using the functional equation we can formulate the other case as follows:

CASE II. If $\rho = \beta + i\gamma = \frac{1}{2} + \delta + i\gamma$ is a zero of an $\mathscr{L}(s, \chi, q)$ function with $\chi(l) \neq 1$ for which

$$|\gamma| \leq \log_2 Y, \tag{49.6.6}$$

then

$$|\delta| \leq 42\frac{\log_3 Y}{\log_2 Y}. \tag{49.6.7}$$

Lemma 49.1. *Suppose that all zeros $\rho = \beta + i\gamma = \frac{1}{2} + \delta + i\gamma$ of all $\mathscr{L}(s, \chi, q)$ functions [with $\chi(l) \neq 1$] for which*

$$|\gamma| \leq Z \tag{49.6.8}$$

satisfy

(49.6.9) $$|\delta| \le 42\frac{\log Z}{Z}$$

where Z is any number such that

(49.6.10) $$Z > e_2(q^{c_4}).$$

Then there exist numbers x′ and x″ such that

(49.6.11) $$\Delta_1(x', q, 1, l) > \frac{2\sqrt{x'}}{\log x'}$$

and

(49.6.12) $$\Delta_1(x'', q, 1, l) < -\frac{2\sqrt{x''}}{\log x''}$$

where

(49.6.13) $$Z^{\exp(-q^{c_5})} \le x', x'' \le Z.$$

Furthermore

(49.6.14) $$V_1(Z, q, 1, l) > \exp(-q^{c_7})\log Z.$$

Choosing

(49.6.15) $$Z = \log_2 Y,$$

Lemma 49.1 obviously proves Theorem 49.2 and 49.3 in Case II (for $i = 1$). (The analogue of Lemma 49.1 holds for $i = 3$ too.)

49.7

Now we prove some relations like those of 41.17–41.18 which permit us to investigate instead of $\Delta_1(x, q, 1, l)$, the more manageable $\Delta_4(x, q, 1, l)$. We use the following notations:

(49.7.1) $$\tilde{\Delta}_i(x) = \begin{cases} \dfrac{\Delta_i(x, q, 1, l)}{\sqrt{x}/\log x} & \text{for} \quad i = 1, 2 \\ \dfrac{\Delta_i(x, q, 1, l)}{\sqrt{x}} & \text{for} \quad i = 3, 4 \end{cases}$$

and

$$\Delta_5(u) = \int_2^u \Delta_4(v, q, 1, l)\, dv. \tag{49.7.2}$$

We assert that

$$\Delta_5(u) = O(u^{3/2}\log q) \quad \text{for} \quad u \le Z. \tag{49.7.3}$$

The finite form of the exact prime number formula for arithmetic progressions implies that for $2 \le u \le Z$,

$$\Delta_5(u) = -\frac{1}{\varphi(q)} \sum_{\chi} [1 - \bar{\chi}(l)] \sum_{|\gamma| \le Z} \frac{u^{\rho+1}}{\rho(\rho+1)} + O[u\log(uq)]. \tag{49.7.4}$$

By (48.8.3) and (49.6.9), the zeros of an $\mathscr{L}(s, \chi, q)$ with $\chi(l) \ne 1$ satisfy

$$\left| \sum_{|\gamma| \le Z} \frac{u^{\rho+1}}{\rho(\rho+1)} \right| \le u^{3/2 + 42\log Z/Z} \sum_{|\gamma| \le Z} \frac{1}{|\rho|^2} = O(u^{3/2}\log q). \tag{49.7.5}$$

As in Lemma 41.3 we get, by partial summation,

$$\tilde{\Delta}_2(x) = \tilde{\Delta}_4(x) + \frac{\Delta_5(x)}{x^{3/2}\log x} + \frac{\log x}{\sqrt{x}} \left\{ \int_2^x \Delta_5(u) \frac{\log u + 2}{u^2 \log^3 u}\, du + O(1) \right\}. \tag{49.7.6}$$

From (49.7.3) and (49.7.6) we get by easy computation

$$\tilde{\Delta}_2(x) = \tilde{\Delta}_4(x) + O\left(\frac{\log q}{\log x} \right) \quad \text{for} \quad 2 \le x \le Z. \tag{49.7.7}$$

Since $\tilde{\Delta}_1(x) = \tilde{\Delta}_2(x) + O(1)$ we obtain from this

$$|\tilde{\Delta}_1(x) - \tilde{\Delta}_4(x)| \le c_{11}\log q. \tag{49.7.8}$$

49.8

Now we study the oscillatory behavior of $\tilde{\Delta}_4(x)$. The finite exact prime number formula

$$\sum_{\substack{n \le x \\ n \equiv l(q)}} \Lambda(n) = \frac{x}{\varphi(q)} - \frac{1}{\varphi(q)} \sum_{\chi} \bar{\chi}(l) \sum_{|\gamma| \le Z} \frac{x^\rho}{\rho} + O[\log^2(qx)] \tag{49.8.1}$$

(see Prachar [1], Satz 4.4, p. 228) if $x \le Z$, implies for

$$\log q \le v \le \log Z \tag{49.8.2}$$

the relation

$$\tilde{\Delta}_4(e^v) = -\frac{1}{\varphi(q)}\sum_{\chi}[1-\bar{\chi}(l)]\sum_{|\gamma|\le Z}\frac{e^{\delta v}\cdot e^{i\gamma v}}{\rho} + O(1). \tag{49.8.3}$$

Now, by (49.6.8), (49.6.9), and (49.8.2), we have

$$e^{\delta v} = 1 + O\left(\frac{\log^2 Z}{Z}\right), \tag{49.8.4}$$

and thus, using (48.8.3) and (49.6.9) for the zeros of an $\mathscr{L}(s,\chi,q)$ with $\chi(l) \ne 1$, we get the inequalities

$$\left|\sum_{|\gamma|\le Z}\frac{e^{\delta v}\cdot e^{i\gamma v}}{\rho} - \sum_{|\gamma|\le Z}\frac{e^{i\gamma v}}{\rho}\right| = O\left(\frac{\log^2 Z}{Z}\sum_{|\gamma|\le Z}\frac{1}{|\rho|}\right) = O\left(\frac{\log^4 Z}{Z}\right), \tag{49.8.5}$$

$$\left|\sum_{|\gamma|\le 2}\frac{e^{i\gamma v}}{\rho}\right| \le \sum_{|\gamma|\le 2}\frac{1}{|\rho|} \le \sum_{|\gamma|\le 2} 3 = O(\log q), \tag{48.8.6}$$

and

$$\left|\sum_{2<|\gamma|\le Z}\frac{e^{i\gamma v}}{\rho} - \sum_{2<|\gamma|\le Z}\frac{e^{i\gamma v}}{i\gamma}\right| \le \sum_{2<|\gamma|\le Z}\frac{1}{\gamma^2} = O(\log q). \tag{49.8.7}$$

So (49.8.3) and (49.8.5)–(49.8.7) together yield

$$\left|\tilde{\Delta}_4(e^v) + \frac{1}{\varphi(q)}\sum_{\chi}[1-\bar{\chi}(l)]\sum_{2<|\gamma|\le Z}\frac{e^{i\gamma v}}{i\gamma}\right| < c_{12}\log q. \tag{49.8.8}$$

We introduce the notation

$$G(v) = \frac{1}{\varphi(q)}\sum_{\chi}[1-\bar{\chi}(l)]\sum_{2<|\gamma|\le Z}\frac{e^{i\gamma v}}{i\gamma}. \tag{49.8.9}$$

It is sufficient to study $G(v)$ because the relations

$$G(v') < -c_{13}\log q \tag{49.8.10}$$

and

$$G(v'') > c_{13}\log q \tag{49.8.11}$$

(where $\log q \le v^{(i)} \le \log Z$) according to (49.7.8) and (49.8.8) imply

$$\tilde{\Delta}_1(e^{v'}) > 2\log q > 2 \tag{49.8.12}$$

and

$$\tilde{\Delta}_1(e^{v''}) < -2\log q < -2 \tag{49.8.13}$$

if $c_{13} = c_{11} + c_{12} + 2$.

49.9

Let $A \ge \max(c_0, q)$ be some number to be chosen later. The constants c_ν for $14 \le \nu \le 17$ (just as the previous constants for $1 \le \nu \le 13$) do not depend on A. Form a mean value of the function $G(v)$ by

(49.9.1)

$$I(\omega) \doteq \frac{1}{2\pi}\int_{-A/4}^{A/4} G\left(\omega + \frac{y}{A}\right)\left(\frac{\sin(y/2)}{y/2}\right)^2 dy$$

$$= \frac{1}{\varphi(q)}\sum_{\chi}[1 - \bar{\chi}(l)] \sum_{2<|\gamma|\le Z} \frac{e^{i\gamma\omega}}{i\gamma}\cdot\frac{1}{2\pi}\int_{-A/4}^{A/4}\left(\frac{\sin(y/2)}{y/2}\right)^2 \cdot e^{i(\gamma/A)y}\,dy.$$

Applying (41.20.8) we get

(49.9.2)

$$\int_{-A/4}^{A/4}\left(\frac{\sin(y/2)}{y/2}\right)^2 \cdot e^{i(\gamma/A)y}\,dy = \int_{-\infty}^{+\infty}\left(\frac{\sin(y/2)}{y/2}\right)^2 \cdot e^{i(\gamma/A)y}\,dy + O\left(\frac{1}{|\gamma|}\right)$$

and so by (48.8.3),

$$I(\omega) = \frac{1}{\varphi(q)}\sum_{\chi}[1 - \bar{\chi}(l)] \sum_{2<|\gamma|\le Z} \frac{e^{i\gamma\omega}}{i\gamma} \cdot\frac{1}{2\pi}\int_{-\infty}^{\infty}\left(\frac{\sin(y/2)}{y/2}\right)^2 \cdot e^{i(\gamma/A)y}\,dy + O(\log q). \tag{49.9.3}$$

Now use the integral formula (41.20.1) with the notation

$$D(\omega) = \frac{1}{\varphi(q)} \sum_{\chi} [1 - \bar{\chi}(l)] \sum_{2<|\gamma|<A} \frac{e^{i\gamma\omega}}{i\gamma}\left(1 - \frac{|\gamma|}{A}\right). \tag{49.9.4}$$

Then

$$|I(\omega) - D(\omega)| \le c_{14} \log q. \tag{49.9.5}$$

If ρ is a zero of $\mathscr{L}(s, \chi)$ then $\bar{\rho}$ is a zero of $\mathscr{L}(s, \bar{\chi})$, so it is easy to see that

$$\begin{aligned} D(\omega) &= \frac{2}{\varphi(q)} \sum_{\chi} [1 - \operatorname{Re} \bar{\chi}(l)] \sum_{2<\gamma<A} \frac{\sin(\gamma\omega)}{\gamma}\left(1 - \frac{\gamma}{A}\right) \\ &\quad - \frac{2}{\varphi(q)} \sum_{\chi} \operatorname{Im} \bar{\chi}(l) \sum_{2<\gamma<A} \frac{\cos(\gamma\omega)}{\gamma}\left(1 - \frac{\gamma}{A}\right) \\ &\doteq D_1(\omega) - D_2(\omega). \end{aligned} \tag{49.9.6}$$

49.10

As in Section 41.21, we estimate $D(\omega)$ by the choice $\omega = \pm 1/A$. We use the well-known relation (see, e.g., Davenport [1], Section 16)

$$N(t, \chi) \doteq \sum_{\substack{\rho_\chi \\ 0<\beta<1 \\ 0\le\gamma\le t}} 1 = \frac{t}{2\pi} \log \frac{q^* t}{2\pi e} + O[\log(q^* t)] \tag{49.10.1}$$

where $t \ge 2$, and we assume that χ is induced by the primitive character χ^* mod q^*. Thus, for the contribution of a single χ to $D_2(1/A)$, we have

$$\begin{aligned} U(\chi) &\doteq \sum_{2<\gamma<A} \frac{\cos(\gamma/A)}{\gamma}\left(1 - \frac{\gamma}{A}\right) \\ &= -\int_2^A N(t, \chi) \frac{d}{dt}\left[\left(1 - \frac{t}{A}\right)\frac{\cos(t/A)}{t}\right] dt + O(\log q). \end{aligned} \tag{49.10.2}$$

If χ is a complex character, the modulus q^* belonging to χ and $\bar{\chi}$ coincide, and

$\operatorname{Im}\bar{\chi}(l) = -\operatorname{Im}\chi(l)$. Hence from (49.10.1) and (49.10.2),

$$(49.10.3)\quad |U(\chi)\operatorname{Im}\bar{\chi}(l) + U(\bar{\chi})\operatorname{Im}\chi(l)|$$
$$\le \int_2^A c_{15}\log(q^*t)\left|\frac{d}{dt}\left(\left(1-\frac{t}{A}\right)\frac{\cos(t/A)}{t}\right)\right|dt + O(\log q)$$
$$< c_{16}\log q.$$

Since for real characters $U(\chi) = 0$, $D_2(1/A)$ has the upper bound

$$(49.10.4)\qquad \left|D_2\left(\frac{1}{A}\right)\right| < c_{16}\log q.$$

From the simple inequality

$$(49.10.5)\qquad \sin x > \frac{x}{2} \quad \text{for} \quad 0 < x < 1$$

and (49.10.1), $D_1(1/A)$ has the lower estimate

$$(49.10.6)\quad D_1\left(\frac{1}{A}\right) \ge \frac{2}{\varphi(q)}\sum_{\chi}[1 - \operatorname{Re}\bar{\chi}(l)]\sum_{2<\gamma<A/2}\frac{\gamma/2A}{\gamma}\cdot\frac{1}{2}$$
$$= \frac{1}{2A\varphi(q)}\sum_{\chi}[1 - \operatorname{Re}\bar{\chi}(l)]\left[N\left(\frac{A}{2},\chi\right) - N(2,\chi)\right]$$
$$> \frac{c_{17}A\log A}{2\varphi(q)\cdot A}\sum_{\chi}[1 - \operatorname{Re}\bar{\chi}(l)]$$
$$= \frac{c_{17}\log A}{2}.$$

Now we choose the value A by

$$(49.10.7)\qquad A = \max\left\{c_0, q, q^{(2/c_{17})(c_{13}+c_{14}+c_{16}+c)}\right\}$$

with an absolute constant c to be given later. Thus from (49.10.4) and (49.10.6) we get

$$(49.10.8)\qquad D\left(\frac{1}{A}\right) > (c_{13} + c_{14} + c)\log q$$

and

$$(49.10.9)\qquad D\left(-\frac{1}{A}\right) < -(c_{13} + c_{14} + c)\log q.$$

49.11

We use Lemma 34.3 to obtain numbers ω_ν' and ω_ν'' for which

$$\left|D(\omega_\nu') - D\left(-\frac{1}{A}\right)\right| < c \tag{49.11.1}$$

and

$$\left|D(\omega_\nu'') - D\left(\frac{1}{A}\right)\right| < c. \tag{49.11.2}$$

In particular, apply Lemma 34.3 to the numbers

$$\frac{\gamma}{2\pi} \qquad (2 < \gamma < A) \tag{49.11.3}$$

where $\rho = \beta + i\gamma$ are the zeros of the $\mathscr{L}(s, \chi, q)$ functions with $\chi(l) \neq 1$. By (49.10.1) and (49.10.7) the number of terms is

$$n < c_{18} A \log(Aq) < q^{c_{19}} \tag{49.11.4}$$

[and obviously ≥ 1, again by (49.10.1)]. The role of q in Lemma 34.3 will be played by $[A] + 1$, and we choose

$$M = \left[\frac{\log Z - 1}{([A] + 1)^n}\right] > \exp(-q^{c_{20}}) \log Z. \tag{49.11.5}$$

[Here we used (49.10.7) and (49.11.4).]
Thus there are positive integers

$$n_\nu \leq \log Z - 1 \qquad \nu = 1, 2, \ldots, M. \tag{49.11.6}$$

such that for all γ with $2 < \gamma < A$

$$\left\| \frac{n_\nu \gamma}{2\pi} \right\| < \frac{1}{A}, \tag{49.11.7}$$

where $\|x\|$ denotes the distance of x from the nearest integer.
Choosing

$$\omega_\nu' = n_\nu - \frac{1}{A}, \qquad \omega_\nu'' = n_\nu + \frac{1}{A} \tag{49.11.8}$$

we have

$$\cos(\omega_\nu'' \gamma) - \cos\frac{\gamma}{A} = O\left(\frac{1}{A}\right) \tag{49.11.9}$$

and

$$\sin\left(\omega_\nu''\gamma\right) - \sin\frac{\gamma}{A} = O\left(\frac{1}{A}\right). \tag{49.11.10}$$

From this, by (49.9.6), (49.10.7), and (49.10.1),

$$\left|D\left(\omega_\nu''\right) - D\left(\frac{1}{A}\right)\right| \le 2\cdot\frac{2}{\varphi(q)}\sum_\chi 2\sum_{2<\gamma<A}\frac{c_{21}}{A\gamma} \tag{49.11.11}$$

$$< \frac{8c_{21}}{\varphi(q)}\cdot\varphi(q)\cdot\frac{1}{A}\cdot c_{22}\log^2 A < c_{23} \stackrel{\text{def}}{=} c$$

(where $c_{23} = c$ does not depend on A), and we similarly obtain (49.11.1).

49.12

Thus (49.9.5), (49.10.8), (49.10.9), (49.11.1), and (49.11.2) together give

$$I\left(\omega_\nu'\right) < -c_{13}\log q \tag{49.12.1}$$

and

$$I\left(\omega_\nu''\right) > c_{13}\log q. \tag{49.12.2}$$

Now, by the relation

$$\frac{1}{2\pi}\int_{-\infty}^{+\infty}\left(\frac{\sin(y/2)}{y/2}\right)^2 dy = 1, \tag{49.12.3}$$

formulas (49.9.1), (49.12.1), and (49.12.2) imply the existence of v_ν' and v_ν'' ($\nu = [\log q] + 2, \dots, M$) with

$$v_\nu^{(i)} \in \left[\omega_\nu^{(i)} - \tfrac{1}{4}, \omega_\nu^{(i)} + \tfrac{1}{4}\right] \subset \left[n_\nu - \tfrac{1}{3}, n_\nu + \tfrac{1}{3}\right] \subset [\log q, Z] \tag{49.12.4}$$

for which (49.8.10) and (49.8.11) and hence (49.8.12) and (49.8.13) hold. Thus $\tilde{\Delta}_1(x)$, and with it $\Delta_1(x, q, 1, l)$, certainly changes sign in all the $M - [\log q] - 1$ disjoint intervals

$$[e^{n_\nu - 1/3}, e^{n_\nu + 1/3}] \qquad (\nu = [\log q] + 2, \dots, M), \tag{49.12.5}$$

which by (49.11.6) are contained in $[1, Z]$; and so $\Delta_1(x, q, 1, l)$ has at least $M - [\log q] - 1$ sign changes in $[1, Z]$. Recalling (49.11.5), this yields (49.6.14), which completes the proof of Theorem 49.3. ■

49.13

On the other hand, applying Dirichlet's approximation theorem in the original form yields a number n_0 with

$$\frac{\log Z - 1}{\exp(q^{c_{20}})} \leq n_0 \leq \log Z - 1 \tag{49.13.1}$$

such that (49.11.7) holds. Then formulas (49.11.7)–(49.12.4) remain valid, and there exist v' and v'' with

$$e^{v'}, e^{v''} \in \left[e^{n_0 - 1/3}, e^{n_0 + 1/3}\right] \subset \left[Z^{\exp(-q^{c_5})}, Z\right] \tag{49.13.2}$$

for which (49.8.12)–(49.8.13), and, by (49.7.1), also the inequalities (49.6.11)–(49.6.12) hold. This proves the other part of Lemma 49.1 and so completes the proof of Theorem 49.2. ■

50

COMPARATIVE PRIME NUMBER THEORY III

Comparison of Two Progressions Belonging to Quadratic Nonresidues

50.1

In the previous section we obtained rather far-reaching results on the comparison of the distribution of primes $\equiv 1$ and $\equiv l \bmod q$. Now we return to the general case, discussed in Section 48; but to obtain a result in the most important case $\Delta_1(x, q, l_1, l_2) = \pi(x, q, l_1) - \pi(x, q, l_2)$ we need another conjecture on the $\mathscr{L}$-zeros, too. Namely, in addition to Haselgrove's condition (48.3.3), we use a "finite" form of the Riemann–Piltz conjecture. In other words, we assume that no $\mathscr{L}(s, \chi) \bmod q$ vanishes in the domain

$$\sigma > \tfrac{1}{2}, \qquad |t| \le c_1 q^{10}, \tag{50.1.1}$$

with a sufficiently large absolute constant c_1.

Using this conjecture we can give a far better localization for the sign changes of $\Delta_4(x, q, l_1, l_2)$ than that given by Theorem 48.1 [for the notation see (48.2.2)], and our argument will also be valid when $i = 2$.

Theorem 50.1. *Assuming both the truth of* (50.1.1) *and the Haselgrove condition* (48.3.3), *we have for*

$$T > \max\left[e_2\left(c_2 q^{20}\right), e_2\left(\frac{50}{A(q)}\right)\right] \tag{50.1.2}$$

the inequality

$$\max_{T^{1/3} \le x \le T} \Delta_i(x, q, l_1, l_2) > \sqrt{T} \exp\left(-35 \frac{\log T \log_3 T}{\log_2 T}\right) \tag{50.1.3}$$

when $i = 2$ *and* $i = 4$.

Since l_1 and l_2 are arbitrary, we can change them to get the analogous inequality with min instead of max, and thus $\Delta_i(x, q, l_1, l_2)$ has a sign change in $[T^{1/3}, T]$ whenever T satisfies (50.1.2). This immediately implies a relatively good lower estimate for the number of sign changes $V_i(Y, q, l_1, l_2)$ of $\Delta_i(x, q, l_1, l_2)$ in the interval $[1, Y]$.

Corollary I. *If both assumptions* (50.1.1) *and the Haselgrove condition are satisfied, then for*

$$Y > \max\left\{ e_2\left(c_3 q^{20}\right), e_2\left[\frac{100}{A(q)}\right]\right\} \tag{50.1.4}$$

the inequality

$$V_i(Y, q, l_1, l_2) > \frac{1}{2 \log 3} \log_2 Y \tag{50.1.5}$$

holds when $i = 2$ *and* $i = 4$.

50.2

These theorems are seemingly unrelated to the oscillatory behavior of $\Delta_1(x, q, l_1, l_2)$ and $\Delta_3(x, q, l_1, l_2)$. But if l_1 and l_2 are quadratic nonresidues mod q then prime squares cannot appear in the corresponding sums $\Delta_2(x, q, l_1, l_2)$ and $\Delta_4(x, q, l_1, l_2)$. Therefore we have for $i = 1, 3$

$$\Delta_i(x, q, l_1, l_2) = \Delta_{i+1}(x, q, l_1, l_2) + O\left[q x^{1/3}\right], \tag{50.2.1}$$

and so Theorem 50.1 and Corollary I are, with minor changes, valid when $i = 1$ and $i = 3$ also. Thus we get the following.

Corollary II. *If* l_1 *and* l_2 *are quadratic nonresidues* mod q *and the assumptions of Theorem* 50.1 *are satisfied, then for* $i = 1$ *and* $i = 3$ *we have the inequality*

$$\max_{T^{1/3} \le x \le T} \Delta_i(x, q, l_1, l_2) > \sqrt{T} \exp\left(-36 \frac{\log T \log_3 T}{\log_2 T}\right). \tag{50.2.2}$$

The analogue of Corollary I is as follows:

Corollary III. *If l_1 and l_2 are quadratic nonresidues* mod q *and the assumptions of Corollary I are satisfied, then the inequality*

$$V_i(Y, q, l_1, l_2) > \frac{1}{2\log 3}\log_2 Y \tag{50.2.3}$$

holds for $i = 1$ and $i = 3$.

The preceding results also contain an answer for the question of the first sign change and for whether there are an infinity of sign changes.

Corollary IV. *If for the functions $\mathscr{L}(s, \chi)$* mod q $(\chi \neq \chi_0)$ *the Haselgrove condition* (48.3.3) *and the "finite Riemann–Piltz conjecture"* (50.1.1) *are true, then for $i = 2$ and $i = 4$ all the functions $\Delta_i(x, q, l_1, l_2)$ and for $i = 1$ and $i = 3$ those belonging to quadratic nonresidues l_1 and l_2* mod q *change their sign infinitely often in $1 \le x < \infty$; their first sign change occurs in the interval*

$$1 \le x \le \max\left\{e_2\left(c_2 q^{20}\right), e_2\left[\frac{50}{A(q)}\right]\right\}. \tag{50.2.4}$$

50.3

The number of quadratic residue classes mod q is $2^{-\nu(q)+j} \cdot \varphi(q)$, where $j = -1, 0$, or 1 and $\nu(q)$ denotes the number of distinct prime divisors of q. We may omit from the positive integers a set of density zero, such that for the remaining q, $\lim_{q\to\infty}\nu(q) = \infty$. Thus Corollary IV asserts that if the Haselgrove condition and the finite Riemann–Piltz conjecture are true for almost all natural numbers q (i.e., apart from a set of density zero) then for almost all moduli q, except $o[\varphi(q)]$ bad residue classes (namely, the quadratic residues), the function $\pi(x, q, l_1) - \pi(x, q, l_2)$ changes sign infinitely often. Hence almost all the functions $\Delta_1(x, q, l_1, l_2)$ [and $\Delta_3(x, q, l_1, l_2)$] change sign infinitely often.

However, it remains an open question as to whether, assuming some natural conditions on the $\mathscr{L}$-zeros, one can insure that all functions $\Delta_1(x, q, l_1, l_2)$ change sign infinitely often (or at least once). It is proved in this section that when $i = 2$ and $i = 4$ such a conclusion is valid. Suppose both l_1 and l_2 are quadratic residues, then the number of solutions of the two congruences

$$x^2 \equiv l_1 \pmod{q} \quad \text{and} \quad x^2 \equiv l_2 \pmod{q} \tag{50.3.1}$$

is equal. Using the prime number theorem of the arithmetical progressions in the form

$$(50.3.2)\qquad \pi(x,q,l)=\frac{\operatorname{li} x}{\varphi(q)}+O\left[x\exp(-c\sqrt{\log x})\right]$$

we easily obtain

$$(50.3.3)\quad \Delta_1(x,q,l_1,l_2)=\Delta_2(x,q,l_1,l_2)+O\left[\sqrt{x}\,\exp(-c'\sqrt{\log x})\right].$$

Thus applying Theorem 50.1 directly we get nothing; however, a more detailed analysis of this case leads to a result analogous to Corollary II. This is discussed in the next section. If l_1 and l_2 are of opposite quadratic character, for example, if l_1 is a quadratic residue mod q, and l_2 a quadratic nonresidue, we denote the number of solutions of the congruence $x^2 \equiv l_1 \pmod q$ by $N(q)$. Then we have the relation

$$(50.3.4)\qquad \Delta_2(x,q,l_1,l_2)-\Delta_1(x,q,l_1,l_2)\sim\frac{N(q)}{\varphi(q)}\cdot\frac{\sqrt{x}}{\log x}.$$

At present we are unable to produce a sufficiently large oscillation for the analytically manageable $\Delta_2(x,q,l_1,l_2)$ that would induce sign changes for $\Delta_1(x,q,l_1,l_2)$ (except in the already settled case when l_1 or l_2 is equal to 1).

Thus the question whether $\pi(x,q,l_1)-\pi(x,q,l_2)$ changes sign infinitely often if l_1 and l_2 are of opposite quadratic character may be considered as the most important open problem of comparative prime number theory. Even for a specific prime q this is a deep problem. Very special computational arguments were used by H. M. Stark [1] to settle the simplest such cases $\pi(x,5,2)-\pi(x,5,4)$ and $\pi(x,5,3)-\pi(x,5,4)$. But to decide this for every q seems beyond reach at present.

All of the statements proved in this section for the functions $\Delta_i(x,q,l_1,l_2)$ also hold for

$$(50.3.5)\qquad \psi(x,q,l)-\frac{x}{\varphi(q)},\qquad \psi(x,q,l)-\frac{\psi(x)}{\varphi(q)}$$

and

$$(50.3.6)\qquad \Pi(x,q,l)-\frac{\operatorname{li} x}{\varphi(q)},\qquad \Pi(x,q,l)-\frac{\Pi(x)}{\varphi(q)}.$$

But (though probably true in full generality) we can prove the corresponding statements for

$$(50.3.7)\qquad \Theta(x,q,l)-\frac{x}{\varphi(q)}$$

and

$$\pi(x, q, l) - \frac{\mathrm{li}\, x}{\varphi(q)} \tag{50.3.8}$$

only when l is a quadratic nonresidue mod q (or $l = 1$).

For the functions

$$\Theta(x, q, l) - \frac{\Theta(x)}{\varphi(q)} \tag{50.3.9}$$

and

$$\pi(x, q, l) - \frac{\pi(x)}{\varphi(q)} \tag{50.3.10}$$

our arguments do not yield any results (even if l is a quadratic nonresidue) except in the case $l = 1$ [see (49.2.8)–(49.2.10)].

50.4

The proof of Theorem 50.1 is rather intricate, although it is somewhat similar to that of Theorem 48.1. The main difficulty is again that the coefficients of the critical power sum

$$\frac{1}{\varphi(q)} \sum_{\chi} [\bar{\chi}(l_1) - \bar{\chi}(l_2)] \sum_{\rho}{}' \left(\frac{e^{y\rho}}{\rho} \right)^{\nu} \tag{50.4.1}$$

do not have a nonnegative real part; thus the one-sided theorems are not directly applicable. However, transforming the sum (50.4.1) to the form

$$\frac{1}{\varphi(q)} \sum_{\chi} [\bar{\chi}(l_1) - \bar{\chi}(l_2)] \sum_{\rho}{}' \frac{\eta_1^{\rho}}{\rho} \left(\frac{e^{y\rho}}{\rho} \right)^{\nu} \tag{50.4.2}$$

a (very weak) lower bound on the absolute value of the real power sum

$$\frac{1}{\varphi(q)} \sum_{\chi} [\bar{\chi}(l_1) - \bar{\chi}(l_2)] \sum_{\rho}{}' \frac{\eta_1^{\rho}}{\rho} \tag{50.4.3}$$

can be proved by an appropriate choice of η_1 (see Lemma 50.2).

But we also need another partial sum of (50.4.3), and the difference must be estimated carefully enough so that the positive lower bound for (50.4.3) is not destroyed [see formula (48.9.16)]. To meet this new difficulty we apply an appropriate two-sided theorem for the sum

$$\frac{1}{\varphi(q)}\sum_{\chi}[\bar{\chi}(l_1)-\bar{\chi}(l_2)]\sum_{\rho}{}'\frac{\eta_1^{\rho}}{\rho}\left(\frac{\eta_2^{\rho}}{\rho}\right)^{\nu_0} \tag{50.4.4}$$

and thus obtain an already satisfactory lower estimate (see Lemma 50.3) with the aid of the weak lower bound for (50.4.3). Having done this, we shall be able to realize our original idea. Making use of the estimate of (50.4.4) we apply the one-sided Corollary II of Section 13 to the sum

$$\frac{1}{\varphi(q)}\sum_{\chi}[\bar{\chi}(l_1)-\bar{\chi}(l_2)]\sum_{\rho}{}'\frac{\eta_1^{\rho}}{\rho}\left(\frac{\eta_2^{\rho}}{\rho}\right)^{\nu_0}\left(\frac{e^{y\rho}}{\rho}\right)^{\nu}; \tag{50.4.5}$$

the proof is then similar to that of Theorem 48.1.

So here we use two power sum theorems. We mention that in the course of the proof of Theorem 48.1 two power sum theorems were also applied (a two-sided and a one-sided one); but one of them was incorporated in Knapowski's Lemma 48.1.

In the proof we use the abbreviation

$$L \doteq c_1 q^{10}, \tag{50.4.6}$$

and we also need to know that if c_1 is chosen sufficiently large, then L is sufficiently large (even compared with q^{10}). On the other hand, all of the constants c_4–c_8 appearing in the proof and those constants implied by the O symbols do not depend on L, or on c_1.

50.5

To estimate the sum (50.4.3) by an appropriate choice of η_1, observe that the sum is essentially $\Delta_4(\eta_1, q, l_1, l_2)$, and this function cannot be too small everywhere, since at the appearance of any prime P such that $P \equiv l_1$ or $P \equiv l_2 \pmod q$, the function must have a jump of $\log P$. This is developed precisely in the lemmas that follow.

Lemma 50.1. *Assuming* (50.1.1), *there exists a prime* $P \equiv l_1 \pmod q$ *satisfying*

$$\tfrac{2}{3}(L)^{1/4} \le P \le (L)^{1/4} \tag{50.5.1}$$

such that $P-1$ *is not a power of* 2.

Proof. Apply the finite exact prime number formula for arithmetic progressions:

$$(50.5.2)\quad \psi(x,q,l_1)=\frac{x}{\varphi(q)}-\frac{1}{\varphi(q)}\sum_{\chi}\bar{\chi}(l_1)\sum_{|\gamma|\le x}\frac{x^{\rho}}{\rho}+O\left[\log^2(qx)\right]$$

(see Prachar [1], Satz 4.4, p. 228).

By (50.1.1) this implies

$$(50.5.3)\quad \psi(x,q,l_1)=\frac{x}{\varphi(q)}+O\left(\sqrt{x}\log^2 x\right)\quad \text{for}\quad q\le x\le c_1q^{10},$$

or in the equivalent form [see (48.2.1)]

$$(50.5.4)\qquad \Theta(x,q,l_1)=\frac{x}{\varphi(q)}+O\left(\sqrt{x}\log^2 x\right).$$

Apply this for $x=L^{1/4}$ and for $2x/3$, and recalling (50.4.6), we obtain

$$(50.5.5)\quad \Theta(x,q,l_1)-\Theta\left(\frac{2x}{3},q,l_1\right)=\frac{x}{3\varphi(q)}+O\left(\sqrt{x}\log^2 x\right)>\log x,$$

which proves our assertion. We use Lemma 50.1 to prove the following.

Lemma 50.2. *Assume* (50.1.1). *Then*

$$(50.5.6)\qquad \max_{L^{1/4}/2\le\omega\le L^{1/4}}\left|\frac{1}{\varphi(q)}\sum_{\chi}\left[\bar{\chi}(l_1)-\bar{\chi}(l_2)\right]\sum_{|\gamma|\le L^{1/3}}\frac{\omega^{\rho}}{\rho}\right|>2.$$

We use the following formula, valid for positive integers x ($x\ge 2$) and a character χ (mod q) which is induced by the primitive χ' (mod q') (see Prachar [1], Satz 4.4, p. 228):

(50.5.7)

$$\sum_{n\le x}{}'\,\Lambda(n)\chi(n)=E_0(\chi)x-\sum_{|\gamma|\le y}\frac{x^{\rho}}{\rho}-d_0(\chi')-\nu_0(\chi')\log x$$

$$-\sum_{\substack{p^{\alpha}\le x\\ p\mid q,\,p\nmid q'}}{}'\,\chi'(p^{\alpha})\log p+O\left(\frac{x\log^2(qxy)}{y}\right).$$

Here $\nu_0(\chi)=0$ or 1, the dash means that the term corresponding to $n=x$ must be taken with coefficient $\frac{1}{2}$, and

$$(50.5.8)\qquad E_0(\chi)=\begin{cases}1, & \text{for}\quad \chi=\chi_0\\ 0, & \text{for}\quad \chi\ne\chi_0.\end{cases}$$

Apply formula (50.5.7) with the choice $y = L^{1/3}$ for $x = P$ and $x = P - 1$ (P is defined in Lemma 50.1). Subtract the two equalities. Since P is a prime $\equiv l_1 (\mathrm{mod}\, q)$ and $P - 1$ is not a power of 2, that is, $\Lambda(P-1) = 0$, (50.4.6) implies that

$$(50.5.9)\quad \tfrac{1}{2}\chi(l_1)\log P = E_0(\chi) - \sum_{|\gamma| \le L^{1/3}} \frac{P^\rho}{\rho} + \sum_{|\gamma| \le L^{1/3}} \frac{(P-1)^\rho}{\rho} + \vartheta$$

with $|\vartheta| \le 1$, if c_1 is sufficiently large.

Multiply by

$$\frac{1}{\varphi(q)}[\bar{\chi}(l_1) - \bar{\chi}(l_2)]$$

and sum over all characters $\chi \bmod q$ to obtain

$$(50.5.10)\quad \tfrac{1}{2}\log P = \frac{-1}{\varphi(q)}\sum_{\chi}[\bar{\chi}(l_1) - \bar{\chi}(l_2)]$$

$$\left(\sum_{|\gamma| \le L^{1/3}} \frac{P^\rho}{\rho} - \sum_{|\gamma| \le L^{1/3}} \frac{(P-1)^\rho}{\rho}\right) + 2\vartheta'$$

(with $|\vartheta'| \le 1$). Identity (50.5.10) clearly implies that the sum appearing in (50.5.6) is absolutely

$$(50.5.11)\quad \ge \tfrac{1}{2}\left(\tfrac{1}{2}\log P - 2\right) > 2 \quad \text{for} \quad \omega = P \quad \text{or} \quad \omega = P - 1.$$

50.6

Let $\omega = \omega_0$ be a value for which (50.5.6) is realized; then we have

$$(50.6.1)\qquad \omega_0 \le L^{1/4}.$$

Lemma 50.3. *Assuming the truth of* (50.1.1), *there exists an integer* ν_0 *with*

$$(50.6.2)\qquad 0.9\frac{L^2}{\log L} \le \nu_0 \le \frac{L^2}{\log L}$$

such that

$$(50.6.3)\quad S \doteq \left|\frac{1}{\varphi(q)}\sum_{\chi}[\bar{\chi}(l_1) - \bar{\chi}(l_2)] \sum_{|\gamma| \le L} \frac{\omega_0^\rho}{\rho}\left(\frac{L^\rho}{\rho}\right)^{\nu_0}\right| > \exp\left(\frac{L^2}{30}\right).$$

Let $\rho_1 = \frac{1}{2} + i\gamma_1$ be the zero with the minimal absolute value of its imaginary part. Then Siegel's theorem (49.1.5) assures

$$|\gamma_1| \le c_4 < L. \tag{50.6.4}$$

Let $\rho_2 = \frac{1}{2} + i\gamma_2$ be the zero with the greatest imaginary part $\le L^{1/3}$ and $\rho_3 = \frac{1}{2} + i\gamma_3$ the zero with the smallest imaginary part $\ge L^{1/3} + 1$.

We write the sum under consideration in the form

(50.6.5)

$$S = \left(\frac{\sqrt{L}}{|\rho_1|}\right)^{\nu_0} \left\{ \frac{1}{\varphi(q)} \sum_{\chi} [\bar{\chi}(l_1) - \bar{\chi}(l_2)] \sum_{|\gamma| \le L} \frac{\omega_0^{\rho}}{\rho} \left(\frac{L^{\rho - 1/2}}{\rho} |\rho_1| \right)^{\nu_0} \right\}.$$

Now we apply the two-sided theorem of S. Knapowski [mentioned without proof in (8.6)] with

$$z_j = \frac{L^{\rho-1/2}}{\rho}|\rho_1| \tag{50.6.6}$$

$$b_j = \frac{1}{\varphi(q)}[\bar{\chi}(l_1) - \bar{\chi}(l_2)]\frac{\omega_0^{\rho}}{\rho} \tag{50.6.7}$$

$$m = \left[0.9\frac{L^2}{\log L}\right] \tag{50.6.8}$$

$$z_\lambda = \frac{L^{\rho_2 - 1/2}}{\rho_2}|\rho_1| \tag{50.6.9}$$

$$z_{\lambda_1} = \frac{L^{\rho_3 - 1/2}}{\rho_3}|\rho_1|. \tag{50.6.10}$$

(The condition $\max|z_j| = 1$ is clearly satisfied.) By (48.8.3) and (50.4.6) the number of terms is

$$n = O[\varphi(q) L \log[q(L+2)]] < L^{1.15}. \tag{50.6.11}$$

By (50.6.8)–(50.6.11) we have

$$|z_\lambda| = \frac{|\rho_1|}{|\rho_2|} > \frac{\frac{1}{2}}{L^{1/3} + 1} > 4L^{-0.4} > \frac{4n}{m} > \frac{4n}{m+n} \tag{50.6.12}$$

and

$$(50.6.13)\qquad |z_\lambda| - |z_{\lambda_1}| = |\rho_1|\left(\frac{1}{|\rho_2|} - \frac{1}{|\rho_3|}\right)$$

$$\geq \frac{1}{2}\left(\frac{1}{L^{1/3}+\frac{1}{2}} - \frac{1}{L^{1/3}+1}\right)$$

$$> \frac{1}{5L^{2/3}} > 2L^{-3/4} > \frac{2n}{m} > \frac{2n}{m+n}.$$

From the definition of z_λ and z_{λ_1}, we get

$$(50.6.14)\qquad B_1 \doteq \min_{\lambda \leq j \leq \lambda_1} |b_1 + b_2 + \cdots + b_j|$$

$$= \left|\frac{1}{\varphi(q)} \sum_{\chi} [\bar{\chi}(l_1) - \bar{\chi}(l_2)] \sum_{|\gamma|<H}{}^{*} \frac{\omega_0^\rho}{\rho}\right|$$

(where Σ^* means that some zeros with $|\gamma| = H$ may be contained in the sum) and H satisfies

$$(50.6.15)\qquad L^{1/3} \leq H \leq |\gamma_3| \leq L^{1/3} + O(\log L)$$

by (49.10.1) and (50.4.6).

Thus, replacing the inner sum by

$$(50.6.16)\qquad \sum_{|\gamma| \leq L^{1/3}} \frac{\omega_0^\rho}{\rho},$$

(48.8.3) and (50.6.1) show that the total error is

$$(50.6.17)\qquad O\left(\log^2 L \cdot \sqrt{\omega_0} \cdot \frac{1}{L^{1/3}}\right) < 1.$$

Thus Lemma 50.2 implies immediately

$$(50.6.18)\qquad B_1 > 1.$$

From (50.6.4)–(50.6.13) and (50.6.18) there exists a ν_0 in the interval $[m+1, m+n]$ that is contained—by (50.6.8) and (50.6.11)—in the interval

(50.6.2), and for which our sum S in (50.6.5), that is, in (50.6.3), satisfies

$$S > \left(\frac{\sqrt{L}}{|\rho_1|}\right)^{\nu_0} \cdot \left(\frac{1}{L^2}\right)^{L^{1.15}} (2L^{-0.4})^{L^2/\log L} \tag{50.6.19}$$

$$> \left(\frac{\sqrt{L}}{c_4+1}\right)^{0.9L^2/\log L} \exp(-L^{1.2})\exp(-0.4L^2)$$

$$> \exp(0.44L^2 - L^{1.2} - 0.4L^2) > \exp\left(\frac{L^2}{30}\right).$$

This proves Lemma 50.3.

50.7. PROOF OF THEOREM 50.1

We again consider only the more complicated case $i = 2$. Let ω_0, L, ν_0 be defined as before [see (50.4.6), (50.6.1), (50.6.2)], and let T satisfy (50.1.2). Define T_1 by

$$T_1 \doteq \frac{T}{\omega_0 L^{\nu_0}} \left(> \frac{T}{e^{2L^2}} > \frac{T}{\log T}\right). \tag{50.7.1}$$

[The inequalities hold by (50.4.6), (50.6.1), (50.6.2), and (50.1.2) if $c_2 \geq 2c_1^2$]. Let y be a number, to be chosen later, in the interval

$$\tfrac{1}{20}\log_2 T_1 \leq y \leq \tfrac{1}{10}\log_2 T_1, \tag{50.7.2}$$

and let ν be an integer restricted at present only by

$$\frac{\log T_1}{y} - \log^{0.9} T_1 \leq \nu \leq \frac{\log T_1}{y}. \tag{50.7.3}$$

Let

$$Y \doteq \omega_0 L^{\nu_0} e^{\nu y} \qquad (\leq T). \tag{50.7.4}$$

As in Section 49.3, define

$$F(s) \doteq \frac{1}{\varphi(q)} \sum_{\chi} [\overline{\chi}(l_1) - \overline{\chi}(l_2)] \frac{\mathcal{L}'}{\mathcal{L}}(s, \chi) \tag{50.7.5}$$

and

$$J(\nu) \doteq \frac{1}{2\pi i}\int_{(2)}\left(\frac{e^{ys}}{s}\right)^{\nu}\frac{(\omega_0 L^{\nu_0})^s}{s^{\nu_0+1}}F(s)\,ds \tag{50.7.6}$$
$$= \frac{1}{2\pi i}\int_{(2)}\frac{Y^s}{s^{\nu_0+\nu+1}}F(s)\,ds.$$

If we use Dirichlet-series representation of $F(s)$ and the integral formula (47.2.5), reasoning similar to that of (49.3.3)–(49.3.5) shows that

$$J(\nu) = \frac{1}{(\nu+\nu_0)!}\int_1^Y \Delta_2(x, q, l_1, l_2)\frac{d}{dx}\left\{\log x\cdot\log^{\nu+\nu_0}\frac{Y}{x}\right\}dx. \tag{50.7.7}$$

Since

$$\frac{d}{dx}\left\{\log x\cdot\log^{\nu+\nu_0}\frac{Y}{x}\right\} \gtreqless 0 \quad\text{for}\quad x \lesseqgtr Y^{1/(\nu+\nu_0+1)}, \tag{50.7.8}$$

we split $J(\nu)$ in two parts

$$J_1(\nu) = \frac{1}{(\nu+\nu_0)!}\int_1^{Y^{1/(\nu+\nu_0+1)}}, \qquad J_2(\nu) = \frac{1}{(\nu+\nu_0)!}\int_{Y^{1/(\nu+\nu_0+1)}}^Y. \tag{50.7.9}$$

A trivial estimate of $J_1(\nu)$ using (50.7.8) yields

$$|J_1(\nu)| \le \max_{x\le Y^{1/(\nu+\nu_0+1)}}|\Delta_2(x, q, l_1, l_2)|\frac{1}{(\nu+\nu_0)!} \tag{50.7.10}$$
$$\times\int_1^{Y^{1/(\nu+\nu_0+1)}}\frac{d}{dx}\left\{\log x\cdot\log^{\nu+\nu_0}\frac{Y}{x}\right\}dx$$
$$\le Y^{1/(\nu+\nu_0+1)}\cdot\frac{(\log Y)^{\nu+\nu_0+1}}{(\nu+\nu_0+1)!}\cdot\left(\frac{\nu+\nu_0}{\nu+\nu_0+1}\right)^{\nu+\nu_0}$$

Also by (50.7.8) we have

(50.7.11)

$$-J_2(\nu) \le \max_{Y^{1/(\nu+\nu_0+1)}\le x\le Y}\Delta_2(x, q, l_1, l_2)\frac{1}{(\nu+\nu_0)!}$$
$$\times\int_{Y^{1/(\nu+\nu_0+1)}}^Y -\frac{d}{dx}\left\{\log x\cdot\log^{\nu+\nu_0}\frac{Y}{x}\right\}dx$$
$$= \max_{Y^{1/(\nu+\nu_0+1)}\le x\le Y}\Delta_2(x, q, l_1, l_2)\cdot\frac{(\log Y)^{\nu+\nu_0+1}}{(\nu+\nu_0+1)!}\left(\frac{\nu+\nu_0}{\nu+\nu_0+1}\right)^{\nu+\nu_0}$$

This together with (50.7.10) gives

$$(50.7.12)\quad \max_{x\le Y}\Delta_2(x,q,l_1,l_2) \ge -J(\nu)\left(\frac{\nu+\nu_0+1}{\nu+\nu_0}\right)^{\nu+\nu_0}\frac{(\nu+\nu_0+1)!}{(\log Y)^{\nu+\nu_0+1}} - Y^{1/(\nu+\nu_0+1)}.$$

By (50.7.1)–(50.7.4) we obtain

$$(50.7.13)\qquad Y^{1/(\nu+\nu_0+1)} < T^{1/\nu} < T^{2y/\log T_1} < e^{3y} < \log T$$

and

$$(50.7.14)\quad \frac{(\nu+\nu_0+1)!}{(\log Y)^{\nu+\nu_0+1}} > \left[\frac{(\nu+\nu_0+1)/4}{\log T}\right]^{\nu+\nu_0+1} > \left(\frac{\nu}{5\log T_1}\right)^{(21/20)\nu} > \frac{1}{(10y)^{(21/20)\nu}} > \exp\left(-21\frac{\log T\log_3 T}{\log_2 T}\right).$$

From (50.7.12)–(50.7.14), if for a $\nu=\nu_2$, $J(\nu_2)<0$, then

$$(50.7.15)\quad \max_{x\le Y}\Delta_2(x,q,l_1,l_2) \ge -J(\nu_2)\exp\left(-21\frac{\log T\cdot\log_3 T}{\log_2 T}\right) - \log T.$$

50.8

Cauchy's integral theorem gives

$$(50.8.1)\qquad J(\nu) = \frac{1}{\varphi(q)}\sum_{\chi}[\bar\chi(l_1)-\bar\chi(l_2)]\sum_{\rho}{}' \frac{Y^\rho}{\rho^{\nu+\nu_0+1}} + \frac{1}{2\pi i}\int_{(V)}\frac{Y^s}{s^{\nu+\nu_0+1}}F(s)\,ds.$$

where V is the infinite broken line appearing in Lemma G.1 of Appendix G, and Σ' means that the summation refers to zeros to the right of V.

Using Lemma G.1 of Appendix G, and (50.1.2), (50.7.1)–(50.7.4), the absolute value of the integral in (50.8.1) has the upper bound

$$(50.8.2)\qquad c_5 Y^{1/100}\cdot(200)^{\nu+\nu_0+1}\cdot q\log^2 q < T^{1/50}.$$

The contribution of zeros with

$$|\gamma| > \log^{1/9} T_1 \tag{50.8.3}$$

to the infinite sum appearing in $J(\nu)$ is, by (48.8.3) and (50.7.1)–(50.7.4), absolutely

$$< c_6 Y \frac{\log\{q(\log^{1/9} T_1 + 2)\}}{(\log^{1/9} T_1 - 1)^{\nu+\nu_0}} < 1. \tag{50.8.4}$$

[See (48.8.3)–(48.8.5).]

50.9

Now we estimate from below the remaining finite power sum $Z(\nu)$ containing the $\mathscr{L}$-zeros to the right of V with $|\gamma| \le \log^{1/9} T_1$, by a proper choice of ν; we will also need Lemma 50.3 and the one-sided Corollary II of Section 13. Let $\rho_4 = \beta_4 + i\gamma_4$ be the zero in $Z(\nu)$ for which $|e^{y\rho}/\rho|$ is maximal. Then $Z(\nu)$ takes the form

$$Z(\nu) = \left(\frac{e^{y\beta_4}}{|\rho_4|}\right)^{\nu} \sum_{\chi} \frac{1}{\varphi(q)} [\bar{\chi}(l_1) - \bar{\chi}(l_2)] \cdot \sum_{|\gamma| \le \log^{1/9} T}' \frac{(\omega_0 L^{\nu_0})^{\rho}}{\rho^{\nu_0+1}} \cdot \left(\frac{e^{y(\rho-\beta_4)}}{\rho} |\rho_4|\right)^{\nu}. \tag{50.9.1}$$

Apply Corollary II of Section 13 with

$$z_j = \frac{e^{y(\rho-\beta_4)}}{\rho} |\rho_4| \qquad (j = 1, 2, \ldots, n) \tag{50.9.2}$$

$$b_j = \frac{1}{\varphi(q)} [\bar{\chi}(l_1) - \bar{\chi}(l_2)] \frac{(\omega_0 L^{\nu_0})^{\rho}}{\rho^{\nu_0+1}}. \tag{50.9.3}$$

The condition $\max|z_j| = 1$ is obviously fulfilled. By Corollary I of Appendix H, (50.1.2), and (50.7.1), there exists a y in the interval (50.7.2) for which

$$\pi \ge |\operatorname{arc} z_j| \ge \frac{1}{8n} \overset{\text{def}}{=} \kappa. \tag{50.9.4}$$

Let

$$m = \left[\frac{\log T_1}{y} - \log^{0.9} T_1\right]. \tag{50.9.5}$$

By (49.10.1), (50.1.2), and (50.7.1) the number n of the z_j is

$$(50.9.6)\quad 1 \le n < c_7 \varphi(q) \log^{1/9} T_1 \log\{q(\log^{1/9} T_1 + 2)\} < \log^{1/8} T_1,$$

and so, by (50.9.4)–(50.9.6), (50.1.2), and (50.7.1),

$$(50.9.7)\quad \left[m+1, m+n\left(3+\frac{\pi}{\kappa}\right)\right] \subset \left[\frac{\log T_1}{y} - \log^{0.9} T_1, \frac{\log T_1}{y}\right].$$

Let $\rho_5 = \frac{1}{2} + i\gamma_5$ be any zero in $Z(\nu)$ with the maximal absolute value of the imaginary part and

$$(50.9.8)\qquad |\gamma| \le \frac{L}{3},$$

and let

$$(50.9.9)\qquad z_{h_1} = \frac{e^{y(\rho_5 - \beta_4)}}{\rho_5} |\rho_4|.$$

We choose $h_2 = n$. Then by (50.9.8), (50.1.1), (50.1.2), (50.4.6), (50.7.1), and (50.7.2) we get

$$(50.9.10)\quad \frac{|z_{h_1}|}{2} \ge \frac{e^{-y/2}}{2(1+L/3)} \cdot \frac{1}{200} > \frac{(\log T_1)^{-1/20}}{200L} > (\log T_1)^{-1/10}.$$

We use the definition of z_{h_1} in (50.9.8)–(50.9.9), the relation (48.8.3), and (50.4.6), (50.6.1), and (50.6.2). For the sum B in (13.8.4) Lemma 50.3. gives the following lower estimate [since the sum in (50.6.3) is real and the summation in B is extended over all zeros with $|\gamma| < L/3$]:

$$(50.9.11)\qquad |B| \ge S - \frac{1}{\varphi(q)} \sum_{\chi}{}^{2} \sum_{|\gamma| \ge L/3} \frac{\omega_0 L^{\nu_0}}{|\rho|^{\nu_0+1}}$$

$$\ge \exp\left(\frac{L^2}{30}\right) - c_8 \frac{\log(qL)\,\omega_0 L^{\nu_0}}{(L/3)^{\nu_0}}$$

$$\ge \exp\left(\frac{L^2}{30}\right) - e^{2\nu_0} > 1.$$

50.10

Thus we can already apply the one-sided Corollary II of Section 13 for the real power sum $Z(\nu)$ in (50.9.1). This yields a ν_2 which, by (50.9.7), is contained in

(50.7.3). Then, by (50.9.1)–(50.9.11), (50.1.2), (50.4.6), and (50.7.1)–(50.7.4),

$$(50.10.1) \qquad -Z(\nu_2) > \left(\frac{e^{y\beta_4}}{|\rho_4|}\right)^{\nu_2}\left(\frac{e^{y(1/2-\beta_4)}}{2|\rho_5|}|\rho_4|\right)^{\nu_2}\cdot\left(\log^{-1/10}T_1\right)^{n(12+\pi/\kappa)}$$

$$> \frac{e^{(1/2)y\nu_2}}{L^{\log T_1/y}}\exp\left(-\log^{1/3}T_1\right)$$

$$> \sqrt{T_1}\cdot\exp\left(-\log_2 T_1\cdot\log^{0.9}T_1 - 20\log L\cdot\frac{\log T_1}{\log_2 T_1}\right)$$

$$> \sqrt{T_1}\cdot\exp\left(-11\frac{\log T_1\log_3 T_1}{\log_2 T_1}\right)$$

$$> \sqrt{T}\cdot\exp\left(-12\frac{\log T\log_3 T}{\log_2 T}\right).$$

Recalling (50.8.2) and (50.8.4), this implies

$$(50.10.2) \qquad -J(\nu_2) > \sqrt{T}\cdot\exp\left(-13\frac{\log T\log_3 T}{\log_2 T}\right)$$

which together with (50.7.15) proves (50.1.3).

51

COMPARATIVE PRIME NUMBER THEORY IV

Comparison of Two Progressions Belonging to Quadratic Residues

51.1

In this section we discuss oscillatory properties of the functions $\pi(x, q, l_1) - \pi(x, q, l_2)$ and $\Theta(x, q, l_1) - \Theta(x, q, l_2)$ when both l_1 and l_2 are quadratic residues. This is the most complicated case. That the difficulties are not apparent from the length of the proof is a result of our being able to use many formulas from the previous section—among them the rather difficult Lemma 50.3—with no change.

In addition to the Haselgrove condition (48.3.3), we again need the finite Riemann–Piltz conjecture, namely, that no $\mathscr{L}(s, \chi) \bmod q$ vanishes in the domain

$$\sigma > \tfrac{1}{2}, \qquad |t| \leq c_1 q^{10} \tag{51.1.1}$$

where c_1 is a sufficiently large absolute constant (the lower restrictions are implied by Section 50). We use the notation of (48.2.2) in the next result.

Theorem 51.1. *If for the $\mathscr{L}$-functions* mod q *the Haselgrove condition* (48.3.3) *and the finite Riemann–Piltz conjecture* (51.1.1) *are assumed true, and l_1 and l_2 are both quadratic residues* mod q, *then for*

$$T > \max\left\{ e_2\left(c_2 q^{20}\right), e_2\left[\frac{50}{A(q)}\right]\right\} \tag{51.1.2}$$

the inequality

$$\max_{T^{1/3}\le x\le T} \Delta_i(x,q,l_1,l_2) > \sqrt{T}\cdot\exp\left(-35\frac{\log T\cdot\log_3 T}{\log_2 T}\right) \tag{51.1.3}$$

holds when $i = 1$ *and* $i = 3$.

(Interchanging l_1 and l_2, we get the analogous inequality with min instead of max.) This Theorem answers Problems 1–3 of Section 48, too.

Corollary I. *With the assumptions of Theorem* 51.1, *the functions* $\Delta_1(x,q,l_1,l_2)$ *and* $\Delta_3(x,q,l_1,l_2)$ *change sign infinitely often in* $1 \le x < \infty$. *For their first sign change we have*

$$1 \le x \le \max\left\{e_2\left(c_2 q^{20}\right), e_2\left[\frac{50}{A(q)}\right]\right\}. \tag{51.1.4}$$

Also, the quantitative version of Corollary I follows after an easy computation from Theorem 51.1.

Corollary II. *With the suppositions of Theorem* 51.1 *the number* $V_i(Y,q,l_1,l_2)$ *of sign changes of* $\Delta_i(x,q,l_1,l_2)$ *in* $[1,Y]$ $(i = 1,3)$ *satisfies the inequality*

$$V_i(Y,q,l_1,l_2) > \frac{1}{2\log 3}\log_2 Y \tag{51.1.5}$$

if only

$$Y > \max\left\{e_2\left(c_3 q^{20}\right), e_2\left[\frac{100}{A(q)}\right]\right\}. \tag{51.1.6}$$

51.2

We restrict ourselves to the proof of the case $i = 1$. Let ω_0 be the number for which the maximum in (50.5.6) is attained, and let L and ν_0 have the same meaning as in Lemma 50.3 [see formulas (50.4.6) and (50.6.1)–(50.6.3)]. Let T_1, y, ν, Y be defined as in (50.7.1)–(50.7.4). Denote the solutions of the congruences $x^2 \equiv l_1 (\mathrm{mod}\ q)$ and $x^2 \equiv l_2 (\mathrm{mod}\ q)$ by $\alpha_1', \dots, \alpha_R'$ and $\alpha_1'', \dots, \alpha_R''$, resp. (We have made use of the simple fact that their number is equal.) We begin with the integral

(51.2.1)

$$J(\nu) = \frac{1}{2\pi i}\int_{(2)} \frac{Y^s}{s^{\nu+\nu_0+1}}\cdot\frac{1}{\varphi(q)}\left\{\sum_{\chi}[\bar\chi(l_1) - \bar\chi(l_2)]\frac{\mathscr{L}'}{\mathscr{L}}(s,\chi)\right.$$
$$\left. + \sum_{j=1}^{R}\sum_{\chi}\left[\bar\chi(\alpha_j'') - \bar\chi(\alpha_j')\right]\frac{\mathscr{L}'}{\mathscr{L}}(2s,\chi)\right\} ds.$$

Using the integral formula (47.2.5) we get

(51.2.2)

$$J(\nu) = \sum_{\substack{n \le Y \\ n \equiv l_2(q)}} \Lambda(n) \frac{\log^{\nu+\nu_0}(Y/n)}{(\nu+\nu_0)!} - \sum_{j=1}^{R} \sum_{\substack{n \le \sqrt{Y} \\ n \equiv \alpha_j''(q)}} \Lambda(n) \frac{\log^{\nu+\nu_0}(Y/n^2)}{(\nu+\nu_0)!}$$

$$-\left\{ \sum_{\substack{n \le Y \\ n \equiv l_1(q)}} \Lambda(n) \frac{\log^{\nu+\nu_0}(Y/n)}{(\nu+\nu_0)!} - \sum_{j=1}^{R} \sum_{\substack{n \le \sqrt{Y} \\ n \equiv \alpha_j'(q)}} \Lambda(n) \frac{\log^{\nu+\nu_0}(Y/n^2)}{(\nu+\nu_0)!} \right\}.$$

Observe that the contribution of the prime squares to the first sum and that of the primes to the second one cancel, similarly for the third and fourth sums. Hence

(51.2.3)
$$J(\nu) = \sum_{\substack{p \le Y \\ p \equiv l_2(q)}} \log p \frac{\log^{\nu+\nu_0}(Y/p)}{(\nu+\nu_0)!}$$

$$- \sum_{\substack{p \le Y \\ p \equiv l_1(q)}} \log p \frac{\log^{\nu+\nu_0}(Y/p)}{(\nu+\nu_0)!}$$

$$+ O\left\{ \left[Y^{1/3} + qY^{1/4} \right] \frac{\log^{\nu+\nu_0} Y}{(\nu+\nu_0)!} \right\}.$$

Then using (51.1.2), as in (50.7.7)–(50.7.15) we get that if $J(\nu_2) < 0$ for $\nu = \nu_2$ then

(51.2.4)
$$\max_{x \le Y} \Delta_1(x, q, l_1, l_2) \ge -J(\nu_2) \exp\left(-21 \frac{\log T \cdot \log_3 T}{\log_2 T} \right) + O(T^{1/3}).$$

Since $J(\nu)$ in (51.2.1) can be written in the form

(51.2.5)

$$J(\nu) = \frac{1}{2\pi i} \int_{(2)} \frac{Y^s}{s^{\nu+\nu_0+1}} \cdot \frac{1}{\varphi(q)} \left\{ \sum_{\chi} [\bar{\chi}(l_1) - \bar{\chi}(l_2)] \frac{\mathscr{L}'}{\mathscr{L}}(s, \chi) \right\} ds$$

$$+ \frac{1}{4\pi i} \int_{(4)} \frac{Y^{s/2}}{(s/2)^{\nu+\nu_0+1}} \cdot \frac{1}{\varphi(q)} \left\{ \sum_{j=1}^{R} \sum_{\chi} [\bar{\chi}(\alpha_j'') - \bar{\chi}(\alpha_j')] \frac{\mathscr{L}'}{\mathscr{L}}(s, \chi) \right\} ds$$

it is possible to shift the line of integration in both integrals to the line V defined in Lemma G.1 of Appendix G. Thus

(51.2.6)

$$J(\nu) = \frac{1}{\varphi(q)} \sum_{\chi} [\bar{\chi}(l_1) - \bar{\chi}(l_2)] \sum_{\rho}{}' \frac{Y^{\rho}}{\rho^{\nu+\nu_0+1}}$$

$$+ \frac{1}{2\varphi(q)} \sum_{j=1}^{R} \sum_{\chi} [\bar{\chi}(\alpha_j'') - \bar{\chi}(\alpha_j')] \sum_{\rho}{}' \frac{Y^{\rho/2}}{(\rho/2)^{\nu+\nu_0+1}}$$

$$+ \frac{1}{2\pi i} \int_{(V)} \frac{Y^s}{s^{\nu+\nu_0+1}} \cdot \frac{1}{\varphi(q)} \left\{ \sum_{\chi} [\bar{\chi}(l_1) - \bar{\chi}(l_2)] \frac{\mathcal{L}'}{\mathcal{L}}(s, \chi) \right\} ds$$

$$+ \frac{1}{4\pi i} \int_{(V)} \frac{Y^{s/2}}{(s/2)^{\nu+\nu_0+1}}$$

$$\cdot \frac{1}{\varphi(q)} \left\{ \sum_{j=1}^{R} \sum_{\chi} [\bar{\chi}(\alpha_j'') - \bar{\chi}(\alpha_j')] \frac{\mathcal{L}'}{\mathcal{L}}(s, \chi) \right\} ds$$

where the dash means that the summation is taken over the zeros to the right of V. If we make use of Lemma G.1 of Appendix G, (51.1.2) and (50.7.1)–(50.7.4), the last two integrals are absolutely

(51.2.7) $$< c_4 \cdot Y^{1/100} \cdot 400^{\nu+\nu_0+1} \cdot q \cdot q \log^2 q < T^{1/50}.$$

For the contribution of zeros with

(51.2.8) $$|\gamma| > \log^{1/9} T_1$$

to the first two sums in $J(\nu)$, (48.8.3) and (50.7.1)–(50.7.4) imply the upper bound

(51.2.9) $$c_5 q Y \frac{\log\{q(\log^{1/9} T_1 + 2)\}}{\left(\frac{1}{2}\log^{1/9} T_1 - 1\right)^{\nu+\nu_0}} < 1.$$

51.3

We now apply Corollary II of Section 13. Let $\rho_4 = \beta_4 + i\gamma_4$ be the zero with $|\gamma| \le \log^{1/9} T_1$ for which $\left|\frac{e^{y\rho}}{\rho}\right|$ is maximal (as in 50.9). Then we can write our remaining sum in the form

(51.3.1)

$$Z(\nu)=\left(\frac{e^{y\beta_4}}{|\rho_4|}\right)^{\nu}\left\{\frac{1}{\varphi(q)}\sum_{\chi}[\bar{\chi}(l_1)-\bar{\chi}(l_2)]\sum_{|\gamma|\le\log^{1/9}T_1}{}'\frac{(\omega_0L^{\nu_0})^{\rho}}{\rho^{\nu_0+1}}\right.$$

$$\cdot\left(\frac{e^{y(\rho-\beta_1)}}{\rho}|\rho_4|\right)^{\nu}+\frac{1}{2\varphi(q)}\sum_{j=1}^{R}\sum_{\chi}[\bar{\chi}(\alpha_j'')-\bar{\chi}(\alpha_j')]$$

$$\left.\sum_{|\gamma|\le\log^{1/9}T_1}\frac{(\omega_0L^{\nu_0})^{\rho/2}}{(\rho/2)^{\nu_0+1}}\left(\frac{e^{y(\rho/2-\beta_4)}}{\rho/2}|\rho_4|\right)^{\nu}\right\}.$$

We introduce the terms: "z_j of the first class" for the numbers

$$\frac{e^{y(\rho-\beta_4)}}{\rho}|\rho_4|. \tag{51.3.2}$$

and "z_j of the second class" for the numbers

$$\frac{e^{y(\rho/2-\beta_4)}}{\rho/2}|\rho_4|. \tag{51.3.3}$$

Correspondingly, the "b_j of the first class" will be the numbers

$$\frac{1}{\varphi(q)}[\bar{\chi}(l_1)-\bar{\chi}(l_2)]\frac{(\omega_0L^{\nu_0})^{\rho}}{\rho^{\nu_0+1}} \tag{51.3.4}$$

and the "b_j of the second class" the numbers

$$\frac{1}{2\varphi(q)}[\bar{\chi}(\alpha_j'')-\bar{\chi}(\alpha_j')]\frac{(\omega_0L^{\nu_0})^{\rho/2}}{(\rho/2)^{\nu_0+1}}. \tag{51.3.5}$$

Let $\rho_5=\frac{1}{2}+i\gamma_5$ be the zero with the maximal absolute value of its imaginary part satisfying

$$|\gamma|>\frac{L}{3} \tag{51.3.6}$$

and let $\rho_6=\frac{1}{2}+i\gamma_6$ be any zero with the minimal absolute value of its imaginary part satisfying

$$\frac{3L}{4}\le|\gamma|\qquad(<L) \tag{51.3.7}$$

by (49.10.1) and (50.4.6). Let

$$z_{h_1}=\frac{e^{y(\rho_5-\beta_4)}}{\rho_5}|\rho_4| \tag{51.3.8}$$

and

$$(51.3.9) \qquad z_{h_2} = \frac{e^{y(\rho_6 - \beta_4)}}{\rho_6} |\rho_4|.$$

Well-known results on the zeros of $\mathscr{L}$-functions (see Prachar [1], p. 295, Satz 6.2) and (51.1.2) yield the estimate

$$(51.3.10) \qquad \beta < 1 - \frac{c_6}{\max\left(\log q, \log^{3/4}(|\gamma| + 3)\log_2^{3/4}(|\gamma| + 3)\right)}$$

$$< 1 - \frac{c_7}{(\log_2 T_1)^{4/5}}$$

for the zeros with $|\gamma| \le \log^{1/9} T_1$. Formulas (50.7.1), (50.7.2), (50.4.6), and (51.1.2) imply for the z_j of the second class, the inequality

$$(51.3.11) \qquad \frac{|z_j|}{|z_{h_2}|} = \frac{e^{y(\beta/2 - 1/2)}}{|\rho|/2} |\rho_6| < \exp\left(- \frac{c_7 y}{2(\log_2 T_1)^{4/5}} \right) 400(L + 1) < 1$$

and so all z_j of the second class have smaller absolute value than z_{h_2}. Thus, on one hand, the condition $\max|z_j| = 1$ is satisfied, on the other hand all the z_j with $j \le h_2$ belong to the first class. So the sum B in (13.8.4) contains only z_j of the first class. All z_j of this class with $|\gamma| < L/3$ definitely appear in B in virtue of (51.3.6) and (51.3.8). Hence the lower estimate for B given in (50.9.11) is valid without any change in the present situation.

The definitions and relations remain unchanged for m, n and κ given in (50.9.4)–(50.9.7) (only the values of the appearing constants c_ν are different). Since the definition of z_{h_1} is the same in (51.3.8) as in (50.9.9), the inequality (50.9.10) for z_{h_1} is valid. Formula (13.8.3) also holds by (51.3.6)–(51.3.9).

Thus the assumptions of Corollary II of Section 13 are satisfied, and we can apply it here. Observe that for the numbers m, n, κ, B, $|z_{h_1}|$ appearing in the final inequalities (13.8.5)–(13.8.6) and that for the implicitly contained numbers y, ν, T_1, T, exactly the same definitions and estimates hold as given in Section 50. Thus the estimate (50.10.1) holds for our real power sum $Z(\nu)$ in (51.3.1) without any change, that is, by an appropriate choice of ν_2 satisfying (50.7.3) we have the estimate (50.10.1). By (51.2.6)–(51.2.9) this implies the inequality

$$(51.3.12) \qquad -J(\nu_2) > \sqrt{T} \cdot \exp\left(-13 \frac{\log T \cdot \log_3 T}{\log_2 T} \right).$$

By (51.2.4) and the fact that $Y \le T$ [see (50.7.4)], this completes the proof of Theorem 51.1. ■

52

COMPARATIVE PRIME NUMBER THEORY V

Comparison of the Progressions $\equiv 1$, $\equiv l$ and $\equiv l^2 \pmod q$

52.1

In Sections 48–51 we found that generally we were unable to prove that the functions

$$\pi(x, q, l_1) - \pi(x, q, l_2)$$

and

$$\Theta(x, q, l_1) - \Theta(x, q, l_2)$$

have any sign change if l_1 and l_2 are of opposite quadratic character, except in the case discussed in Section 49 when l_1 or l_2 is equal to 1. Now we show that slight modifications of the proofs of Section 49 do yield such theorems for the expressions

$$\tfrac{3}{4}\pi(x, q, 1) - \pi(x, q, l) + \tfrac{1}{4}\pi(x, q, l^2)$$

or

$$\tfrac{3}{4}\Theta(x, q, 1) - \Theta(x, q, l) + \tfrac{1}{4}\Theta(x, q, l^2).$$

Here the deeper case is when l is a quadratic nonresidue: in this case the preceding functions take large positive values infinitely often when $1 \le x < \infty$.

This means that in the two quadratic residue classes 1 and l^2—with the weights $\frac{3}{4}$ and $\frac{1}{4}$—there are for many x more primes $\leq x$ than in the nonresidue class l. We must note that we cannot choose the coefficients $\frac{3}{4}$ and $\frac{1}{4}$ freely; their values are required from the proof.

Define, as in (48.2.2), the functions

$$(52.1.1)\quad \Delta_1^*(x,q,1,l) \doteq \tfrac{3}{4}\pi(x,q,1) - \pi(x,q,l) + \tfrac{1}{4}\pi(x,q,l^2)$$

$$\Delta_2^*(x,q,1,l) \doteq \tfrac{3}{4}\Pi(x,q,1) - \Pi(x,q,l) + \tfrac{1}{4}\Pi(x,q,l^2)$$

$$\Delta_3^*(x,q,1,l) \doteq \tfrac{3}{4}\Theta(x,q,1) - \Theta(x,q,l) + \tfrac{1}{4}\Theta(x,q,l^2)$$

$$\Delta_4^*(x,q,1,l) \doteq \tfrac{3}{4}\psi(x,q,1) - \psi(x,q,l) + \tfrac{1}{4}\psi(x,q,l^2)$$

Let $V_i{}^*(Y,q,1,l)$ denote the number of sign changes of $\Delta_i^*(x,q,1,l)$ in $[1,Y]$.

Theorem 52.1. *All the results of Section* 49 (*i.e., Theorems* 49.1–49.3 *and Corollaries I–VI*) *are valid for the functions* $\Delta_i^*(x,q,1,l)$ *in place of* $\Delta_i(x,q,1,l)$, *and for* $V_i{}^*(Y,q,1,l)$ *in place of* $V_i(Y,q,1,l)$.*

Theorem 52.1 can also be considered as a generalization of the results of Section 49, since if $l^2 \equiv 1 \pmod q$,

$$(52.1.2)\qquad \Delta_i^*(x,q,1,l) = \Delta_i(x,q,1,l).$$

So in the proof of Theorem 52.1 we can assume that $l^2 \not\equiv 1 \pmod q$.

52.2

The proof is so similar to the proofs of Theorems 49.1–49.3 that we only point out the differences.

We work with the function

$$(52.2.1)\quad F^*(s) \doteq \frac{1}{\varphi(q)}\sum_{\chi}\left[\frac{3}{4} - \bar{\chi}(l) + \frac{1}{4}\bar{\chi}(l^2)\right]\frac{\mathscr{L}'}{\mathscr{L}}(s,\chi)$$

$$= -\left[\frac{3}{4}\sum_{n\equiv 1(q)}\frac{\Lambda(n)}{n^s} - \sum_{n\equiv l(q)}\frac{\Lambda(n)}{n^s} + \frac{1}{4}\sum_{n\equiv l^2(q)}\frac{\Lambda(n)}{n^s}\right]$$

instead of with the $F(s)$ of (49.3.3).

*Theorem 52.1 is based on an unpublished manuscript of S. Knapowski–P. Turán, where the proof is sketched.

It is very important here that the coefficients

$$b_j^* = \frac{1}{\varphi(q)}\left[\frac{3}{4} - \bar{\chi}(l) + \bar{\chi}(l^2)\right] \tag{52.2.2}$$

have a nonnegative real part because

$$\frac{3}{4} - \cos\varphi + \frac{1}{4}\cos 2\varphi = \frac{(1-\cos\varphi)^2}{2}. \tag{52.2.3}$$

So the only change in the proof of Theorem 49.1 is that instead of (49.5.4) we have the weaker lower bound

$$\operatorname{Re}\left(b_1^* + \cdots + b_\mu^*\right) \geq \frac{1}{\varphi(q)} \frac{(1-\cos[2\pi/\varphi(q)])^2}{2} > \frac{1}{q^5}, \tag{52.2.4}$$

but the crucial estimate in (49.5.7) remains valid without any change.

The proofs of Theorems 49.2 and 49.3 are carried through as before until we get to $D^*(\omega)$ in (49.9.6). Now we have

(52.2.5)

$$\begin{aligned} D^*(\omega) &= \frac{2}{\varphi(q)}\sum_\chi\left[\frac{3}{4} - \operatorname{Re}\bar{\chi}(l) + \frac{1}{4}\operatorname{Re}\bar{\chi}(l^2)\right]\sum_{2<\gamma<A}\left(1-\frac{\gamma}{A}\right)\frac{\sin(\gamma\omega)}{\gamma} \\ &\quad - \frac{2}{\varphi(q)}\sum_\chi\left[\operatorname{Im}\bar{\chi}(l) - \tfrac{1}{4}\operatorname{Im}\bar{\chi}(l^2)\right]\sum_{2<\gamma<A}\left(1-\frac{\gamma}{A}\right)\frac{\cos(\gamma\omega)}{\gamma} \\ &= \frac{2}{\varphi(q)}\sum_\chi\left[\frac{3}{4} - \operatorname{Re}\bar{\chi}(l) + \frac{1}{4}\operatorname{Re}\bar{\chi}(l^2)\right]\sum_{2<\gamma<A}\left(1-\frac{\gamma}{A}\right)\frac{\sin(\gamma\omega)}{\gamma} \\ &\quad - \frac{2}{\varphi(q)}\sum_\chi \operatorname{Im}\bar{\chi}(l)\sum_{2<\gamma<A}\left(1-\frac{\gamma}{A}\right)\frac{\cos(\gamma\omega)}{\gamma} \\ &\quad + \frac{1}{2\varphi(q)}\sum_\chi \operatorname{Im}\bar{\chi}(l^2)\sum_{2<\gamma<A}\left(1-\frac{\gamma}{A}\right)\frac{\cos(\gamma\omega)}{\gamma} \\ &\doteq D_1^*(\omega) - D_2^*(\omega) + D_3^*(\omega). \end{aligned}$$

Here $D_2^*(\omega)$ is exactly the same as $D_2(\omega)$ in (49.9.6), so the estimate (49.10.4) remains valid for $D_2^*(1/A)$. Furthermore, this also holds for $D_3^*(1/A)$ since it differs from $D_2^*(1/A) = D_2(1/A)$ only by a constant factor $\frac{1}{4}$; of course $l \not\equiv 1(q)$ is replaced by $l^2 \not\equiv 1(q)$, but $l \not\equiv 1(q)$ was arbitrary in Section 49. In the estimate of $D_1^*(1/A)$ the coefficients

$$\frac{3}{4} - \operatorname{Re}\bar{\chi}(l) + \frac{1}{4}\operatorname{Re}\bar{\chi}(l^2) = \frac{[1 - \operatorname{Re}\bar{\chi}(l)]^2}{2} \tag{52.2.5}$$

are nonnegative, and thus we get, just as in (49.10.6)

$$D_1^*\left(\frac{1}{A}\right) > \frac{c_{17} A \log A}{2A\varphi(q)} \sum_{\chi} \left[\frac{3}{4} - \operatorname{Re}\bar{\chi}(l) + \frac{1}{4}\operatorname{Re}\bar{\chi}(l^2)\right] \tag{52.2.6}$$

$$= \frac{3c_{17}\log A}{8} \doteq \frac{c_{17}^*\log A}{2}.$$

Change c_{17} to c_{17}^*, and the remaining part of the proof of Lemma 49.1 remains valid. This yields the modified forms of Theorems 49.2–49.3 and completes the proof of Theorem 52.1. ■

53

COMPARATIVE PRIME NUMBER THEORY VI

A Modification of Chebyshev's Assertion

53.1

We return to the original assertion of Chebyshev [1]

$$(53.1.1) \qquad \lim_{x\to\infty} \sum_{p>2} (-1)^{(p-1)/2} e^{-p/x} = -\infty$$

which states that there are more primes $\equiv 3$ (4) than are $\equiv 1$ (4) in the above "Abelian" sense. As we mentioned (see 48.1) Hardy and Littlewood [1] and Landau [1, 2] showed that this is equivalent to the very deep unproved conjecture

$$(53.1.2) \qquad \mathscr{L}(s, \chi_1) = \sum_{n=0}^{\infty} \frac{(-1)^n}{(2n+1)^s} \neq 0 \quad \text{for} \quad \sigma > \tfrac{1}{2}.$$

The equivalence may also be stated for the more manageable form:

$$(53.1.3) \qquad \lim_{x\to\infty} \sum_{p>2} (-1)^{(p-1)/2} \log p \cdot e^{-p/x} = -\infty.$$

We naturally ask why (53.1.2) and (53.1.3) are equivalent, and if we assume (53.1.2), why are the primes $\equiv 3$ (mod 4) more numerous than those $\equiv 1$ (mod 4). The situation can be briefly described as follows.

The role of the residue classes $\equiv 1 \pmod 4$ and $\equiv 3 \pmod 4$ will be symmetric if we also allow the prime powers, that is, for the function

$$(53.1.4)\qquad \sum_{n\,\text{odd}} (-1)^{(n-1)/2}\Lambda(n)e^{-n/x}$$

no analogue of (53.1.3) exists whether or not (53.1.2) is true. Now if $\mathscr{L}(s, \chi_1)$ has a zero $\rho_0 = \beta_0 + i\gamma_0$ with $\beta_0 > \frac{1}{2}$, then (53.1.4) oscillates with order larger than

$$(53.1.5)\qquad x^{\beta_0 - \varepsilon},$$

and thus the contribution of prime powers, being $O(\sqrt{x})$, does not count. But if (53.1.2) is true, then the oscillation of (53.1.4), being smaller than

$$(53.1.6)\qquad c_0\sqrt{x},$$

is majorized by the large positive contribution of

$$(53.1.7)\qquad \sum_{\substack{n=p^\alpha,\ \alpha\ge 2\\ p>2}} (-1)^{(n-1)/2}\Lambda(n)e^{-n/x}$$

which is asymptotic to

$$(53.1.8)\qquad c_1\sqrt{x} \quad \text{with} \quad c_1 > c_0.$$

And so

$$(53.1.9)\qquad \begin{aligned} &\sum_{p>2} (-1)^{(p-1)/2}\log p \cdot e^{-p/x} \\ &\quad = \sum_{n\,\text{odd}} (-1)^{(n-1)/2}\Lambda(n)e^{-n/x} \\ &\qquad - \sum_{\substack{n=p^\alpha,\ \alpha\ge 2\\ p>2}} (-1)^{(n-1)/2}\Lambda(n)e^{-n/x} \\ &\quad < -c_2\sqrt{x} \end{aligned}$$

with an absolute constant $c_2 > 0$.

For general q the situation becomes more difficult. In the case $q = 8$ it was possible (see Turán [8]) to show that for $l = 3, 5$, or 7 the relation

$$(53.1.10)\qquad \lim_{x\to\infty} \left\{ \sum_{p\equiv 1(8)} \log p \cdot e^{-p/x} - \sum_{p\equiv l(8)} \log p \cdot e^{-p/x} \right\} = -\infty$$

holds if and only if no function $\mathscr{L}(s, \chi)$ mod 8 with $\chi \ne \chi_0$ vanishes for $\sigma > \frac{1}{2}$.

The proof of this follows the line of the Hardy–Littlewood–Landau arguments and relies on numerical data concerning the low zeros of the $\mathscr{L}$-functions mod 8. But for general q these arguments work only if we prove that the analogue of (53.1.3) implies the analogue of (53.1.2), but not conversely.

However, we may modify the problem somewhat by replacing the "kernel" function

$$e^{-p/x} \quad \text{by} \quad e^{-(1/r)\log^2(p/x)} \tag{53.1.11}$$

where $r = r(x)$ is some function for which we assume only that $c_3 \le r \le \log x$. Then the situation becomes more pleasant. It suffices to use only the Haselgrove condition (48.3.3)—its necessity is demonstrated in the arguments of Section 48.3—to prove the complete analogue of the Hardy–Littlewood–Landau theorem. In the following three sections we show another reason for the change of the kernel function: there we prove some theorems concerning the accumulation of primes in (relatively) short intervals in various residue classes. These results are the consequences of some Abel-type theorems.

53.2

We introduce the notation

$$\varepsilon(n, q, l_1, l_2) = \begin{cases} 1 & \text{if} \quad n \equiv l_1 (\operatorname{mod} q) \\ -1 & \text{if} \quad n \equiv l_2 (\operatorname{mod} q) \\ 0 & \text{otherwise} \end{cases} \tag{53.2.1}$$

for the next result.

Theorem 53.1. *Suppose, for a modulus q, that the Haselgrove condition* (48.3.3) *is true and that for a quadratic nonresidue l_1 and a residue l_2* (mod q) *no $\mathscr{L}(s, \chi, q)$ with $\chi(l_1) \neq \chi(l_2)$ vanishes for $\sigma > \frac{1}{2}$. Then the inequality*

$$\sum_p \varepsilon(p, q, l_1, l_2) \log p \exp\left(-\frac{\log^2(p/x)}{4k}\right) > \sqrt{x} \tag{53.2.2}$$

holds whenever

$$(\log q \le) c_4 \frac{\log q}{A^2(q)} \le k \le \tfrac{1}{10} \log x, \qquad x \ge \max(c_5, e^q) \tag{53.2.3}$$

where c_4 and c_5 are sufficiently large absolute constants.

The inequality asserts that there are more primes $\equiv l_1$ (mod q) than are $\equiv l_2$ (mod q) in the given sense in every interval of the form

$$\left[x \exp(-3\sqrt{k \log x}),\, x \exp(3\sqrt{k \log x})\right] \tag{53.2.4}$$

when x is sufficiently large. In the case of the original problem, the preponderance can be established only in larger intervals of the form

$$[x^{1/2-\varepsilon}, x\log x]. \tag{53.2.5}$$

This is another reason why the kernel function $\exp[-(1/r)\log^2(p/x)]$ is more advantageous in proving "accumulation theorems."

The other case is settled by the next result.

Theorem 53.2. *Assume that the Haselgrove condition is satisfied for* q, $(l, q) = 1$, *and there is a zero* $\rho_0 = \beta_0 + i\gamma_0$ $(\beta_0 > \frac{1}{2})$ *of an* $\mathscr{L}(s, \chi_1)$ mod q *such that* $\chi_1(l) \neq 1$. *Then for all* T *satisfying*

$$T > \max\left(c_6, \exp\left[\frac{1}{A(q)}\right]^{30}, \exp(q^{50}), \exp\left\{\left(3\frac{1+\gamma_0^2}{\beta_0 - \frac{1}{2}}\right)^{7/2}\right\}\right) \tag{53.2.6}$$

there exist k_1 *and* k_2 *with*

$$\log^{5/7}T \le k_j \le \log^{5/7}T + \log^{1/3}T \tag{53.2.7}$$

and x_1 *and* x_2 *with*

$$T \le x_j \le T\exp(\log^{3/4}T) \tag{53.2.8}$$

such that the inequalities

$$\sum_p \varepsilon(p, q, l, 1)\log p \exp\left(-\frac{\log^2(p/x_1)}{4k_1}\right) > x_1^{\beta_0}\exp[-2(1+\gamma_0^2)\log^{5/7}x_1] \tag{53.2.9}$$

and

$$\sum_p \varepsilon(p, q, l, 1)\log p \exp\left(-\frac{\log^2(p/x_2)}{4k_2}\right) < -x_2^{\beta_0}\exp[-2(1+\gamma_0^2)\log^{5/7}x_2] \tag{53.2.10}$$

hold.

The proof shows that if we replace $\log p$ by $\Lambda(n)$ the same theorem also holds for $\beta_0 = \frac{1}{2}$ (naturally the last term in (53.2.6) must be omitted). Thus, by Siegels's theorem (49.1.5), for suitable k_j and x_j $(j = 1, 2)$ satisfying (53.2.7)–(53.2.8) the inequalities

(53.2.11)

$$\sum_{n=1}^{\infty} \varepsilon(n, q, l, 1)\Lambda(n)\exp\left(-\frac{\log^2(n/x_1)}{4k_1}\right) > \sqrt{x_1}\cdot\exp(-c_7\log^{5/7}x_1)$$

and

(53.2.12)

$$\sum_{n=1}^{\infty} \varepsilon(n, q, l, 1)\Lambda(n)\exp\left(-\frac{\log^2(n/x_2)}{4k_2}\right) < -\sqrt{x_2}\exp(-c_7\log^{5/7}x_2)$$

hold assuming only the Haselgrove condition and independent of the Riemann–Piltz hypothesis.

From Theorems 53.1 and 53.2 we get the analogue of the Hardy–Littlewood–Landau theorem.

Corollary I. *Suppose that q satisfies the Haselgrove condition. Then the relation*

$$\lim_{x\to\infty} \sum_{p} \varepsilon(p, q, l, 1)\log p \exp\left(-\frac{\log^2(p/x)}{4k}\right) = +\infty \tag{53.2.13}$$

is true for all quadratic nonresidues l (mod q) *and for all k satisfying*

$$c_4\frac{\log q}{A^2(q)} \le k \le \tfrac{1}{10}\log x \tag{53.2.14}$$

if and only if none of the $\mathcal{L}(s, \chi, q)$ functions with $\chi \ne \chi_0$ vanishes for $\sigma > \frac{1}{2}$.

Proof. It is enough to observe that for every χ (mod q) with $\chi \ne \chi_0$ there exists a quadratic nonresidue l (mod q) for which $\chi(l) \ne 1$. [Namely, if $\chi(l) = 1$ for all nonresidues, b is an arbitrary residue, and l is an arbitrary nonresidue, then bl being a nonresidue implies that $\chi(b) = \chi(bl)/\chi(l) = 1$. Hence $\chi(b) = 1$ for all residues b, too, which implies $\chi = \chi_0$.] Thus if the Riemann–Piltz hypothesis is false, then Theorem 53.2 is applicable. If it is true, then Theorem 53.1 with $l_2 = 1$ proves the corollary. ■

53.3

In the proofs of Theorems 53.1–53.2 we use the abbreviations

$$F(s) \doteq \frac{1}{\varphi(q)}\sum_{\chi}[\bar\chi(l_2) - \bar\chi(l_1)]\frac{\mathcal{L}'}{\mathcal{L}}(s, \chi) \tag{53.3.1}$$

$$\mu \doteq \log x \tag{53.3.2}$$

$$J(k, x) \doteq \frac{1}{2\sqrt{k\pi}}\sum_{n}\varepsilon(n, q, l_1, l_2)\Lambda(n)\exp\left(-\frac{\log^2(n/x)}{4k}\right). \tag{53.3.3}$$

From the integral formula (41.8.2) we obtain

$$J(k,x) = \frac{1}{2\pi i}\int_{(2)} e^{ks^2+\mu s}F(s)\,ds. \tag{53.3.4}$$

Shift the line of integration to $\sigma = -\frac{1}{4}$. Then

$$\begin{aligned} J(k,x) &= \frac{1}{\varphi(q)}\sum_{\chi}[\bar\chi(l_2) - \bar\chi(l_1)]\sum_{\rho} e^{k\rho^2+\mu\rho} \\ &\quad + \frac{1}{2\pi i}\int_{(-1/4)} e^{ks^2+\mu s}F(s)\,ds. \end{aligned} \tag{53.3.5}$$

But well-known estimates on $\mathscr{L}$-functions imply that if $1 \le k \le \mu$, the last integral is $O(\log q)$, namely, for $\sigma = -\frac{1}{4}$, $|\mathscr{L}'/\mathscr{L}(s,\chi)| = O[\log q(|t| + 2)]$, which follows easily from G.1.2 of Appendix G and from (48.8.3). Thus

$$J(k,x) = \frac{1}{\varphi(q)}\sum_{\chi}[\bar\chi(l_2) - \bar\chi(l_1)]\sum_{\rho} e^{k\rho^2+\mu\rho} + O(\log q). \tag{53.3.6}$$

We also use the fact that if $1 \le k \le \mu/10$, $x = e^{\mu} \to \infty$, and if $g(n)$ is a function for which

$$\sum_{n\le r} g(n) \doteq G(r) = [1 + o(1)]\,cr^B \tag{53.3.7}$$

with $\frac{1}{10} \le B \le 1$ then

$$\begin{aligned} &\frac{1}{2\sqrt{\pi k}}\sum_{n=1}^{\infty} g(n)\exp\left(-\frac{\log^2(n/x)}{4k}\right) \\ &= \frac{1}{2\sqrt{\pi k}}\int_1^{\infty} G(r)\exp\left(-\frac{(\log r - \mu)^2}{4k}\right)\frac{\log r - \mu}{2kr}\,dr \\ &= \int_{-\mu}^{\infty}[1 + o(1)]\,c\exp\left\{B(\mu + y) - \frac{y^2}{4k}\right\}\frac{y}{4k\sqrt{\pi k}}\,dy \\ &= ce^{kB^2+\mu B}\int_{-\mu}^{\infty}[1 + o(1)]\exp\left\{-\left(\frac{y^2}{4k} - By + kB^2\right)\right\}\frac{y}{4k\sqrt{\pi k}}\,dy \\ &= ce^{kB^2+\mu B}\int_{-\mu/2\sqrt{k} - B\sqrt{k}}^{\infty}[1 + o(1)]\,e^{-t^2}\cdot\frac{B\sqrt{k} + t}{\sqrt{\pi k}}\,dt \\ &= [1 + o(1)]\,cBe^{kB^2+\mu B}. \end{aligned} \tag{53.3.8}$$

Naturally if

$$\sum_{n \le r} |g(n)| \le [1 + o(1)]\, cr^{B}, \tag{53.3.9}$$

then

$$\left| \frac{1}{2\sqrt{\pi k}} \sum_{n} g(n) \exp\left(-\frac{\log^2(n/x)}{4k} \right) \right| \le [1 + o(1)]\, cBe^{kB^2 + \mu B}. \tag{53.3.10}$$

53.4. PROOF OF THEOREM 53.1

Observe that for the zeros in (53.3.6) we have

$$\beta = \tfrac{1}{2}, \qquad |\gamma| \ge A(q), \tag{53.4.1}$$

and thus by (48.8.3) and (53.2.3) the infinite power sum in (53.3.6) is absolutely

(53.4.2)

$$< \sum_{n=1}^{\infty} c_8 \log\{q(n+1+2)\} \exp\left[\frac{\mu}{2} + \frac{k}{4} - k \max\left((n-1)^2, A^2(q) \right) \right]$$

$$< c_9 \log q \exp\left[\frac{\mu}{2} + \frac{k}{4} - kA^2(q) \right] \le c_9 \frac{\log q}{q^{c_4}} e^{\mu/2 + k/4}.$$

On the other hand, since l_1 is a quadratic residue and l_2 is a nonresidue mod q,

(53.4.3)

$$\sum_{\substack{n = p^{\alpha},\, \alpha \ge 2 \\ n \le r}} \varepsilon(n, q, l_1, l_2) \Lambda(n)$$

$$= \sum_{\substack{p^2 \equiv l_1(q) \\ p \le \sqrt{r}}} \log p + \sum_{\substack{n = p^{\alpha},\, \alpha \ge 3 \\ n \le r}} \varepsilon(n, q, l_1, l_2) \Lambda(n)$$

$$= [1 + o(1)]\, N(q) \frac{\sqrt{r}}{\varphi(q)} + O\left(\sqrt[3]{r} \right) = [1 + o(1)] \frac{N(q)}{\varphi(q)} \sqrt{r}$$

where $N(q)$ denotes the number of solutions of the congruence $x^2 \equiv l_1(q)$. Thus (53.3.7)–(53.3.8) imply

(53.4.4)

$$\frac{1}{2\sqrt{\pi k}} \sum_{n = p^{\alpha},\, \alpha \ge 2} \varepsilon(n, q, l_1, l_2) \Lambda(n) \exp\left(-\frac{\log^2(n/x)}{4k} \right)$$

$$= [1 + o(1)] \frac{N(q)}{2\varphi(q)} e^{\mu/2 + k/4}.$$

Now from (53.2.3), (53.3.3), (53.3.6), (53.4.2), and (53.4.4) we get (53.2.2) if c_4 is sufficiently large.

53.5

To prove Theorem 53.2 it will be sufficient to show that with suitable $k_1, k_2, x_1 = e^{\mu_1}, x_2 = e^{\mu_2}$ satisfying (53.2.7)–(53.2.8), the inequalities

$$(53.5.1)\qquad \frac{1}{\varphi(q)}\sum_{\chi}[1-\bar{\chi}(l)]\sum_{\rho}e^{k_1\rho^2+\mu_1\rho} > 2x_1^{\beta_0}\exp\{-2(1+\gamma_0^2)\log^{5/7}x_1\}$$

and

(53.5.2)

$$\frac{1}{\varphi(q)}\sum_{\chi}[1-\bar{\chi}(l)]\sum_{\rho}e^{k_2\rho^2+\mu_2\rho} < -2x_2^{\beta_0}\exp\{-2(1+\gamma_0^2)\log^{5/7}x_2\}$$

hold.

Namely, using the fact that for $c', 2 \le c' \le \varphi(q)$,

$$(53.5.3)\qquad \sum_{\substack{n=p^\alpha,\,\alpha\ge 2\\ n\le r}}|\varepsilon(n,q,l,1)|\Lambda(n) = [1+o(1)]\frac{c'}{\varphi(q)}\sqrt{r},$$

(53.3.7)–(53.3.8) imply

$$(53.5.4)\qquad \left|\frac{1}{2\sqrt{\pi k}}\sum_{n=p^\alpha,\,\alpha\ge 2}\varepsilon(n,q,l,1)\Lambda(n)\cdot\exp\left(-\frac{\log^2(n/x)}{4k}\right)\right| < \frac{2}{3}\cdot e^{\mu/2+k/4} = \frac{2}{3}\sqrt{x}\cdot e^{k/4}.$$

By (53.2.6)–(53.2.8),

(53.5.5)

$$\frac{k}{4} + 2(1+\gamma_0^2)\log^{5/7}x < 3(1+\gamma_0^2)\log^{5/7}x < (\beta_0 - \tfrac{1}{2})\log^{2/7}T\cdot\log^{5/7}x < (\beta_0-\tfrac{1}{2})\log x,$$

that is,

$$(53.5.6)\qquad e^{k/4} < x^{\beta_0-1/2}\cdot\exp[-2(1+\gamma_0^2)\log^{5/7}x].$$

So by (53.5.4) and (53.5.6) the inequalities (53.5.1)–(53.5.2) imply (53.2.9) and (53.2.10) when (53.3.3) and (53.3.6) are also taken into account.

The contribution of zeros to (53.5.1)–(53.5.2) with

$$(53.5.7) \qquad |\gamma| > 2\log^{1/7}T$$

is again, by (48.8.3) and (53.2.6)–(53.2.8), absolutely

$$(53.5.8) \qquad < \sum_{n=[2\log^{1/7}T]}^{\infty} c_9 \log[q(n+1+2)]\exp[\mu + k(1-n^2)]$$

$$< c_{10}\log(q\log^{1/7}T)\exp\{\mu + k(1-[2\log^{1/7}T]^2)\} < 1.$$

Thus for the remaining finite real power sum it is sufficient to show the estimates

$$(53.5.9) \qquad \frac{1}{\varphi(q)}\sum_{\chi}[1-\bar{\chi}(l)] \sum_{|\gamma|\le 2\log^{1/7}T} e^{k_1\rho^2+\mu_1\rho}$$

$$> 3x_1^{\beta_0}\exp\{-2(1+\gamma_0^2)\log^{5/7}x_1\}$$

and

$$(53.5.10) \qquad \frac{1}{\varphi(q)}\sum_{\chi}[1-\bar{\chi}(l)] \sum_{|\gamma|\le 2\log^{1/7}T} e^{k_2\rho^2+\mu_2\rho}$$

$$< -3x_2^{\beta_0}\exp\{-2(1+\gamma_0^2)\log^{5/7}x_2\}.$$

53.6

Now apply Corollary II of Appendix H to obtain a ξ with

$$(53.6.1) \qquad \log^{2/7}T \le \xi \le \log^{2/7}T + \frac{1}{A(q)} \le \log^{2/7}T + \log^{1/30}T$$

such that for the n zeros with

$$(53.6.2) \qquad |\gamma| \le 2\log^{1/7}T$$

the inequality

$$(53.6.3) \qquad (\pi \ge)\ \left|\operatorname{arc} e^{\rho^2+\xi\rho}\right| > \frac{1}{4n} \doteq \kappa$$

holds. Let k be any integer, to be chosen later, satisfying (53.2.7) and let

$$(53.6.4) \qquad \log x = \mu = k\xi.$$

Then (53.6.1) and (53.2.7) imply (53.2.8).

Now we apply the one-sided Corollary I of Section 13 with

$$b_j = \frac{1}{\varphi(q)}[1 - \bar{\chi}(l)] \tag{53.6.5}$$

$$z_j = e^{\rho^2 + \xi\rho} \tag{53.6.6}$$

$$m = \left[\log^{5/7} T\right]. \tag{53.6.7}$$

By (48.8.3) and (53.2.6) the number of terms is

$$1 \le n < c_{11}\varphi(q)\log^{1/7} T \log\left[q(2\log^{1/7} T + 2)\right] < \frac{\log^{1/6} T}{4} \tag{53.6.8}$$

and

$$\operatorname{Re}(b_1 + \cdots + b_j) > \frac{1}{\varphi(q)} \cdot \frac{1}{q^2} > \frac{1}{q^3}. \tag{53.6.9}$$

Thus (53.6.3), (53.6.7), and (53.6.8) imply the existence of a k_1 and k_2 with

$$k_j \in \left[m + 1, m + n\left(3 + \frac{\pi}{\kappa}\right)\right] \subset \left[\log^{5/7} T, \log^{5/7} T + \log^{1/3} T\right] \tag{53.6.10}$$

such that for the real power sum $Z(k_j, \mu_j) = Z(k_j, k_j\xi) = Z_\xi(k_j)$ in (53.5.9)–(53.5.10) the inequalities

$$\begin{aligned} Z_\xi(k_1) &> \frac{q^{-3}}{\frac{3}{4}\log^{1/6} T}\left(\frac{1}{\log T}\right)^{(\log^{1/6} T)/2} \cdot e^{k_1(\beta_0^2 - \gamma_0^2) + \mu_1\beta_0} \\ &> 3x_1^{\beta_0}\exp\left(-k_1\gamma_0^2 - \log^{1/5} T\right) \\ &> 3x_1^{\beta_0}\exp\left[-2(1 + \gamma_0^2)\log^{5/7} x_1\right] \end{aligned} \tag{53.6.11}$$

hold.

Similarly,

$$Z_\xi(k_2) < -3x_2^{\beta_0}\exp\left[-2(1 + \gamma_0^2)\log^{5/7} x_2\right]. \tag{53.6.12}$$

Thus the inequalities (53.5.9)–(53.5.10) are valid, and Theorem 53.2 is proven.

■

54

COMPARATIVE PRIME NUMBER THEORY VII

Accumulation Theorems for Residue Classes Representing Quadratic Nonresidues

54.1

In the present section we compare the number of primes (in the Abelian sense) in two residue classes mod q that are both quadratic nonresidues. As one expects in this case, no constant preponderance of primes in one or the other class exists. Now the Haselgrove condition (48.3.3) is not by itself sufficient for our purpose, and we also must assume the finite Riemann–Piltz hypothesis, in a somewhat stronger form than given in Sections 50–51. Namely, we assume that no function $\mathscr{L}(s, \chi, q)$ vanishes in the domain

(54.1.1) $$\sigma > \frac{1}{2}, \qquad |t| \leq D$$

where D satisfies

(54.1.2) $$D \geq \max\left[\frac{5}{A(q)}, c_1 q^{10}\right]$$

and c_1 is a sufficiently large absolute constant. (Some lower restrictions for c_1 are implied in Section 50.)

Theorem 54.1. *If for the $\mathscr{L}$-functions* mod q *the Haselgrove condition* (48.3.3) *and the finite Riemann–Piltz conjecture* (54.1.1)–(54.1.2) *hold, then for all*

(54.1.3) $$T > \max\left[c_2, \exp(2D^6)\right]$$

and for all quadratic nonresidues l_1 *and* l_2 mod q *there exists a* k *with*

$$\frac{4\log T}{D^2} \le k \le \frac{4\log T}{D^2} + D^3 \tag{54.1.4}$$

and an x *with*

$$T \le x \le T^{1+1/D} \tag{54.1.5}$$

such that [*in the notation* (53.2.1)] *the inequality*

$$\sum_p \varepsilon(p, q, l_1, l_2)\log p \cdot \exp\left(-\frac{\log^2(p/x)}{4k}\right) > T^{1/2-1/D} \tag{54.1.6}$$

hold.

Interchanging l_1 and l_2 gives an analogous inequality with $>$ replaced by $<$.

It is then relatively easy to deduce that there are "short" intervals in which there are much more primes $\equiv l_1(\bmod q)$ than $\equiv l_2(\bmod q)$ (and conversely). This "accumulation theorem" is stated precisely as follows.

Theorem 54.2. *Under the conditions of Theorem* 54.1 *there is an interval*

$$[U_1, U_2] \subset [T^{1-8/D}, T^{1+8/D}] \tag{54.1.7}$$

such that

$$\sum_{\substack{p \equiv l_1(q) \\ U_1 < p \le U_2}} 1 - \sum_{\substack{p \equiv l_2(q) \\ U_1 < p \le U_2}} 1 > T^{1/2-2/D}. \tag{54.1.8}$$

54.2

Recalling (54.1.1)–(54.1.2) and the notation $L = c_1 q^{10}$ in (50.4.6), Lemma 50.1 implies the existence of a prime $P \equiv l_1 \pmod q$ such that

$$\tfrac{2}{3}L^{1/4} \le P \le L^{1/4}. \tag{54.2.1}$$

Using this prime P, we can prove a result similar to Lemma 50.2.

Lemma 54.1. *If* $(l_1, q) = (l_2, q) = 1$ *and* $P \equiv l_1 \pmod q$ *is a prime satisfying* (54.2.1), *then for*

$$k_0 = \frac{1}{P^2\log^2 P}, \qquad \mu_0 = \log P \tag{54.2.2}$$

we have

$$(54.2.3)\qquad S \doteq \frac{1}{\varphi(q)} \sum_{\chi} [\bar{\chi}(l_2) - \bar{\chi}(l_1)] \sum_{\rho} e^{k_0\rho^2 + \mu_0\rho} > P.$$

Proof. Apply (53.3.1)–(53.3.5), and an easy computation for the integral in the analogue of (53.3.5) yields the estimate $O(P^{3/4}\log^2 P)$. Then, as in (53.3.6),

(54.2.4)

$$S = \frac{1}{2\sqrt{\pi k_0}} \sum_n \varepsilon(n, q, l_1, l_2)\Lambda(n)\exp\left(-\frac{\log^2(n/P)}{4k_0}\right) + O(P^{3/4}\log^2 P).$$

The total contribution of the n with $n \neq P$ to the inner sum is

$$(54.2.5)\qquad < \sum_{\substack{n=1\\ n\neq P}}^{\infty} \log n \exp\left(-\frac{P^2\log^2 P}{4}\log^2\frac{n}{P}\right) < \exp\left(-\frac{\log^2 P}{5}\right).$$

The contribution of $n = P$ is obviously $\log P$, and so

$$(54.2.6)\quad S > \frac{P\log P}{2\sqrt{\pi}}\left[\log P - \exp\left(-\frac{\log^2 P}{5}\right)\right] + O(P^{3/4}\log^2 P) > P.$$

(We used the fact that if c_1 is sufficiently large, L and hence P are also sufficiently large, even compared to q.) ■

54.3

To prove Theorem 54.2 it is sufficient to show that

$$(54.3.1)\qquad \frac{1}{2\sqrt{\pi k}} \sum_n \varepsilon(n, q, l_1, l_2)\Lambda(n)\exp\left(-\frac{\log^2(n/x)}{4k}\right) > 2T^{1/2-1/D},$$

because l_1 and l_2 are quadratic nonresidues and therefore

$$(54.3.2)\qquad \sum_{\substack{n=p^\alpha,\,\alpha\geq 2\\ n\leq r}} |\varepsilon(n, q, l_1, l_2)|\Lambda(n) = O(r^{1/3}).$$

Thus, using (53.3.9)–(53.3.10), we get

(54.3.3)

$$\frac{1}{2\sqrt{\pi k}}\left|\sum_{n=p^\alpha,\,\alpha\geq 2} \varepsilon(n, q, l_1, l_2)\Lambda(n)\exp\left(-\frac{\log^2(n/x)}{4k}\right)\right|$$

$$= O(x^{1/3}\cdot e^{k/9}) < T^{1/2-1/D}.$$

In view of (53.3.1)–(53.3.6), proving (54.3.1) amounts to showing

$$\frac{1}{\varphi(q)}\sum_{\chi}[\bar{\chi}(l_2)-\bar{\chi}(l_1)]\sum_{\rho}e^{k\rho^2+\mu\rho} > 3T^{1/2-1/D} \tag{54.3.4}$$

(with $\mu = \log x$).

The contribution of zeros with

$$|\gamma| > D \tag{54.3.5}$$

is by (48.8.3), (54.1.3)–(54.1.5), absolutely

(54.3.6)

$$< \sum_{n=[D]}^{\infty} c_3 \log[q(n+1+2)]\exp\left\{\frac{4\log T}{D^2}(1-n^2)+\log T\left(1+\frac{1}{D}\right)\right\} < 1,$$

and so we have only to prove

$$\frac{1}{\varphi(q)}\sum_{\chi}[\bar{\chi}(l_2)-\bar{\chi}(l_1)]\sum_{|\gamma|\le D}e^{k\rho^2+\mu\rho} > 4T^{1/2-1/D} \tag{54.3.7}$$

for an appropriate choice of k and $\mu = \log x$ satisfying (54.1.4)–(54.1.5).

54.4

We again estimate a power sum similar to that of (53.5.9)–(53.5.10), but here the coefficients $[1/\varphi(q)][\bar{\chi}(l_2)-\bar{\chi}(l_1)]$ no longer have a nonnegative real part. This is the reason why Lemma 54.1 was necessary. With the notation (54.2.2), let

$$k \doteq k' + k_0, \qquad \mu \doteq \mu' + \mu_0 \doteq k'\xi + \mu_0 \tag{54.4.1}$$

where k' is an integer to be chosen later for which

$$\frac{4\log T}{D^2} \le k' \le \frac{4\log T}{D^2} + D^3 - 1 \tag{54.4.2}$$

and ξ is a number, given by Corollary II of Appendix H satisfying

$$\xi \in \left[\frac{D^2}{4}, \frac{D^2}{4} + \frac{1}{A(q)}\right] \subseteq \left[\frac{D^2}{4}, \frac{D^2}{4} + \frac{D}{5}\right] \tag{54.4.3}$$

such that for the n zeros with $|\gamma| \le D$

$$(\pi \ge)\ |\text{arc}\, e^{\rho^2+\xi\rho}| > \frac{1}{4n} \doteq \kappa. \tag{54.4.4}$$

By (54.1.2), (54.1.3), (54.2.2), and (54.4.2)–(54.4.3) it is easy to see that k and μ in (54.4.1) satisfy the inequalities prescribed in (54.1.4)–(54.1.5).

54.5

Let $\rho_1 = \frac{1}{2} + i\gamma_1$ be the zero with minimal absolute value of γ. By Siegel's theorem (49.1.5), $|\gamma| = O(1) < D$. Now by (54.4.1), our critical sum (54.3.7) takes the form

$$Z(k') = \exp\left[\left(\frac{1}{4} - \gamma_1^2 + \frac{\xi}{2}\right)k'\right] \cdot \sum b_j z_j^{k'} \tag{54.5.1}$$

with the more convenient coefficients

$$b_j = \frac{1}{\varphi(q)}\left[\bar{\chi}(l_2) - \bar{\chi}(l_1)\right] e^{k_0\rho^2 + \mu_0\rho} \tag{54.5.2}$$

and

$$z_j = \exp\left[\rho^2 + \xi\rho - \left(\frac{1}{4} - \gamma_1^2 + \frac{\xi}{2}\right)\right]. \tag{54.5.3}$$

The condition $\max|z_j| = 1$ is satisfied; the number of terms is

$$1 \le n < c_4\varphi(q) D \log[q(D+2)] < D^{1.2} \tag{54.5.4}$$

by (48.8.3) and (54.1.2). Thus choosing

$$m = \left[\frac{4\log T}{D^2}\right], \tag{54.5.5}$$

(54.4.4) and (54.5.4) imply that

$$\left[m+1, m+n\left(3+\frac{\pi}{\kappa}\right)\right] \subset \left[\frac{4\log T}{D^2}, \frac{4\log T}{D^2} + D^3 - 1\right]. \tag{54.5.6}$$

We apply the one-sided Corollary II of Section 13 where z_{h_1} is the number belonging to the zero $\rho_2 = \frac{1}{2} + i\gamma_2$ with the largest imaginary part among those satisfying

$$|\gamma| \le P\log^2 P \left(< \frac{\sqrt{D}}{5}\right) \tag{54.5.7}$$

and choose $h_2 = n$.

Then

$$|z_{h_1}| = e^{\gamma_1^2 - \gamma^2} > e^{-\gamma^2} > e^{-D/25} > 2e^{-D/20}. \tag{54.5.8}$$

Since $|z_j| = e^{\gamma_1^2 - \gamma^2}$, the numbers $|z_j|$ decrease with increasing values of $|\gamma|$, and thus all zeros with $|\gamma| < P\log^2 P$ are contained in the critical sum B of (13.8.4). Using Lemma 54.1, and observing further that by (54.2.1) P is large enough (even compared to q), the real sum B has the lower estimate

(54.5.9)

$$
\begin{aligned}
B &> S - \frac{1}{\varphi(q)}\sum_{\chi} 2 \sum_{|\gamma| \geq P\log^2 P} \exp\left(\frac{1}{P^2\log^2 P}(1-\gamma^2) + \log P\right) \\
&\geq P - \sum_{n=[P\log^2 P]} c_5 \log[q(n+1+2)] \\
&\quad \cdot \exp\left[\frac{1}{P^2\log^2 P}(1-n^2) + \log P\right] \\
&\geq P - \exp\left(-\frac{\log^2 P}{2}\right) > P - 1 > 1.
\end{aligned}
$$

From (54.1.2), (54.1.3), (54.4.4), and (54.5.1)–(54.5.9) and Corollary II of Section 13, there exists a k' satisfying (54.4.2) such that

(54.5.10) $$\sum b_j z_j^{k'} > (e^{-D/20})^{5D^{-2}\log T} = \exp\left(-\frac{\log T}{4D}\right) > 4\exp\left(-\frac{\log T}{2D}\right).$$

Furthermore, $|\gamma| < c_6$ by Siegel's theorem (49.1.5) so (54.4.2) and (54.4.3) yield

(54.5.11) $$\left(\frac{1}{4} - \gamma_1^2 + \frac{\xi}{2}\right)k' > -c_6^2 \cdot \frac{5\log T}{D^2} + \frac{\log T}{2} > \frac{\log T}{2} - \frac{\log T}{2D}.$$

Then (54.5.10) and (54.5.11) give the following lower bound for the $Z(k')$ in (54.5.1):

(54.5.12) $$Z(k') > T^{1/2 - 1/2D} \cdot 4T^{-1/2D} = 4T^{1/2 - 1/D}$$

which proves (54.3.7) and hence also (54.1.6).

54.6

Theorem 54.2 can be deduced from Theorem 54.1 and the following lemma.

Lemma 54.2. *If $1 \leq k \leq \mu/12$ and $\mu > c_7$, then for arbitrary l_1, l_2, q there are U_1 and U_2 satisfying*

(54.6.1) $$[U_1, U_2] \subseteq \left[e^{\mu - 3\sqrt{\mu k}}, e^{\mu + 3\sqrt{\mu k}}\right]$$

such that the inequality

(54.6.2)
$$\sum_{U_1<p\le U_2} \varepsilon(p,q,l_1,l_2)$$
$$> \frac{1}{\mu+1}\left\{\sum_p \varepsilon(p,q,l_1,l_2)\log p \exp\left[-\frac{(\log p-\mu)^2}{4k}\right]-3\right\}$$

holds.

Proof. Let

(54.6.3)
$$h(r)=\sum_{p\le r}\varepsilon(p,q,l_1,l_2).$$

By partial summation we get

(54.6.4)
$$\begin{aligned} H &\doteq \sum_p \varepsilon(p,q,l_1,l_2)\log p \exp\left[-\frac{(\log p-\mu)^2}{4k}\right] \\ &= \int_1^\infty \log r \exp\left[-\frac{(\log r-\mu)^2}{4k}\right] dh(r) \\ &= \int_1^\infty [-h(r)]\frac{d}{dr}\left\{\exp\left[-\frac{(\log r-\mu)^2}{4k}\right]\log r\right\} dr \\ &= \int_1^\infty \left[-\frac{h(r)}{r}\right]\exp\left[-\frac{(\log r-\mu)^2}{4k}\right]\left[1-\frac{\log r(\log r-\mu)}{2k}\right] dr. \end{aligned}$$

Split the integral H into three parts:

(54.6.5)
$$H_1=\int_1^{e^{\mu-3\sqrt{\mu k}}},\qquad H_2=\int_{e^{\mu-3\sqrt{\mu k}}}^{e^{\mu+3\sqrt{\mu k}}},\qquad H_3=\int_{e^{\mu+3\sqrt{\mu k}}}^{\infty},$$

and we can estimate H_1 and H_3 using the trivial estimate $|h(r)|\le r$. Introduce the new variable $y=\log r-\mu$, and since $3\sqrt{\mu k}>10k$ we have

(54.6.6)
$$\begin{aligned} |H_3| &\le \int_{3\sqrt{\mu k}}^\infty \exp\left(-\frac{y^2}{4k}+\mu+y\right)\left|1-\frac{(\mu+y)y}{2k}\right| dy \\ &< \int_{3\sqrt{\mu k}}^\infty \exp\left(-\frac{y^2}{4k}+\mu+y\right)\mu y^2\, dy \\ &< \mu e^\mu \int_{3\sqrt{\mu k}}^\infty \exp\left\{-\frac{y^2}{8k}-\frac{y}{4}\left(\frac{y}{2k}-5\right)\right\} y dy < \mu e^\mu \int_{3\sqrt{\mu k}}^\infty \exp\left(-\frac{y^2}{8k}\right) y dy \\ &= 4k\mu e^\mu \int_{3\sqrt{\mu}/\sqrt{8}}^\infty e^{-v^2} 2v\, dv = 4k\mu e^\mu\cdot e^{-(9/8)\mu}<1. \end{aligned}$$

Similarly we have

$$|H_1| < 1. \tag{54.6.7}$$

From (54.6.4) we see that

$$\frac{d}{dr}\left\{\exp\left[-\frac{(\log r-\mu)^2}{4k}\right]\log r\right\} \gtreqqless 0 \quad \text{if} \quad e \le r \lesseqqgtr r_0 \tag{54.6.8}$$

where r_0 is a given number such that

$$0 < \log r_0 - \mu = \frac{2k}{\log r_0} < \frac{2k}{\mu} < 1 < 3\sqrt{\mu k}\,. \tag{54.6.9}$$

Thus

(54.6.10)

$$\begin{aligned} H_2 \le &- \min_{e^{\mu-3\sqrt{\mu k}} \le r \le r_0} h(r) \int_{e^{\mu-3\sqrt{\mu k}}}^{r_0} \frac{d}{dr}\left\{\exp\left[-\frac{(\log r-\mu)^2}{4k}\right]\log r\right\} dr \\ &+ \max_{r_0 \le r \le e^{\mu+3\sqrt{\mu k}}} h(r) \int_{r_0}^{e^{\mu+3\sqrt{\mu k}}} -\frac{d}{dr}\left\{\exp\left[-\frac{(\log r-\mu)^2}{4k}\right]\log r\right\} dr \\ \le &\left[\max_{r_0 \le r \le e^{\mu+3\sqrt{\mu k}}} h(r) - \min_{e^{\mu-3\sqrt{\mu k}} \le r \le r_0} h(r)\right] \\ &\cdot \exp\left[-\frac{(\log r_0-\mu)^2}{4k}\right]\log r_0 + 2e^{\mu+3\sqrt{\mu k}}\left(\mu + 3\sqrt{\mu k}\right)e^{-(9/4)\mu} \\ \le &\left[\max_{r_0 \le r \le e^{\mu+3\sqrt{\mu k}}} h(r) - \min_{e^{\mu-3\sqrt{\mu k}} \le r \le r_0} h(r)\right](\mu+1) + 1. \end{aligned}$$

If the maximum is obtained for U_2 and the minimum for U_1, then from (54.6.3)–(54.6.7) and (54.6.10) we have the inequality

$$\sum_{U_1 < p \le U_2} \varepsilon(p, q, l_1, l_2) > \frac{H-3}{\mu+1}, \tag{54.6.11}$$

which proves the lemma. ■

From (54.1.3)–(54.1.5) with $\mu = \log x$ we get

$$3\sqrt{\mu k} < \frac{7\log T}{D}. \tag{54.6.12}$$

This, together with (54.1.5) and (54.6.1), implies that

$$(54.6.13)\qquad [U_1, U_2] \subseteq \left[e^{\mu - 3\sqrt{\mu k}}, e^{\mu + 3\sqrt{\mu k}}\right] \subset [T^{1-7/D}, T^{1+8/D}].$$

Hence (54.1.6) and (54.6.2) prove (54.1.8) because

(54.6.14)

$$\frac{T^{1/2-1/D} - 3}{\mu + 1} > \frac{T^{1/2-1/D}}{2\log T} > \frac{T^{1/2-1/D}}{e^{\sqrt{\log T}}} > \frac{T^{1/2-1/D}}{e^{(1/D)\log T}} = T^{1/2-2/D}$$

by (54.1.4) and (54.1.5).

55

COMPARATIVE PRIME NUMBER THEORY VIII

Accumulation Theorems for Residue Classes Representing Quadratic Residues

55.1

In this section we show that the previous results can be extended to the case when l_1 and l_2 have the same quadratic character.

Theorem 55.1. *Theorem* 54.1 *is valid if* l_1 *and* l_2 *are both quadratic residues* mod q.

Theorem 55.2. *Theorem* 54.2 *is true when* l_1 *and* l_2 *are both quadratic residues* mod q.

In contrast to the analogous theorems discussed in Sections 50–51, the extension of Theorem 54.1 requires only a slight modification in the proof.

Theorem 55.2 follows immediately from Theorem 55.1 and Lemma 54.2 [the simple arguments given in (54.6.12)–(54.6.14) remain valid].

55.2

Denote the solutions of the congruences

$$x^2 \equiv l_1 \pmod q, \qquad x^2 \equiv l_2 \pmod q \tag{55.2.1}$$

by

$$\alpha_1', \ldots, \alpha_R' \quad \text{and} \quad \alpha_1'', \ldots, \alpha_R'' \text{ resp.} \tag{55.2.2}$$

(their number being equal). Now as in 51.2, we consider the function

$$\tilde{F}(s) \doteq \frac{1}{\varphi(q)}\left\{\sum_{\chi}[\bar{\chi}(l_2) - \bar{\chi}(l_1)]\frac{\mathscr{L}'}{\mathscr{L}}(s,\chi) + \sum_{j=1}^{R}\sum_{\chi}[\bar{\chi}(\alpha_j') - \bar{\chi}(\alpha_j'')]\frac{\mathscr{L}'}{\mathscr{L}}(2s,\chi)\right\} \tag{55.2.3}$$

and the corresponding function

(55.2.4)

$$\tilde{J}(k,x) \doteq \frac{1}{2\sqrt{\pi k}}\left\{\sum_{n}\varepsilon(n,q,l_1,l_2)\Lambda(n)\exp\left(-\frac{\log^2(n/x)}{4k}\right) - \sum_{n}\varepsilon(n^2,q,l_1,l_2)\Lambda(n)\exp\left(-\frac{\log^2(n^2/x)}{4k}\right)\right\}.$$

Following the argument of Section 53.3 we get

$$\tilde{J}(k,x) = \frac{1}{2\pi i}\int_{(2)} e^{ks^2+\mu s}\tilde{F}(s)\,ds \tag{55.2.5}$$

(with $\mu = \log x$) and from this (as in 53.3.6)

$$\tilde{J}(k,x) = \frac{1}{\varphi(q)}\left\{\sum_{\chi}[\bar{\chi}(l_2) - \bar{\chi}(l_1)]\sum_{\rho}e^{k\rho^2+\mu\rho} + \frac{1}{2}\sum_{j=1}^{R}\sum_{\chi}[\bar{\chi}(\alpha_j') - \bar{\chi}(\alpha_j'')]\sum_{\rho}\exp\left[k(\rho/2)^2 + \mu\rho/2\right]\right\} + O(\log q). \tag{55.2.6}$$

The contribution of the prime squares to the first sum in (55.2.4) and that of the primes to the second sum cancel. For the remaining terms, except the primes in the first sum, we have

(55.2.7)
$$\sum_{\substack{n=p^\alpha,\,\alpha\ge 3\\ n\le r}} |\varepsilon(n,q,l_1,l_2)|\Lambda(n) + \sum_{\substack{n=p^\alpha,\,\alpha\ge 2\\ n^2\le r}} |\varepsilon(n^2,q,l_1,l_2)|\Lambda(n) = O(r^{1/3}),$$

and thus by (53.3.9)–(53.3.10) and (54.1.3)–(54.1.5) we get the estimate

(55.2.8)
$$\frac{1}{2\sqrt{\pi k}}\left|\sum_{n=p^\alpha,\,\alpha\ge 3} \varepsilon(n,q,l_1,l_2)\Lambda(n)\exp\left(-\frac{\log^2(n/x)}{4k}\right) - \sum_{n=p^\alpha,\,\alpha\ge 2} \varepsilon(n^2,q,l_1,l_2)\Lambda(n)\exp\left(-\frac{\log^2(n^2/x)}{4k}\right)\right|$$
$$= O(e^{k/9+\mu/3}) = O(T^{2/5}).$$

From (55.2.4) and (55.2.8) we immediately obtain

(55.2.9)
$$\tilde{J}(k,x) = \frac{1}{2\sqrt{\pi k}}\sum_p \varepsilon(p,q,l_1,l_2)\log p\exp\left(-\frac{\log^2(p/x)}{4k}\right) + O(T^{2/5}).$$

Then by (54.1.1)–(54.1.5) and (48.8.3), for any $\mathscr{L}(s,\chi)$ mod q,

(55.2.10)
$$\left|\sum_\rho \exp\left[k\left(\frac{\rho}{2}\right)^2 + \mu\frac{\rho}{2}\right]\right| \le \sum_\rho \exp\left[k\left(\frac{\beta^2}{4} - \frac{\gamma^2}{4}\right) + \frac{\mu\beta}{2}\right]$$
$$\le \sum_{|\gamma|\le D} e^{k/16+\mu/4} + \sum_{|\gamma|>D} \exp\left[k\left(\frac{1}{4} - \frac{\gamma^2}{4}\right) + \frac{\mu}{2}\right]$$
$$\le c_3 D\log[q(D+2)]\,e^{k/16+\mu/4} + \sum_{n=[D]}^{\infty} c_3\log[q(n+1+2)]\exp\left[k\left(\frac{1}{4} - \frac{n^2}{4}\right) + \frac{\mu}{2}\right]$$
$$< T^{1/3} + c_4\log(qD)\exp\left[\frac{k}{4}(1-[D]^2) + \frac{\mu}{2}\right] < T^{1/3} + 1.$$

Thus the second sum in (55.2.6) is absolutely

$$(55.2.11) \qquad < \frac{1}{\varphi(q)} \cdot \frac{1}{2} \cdot \varphi(q) \cdot \varphi(q) \cdot 2(T^{1/3} + 1) < T^{2/5}.$$

However for the first part of the sum in (55.2.6) and suitable k and μ satisfying (54.1.4)–(54.1.5), we proved the estimate

$$(55.2.12) \qquad \frac{1}{\varphi(q)} \left\{ \sum_{\chi} [\bar{\chi}(l_2) - \bar{\chi}(l_1)] \sum_{\rho} e^{k\rho^2 + \mu\rho} \right\} > 3T^{1/2 - 1/D}$$

in the previous section [see (54.3.4)]. With (55.2.6), (55.2.11), and (55.2.9), this clearly proves (54.1.6). Hence Theorems 55.1 and 55.2 are proven. ■

56

COMPARATIVE PRIME NUMBER THEORY IX

Accumulation Theorems for Residue Classes $\equiv 1$ and $\equiv l \bmod q$

56.1

In Sections 54–55 we established that if l_1 and l_2 have the same quadratic character than there are frequent short intervals containing more primes $\equiv l_1 \pmod q$ than $\equiv l_2 \pmod q$, and conversely. We deduced these results from the fact that the parameters k and x can be suitably chosen in the modified Abel means, with the weights $\exp[-(\log^2(n/x))/4k]$, so that there are much more primes $\equiv l_1 \pmod q$ than $\equiv l_2 \pmod q$ in the above sense; conversely for other values of k and x. However, assuming the Riemann–Piltz conjecture and Haselgrove's condition, if l_1 is a quadratic residue and l_2 is a nonresidue, then Theorem 53.1 implies that there are more primes $\equiv l_2 \pmod q$ than $\equiv l_1 \pmod q$ as measured by a modified Abel mean for an arbitrary choice of the parameters k and x satisfying (53.2.3). Thus we ask if there are any accumulation theorems, like Theorems 54.2 and 55.2, possible when l_1 and l_2 are of opposite quadratic character, and, if so, how would they be proved. In this section we show that such a theorem exists in the easiest case $l_1 = 1$, and the proof follows from a suitable combination of the results proved in the previous sections. If the Riemann–Piltz conjecture is not true for q [by which we mean there exists an $\mathscr{L}(s, \chi_1, q)$ with $\chi_1(l_2) \neq 1$ which has a zero in $\sigma > \frac{1}{2}$, $|t| \neq 0$], then Theorem 53.2 gives the desired result; if it is true then the result follows by way of the arguments of Littlewood, Ingham, and Skewes, used in Theorem 49.2.

Theorem 56.1. *If the Haselgrove condition* (48.3.3) *is true for the modulus q and*

$$Y > \max\left\{c_1, e_3(q^{c_2}), \exp\left[2\left(\frac{1}{A(q)}\right)^{30}\right]\right\} \tag{56.1.1}$$

then there are numbers U_1, U_2, U_3, U_4 *with*

$$(\log Y)^{\exp(-q^{c_3})} \leq U_2\exp\left(-8\log^{6/7}U_2\right) \leq U_1 < U_2 \leq Y \tag{56.1.2}$$

and

$$(\log Y)^{\exp(-q^{c_3})} \leq U_4\exp\left(-8\log^{6/7}U_4\right) \leq U_3 < U_4 \leq Y \tag{56.1.3}$$

such that the inequalities

$$\sum_{\substack{U_1 < p \leq U_2 \\ p \equiv 1(q)}} 1 - \sum_{\substack{U_1 < p \leq U_2 \\ p \equiv l(q)}} 1 > \frac{\sqrt{U_2}}{\log U_2} \tag{56.1.4}$$

and

$$\sum_{\substack{U_3 < p \leq U_4 \\ p \equiv 1(q)}} 1 - \sum_{\substack{U_3 < p \leq U_4 \\ p \equiv l(q)}} 1 < -\frac{\sqrt{U_4}}{\log U_4} \tag{56.1.5}$$

hold.

We remark that if the Riemann–Piltz conjecture is true, then the inequalities (56.1.4)–(56.1.5) are nearly the best possible, the corresponding difference being $O(\sqrt{U_2}\log U_2)$ and $O(\sqrt{U_4}\log U_4)$, respectively.

Here the essential point is that for a quadratic nonresidue $l(\bmod q)$, inequality (56.1.4) holds. The interesting special case $q = 4$, $l = 3$ (where the Haselgrove condition is satisfied) can be formulated as follows.

Corollary I. *For* $Y > c_4$ *there are numbers* U_1, U_2, U_3, U_4 *with*

$$(\log Y)^{c_5} \leq U_2\exp\left(-8\log^{6/7}U_2\right) \leq U_1 < U_2 \leq Y \tag{56.1.6}$$

and

$$(\log Y)^{c_5} \leq U_4\exp\left(-8\log^{6/7}U_4\right) \leq U_3 < U_4 \leq Y \tag{56.1.7}$$

such that the inequalities

$$\sum_{\substack{U_1 < p \le U_2 \\ p \equiv 1(4)}} 1 - \sum_{\substack{U_1 < p \le U_2 \\ p \equiv 3(4)}} 1 > \frac{\sqrt{U_2}}{\log U_2} \tag{56.1.8}$$

and

$$\sum_{\substack{U_3 < p \le U_4 \\ p \equiv 1(4)}} 1 - \sum_{\substack{U_3 < p \le U_4 \\ p \equiv 3(4)}} 1 < -\frac{\sqrt{U_4}}{\log U_4} \tag{56.1.9}$$

hold.

A remarkable consequence of this corollary is the existence of consecutive primes p_ν and $p_{\nu+1}$ both $\equiv 1 \pmod 4$ satisfying the inequality

$$(\log Y)^{c_5} \le p_\nu < p_{\nu+1} \le Y. \tag{56.1.10}$$

The weaker assertion that

$$p_\nu \equiv p_{\nu+1} \equiv 1 \pmod 4 \tag{56.1.11}$$

for infinitely many pairs of consecutive primes could have been derived from Littlewood's theorem

$$\overline{\lim_{x \to \infty}} \; [\pi(x,4,1) - \pi(x,4,3)] = +\infty \tag{56.1.12}$$

but not from easier results; in particular, no arithmetical approach can prove (56.1.11) at present. The natural conjecture that for arbitrarily large ω there are infinitely many ν such that

$$p_\nu \equiv p_{\nu+1} \equiv \cdots \equiv p_{\nu+\omega} \equiv 1 \pmod 4 \tag{56.1.13}$$

is, at present, beyond our grasp, even for $\omega = 2$.

Corollary I also shows that there are at least $(\log Y)^{c_6}$ prime pairs with

$$p_\nu \equiv p_{\nu+1} \equiv 1 \pmod 4 \tag{56.1.14}$$

and

$$p_\nu < p_{\nu+1} \le Y. \tag{56.1.15}$$

In the corresponding case $p_\nu \equiv p_{\nu+1} \equiv 3 \pmod 4$ much more can be said (and the proof is even easier). Though it follows from the corollary only that the same results hold for these pairs, (53.2.11) and Lemma 54.2 show that in every interval of the form

(56.1.16) $$[Y\exp(-8\log^{6/7}Y), Y] \quad \text{for} \quad Y > c_7$$

there are at least

(56.1.17) $$\sqrt{Y}\cdot\exp(-5\log^{6/7}Y)$$

prime pairs p_ν, $p_{\nu+1}$ such that

(56.1.18) $$p_\nu \equiv p_{\nu+1} \equiv 3 \pmod 4.$$

56.2

The proof of Theorem 56.1 has two cases:

CASE I. There exists an $\mathscr{L}(s, \chi_1) \bmod q$ with $\chi_1(l) \neq 1$ which has a zero $\rho_0 = \beta_0 + i\gamma_0$ satisfying

(56.2.1) $$\beta_0 > \frac{1}{2} + \frac{1}{\log^{1/11}Y}, \qquad |\gamma_0| < \log^{1/11}Y.$$

In this case,

(56.2.2) $$\left(3\frac{1+\gamma_0^2}{\beta_0 - \frac{1}{2}}\right)^{7/2} < (4\log^{3/11}Y)^{7/2} = 128\log^{21/22}Y,$$

and thus, recalling (53.2.6) and (56.1.1), we apply Theorem 53.2 with the choice

(56.2.3) $$T = Y\exp(-4\log^{6/7}Y).$$

Since for any x satisfying (53.2.8), formulas (56.2.1), and (56.2.3) show that

(56.2.4) $$\beta_0 - \frac{1}{2} > \frac{1}{\log^{1/11}Y} > \frac{2}{\log^{1/7}x} + \frac{2(1+\gamma_0^2)}{\log^{2/7}x},$$

that is,

(56.2.5) $$x^{\beta_0}\exp\{-2(1+\gamma_0^2)\log^{5/7}x\} > \sqrt{x}\cdot\exp(2\log^{6/7}x),$$

we obtain

(56.2.6) $$\sum_p \varepsilon(p, q, 1, l)\log p\cdot\exp\left(-\frac{\log^2(p/x)}{4k}\right) > \sqrt{x}\cdot\exp(2\log^{6/7}x)$$

by (53.2.10).

With $\mu = \log x$ we have

$$(56.2.7) \qquad 3\sqrt{\mu k} < \tfrac{7}{2}\log^{6/7}T \le \tfrac{7}{2}\log^{6/7}x,$$

by (53.2.7)–(53.2.8).
Then Lemma 54.2 gives us numbers U_1 and U_2 with

$$(56.2.8) \qquad [U_1, U_2] \subset \left[x\exp\left(-\tfrac{7}{2}\log^{6/7}x\right), x\exp\left(\tfrac{7}{2}\log^{6/7}x\right)\right]$$

for which, by (56.2.6) and (54.6.2), the inequality

$$(56.2.9) \qquad \sum_{U_1 < p \le U_2} \varepsilon(p, q, 1, l) > \frac{1}{\log x + 1}\left[\sqrt{x}\cdot\exp(2\log^{6/7}x) - 3\right] > \sqrt{U_2}$$

holds.

By (53.2.8), (56.2.3), and (56.2.8), U_1 and U_2 satisfy the restriction (56.1.2), and so (56.2.9) proves (56.1.4) in Case I. Naturally the proof of (56.1.5) goes unchanged in Case I.

From the functional equation, the second case may be formulated as follows:

Case II. None of the functions $\mathscr{L}(s, \chi) \bmod q$ with $\chi(l) \ne 1$ vanish in the domain

$$(56.2.10) \qquad \frac{1}{2} > \left|\sigma - \frac{1}{2}\right| > \frac{1}{Z}, \qquad |t| < Z$$

where

$$(56.2.11) \qquad Z = \log^{1/11}Y.$$

The conditions of Lemma 49.1 are satisfied by (56.1.1) and (56.2.10)–(56.2.11). Hence there exists a U_2 with

$$(56.2.12) \qquad Z^{\exp(-q^{c_8})} \le U_2 \le Z$$

such that

$$(56.2.13) \qquad \pi(U_2, q, 1) - \pi(U_2, q, l) > \frac{2\sqrt{U_2}}{\log U_2}.$$

The finite exact prime number formula

$$(56.2.14) \quad \psi(x, q, l) = \frac{x}{\varphi(q)} - \frac{1}{\varphi(q)}\sum_{\chi}\bar{\chi}(l)\sum_{|\gamma| < x}\frac{x^{\rho}}{\rho} + O\left[\log^2(qx)\right]$$

(see Prachar [1], Satz 4.4, p. 228) together with (48.8.3) and (56.2.10) imply

$$\psi(x, q, l) = \frac{x}{\varphi(q)} + O\left[\sqrt{x} \cdot \log^2(qx)\right] \quad \text{for} \quad x \leq Z; \tag{56.2.15}$$

and thus

$$|\psi(x, q, l) - \psi(x, q, 1)| = O\left[\sqrt{x} \cdot \log^2(qx)\right]. \tag{56.2.16}$$

By partial summation we get the analogous estimate

$$|\pi(x, q, l) - \pi(x, q, 1)| = O\left[\sqrt{x} \cdot \log^2(qx)\right]. \tag{56.2.17}$$

Now choose

$$U_1 = \frac{U_2}{\log^8 U_2} (> q), \tag{56.2.18}$$

and, from (56.2.17),

$$|\pi(U_1, q, l) - \pi(U_1, q, 1)| = O\left(\frac{\sqrt{U_2}}{\log^2 U_2}\right). \tag{56.2.19}$$

Combine this with (56.2.13), and we immediately obtain (56.1.4) in Case II. [Formula (56.1.2) is satisfied by (56.2.11), (56.2.12), and (56.2.18).] Formulas (56.1.3)–(56.1.5) can be proved in like manner in Case II, and this completes the proof of Theorem 56.1.* ■

*The special case $q = 4$ of this theorem is contained in a weaker form in Knapowski–Turán [15], with $\log_3 Y$ instead of $(\log Y)^{\exp(-q^{c_5})}$. The improvement and generalization proved here is due to J. Pintz based essentially on the work of Knapowski–Turán [20], which contains the proof of (56.1.14)–(56.1.15).

57

THE ESTIMATION OF DEDEKIND'S ZETA FUNCTIONS

57.1

Power sums occur, although in a different way, in the theory of Dedekind's zeta functions

$$\zeta_K(s) = \sum_{a \neq 0} \frac{1}{N^s(a)}$$

belonging to an algebraic number field K. [The variable a runs through the nonzero integral ideals of K, and $N(a)$ is its absolute norm.] The problem is to estimate $\zeta_K(s)$ explicitly in terms of the characteristics of K: the degree n and the discriminant Δ. The first application of the second main theorem to this problem is due to A. V. Sokolovski [1]. By a refinement of his proof, K. M. Bartz obtained the following quantitative estimate [1]:

$$|\zeta_K(s)| < e^{cn^6|\Delta^3|}\log|t| \qquad (s = 6 + it)$$

for

$$|t| \geq 1,1, \qquad \sigma > 1 - \frac{c'}{n^5}\left(\frac{\log\log|t|}{\log|t|}\right)^{2/3}$$

where c and c' are effectively computable numerical constants. In the standard way this estimate implies explicit remainder terms in the prime ideal theorem. Similar estimates in Theorem 38.2 give us explicit density results for $\zeta_K(s)$

(published by W. Staś [9]) that might be useful in removing the unproved assumption in Hooley's conditional proof of Artin's conjecture.

57.2

We shall be contented with an outline of the proof, refer to Bartz [1] for details.

For $\sigma > 1 - 1/n + 1$, $|t| > 1$ we have the classical formula

$$\zeta_K(s) = \sum_{N(a) \leq |t|^{n+1}} \frac{1}{N^s(a)} + O(1),$$

and recent developments make it possible to express $O(1)$ explicitly in terms of n and Δ. To express the terms in a manageable analytic form, we split the sum in the usual way according to the ideal classes:

$$\sum_{N(a) \leq |t|^{n+1}} \frac{1}{N^s(a)} = \sum_{j=1}^{h} N^s(a_j) \sum_{\substack{\alpha \\ a_j | \alpha \\ 0 < |N(\alpha)| \leq N(a_j)|t|^{n+1}}} \frac{1}{|N(\alpha)|^s}$$

where $a_1, \ldots, a_n$ is a complete system of integral representatives of the ideal classes. In the inner sum α runs through elements of a_j, but only one α is taken from each set of associates. a_j can be chosen with small norm, and h, the number of ideal classes, can be estimated by n and Δ. So it remains to consider the inner sum where, selecting a basis $\beta_1, \ldots, \beta_n$ for a_j, $a_j | \alpha$ is equivalent to

$$\alpha = r_1\beta_1 + \cdots + r_n\beta_n$$

with rational integers r_K. A suitable way of choosing exactly one from each set of associates in the inner sum is to let $(r_1, \ldots, r_n)$ run through those n-tuples such that α (in its geometric representation in the n-dimensional real Euclidean space) belongs to the fundamental domain V of K belonging to a fixed system of fundamental units. It is more convenient to replace V by $\overline{V}$, the set obtained by multiplying elements of V by roots of unity belonging to K. This means that each term occurs m times, where m is the number of roots of unity in K.

57.3

We split the sum further according to the magnitude of $\max(|r_1|, \ldots, |r_n|)$ into sums of the form

$$\sum_{\substack{r_1, \ldots, r_n \\ \alpha \in \overline{V} \\ |N(\alpha)| \leq N(a_j)t^{n+1} \\ x < \max(|r_1|, \ldots, |r_n|) \leq 2x}} \frac{1}{|N(\alpha)|^s}$$

In this summation, a given r_k, with the r_j fixed for $j \neq k$, runs through the union of a bounded number of intervals. So, using partial summation, it suffices to consider sums of the form

$$\sum_{\substack{a < r_k \leq b \\ x < \max(|r_1|, \ldots, |r_n|) \leq 2x}} e^{iF(r_1, \ldots, r_k, \ldots, r_n)} \tag{57.3.1}$$

for fixed r_j ($j \neq k$) where one has the analytic expression

$$F(r_1, \ldots, r_n) = -t \log |N(\alpha)| = -t \sum_{l=1}^{n} \log |r_1 \beta_1^{(l)} + \cdots + r_n \beta_n^{(l)}|.$$

(In general, $\alpha^{(l)}$ stands for conjugates.) Approximating by Taylor's formula reduces the problem to an estimation of sums of the form

$$\sum_{r=1}^{R} e^{if(r)}$$

where

$$f(r) = d_1 r + \cdots + d_m r^m \tag{57.3.2}$$

with

$$d_\nu = \frac{1}{\nu!} \frac{\partial^\nu F}{\partial r_k^\nu} = \frac{(-1)^\nu}{\nu!} t \sum_{l=1}^{n} \left[\frac{\beta_k^{(l)}}{r_1 \beta_1^{(l)} + \cdots + r_n \beta_n^{(l)}} \right]^\nu. \tag{57.3.3}$$

57.4

Here Vinogradov's estimate can be applied. It yields a nontrivial bound provided that $|d_m|$ is neither too small nor too large.

Then

$$A_1 \max(|r_1|, \ldots, |r_n|) \leq \left| \frac{r_1 \beta_1^{(l)} + \cdots + r_n \beta_n^{(l)}}{\beta_k^{(l)}} \right| \leq A_2 \max(|r_1|, \ldots, |r_n|) \tag{57.4.1}$$

for $l = 1, \ldots, n$ where A_1 and A_2 are explicitly expressible constants. This is shown using results due to Mahler which allow us to choose our basis $\beta_1, \ldots, \beta_n$ with prescribed ranges for the values $|\beta_1^{(l)}|$. The upper bound then follows trivially. The lower bound is obtained from the fact that $\alpha \in \overline{V}$, and V can be defined by a system of fundamental units satisfying appropriate inequalities in terms of n and Δ, by results of Siegel. Since $x < \max(|r_1|, \ldots, |r_n|) \leq 2x$, we immediately obtain an upper bound for $|d_m|$ in (57.3.3). Observing its power sum character, and using the second inequality in (57.4.1), we find that the first main theorem yields a lower bound for a suitable localized value of m.

57.5

If x is small, an improved version of the exponential sum estimation due to Korobov and Vinogradov gives a better result. We replace r_k in (57.3.1) by $u + rv$, $a < u \leq b$, $1 \leq r, v \leq Y$. With Y relatively small, most of the r_k of (57.3.1) are represented exactly Y^2 times in this form. As before, this reduces to estimating sums of the form

$$\sum_{1 \leq r, v \leq Y} e^{if(rv)}$$

with f given in (57.3.2).

Using Hölder's inequality, a straightforward calculation yields

$$(57.5.1) \quad \left(\sum_{1 \leq r \leq Y} \left| \sum_{1 \leq v \leq Y} e^{if(rv)} \right| \right)^{2p} \leq Y^{2p-1} N \sum_{\substack{n_1, \ldots, n_m \\ |n_\nu| < pY^\nu}} \left| \sum_{1 \leq r \leq Y} e^{ig(r)} \right|$$

where

$$g(r) = d_1 n_1 r + \cdots + d_m n_m r^m,$$

and N is the maximal number of solutions of the simultaneous equations

$$x_1^\nu + \cdots + x_r^\nu - x_{p+1} - \cdots - x_{2p}^\nu = n.$$

(This may be estimated nontrivially by Vinogradov's mean value theorem.) Another application of the same procedure to the right-hand side of (57.5.1) gives rise to geometric series with respect to n_ν, and evaluating them we obtain the upper bound

$$Y^{1-1/2p} \left[N \sum_{\substack{n_1, \ldots, n_m \\ |n_\nu| < pY^\nu}} \prod_{\nu=1}^{m} \min\left(2pY^\nu, \frac{\pi}{\|n_\nu d_\nu\|}\right) \right]^{1/2p}$$

for the sum on the right of (57.5.1), where $\| \ \|$ stands for the distance from the nearest multiple of 2π. For larger values of ν, $|n_\nu d_\nu|$ will always be less than 1, so the estimate boils down to finding lower bounds for sufficiently many of the $|d_\nu|$. Recalling their power sum representations (57.3.3), the first main theorem applied to disjoint ν-intervals of length n can supply this. Thus the special features of the main theorems, the strong localization, and relatively good bounds play an even more important role here than in Section 57.4. Owing to its very weak localization, Dirichlet's classical theorem on diophantine approximation would only give exponential growth in n.

58

SOME OPEN PROBLEMS IN CONNECTION WITH PART II

PROBLEM 1

Can the constant $4e$ be replaced by 2 in Corollary IV of Section 18?

PROBLEM 2

Can our power sum methods prove Pólya's theorem, which states that if the exponent of the series

$$f(s) = \sum_{1}^{\infty} a_j e^{-\lambda_j s}$$

satisfies the gap condition

$$\varliminf_{j} (\lambda_{j+1} - \lambda_j) \geq \gamma > 0,$$

then every closed segment of length $2\pi/\gamma$ on the line of convergence contains a singular point of $f(s)$?

PROBLEM 3

Prove the natural analogue of Theorem 21.1 for harmonic entire functions of the three space, and prove the analogue of Pólya's theorem.

PROBLEM 4

Prove the theorems of § 21 for entire functions of infinite order, assuming only Fejér's gap condition (21.8.2) at least for suitable $r_1 < r_2 < \cdots \to \infty$.

PROBLEM 5

Extend the theorem of Pólya-Fuchs to entire functions representable by Dirichlet-series.

PROBLEM 6

Can the algorithm described in Theorem 31.1 be modified so that it approximates a zero of maximal absolute value?

APPENDIXES

A

SOME INEQUALITIES TO LEMMA 5.8

A.1

The inequality

$$\frac{1}{\sin^2 x} - \frac{1}{x^2} \le \frac{\pi^2}{12} \qquad \left(0 < x \le \frac{\pi}{2}\right) \tag{A.1.1}$$

is a simple consequence of the Taylor series and concavity of $\sin x$.

We also need the inequality

$$U(x) = \frac{1}{\sin^2 x} - \frac{1}{x^2} > \frac{3}{10} \quad \text{for} \quad x > 0, x \neq n\pi. \tag{A.1.2}$$

For $x \ge \sqrt{2}$,

$$U(x) \ge 1 - \tfrac{1}{2} = \tfrac{1}{2} > \tfrac{3}{10}.$$

For $0 < x < \sqrt{2}$, we have

$$x^2 < 2,$$

$$\frac{x^2}{20} < \frac{1}{10},$$

$$0 < \sin x < x - \frac{x^3}{3!} + \frac{x^5}{5!} = x - \frac{x^3}{6}\left(1 - \frac{x^2}{20}\right) < x - \frac{3}{20}x^3$$

and hence,

$$U(x) > \frac{1}{\left(x - \frac{3}{20}x^3\right)^2} - \frac{1}{x^2} = \frac{1}{x^2}\left\{\frac{1}{\left(1 - \frac{3}{20}x^2\right)^2} - 1\right\}.$$

$$= \frac{3}{20}\frac{2 - \frac{3}{20}x^2}{\left(1 - \frac{3}{20}x^2\right)^2} = \frac{3}{20}\left\{\frac{1}{\left(1 - \frac{3}{20}x^2\right)^2} + \frac{1}{1 - \frac{3}{20}x^2}\right\} > \frac{3}{10}.$$

Next let $0 < \alpha < \pi$. Then (A.1.2) gives

(A.1.3) $$\frac{1}{\sin^2[(k+1)\alpha/2]} - \frac{1}{[(k+1)\alpha/2]^2} > \frac{3}{10} \quad \left(\text{for } \frac{k+1}{2}\alpha \neq n\pi\right)$$

and since

$$\frac{3}{10} > \frac{\pi^2}{12}\cdot\frac{1}{k+1} \quad \text{for } k \geq 2,$$

(A.1.3) and (A.1.1) imply

$$\frac{1}{\sin^2[(k+1)\alpha/2]} - \frac{1}{[(k+1)\alpha/2]^2} > \frac{1}{k+1}\left(\frac{1}{\sin^2(\alpha/2)} - \frac{1}{(\alpha/2)^2}\right).$$

Thus, for $0 < \alpha < \pi$ and integer $k \geq 2$,

(A.1.4)

$$1 - \left[\frac{\sin\{(k+1)\alpha/2\}}{(k+1)\sin(\alpha/2)}\right]^2 \geq \left(1 - \frac{1}{k+1}\right)\left\{1 - \left[\frac{\sin\{(k+1)\alpha/2\}}{(k+1)\alpha/2}\right]^2\right\}.$$

We also need in Section 5.7 an upper bound for

$$I = \frac{1}{\pi}\int_0^\infty \log\frac{1}{1 - (\sin x/x)^2}\,dx.$$

From the representation

$$I = \frac{1}{2} + \sum_{\nu=2}^{\infty}\frac{1}{\nu\pi}\left\{\int_0^{\sqrt{2}} + \int_{\sqrt{2}}^{\infty}\left(\frac{\sin x}{x}\right)^{2\nu} dx\right\},$$

and since $\sin x \leq x - (3/20)x^3$ for $(0, \sqrt{2})$ and $|\sin x| \leq 1$ for the rest, we get

(A.1.5) $$I \leq \tfrac{3}{2}.$$

B

AN APPROXIMATION LEMMA TO SECTION 20

B.1

Let $s = \sigma + it$ and, in the half-plane $\sigma > 0$, let $f(s)$ be represented by the convergent Dirichlet series

$$\text{(B.1.1)} \qquad \sum_{n=1}^{\infty} a_n e^{-\lambda_n s}$$

$$\lambda_1 < \lambda_2 < \cdots \to \infty.$$

Suppose $f(s)$ is regular for, say

$$\text{(B.1.2)} \qquad \sigma \geq -2\eta, \qquad |t| \leq 3$$

with $0 < \eta < 1/25$. Then we assert the following:

Lemma B.1. *There exist (explicitly calculable) complex numbers*

$$\text{(B.1.3)} \qquad \beta_{n\nu} \qquad 1 \leq \nu \leq n$$

depending only on η *and the* λ_j *so that, for*

$$\text{(B.1.4)} \qquad D\colon -\frac{\eta}{2} \leq \sigma \leq \frac{\eta}{2}, \qquad |t| \leq 1,$$

the inequality

$$\left| f(s) - \sum_{j=1}^{n} \beta_{nj} a_j e^{-\lambda_j s} \right| \leq c e^{-\lambda_n \eta/4} \tag{B.1.5}$$

holds, where $c = c(f, \eta)$ *depends only upon* f *and* η.

B.2

We may suppose, without loss of generality, that

$$\lambda_n \geq 100 \tag{B.2.1}$$

and begin with an investigation of

$$Y_n(s) = \frac{1}{2\pi i} \int_{(\eta)} \frac{e^{\lambda_n z} f(s+z)}{z(1 + 3z\sqrt{\eta})^{1+[\lambda_n]}} dz \tag{B.2.2}$$

$$z = x + iy, \qquad s \in D.$$

The sharp decrease of the kernel justifies termwise integration:

$$Y_n(s) = \sum_{j=1}^{\infty} a_j e^{-\lambda_j s} \frac{1}{2\pi i} \int_{(\eta)} \frac{e^{(\lambda_n - \lambda_j)z}}{z(1 + 3z\sqrt{\eta})^{1+[\lambda_n]}} dz. \tag{B.2.3}$$

Using "big" half-circles in the right half-plane, we show easily that the integrals belonging to $j > n$ vanish, that is,

$$Y_n(s) = \sum_{j=1}^{n} \beta_{nj} a_j e^{-\lambda_j s} \tag{B.2.4}$$

with

$$\beta_{nj} = \frac{1}{2\pi i} \int_{(\eta)} \frac{e^{(\lambda_n - \lambda_j)z}}{z(1 + 3z\sqrt{\eta})^{1+[\lambda_n]}} dz, \tag{B.2.5}$$

depending on λ_n, λ_j and η only, as required.

B.3

To get another representation of $Y_n(s)$ we replace the segment $[\eta - i, \eta + 1]$ of the path of integration by the broken line consisting of the segments

$$\text{(B.3.1)} \qquad L_2: \quad -\frac{\eta}{2} \le x \le \eta, \qquad y = -1$$

$$L_3: \quad x = -\frac{\eta}{2}, \qquad -1 \le y \le 1$$

$$L_4: \quad -\frac{\eta}{2} \le x \le \eta, \qquad y = 1.$$

Denoting the new path of integration by l, Cauchy's formula implies that

$$\text{(B.3.2)} \qquad Y_n(s) = f(s) + \frac{1}{2\pi i}\int_{(l)} \frac{e^{\lambda_n z} f(s+z)}{z\left(1 + 3z\sqrt{\eta}\right)^{[\lambda_n]+1}}\,dz.$$

l consists of L_2, L_3, L_4, and

$$\text{(B.3.3)} \qquad L_1: x = \eta, \qquad y \le -1$$

$$L_5: x = \eta, \qquad y \ge 1.$$

There is a $c = c(f, \eta)$ such that, for $\eta/2 \le x$

$$|f(z)| \le c(|y| + 1).$$

Hence

$$\text{(B.3.4)} \qquad \left|\frac{1}{2\pi i}\int_{L_5}\right| \le ce^{\lambda_n \eta}\int_1^\infty \frac{dy}{|1 + i \cdot 3\sqrt{\eta}\, y|^{1+[\lambda_n]}}$$

$$< ce^{\lambda_n \eta}\int_1^\infty \frac{dy}{\left(1 + 9\eta y^2\right)^{\lambda_n/2}} < c(\eta)e^{-(\lambda_n/2)\eta}$$

and the same holds for the integral over L_1. By (B.1.2) we have for $s \in D$

$$\text{(B.3.5)} \qquad |f(s+z)| \le M = M(f, \eta) \qquad (z \in L_2 \cup L_3 \cup L_4).$$

Furthermore,

$$\left|\frac{1}{2\pi i}\int_{L_4}\right| \le \frac{M}{2\pi} e^{\lambda_n \eta} \int_{-\eta/2}^{\eta} \frac{dx}{(1+8\eta)^{\lambda_n/2}} < c(\eta) M e^{-\lambda_n \eta/2} \tag{B.3.6}$$

and similarly for the integral over L_2. Finally

$$\left|\frac{1}{2\pi i}\int_{L_3}\right| \le \frac{2M}{\eta\pi} e^{-\lambda_n \eta/2} \frac{5}{(1-(3/2)\eta^{3/2})^{\lambda_n}} < c(M,\eta) e^{-\lambda_n \eta/4} \tag{B.3.7}$$

and (B.2.4), (B.3.2), (B.3.4), (B.3.6) and (B.3.7) prove Lemma B.1.

C

SOME LEMMAS TO SECTION 21

C.1

We shall need the following lemma of Edrei–Fuchs [1].

Lemma C.1. *Let $S(x)$ and $\mu(x)$ be increasing, continuous, positive functions of x for $x > 0$ such that*

$$\int_1^\infty \frac{dy}{\mu(y)} < \infty. \tag{C.1.1}$$

Let $h > 0$ be any fixed constant. Then the set of values x such that

$$S\left\{x + \frac{1}{\mu[S(x)]}\right\} > S(x) + h \tag{C.1.2}$$

has finite measure.

Proof. Let x_0 be the smallest $x \geq 1$ satisfying (C.1.2) (if any) and let

$$\xi_0 = x_0 + \frac{1}{\mu[S(x_0)]}.$$

Then, let x_1 be the smallest $x_1 \geq \xi_0$ satisfying (C.1.2) and set

$$\xi_1 = x_1 + \frac{1}{\mu[S(x_1)]},$$

and so forth. Then

$$S(x_n) \geq S(\xi_{n-1}) > S(x_{n-1}) + h.$$

that is,

$$S(x_n) > nh.$$

Hence $x_n \to \infty$, and the intervals $[x_n, \xi_n]$ together with $[0, 1]$ cover the whole set. Since

$$\xi_n - x_n = \frac{1}{\mu[S(x_n)]} \leq \frac{1}{\mu(nh)}$$

we have

$$\sum_{n=2}^{\infty} (\xi_n - x_n) \leq \sum_{n=2}^{\infty} \frac{1}{\mu(nh)} < \frac{1}{h} \int_h^{\infty} \frac{dx}{\mu(x)} < \infty$$

which completes the proof.

Hence the inequality

$$S\left\{x + \frac{1}{\mu[S(x)]}\right\} \leq S(x) + h \tag{C.1.3}$$

holds for $x > 0$ except perhaps in a set that can be covered by disjoint intervals of finite total length.

Suppose $Q(r)$ is positive and increasing for $r > 1$, and in (C.1.2) choose

$$S(x) = \log Q(e^x), \qquad \eta > 0,$$

$$\mu(x) = x^{1+\eta/2}, \qquad h = \log 2.$$

Lemma C.1 gives the inequality

$$Q\left(r + \frac{r}{\log^{1+\eta/2} Q(r)}\right) \leq Q\left(r \exp\{\log^{-1-\eta/2} Q(r)\}\right) \leq 2Q(r) \tag{C.1.4}$$

except from an r-set of finite logarithmic measure. Applying this with $Q(r) = \log M(r)$ we obtain that, except perhaps from an r-set with finite logarithmic measure, the inequality

$$M\left\{r\left[1 + \frac{1}{[\log\log M(r)]^{1+\eta/2}}\right]\right\} \leq M(r)^2 \tag{C.1.5}$$

holds.

D

A LEMMA TO SECTION 23

D.1

Let ω be a positive integer,

(D.1.1) $$R_1 < R_3, \qquad R_2 < R_3.$$

Let $f(z)$ be regular for $|z| \le R_3$ and its zeros in $|z| \le R_1$

$$\rho_1, \rho_2, \ldots, \rho_N$$

(multiple zeros being repeated). Let

(D.1.2) $$\varphi(z, \rho_j) = R_3 \frac{z - \rho_j}{R_3^2 - \bar{\rho}_j z}$$

and consider the integral

(D.1.3) $$I = \frac{1}{2\pi i} \int_{|\zeta| = R_3} \frac{f(\zeta)^\omega}{\zeta - z} \prod_{j=1}^{N} \left[\frac{\varphi(z, \rho_j)}{\varphi(\zeta, \rho_j)} \right]^\omega d\zeta,$$

with z in $|\zeta| \le R_2$. Cauchy's integral theorem gives

$$I = f(z)^\omega.$$

Thus choosing z on $|\zeta| = R_2$ so that

$$\max_{\varphi} |f(R_2 e^{i\varphi})| = |f(z_0)|$$

we get

$$|M(R_2, f)|^{\omega} \le \frac{M(R_3, f)^{\omega}}{R_3 - R_2} \prod_{j=1}^{N} \max_{|z|=R_2} |\varphi(z, \rho_j)|^{\omega}.$$

Taking ωth roots and letting $\omega \to \infty$ we get

$$M(R_2, f) \le M(R_3, f) \prod_{j=1}^{N} \max_{|z|=R_2} |\varphi(z, \rho_j)|. \tag{D.1.4}$$

Since, for $|z| = R_2, |\rho| \le R_1$,

$$\max_{|z|=R_2} |\varphi(z, \rho_j)| \le \frac{R_2 + R_1}{R_3^2 + R_1 R_2} R_3,$$

we obtained from (D.1.3) the following.

Lemma D.1. *For the number $N(R_1)$ of zeros of $f(z)$ in $|z| \le R_1$ (with multiplicity) the inequality*

$$N(R_1) \le \frac{\log \dfrac{M(R_3, f)}{M(R_2, f)}}{\log \dfrac{R_3^2 + R_1 R_2}{(R_1 + R_2) R_3}}$$

holds (under D.1.1).

E

A LEMMA FOR SECTIONS 33, 38, AND 39, ESSENTIALLY DUE TO LANDAU

E.1

Let $G(s)$ be regular for $|s - s_0| \le R$, $G(s_0) \ne 0$ and assume that

$$\left|\frac{G(s)}{G(s_0)}\right| \le e^M, \tag{E.1.1}$$

there. Let $s_j (j = 1, 2, \ldots)$ be its zeros in $|s - s_0| < R$.

Lemma E.1. *For integer ν and $r < R$*

$$\left|\left[\frac{G'}{G}(s)\right]^{(\nu)}_{s=s_0} + \nu! \sum_{|s_j - s_0| \le r} \frac{1}{(s_j - s_0)^{\nu+1}}\right| \le \frac{7M(\nu+1)!}{r^{\nu+1}} \max\left(1, \frac{1}{\log(R/r)}\right).$$

E.2

Without loss of generality we may assume $s_0 = 0$ and $R = 1$. Define $g(s) = u(s) + iv(s)$ by

$$\frac{G(s)}{G(0)\prod_j [(s - s_j)/(1 - s\bar{s}_j)]} = e^{g(s)}. \tag{E.2.1}$$

Then

$$u(s) \le M, \qquad u(0) \ge 0 \tag{E.2.2}$$

and so

(E.2.3)

$$\begin{aligned}\int_0^{2\pi} |u(e^{i\vartheta})| d\vartheta &= 2\int_{u(e^{i\vartheta})>0} u(e^{i\vartheta})\, d\vartheta - \int_0^{2\pi} u(e^{i\vartheta})\, d\vartheta \\ &\le 2M \cdot 2\pi - 2\pi u(0) \le 4\pi M.\end{aligned}$$

Now,

$$g(s) = \frac{1}{2\pi}\int_0^{2\pi} \frac{e^{i\vartheta}+s}{e^{i\vartheta}-s} u(e^{i\vartheta})\, d\vartheta + \text{const.},$$

and, by (E.2.3)

$$|g(0)| = \left| \frac{2(\nu+1)!}{2\pi} \int_0^{2\pi} e^{-i(\nu+1)\vartheta} u(e^{i\vartheta})\, d\vartheta \right| \le 4(\nu+1)!M.$$

This implies, using (E.2.1),

$$|g^{\nu+1}(0)| = \left| \left[\frac{G'}{G}(s) \right]_{s=0}^{(\nu)} + \nu! \sum_j \left(\frac{1}{s_j^{\nu+1}} - \bar{s}_j^{\nu+1} \right) \right| \le 4(\nu+1)!M.$$

The maximum principle for $u(s)$ gives Jensen's inequality in (E.2.1):

$$\prod_j |s_j| \ge e^{-M}$$

implying

$$\left| \sum_{|s_j| \le r} \bar{s}_j^{\nu+1} \right| \le \sum_{|s_j| \le r} 1 \le \frac{M}{\log(1/r)},$$

and since

$$\left| \frac{1}{s^{\nu+1}} - \bar{s}^{\nu+1} \right| = \frac{1 - |s|^{2(\nu+1)}}{|s|^{\nu+1}} \le \frac{2(\nu+1)\log(1/|s|)}{r^{\nu+1}} \qquad (r \le |s| < 1),$$

also

$$\left| \sum_{|s_j|>r} \left(\frac{1}{s_j^{\nu+1}} - \bar{s}_j^{\nu+1} \right) \right| \le \frac{2(\nu+1)}{r^{\nu+1}} \log \prod_j \frac{1}{|s_j|} \le \frac{2(\nu+1)M}{r^{\nu+1}}.$$

Collecting our estimates we obtain the lemma.

F

AN EXPLICIT FORMULA FOR PRIMES

F.1

In Section 40 we use the following:

Lemma F.1. *If* $s = \sigma + it$, $\sigma > 1$, $\xi > 1$, k *is a positive integer, then*

(F.1.1)

$$\frac{1}{(k+1)!} \sum_{n \geq \xi} \frac{\Lambda(n)}{n^s} \log^{k+1} \frac{n}{\xi} = \frac{\xi^{1-s}}{(s-1)^{k+2}} - \sum_{\rho} \frac{\xi^{\rho - s}}{(s-\rho)^{k+2}} + O\left[\frac{\log(|t|+2)}{\xi^2}\right].$$

The proof of this lemma is very similar to the proof of Riemann–Mangoldt's exact prime number formula, thus a sketch of it will suffice.

We start with the integral

(F.1.2)
$$D \doteq -\frac{1}{2\pi i} \int_{\left(-\frac{\sigma-1}{2}\right)} \frac{\xi^w}{w^{k+2}} \cdot \frac{\zeta'}{\zeta}(s+w)\, dw.$$

Integrating term by term we obtain

(F.1.3)
$$D = \frac{(-1)^{k+2}}{(k+1)!} \sum_{n \geq \xi} \frac{\Lambda(n)}{n^s} \log^{k+1} \frac{n}{\xi}.$$

On the other hand, we can apply Cauchy's integral formula to the parallelogram with vertices

(F.1.4) $$-\frac{\sigma-1}{2} \pm iT_m, \qquad -1-\sigma \pm iT_m$$

where

(F.1.5) $$(3 \leq)\, m \leq T_m \leq m+1$$

and T_m is such that

(F.1.6) $$\left|\frac{\zeta'}{\zeta}(x+iT_m)\right| \leq c\log^2 T_m \quad \text{for} \quad -1 \leq x \leq 2.$$

Thus if we subsequently let $m \to \infty$ we get

(F.1.7)

$$D = \frac{\xi^{1-s}}{(1-s)^{k+2}} - \sum_{\rho} \frac{\xi^{\rho-s}}{(\rho-s)^{k+2}} - \frac{1}{2\pi i}\int_{(-1-\sigma)} \frac{\xi^{w}}{w^{k+2}} \cdot \frac{\zeta'}{\zeta}(s+w)\, dw.$$

The well-known estimate

(F.1.8) $$\left|\frac{\zeta'}{\zeta}(-1+iT)\right| = O[\log(|T|+2)]$$

together with (F.1.3) and (F.1.7) completes the proof. ■

G

A LEMMA FOR DIRICHLET'S $\mathscr{L}$-FUNCTIONS

G.1

The following lemma plays an essential role in Sections 47–52.

Lemma G.1. *In the vertical strip* $1/200 \leq \sigma \leq 1/100$ *there is a broken line* V, *symmetric to the real axis, consisting alternately of horizontal and vertical segments, and ascending monotonically from* $-\infty$ *to* $+\infty$ *such that for all* $\mathscr{L}(s, \chi)$ *functions* mod q *the inequality*

$$\left|\frac{\mathscr{L}'}{\mathscr{L}}(s, \chi)\right| \leq c\varphi(q)\log^2[q(|t| + 2)] \tag{G.1.1}$$

holds at every point $s \in V$.

Proof. We use the fact that if $-\frac{1}{2} \leq \sigma \leq 2$ and $\min(|s|, |1 - s|) > c_1$ then for all $\mathscr{L}(s, \chi)$ mod q

$$\frac{\mathscr{L}'}{\mathscr{L}}(s, \chi) = \sum_{|\gamma - t| \leq 1} \frac{1}{s - \rho} + O\{\log[q(|t| + 2)]\} \tag{G.1.2}$$

where $s = \sigma + it$, $\rho = \beta + i\gamma$ (see Prachar [1], Satz 4.1, p. 225). Let $n \geq 0$ be any integer. Then the total number (with multiplicity) of the zeros of all $\mathscr{L}(s, \chi, q)$ functions with $n \leq |\gamma| \leq n + 1$ is, by (47.3.5), smaller than

$$c_2\varphi(q)\log[q(n + 2)]. \tag{G.1.3}$$

Split the parallelogram with vertices

$$in, \tfrac{1}{2} + in, \tfrac{1}{2} + i(n+1), i(n+1) \tag{G.1.4}$$

and its reflection on the real axis into $[c_2\varphi(q)\log(q(n+2))] + 1$ pairs of narrow horizontal parallelograms with equal height. Then we can choose one pair of symmetrical parallelograms free from zeros of any $\mathscr{L}(s, \chi, q)$ functions. If s is any point of the horizontal central line of one of the two narrow zero-free parallelograms, that is, $s = \sigma \pm it_n$, $0 \leq \sigma \leq \frac{1}{2}$, then for all ρ

$$|s - \rho| > \frac{1}{3c_2\varphi(q)\log[q(n+2)]}. \tag{G.1.5}$$

So by (G.1.2) we get

$$\left|\frac{\mathscr{L}'}{\mathscr{L}}(s, \chi)\right| \leq c_3\varphi(q)\log^2[q(t_n + 2)], \tag{G.1.6}$$

because the number of zeros occurring in (G.1.2) is $O[\log q(t_n + 2)]$ by (47.3.5).

Similarly, we can split the parallelogram with vertices

$$\tfrac{1}{200} + in, \tfrac{1}{100} + in, \tfrac{1}{100} + i(n+2), \tfrac{1}{200} + i(n+2) \tag{G.1.7}$$

and its reflection on the real axis into $[2c_2\varphi(q)\log(q(n+3))] + 1$ pairs of narrow vertical parallelograms with equal width from which we can choose one pair of symmetric ones free from zeros of all $\mathscr{L}$-functions mod q. If s is any point of the vertical central line of one of the two narrow zero-free parallelograms, that is, $s = \sigma_n + it$, $n \leq |t| \leq n + 2$, then for all ρ we have

$$|s - \rho| \geq \frac{1}{1000c_2\varphi(q)\log[q(n+3)]}, \tag{G.1.8}$$

and from this we again get (G.1.6).

Thus we get the required symmetric broken line in the strip $1/200 \leq \sigma \leq 1/100$ proceeding from $-\infty$ to $+\infty$, and consisting of alternate segments of the horizontal lines

$$s = \sigma \pm it_n, \qquad \min(\sigma_{n-1}, \sigma_n) \leq \sigma \leq \max(\sigma_{n-1}, \sigma_n) \tag{G.1.9}$$

and the vertical lines

$$s = \sigma_n \pm it \qquad t_n \leq t \leq t_{n+1}. \tag{G.1.10}$$

H

A LEMMA TO COMPARATIVE PRIME NUMBER THEORY

H.1

We now prove a lemma that makes it possible to check the argument conditions in the applications of one-sided power sum theorems. This lemma is needed in each of Sections 48–56 dealing with comparative prime number theory.

Lemma H. *Let a_j, b_j be real numbers, $j = 1, \dots, N$; $|a_j| \geq 1/M$. Then for every H there exists an y_0 with*

$$y_0 \in [H, H + M] \tag{H.1.1}$$

such that for any integer k and $j = 1, \dots, N$

$$|y_0 a_j + b_j - 2k\pi| > \frac{1}{4N}. \tag{H.1.2}$$

H.2

To prove the lemma it is sufficient to show that for a fixed $j = \nu$ the Lebesgue measure of all the y in (H.1.1) for which there exists an integer k with

$$|y a_\nu + b_\nu - 2k\pi| \leq \frac{1}{4N} \tag{H.2.1}$$

is less than M/N.

(H.1.1) and (H.2.1) restricts k to an interval of length

$$\frac{M|a_\nu| + 1/2N}{2\pi} < \frac{M|a_\nu|}{\pi}, \tag{H.2.2}$$

and so k can assume at most

$$\left[\frac{M|a_\nu|}{\pi}\right] + 1 < 2M|a_\nu| \tag{H.2.3}$$

integer values.

Since for any fixed k (H.2.1) can hold for the y in an interval of length $1/2N|a_\nu|$, the total measure of all y in (H.1.1) for which, with a suitable integer k, (H.2.1) holds is

$$< \frac{2M|a_\nu|}{2N|a_\nu|} = \frac{M}{N}, \tag{H.2.4}$$

by (H.2.3). As we have seen, this estimate proves our lemma. ■

H.3

With specific applications in mind we provide two other formulations of this lemma:

Corollary I. *If we have n complex numbers $\rho = \beta_\rho + i\gamma_\rho$ where $|\gamma_\rho| \geq A$, then for every $R \geq 4/A$ there exists a y with*

$$y \in \left[\frac{R}{2}, R\right] \tag{H.3.1}$$

such that for all ρ the inequalities

$$\pi \geq \left|\operatorname{arc}\frac{e^{i\gamma_\rho y}}{\rho}\right| > \frac{1}{8n} \tag{H.3.2}$$

and

$$\pi \geq \left|\operatorname{arc}\frac{e^{i(\gamma_\rho/2)y}}{\rho/2}\right| > \frac{1}{8n} \tag{H.3.3}$$

hold.

Denoting the ρ and $\rho/2$ by $\lambda_j = \sigma_j + it_j (j = 1, 2, \ldots, 2n)$ and applying Lemma H.1 with

$$\text{(H.3.4)} \quad a_j = t_j, \qquad b_j = -\text{arc}\,\lambda_j, \qquad N = 2n, \qquad M = \frac{2}{A}, \qquad H = \frac{R}{2}$$

proves Corollary I.

Corollary II. *If we have n complex numbers* $\rho = \beta_\rho + i\gamma_\rho$, *where* $|\gamma_\rho| \geq A$, *then for every H we have a* ξ *with*

$$\text{(H.3.5)} \qquad \xi \in \left[H, H + \frac{1}{A}\right]$$

such that for all ρ *the inequality*

$$\text{(H.3.6)} \qquad \pi \geq \left|\text{arc}\, e^{\rho^2 + \xi\rho}\right| > \frac{1}{4n}$$

holds.

Proof. Apply Lemma H.1 with the choice

$$\text{(H.3.7)} \qquad a_j = \gamma_\rho, \qquad b_j = 2\beta_\rho\gamma_\rho, \qquad N = n, \qquad M = \frac{1}{A}.$$

BIBLIOGRAPHY

F. V. Atkinson [1]. On sums of powers of complex numbers, *Acta Math. Hung.*, **XII** (1–2), 185–188 (1961).

F. V. Atkinson [2]. Some further estimates concerning sums of powers of complex numbers, *ibid.*, **XX** (1–2), 193–210 (1969).

A. A. Balkema and R. Tijdeman [1]. Some estimates in theory of exponential sums, *Acta Math. Hung.*, **24**, 115–133 (1973).

G. Birkhoff [1]. Current trends in algebra, *Amer. Math. Monthly*, **80**, 760–781 (1973).

P. E. Blanksby and H. L. Montgomery [1]. Algebraic integers near the unit circle, *Acta Arith.*, **XVIII**, 355–369 (1971).

H. Bohr [1]. Über das Verhalten von $\zeta(s)$ in der Halbebene $\sigma > 1$, *Gött. Nachr.*, 409–428 (1911).

H. Bohr [2]. Sur l'existence des valeurs arbitrairement petites de la fonction $\zeta(s) = \zeta(\sigma + it)$ de Riemann pour $\sigma > 1$, *Oversigt Vidensk. Selsk. Kobenhaven*, 201–208 (1911).

H. Bohr [3]. Again the Kronecker theorem, *J. Lond. Math. Soc.*, **9**, 5–6 (1934).

N. G. de Bruijn [1]. On Turán's first main theorem, *Acta Math. Hung*, **XI** (3–4), 213–216 (1960).

J. D. Buchholtz [1]. Extremal problems for sums of powers of complex numbers, *Acta Math. Hung.*, XVII, 147–153 (1966).

J. D. Buchholtz [2]. Sums of complex numbers, *J. Math. Anal. Appl.*, **17**, 269–279 (1967).

F. Carlson, E. Landau [1]. Neuer Beweis und Verallgemeinerungen des Fabryschen Lückensatzes, *Gött. Nachr.*, 184–188 (1921).

J. W. S. Cassels [1]. On the sums of powers of complex numbers, *Acta Math. Hung.*, **VII** (3–4), 283–290 (1956).

P. L. Chebyshev [1]. Lettre de M. le professeur Tchébychev à M. Fuss, sur un nouveau théorème rélatif aux nombres premiers contenus dans la formes $4n + 1$ et $4n + 3$, *Bull. de la Classe phys.-math. de l'Acad. Imp. des Sciences St. Petersbourg*, **11**, 208 (1853).

J. Coates [1]. On the algebraic approximations of functions IV, *Indag. Math.*, **XXIX**, 205–212 (1977).

S. Dancs [1]. On the sums of powers of complex numbers (in Hungarian), *Mat. Lapok*, **XIII** (1–2), 108–114 (1962).

S. Dancs [2]. On an extremal problem, *Acta Math. Hung.*, **IX** (3–4), 309–313 (1958).

S. Dancs [3]. On generalised sums of powers of complex numbers, *Ann. Univ. Sci. Budapest. Eötvös Sect. Math.*, **VII**, 113–121 (1964).

S. Dancs [4]. Remarks on a paper of P. Turán, *ibid.*, 133–141 (1964).

S. Dancs and P. Turán [1]. On the distribution of values of a class of entire functions I, *Publ. Math. Debrecen.*, **11** (1–4), 257–265 (1964).

S. Dancs and P. Turán [2]. On the distribution of values of a class of entire functions II, *ibid.*, 266–272 (1964).

S. Dancs and P. Turán [3]. Investigations in the power sum theory I, *Ann. Univ. Sci. Budapest. Eötvös Sect. Math.*, **XVI**, 47–52 (1973).

S. Dancs and P. Turán [4]. Investigations in the power sum theory II, *Acta Arith.*, **XXV**, 105–113 (1973).

S. Dancs and P. Turán [5]. Investigations in the power sum theory III, *Ann. Mat. Pura Appl.*, **CIII**, 199–205 (1975).

S. Dancs and P. Turán [6]. Investigations in the power sum theory IV, *Publ. Math. Debrecen.*, **22**, 123–131 (1975).

H. Davenport [1]. *On multiplicative number theory*, Chicago, Markham (1967).

D. G. Dickson [1]. Asymptotic distribution of zeros of exponential sums, *Publ. Math. Debrecen.*, **11** (1–4), 297–300 (1964).

A. Edrei and W. H. J. Fuchs [1]. Bounds for the number of deficient values of certain classes of meromorphic functions, *Proc. Lond. Math. Soc.*, **3** (12), 315–344 (1962).

W. J. Ellison and M. Mendès France [1]. *Les nombres premiers*, Publ. Inst. Math. Univ. Nancago 9, Actualités scient. et industr. 19366, (1975).

P. Erdős and A. Rényi [1]. A probabilistic approach to problems of diophantine approximation, *Illinois J. Math.*, **1**, 303–315 (1957).

P. Erdős and A. Rényi [2]. On a Problem of A. Zygmund, *Studies in Mathematical Analysis and Related Topics*. Stanford University Press, 110–116 (1962).

P. Erdős and P. Turán [1]. On a problem in the theory of uniform distribution I and II, *Indag. Math.*, **X** (5), 3–11, 12–19 (1948).

E. Fogels [1]. On the zeros of $\mathscr{L}$-functions, *Acta Arith.*, **11**, 67–96 (1965).

J. N. Franklin [1]. On numerical solution of characteristic equations in flutter analysis, *J. Assoc. Comp. Mach.*, **5**, 45–51 (1958).

W. H. J. Fuchs [1]. Proof of a conjecture of G. Pólya concerning gap series, *Illinois J. Math.*, **7**, 661–667 (1963).

W. H. J. Fuchs see also Edrei-Fuchs.

D. Gaier [1]. Der allgemeine Lückenumkehrsatz für das Borel-Verfahren., *Math. Z.*, **88**, 410–417 (1965).

D. Gaier [2]. Bemerkungen zum Turánschen Lemma, *Abhandlungen Math. Seminar, Univ. Hamburg*, **35**, 1–7 (1970).

P. X. Gallagher [1]. A large sieve density estimate near $\sigma = 1$, *Invent. Math.*, **11**, 329–338 (1970).

A. O. Gelfond [1]. *Transcendental and algebraic numbers*, Dover Publ. Inc., see in particular Sect. 4, Ch. 3 (1960).

A. O. Gelfond [2]. Sùr les systèmes complets de fonctions analitiques, *Rec. Math. Mosc.*, **4** (46), 149–156 (1938).

J. M. Geysel [1]. On generalised sums of powers of complex numbers, M. C. Report Z. W. 1968–013 Math. Cent. Amsterdam (1968).

G. Halász and P. Turán [1]. On the distribution of roots of Riemann zeta and allied functions I, *J. Number Theory*, **I**, 122–137 (1969).

G. Halász and P. Turán [2]. The same **II**, *Acta Math. Hung.*, **XXI** (3–4), 403–419 (1970).

W. Haneke [1]. Ein neuer Beweis des Linnikschen Nullstellensatzes, *Publ. Math. Debrecen.*, **17** (1–4), 249–258 (1970).

G. H. Hardy and J. E. Littlewood [1]. Contributions to the theory of Riemann zeta-function and the theory of the distribution of primes, *Acta Math.*, **41**, 119–196 (1918).

W. K. Hayman [1]. Angular value distribution of power series with gaps, *Proc. Lond. Math. Soc.*, **XXIV** (3), 590–624 (1972).

E. Hlawka [1]. Zur Theorie der diophantischen Approximation, *Anz. Math.-Naturwiss. Klasse der Österr. Akad. Wiss.*, **4**, 41–48 (1958).

A. E. Ingham [1]. A note on the distribution of primes, *Acta Arith.*, **1** (2), 201–211 (1936).

A. E. Ingham [2]. *The Distribution of Prime Numbers*, Cambridge, University Press (1932).

C. G. J. Jacobi [1]. Observatioulae ad theoriam aequationum pertinentes, *Ges. Werke*, **III**, 28 (1834).

H. Jager [1]. A note on the vanishing of power-sums, *Ann. Univ. Sci. Budapest. Eötvös Sect. Math.*, **X**, 13–16 (1967).

M. Jutila [1]. On two theorems of Linnik concerning the zeros of Dirichlet's *L*-functions, *Ann. Acad. Sci. Fenn. Ser. A. I. Math.*, **458**, 1–32 (1969).

M. Jutila [2]. A new estimate for Linnik's constant, *ibid.*, **471**, 1–8 (1969).

S. Knapowski [1]. On prime numbers in an arithmetical progression, *Acta Arith.*, **IV** (1), 57–70 (1958).

S. Knapowski [2]. On the Möbius function, *ibid.*, **IV** (3), 209–216 (1958).

S. Knapowski [3]. Contributions to the theory of distribution of prime numbers in arithmetical progressions, *ibid.*, **VI** (4), 415–434 (1960).

S. Knapowski [4]. On sign-changes of the difference of $\pi(x)$-Li x, *ibid.*, **VII** (2), 107–120 (1962).

S. Knapowski [5]. Mean-value estimations for the Möbius function I, *ibid.*, 121–130 (1961).

S. Knapowski [6]. Contributions to the theory of distribution of prime numbers in arithmetical progressions II, *ibid.*, **VII** (4), 325–336 (1961).

S. Knapowski [7]. Mean value estimations for the Möbius function II, *ibid.*, 337–344 (1961).

S. Knapowski [8]. Contributions to the theory of distribution of prime numbers in arithmetical progression III, *ibid.*, **VIII** (1), 97–106 (1962).

S. Knapowski [9]. On oscillations of certain means formed from the Möbius series I, *ibid.*, **VIII** (3), 311–320 (1962).

S. Knapowski [10]. On oscillations of certain means formed from the Möbius series II, *ibid.*, **X** (4), 377–386 (1964).

S. Knapowski [11]. On Siegel's theorem, *ibid.*, **XIV** (4), 417–424 (1968).

S. Knapowski [12]. On sign-changes in the remainder term in the prime number formula, *J. Lond. Math. Soc.*, 451–460 (1961).

S. Knapowski [13]. On Linnik's theorem concerning exceptional *L*-zeros, *Publ. Math. Debrecen.*, **9**, 168–178 (1962).

S. Knapowski and W. Staś [1]. A note on a theorem of Hardy and Littlewood, *Acta Arith.*, **VII** (2), 161–166 (1962).

S. Knapowski and W. Staś [2]. Another note on Hardy–Littlewood's theorem, *Acta Arith.*, **VIII** (2), 205–212 (1963).

S. Knapowski and P. Turán [1]. The comparative theory of primes I, *Acta Math. Hung.*, **XIII** (3–4), 299–314 (1962).

S. Knapowski and P. Turán [2]. The same II, *ibid.*, 315–342 (1962).

S. Knapowski and P. Turán [3]. The same III, *ibid.*, 343–364 (1962).

S. Knapowski and P. Turán [4]. The same IV, *ibid.*, **XIV** (1–2), 31–42, (1963).

S. Knapowski and P. Turán [5]. The same V, *ibid.*, 43–64 (1963).

S. Knapowski and P. Turán [6]. The same VI, *ibid.*, 65–78 (1963).

S. Knapowski and P. Turán [7]. The same VII, *ibid.*, **XIV** (3–4), 241–250 (1963).

S. Knapowski and P. Turán [8]. The same VIII, *ibid.*, 251–268 (1963).

S. Knapowski and P. Turán [9]. Further developments in the comparative prime number theory I, *Acta Arith.*, **IX** (1), 23–40 (1964).

S. Knapowski and P. Turán [10]. The same II, *ibid.*, **IX** (3), 293–314 (1964).

S. Knapowski and P. Turán [11]. The same III, *ibid.*, **XI** (1), 115–127 (1965).

S. Knapowski and P. Turán [12]. The same IV, *ibid.*, **XI** (2), 147–162 (1965).

S. Knapowski and P. Turán [13]. The same V, *ibid.*, 193–202 (1965).

S. Knapowski and P. Turán [14]. The same VI, *ibid.*, **XII** (1), 85–96 (1966).

S. Knapowski and P. Turán [15]. The same VII, *ibid.*, **XXI**, 193–201 (1972).

S. Knapowski and P. Turán [16]. On the sign changes of $\pi(x) - \text{li}\, x$, *Topics in Number Theory*, *Coll. Math. Soc. J. Bolyai*, **13**, 153–170 (1976).

S. Knapowski and P. Turán [17]. The same II, *Monatsh. Math.*, **82**, 163–175 (1976).

S. Knapowski and P. Turán [18]. On an assertion of Cebysev, *J. Analyse Math.*, **XIV**, 267–274 (1965).

S. Knapowski and P. Turán [19]. Über einige Fragen der vergleichenden Primzahltheorie. *Abhandl. aus der Zahlentheorie und Analysis*. VEB Deutsch. Verlag der Wiss. Berlin., 159–171 (1968).

S. Knapowski and P. Turán [20]. On prime numbers $\equiv 1$ resp. 3 (mod 4), *Number Theory and Algebra*, Ed. H. Zassenhaus, Academic Press, 157–166 (1977).

J. Komlós, A. Sárközy, E. Szemerédi [1]. On sums of powers of complex numbers (in Hungarian) *Mat. Lapok*, **XV** (4), 337–347 (1964).

T. Kővári [1]. On theorems of G. Pólya and P. Turán, *J. Analyse Math.*, **VI**, 323–331 (1958).

T. Kővári [2]. A gap-theorem for entire functions of infinite order, *Mich. Math. J.*, **12**, 133–140 (1965).

P. Lancaster [1]. *Lambda Matrices and Vibrating Systems*, Pergamon Press (1966).

E. Landau [1]. Über einige ältere Vermutungen und Behauptungen in der Primzahltheorie, *Math. Z.*, **1**, 1–24 (1918).

E. Landau [2]. The same II, *ibid*, 213–219 (1918).

E. Landau. *See also* Carlson–Landau.

J. Ławrynowicz [1]. Calculation of a minimum maximum of complex numbers, *Bull. Soc. Sci. Lettr. Lodz.*, **XI** (2), 1–9 (1960).

J. Ławrynowicz [2]. Remark on power sums of complex numbers, *Acta Math. Hung.*, **XVIII** (3–4), 279–281 (1967).

C. Lech [1]. A note on recurring series, *Ark. Math.*, **2**, 417–421 (1953).

H. Leenmann–Tijdeman [1]. Bounds for the maximum modulus of the first k power sums, *K. Ned. Akad. Wesenschappen. Indag. Math.*, **36** (4), 387–391 (1974).

R. S. Lehman [1]. On the difference $\pi(x)$-li x, *Acta Arith.*, **XI** (4), 397–410 (1966).

D. H. Lehmer [1]. A machine method for solving polynomial equations, *J. Assoc. Comp. Mach.*, **8** (2), 151–162 (1961).

D. N. Lehmer [2]. *List of primes from* 1 *to* 10,006,721, Carnegie Inst. Wash. Publ. 165., Washington, D.C. (1914).

N. Levinson [1]. On a class of non-vanishing functions, *Proc. Lond. Math. Soc.*, **41**, 393–407 (1936).

J. E. Littlewood [1]. Mathematical notes (12): An inequality for a sum of cosines, *J. Lond. Math. Soc.*, **12**, 217–221 (1937).

J. E. Littlewood [2]. Sur la distribution des nombres premiers, *C. R. Acad. Sci. Paris*, **158**, 1869–1872 (1914).

J. E. Littlewood. See *also* G. H. Hardy and J. E. Littlewood.

A. J. Macintyre [1]. Asymptotic paths of integral functions with gap power series, *Proc. Lond. Math. Soc.* **3** (2), 386–398 (1952).

K. Mahler [1]. Eine arithmetische Eigenschaft der Taylorkoeffizienten rationaler Functionen, *Proc. Akad. Amsterdam*, **38**, 51–60 (1935).

K. Mahler [2]. On the Taylor-coefficients of rational functions, *Proc. Cambridge Philos. Soc.*, **52**, 39–48 (1956).

E. Makai [1]. An estimation in the theory of diophantine approximations, *Acta Math. Hung.*, **IX** (3–4), 299–307 (1958).

E. Makai [2]. The first main theorem of P. Turán, *Acta Math. Hung.*, **X** (3–4), 405–411 (1959).

E. Makai [3]. On a maximum problem, *Acta Math. Hung.*, **IX** (1–2), 105–110 (1958).

E. Makai [4]. On a minimum problem, *Ann. Univ. Sci. Budapest. Eötvös Sect. Math.*, (3–4), 177–182 (1964).

E. Makai [5]. On a minimum problem, *Acta Math. Hung.*, **XV** (1–2), 63–66 (1964).

S. Mandelbrojt [1]. *Séries de Fourier et Classes Quasianalytiques*, Paris, Gauthier-Villars (1935).

M. Mendès France. *See* W. J. Ellison and M. Mendès France.

H. L. Montgomery. *See* Blanksby-Montgomery.

L. J. Mordell [1]. Poisson's summation formula in several variables and some applications to the theory of numbers, *Proc. Cambridge Philos. Soc.*, **25**, 412–420 (1929).

A. Ostrowski [1]. Sur une propriété des sommes des racines d'un polynome, *C. R. Acad. Sci. Paris*, Ser. A–B, 263, A46–A48 (1966).

A. Ostrowski [2]. Recherches sur la méthode de Graeffe et les zeros de polynomes et des séries de Laurent, *Acta Math.*, **72**, 99–257 (1940).

R. E. A. C. Paley and N. Wiener [1]. *Fourier transforms in the complex domain*, *Amer. Math. Soc. Colloq. Publ.*, **XIX** (1934).

O. Perron [1]. Über Stabilität und asymptotisches Verhalten der Integrale von Differentialgleichungssystemen, *Math. Z.*, **29**, 129–160 (1928).

J. Pintz [1] On the remainder term of the prime number formula II, On a theorem of Ingham, *Acta Arith.*, **37**, 209–220 (1980).

J. Pintz [2]. The same III, Sign changes of $\pi(x) - \text{li}\, x$, *Stud. Sci. Math. Hung.*, **12**, 345–369 (1977).

J. Pintz [3]. The same IV, Sign changes of $\pi(x) - \text{li}\, x$, *Stud. Sci. Math. Hung.*, **13** 29–42 (1979).

G. Pólya [1]. Untersuchungen über Lücken und Singularitäten von Potenzreihen, *Math. Z.*, **29**, 549–640 (1929).

G. Pólya [2]. Über das Vorzeichen des Restgliedes im Primzahlsatz, *Gött. Nachr.*, 19–27 (1930).

A. J. van der Poorten [1]. Simultaneous algebraic approximation of functions, Ph. D. thesis, University of New South Wales (1968).

A. J. van der Poorten [2]. Generalisation of Turán's main theorems on lower bounds for sums of powers, *Bull. Austral. Math. Soc.*, **2**, 15–37 (1970).

A. J. van der Poorten [3]. Generalisations of Turán's main theorems to binomials and logarithms, *ibid*, 183–195 (1970).

A. J. van der Poorten [4]. On a theorem of S. Dancs and P. Turán, *Acta Math. Hung.*, **XXII** (3–4), 359–364 (1972).

K. Prachar [1]. *Primzahlverteilung*, Springer, Berlin-Göttingen-Heidelberg (1957).

Q. I. Rahman and P. Turán [1]. On a property of rational functions, *Ann. Univ. Sci. Budapest. Eötvös Sect. Math.*, **XVI**, 37–45 (1973).

A. Rényi and P. Turán [1]. On the zeros of polynomials, *Acta Math. Hung.*, **III** (4), 275–282 (1952).

A. Rényi. *See also* Erdős–Rényi.

B. Riemann [1]. Über die Anzahl der Primzahlen unter einer gegebenen Grösse, *Monatsh. Preuss. Akad. Wiss. Berlin*, 671–680 (1859–1860).

J. B. Rosser–L. Schoenfeld [1]. Approximate formulas for some functions of prime numbers, *Illinois J. Math.*, **6**, 64–94 (1962).

A. Sárközy. *See* Komlós–Sárközy–Szemerédi.

E. Schmidt [1]. Über die Anzahl der Primzahlen unter gegebener Grenze, *Math. Ann.*, **57**, 195–204 (1903).

H. N. Shapiro [1]. On a theorem concerning exponential polynomials, *Comm. Pure Appl. Math.*, **12**, 487–500 (1959).

L. Schoenfeld. *See* Rosser–Schoenfeld.

C. L. Siegel [1]. Neuer Beweis des Satzes von Minkowski über lineare Formen, *Math. Ann.*, **87**, 36–38 (1922).

C. L. Siegel [2]. Über Gitterpunkte in Convexen Körpern und ein damit zusammenhängendes Extremalproblem, *Acta Math.*, **65**, 307–323 (1935).

C. L. Siegel [3]. On the zeros of the Dirichlet L-functions, *Ann. Math.*, **46**, 409–422 (1945).

S. Skewes [1]. On the difference $\pi(x) - \text{li } x$ I, *J. Lond. Math. Soc.*, **8**, 277–283 (1933).

S. Skewes [2]. The same II, *Proc. Lond. Math. Soc.*, **5**, 48–70 (1955).

V. T. Sós and P. Turán [1]. On some new theorems in the theory of diophantine approximations, *Acta Math. Hung.*, **VI** (3–4), 241–257 (1955).

A. V. Sokolovskij [1]. Teorema o nullach dzeta funkcii Dedekinda i rastojannie mesdy sosodnimi prostimi idealami, *Acta Arith.*, **XIII** (3), 321–334 (1968).

R. Spira [1]. Calculation of Dirichlet L-function, *Math. Comp.*, **23** (0107), 489–497 (1969).

H. M. Stark [1]. A problem in comparative prime number theory, *Acta Arith.*, **18**, 311–320 (1971).

W. Staś [1]. Über eine Anwendung der Methode von Turán, auf die Theorie des Restgliedes im Primidealsatz, *Acta Arith.*, **V** (2), 179–195 (1959).

W. Staś [2]. Über eine Abschätzung des Restgliedes im Primzahlsatz, *ibid.*, **V** (4) 427–434 (1959).

W. Staś [3]. Über einige Abschätzungen in Idealklassen, *ibid.*, **VI** (1), 1–10 (1960).

W. Staś [4]. Über die Dichte der Nullstellen der Dirichletschen L-Funktionen, *ibid.*, **VI** (3), 313–323 (1961).

W. Staś [5]. Über die Umkehrung eines Satzes von Ingham, *ibid.*, **VI** (4), 435–446 (1961).

W. Staś [6]. Zur Theorie der Möbiusschen μ-funktion, *ibid.*, **VII** (4), 409–416 (1962).

W. Staś [7]. Über eine Reihe von Ramanujan, *ibid.*, **VIII** (3), 261–272 (1963).

W. Staś [8]. Some remarks on a series of Ramanujan, *ibid.*, **X** (4), 359–368 (1965).

M. Szalay [1]. On a theorem of J. W. S. Cassels, (in Hungarian), *Mat. Lapok*, **23**, 109–112 (1972).

M. Szalay [2]. On number theoretical extremal problems, Thesis, (1974).

O. Szász [1]. Singularitäten von Potenzreihen und Dirichletschen Reihen am Rande des Konvergenzbereiches, *Math. Ann.*, **85**, 111–114 (1922).

E. Szemerédi. *See* Komlós–Sárközy–Szemerédi.

R. Tijdeman [1]. On a conjecture of Turán and Erdős, *Indag. Math.*, **XXVIII**, 374–383 (1966).

R. Tijdeman [2]. On the distribution of the values of certain functions, Ph. D. Thesis, University of Amsterdam (1969).

R. Tijdeman [3]. On the number of zeros of general exponential polynomials, *Indag. Math.*, **33** (1), 1–7 (1971).

R. Tijdeman [4]. On the algebraic independence of certain numbers, *ibid.*, 146–162 (1971).

R. Tijdeman [5]. On an inequality of P. Turán, *Acta Math. Hung.*, **XXI** (3–4), 357–363 (1970).

R. Tijdeman. *See also* Balkema–Tijdeman.

E. C. Titchmarsh [1]. *The Theory of the Riemann Zeta-Function*, Oxford, Clarendon Press (1951).

E. Thorp and R. Whitley [1]. The strong maximum modulus theorem for analytic functions into a Banach space, *Proc. Amer. Math. Soc.*, **18**, 640–646 (1967).

P. Turán [1]. Über die Verteilung der Primzahlen I, *Acta Sci. Math.* (*Szeged*), **X**, 81–104 (1941).

P. Turán [2]. On a theorem of Littlewood, *J. Lond. Math. Soc.*, **21**, 268–275 (1946).

P. Turán [3]. Sur la théorie des fonctions quasianalytiques, *C. R. Acad. Sci. Paris*, 1750–1752 (1947).

P. Turán [4]. On the gap theorem of Fabry, *Acta Math. Hung.*, *I*, 21–29 (1947).

P. Turán [5]. On Riemann's hypothesis, *Acad. Sci. URSS Bull.* Ser. Math., **11**, 197–262 (1947).

P. Turán [6]. On a new method in the analysis with applications, *Casopis pro pest. mat. a fys. Rc.*, **74**, 123–131 (1949).

P. Turán [7]. On the remainder term of the prime-number formula I, *Acta Math. Hung.*, **I** (1), 48–63 (1950).

P. Turán [8]. On the remainder term of the prime number formula II, *ibid.*, **I** (3–4), 155–166 (1950).

P. Turán [9]. On approximative solution of algebraic equations, *Publ. Math. Debrecen.*, **II** (1), 28–42 (1951).

P. Turán [10]. On Carlson's theorem in the theory of zetafunction of Riemann, *Acta Math. Hung.*, **II** (1–2), 39–73 (1951).

P. Turán [11]. On a property of lacunary power-series, *Acta Sci. Math.* (*Szeged*), **XIV** (4), 209–218 (1952).

P. Turán [12]. *Über eine neue Methode der Analysis und ihre Anwendungen*, Akad. Kiadó (1953).

P. Turán [13]. On Lindelöf's conjecture, *Acta Math. Hung.*, **V** (3–4), 145–153 (1954).

P. Turán [14]. Revised Chinese edition of [12], Peking (1956).

P. Turán [15]. On the instability of systems of differential equations, *Acta Math. Hung.*, **VI** (3–4), 257–271 (1955).

P. Turán [16]. On the zeros of the zetafunction of Riemann, *J. Indian Math. Soc.*, **XX**, 17–36 (1956).

P. Turán [17]. Über eine neue Methode der Analysis, *Wissenschaftl. Z. Humboldt Universität zu Berlin*, 275–279 (1955–1956).

P. Turán [18]. Über eine Anwendung einer neuen Methode auf die Theorie der Riemannschen Zetafunktion, *ibid.*, 281–285 (1955–1956).

P. Turán [19]. Remark on the preceding paper of J. W. S. Cassels, *Acta Math. Hung.*, **VII** (3–4), 291–294 (1957).

P. Turán [20]. Remark on the theory of quasianalytic function classes, *Publ. Math. Inst. Hung. Acad. Sci.*, **I** (4), 481–487 (1956).

P. Turán [21]. Über lakunären Potenzreihen, *Rev. Math. Pures Appl.*, **I**, 27–32 (1956).

P. Turán [22]. On the so-called density hypothesis in the theory of zeta-function of Riemann, *Acta Arith.*, **IV** (1), 31–56 (1958).

P. Turán [23]. Zur Theorie der Dirichletschen Reihen, *Euler Festschr.*, 322–336 (1959).

P. Turán [24]. On a property of the stable or conditionally stable solutions of systems of nonlinear differential equations, *Ann. Mat.*, **XLVIII**, 333–340 (1959).

P. Turán [25]. A note on the real zeros of Dirichlet *L*-functions, *Acta Arith.*, **V**, 309–314 (1959).

P. Turán [26]. On the distribution of zeros of general exponential polynomials, *Publ. Math. Debrecen.*, **VII**, 130–136 (1960).

P. Turán [27]. On an improvement of some new one-sided theorems of the theory of diophantine approximation, *Acta Math. Hung.*, **XI** (3–4), 299–316 (1960).

P. Turán [28]. On a density theorem of Ju. V. Linnik, *Publ. Math. Inst. Hung. Acad. Sci.*, **VI**, Ser. A. (1–2), 165–179 (1961).

P. Turán [29]. On the eigenvalues of matrices, *Ann. Mat. Pura Appl.*, **IV** (LIV), 397–401 (1961).

P. Turán [30]. On some further one-sided theorems of new type, *Acta Math. Hung.*, **XII** (3–4), 455–468 (1961).

P. Turán [31]. A remark on the heat equation, *J. Analyse Math.*, **XIV**, 443–448 (1965).

P. Turán [32]. On a certain limitation of eigenvalues of matrices, *Aequat. Math.*, **2** (2–3), 184–189 (1969).

P. Turán [33]. On the approximate solution of algebraic equations (in Hungarian), *Commun. Math. Phys. Class, Hung. Acad.*, **XVIII**, 223–236 (1968).

P. Turán [34]. A remark on linear differential equations, *Acta Math. Hung.*, **XX** (3–4), 357–360 (1969).

P. Turán [35]. Zeta roots and prime numbers, *Colloq. Math. Soc. János Bolyai*, **2** (Number theory), Debrecen, 205–216 (1969).

P. Turán [36]. Exponential sums and the Riemann conjecture, *Proc. Symp. Pure Math. XXIV Anal. Number Theory 1973*, 305–314.

P. Turán [37]. On an inequality of Cebysev, *Ann. Univ. Sci. Budapest. Eötvös Sect. Math.*, **XI**, 15–16 (1968).

P. Turán [38]. On an inequality, *ibid.*, **I**, 3–6 (1958).

P. Turán [39]. On a certain problem in the theory of power series with gaps, *Studies in Mathematical Analysis and Related Topics*, Stanford University Press, 404–409 (1962).

P. Turán [40]. On a trigonometrical inequality, *Proc. Conf. Constr. Theory of Functions*, Akad. Kiadó, Budapest, 503–512 (1969).

P. Turán [41]. Investigations in the power sum theory II, (with S. Dancs) *Acta Arith.*, **XXV**, 105–113 (1973).

P. Turán [42]. On the latent roots of λ-matrices, *Comput. Math. Appl.*, **I**, 307–313 (1975).

P. Turán [43]. On some recent results in the analytical theory of numbers, *Proc. Symp. Pure Math. XX. Inst. on Number Theory*, 359–374 (1969).

P. Turán. *See also* Dancs–Turán, Erdős–Turán, Halász–Turán, Knapowski–Turán, Rahman–Turán, Rényi–Turán, T. Sós–Turán.

S. Uchiyama [1]. Complex numbers with vanishing power sums, *Proc. Jap. Acad.*, **33**, 10–12 (1957).

S. Uchiyama [2]. Systems of *n* complex numbers with vanishing power sums, *J. Fac. Sci. Hokkaido Univ.*, **14**, 29–36 (1958).

S. Uchiyama [3]. Sur un problème posé par M. Paul Turán, *Acta Arith.*, **IV** (3), 240–246 (1958).

S. Uchiyama [4]. Sur les sommes de puissances des nombres complexes, *Acta Math. Hung.*, **IX** (3–4), 257–278 (1958).

S. Uchiyama [5]. A note on the second main theorem of P. Turán, *Acta Math. Hung.*, **IX** (3–4), 379–380 (1958).

R. Whitley. *See* Thorp–Whitley.

N. Wiener [1]. A class of gap theorems, *Ann. Pisa*, **2**, 370–372 (1934).

N. Wiener. *See also* Paley–Wiener.

A. Wintner [1]. Asymptotic integration constants, *Amer. J. Math.*, **68**, 553–559 (1946).

H. Wittich [1]. Zur Kennzeichnung linearer Differentialgleichungen mit konstanten Koeffizienten, *Festband zum 70. Geburtstag von R. Nevanlinna,* Springer, 128–134 (1966).

M. Waldschmidt [1]. Independance algébrique de valeurs de la fonction exponentielle, *Bull. Soc. Math. Fr.*, 285–304 (1971).

H. Weyl [1]. Über die Gleichverteilung von Zahlen mod Eins, *Math. Ann.*, **77**, 313–352 (1916).

AUTHOR INDEX